Australian vegetation has interested botanists and naturalists since Europeans first encountered Australia and its plant life. This new edition of *Australian Vegetation* reviews the vegetation of the continent as a whole. In the introductory section, chapters on phytogeography, vegetation history and alien plants set the scene for further sections covering all the major vegetation types. The plant life of extreme Australian habitats is also discussed, and the book closes with a chapter on the conservation of Australian vegetation. Each chapter is written by experts on each particular habitat type.

This book will inform and stimulate the interest of students and professional botanists, especially those fortunate enough to see for themselves the unique vegetation and flora of Australia.

Australian vegetation

EDITED BY R. H. GROVES

Australian Vegetation
Second edition

The Editor is presently at the CSIRO Biological Control Unit,
Montferrier-sur-Lez, France

CAMBRIDGE
UNIVERSITY PRESS

Published by the Press Syndicate of the University of Cambridge
The Pitt Building, Trumpington Street, Cambridge CB2 1RP
40 West 20th Street, New York, NY 10011-4211, USA
10 Stamford Road, Oakleigh, Melbourne 3166, Australia

First published 1981
Second edition 1994

Printed in Great Britain at the University Press, Cambridge

A catalogue record for this book is available from the British Library

Library of Congress cataloguing in publication data

Australian vegetation / edited by R. H. Groves. – 2nd ed.
    p.   cm.
Includes bibliographical references and index.
ISBN 0 521 41420 2 (hb). – ISBN 0 521 42476 3 (pb)
1. Botany – Australia – Ecology.   2. Plant communities – Australia.
3. Phytogeography – Australia.   I. Groves, R. H.
QK431.A9   1994
581.994 – dc20   93-28688   CIP

ISBN 0 521 41420 2 hardback
ISBN 0 521 42476 3 paperback

To Jean

CONTENTS

List of contributors xi   *Preface to First Edition* xiii
Preface to Second Edition xv   *Acknowledgements* xvii
Map to show the natural vegetation of Australia xviii

Part 1   Introduction   1
  1   Phytogeography of the Australian region
     B. A. Barlow   3
  2   Quaternary vegetation history
     J. R. Dodson   37
  3   Alien plants
     P. W. Michael   57

Part 2   Major vegetation types   85
  4   The rainforests of northern Australia
     L. J. Webb & J. G. Tracey   87
  5   Southern rainforests
     J. R. Busby & M. J. Brown   131
  6   Tall open-forests
     D. H. Ashton & P. M. Attiwill   157
  7   Patterns and processes in open-forests of *Eucalyptus* in
     southern Australia
     A. M. Gill   197
  8   Woodlands
     A. N. Gillison   227
  9   *Acacia* open-forests, woodlands and shrublands
     R. W. Johnson & W. H. Burrows   257
  10   *Eucalyptus* scrubs and shrublands
     R. F. Parsons   291
  11   Heathlands
     R. L. Specht   321
  12   Chenopod shrublands
     J. H. Leigh   345

13 Natural and derived grasslands
    J. J. Mott & R. H. Groves   369

Part 3 Vegetation of extreme habitats   393
14 Saltmarsh and mangrove
    P. Adam   395
15 Aquatic vegetation of inland wetlands
    M. A. Brock   437
16 Alpine and subalpine vegetation
    R. J. Williams & A. B. Costin   467
17 Coastal dune vegetation
    P. J. Clarke   501

Part 4 Conservation of vegetation   523
18 Biodiversity and conservation
    R. L. Specht   525

    Index   557

CONTRIBUTORS

P. Adam, Department of Botany, University of New South Wales, PO Box 1, Kensington, NSW 2033

D.H. Ashton, Department of Botany, University of Melbourne, Parkville, Vic. 3052

P.M. Attiwill, Department of Botany, University of Melbourne, Parkville, Vic. 3052

B.A. Barlow, CSIRO Division of Plant Industry, GPO Box 1600, Canberra, ACT 2601

M.A. Brock, Department of Botany, University of New England, Armidale, NSW 2351

M.J. Brown, Forestry Commission, 199 Macquarie St, Hobart, Tas. 7000

W.H. Burrows, Department of Primary Industries, PO Box 6014, Rockhampton Mail Centre, Qld 4702

J.R. Busby, ERIN Unit, Australian Nature Conservation Agency, GPO Box 636, Canberra, ACT 2601

P.J. Clarke, Department of Botany, University of New England, Armidale, NSW 2351

A.B. Costin, PO Box 62, Bodalla, NSW 2545

J.R. Dodson, School of Geography, University of New South Wales, PO Box 1, Kensington, NSW 2033

A.M. Gill, CSIRO Division of Plant Industry, GPO Box 1600, Canberra, ACT 2601

A.N. Gillison, CSIRO Tropical Forest Research Centre, PO Box 780, Atherton, Qld 4883

R.H. Groves, CSIRO Biological Control Unit, Campus International de Baillarguet, 34982 Montferrier-sur-Lez, France

R.W. Johnson, Department of Primary Industries, Botany Branch, Meiers Rd, Indooroopilly, Qld 4068

J.H. Leigh, CSIRO Division of Plant Industry, GPO Box 1600, Canberra, ACT 2601

P.W. Michael, 5 George St, Epping, NSW 2121

J.J. Mott, Department of Agriculture, University of Queensland, St Lucia, Qld 4067

R.F. Parsons, Department of Botany, La Trobe University, Bundoora, Vic. 3083

R.L. Specht, 107 Central Ave, St Lucia, Qld 4067

J.G. Tracey, CSIRO Tropical Forest Research Centre, PO Box 780, Atherton, Qld 4883

L.J. Webb, Environmental Sciences, Griffith University, Nathan, Qld 4111

R.J. Williams, CSIRO Division of Wildlife & Ecology, Private Bag, Winnellie, NT 5789

PREFACE TO THE FIRST EDITION

Australian vegetation has interested both botanists and naturalists since Europeans first encountered Australia and its plant life. The ubiquity yet distinctiveness of species of *Eucalyptus*, the colour and variety of the vegetation around the early settlement at Sydney, the sometimes-bizarre plant forms of southwestern Australia, and the monotony of the inland mallee areas; all these features and others evoked different responses in different people. Banks and Solander in 1770 had their interest quickened and their botanical horizons widened. The very naming of Botany Bay bears witness to this. Charles Darwin in 1836, on the other hand, found the vegetation around Albany, Western Australia, dreary in the extreme.

The vegetation of Australia still evokes an ambivalent response from more general observers, both those from within the country and those from other continents. A *Nothofagus* or a *Eucalyptus regnans* forest experienced on a misty morning in spring in southeastern Australia can evoke awe in some, but the omnipresent eucalypts of coastal Australia dull the senses and seem monotonous to those with a more 'Europocentric' vision.

No matter what the general response to Australian vegetation, modern botanists are usually intrigued and stimulated by it and want to know more about it. Almost 200 years after European settlement we know quite a lot about Australian vegetation, as the following chapters and their reference lists show. The effects of fire on vegetation and individual species, the role of soil nutrients in delimiting vegetation distribution, the selective grazing of animals are just three of the subjects which Australian ecologists have contributed to international botanical science. Other subjects, such as pollination biology, are greatly in need of more research.

This book is aimed at stimulating the interest of graduate students and fellow botanists around the world, especially those fortunate enough to see for themselves the unique Australian vegetation and flora. To professional botanists resident in Australia both audiences can bring new insights and enthusiasms to the study of Australian vegetation. If reading the chapters that follow provides the catalyst for further research on the ecology of Australian vegetation then the efforts of the individual authors will have been worthwhile.

Some of the chapters include distribution maps of vegetation types. A recent map of all major vegetation types, using the most recent terminology, appears on p. xiv.

Readers unfamiliar with Australian geography can find place names in a recent atlas such as *The Reader's Digest Atlas of Australia*. Places mentioned in the text but not listed in that atlas have their latitude and longitude cited after their first mention.

The chapters were completed between September 1978 and August 1979; later research may not have been included.

PREFACE TO THE SECOND EDITION

It has been encouraging to authors and editor alike that there is a continuing need for a reasonably concise book on Australian vegetation. Publication of the first edition was prompted in part by the visit to Australia of numerous botanists to attend the XII International Botanical Congress held at Sydney in August 1981. An introduction to the unique vegetation types of Australia was required by the international audience who attended symposia and enjoyed excursions on that occasion. Now, 12 years later, a second edition is required by a still wider audience of biologists as a reference and as a stimulus to their own understanding of Australian plant ecology. The volume of scientific publications on Australian vegetation has increased considerably over the last twelve years; the chapters in this book seek to reflect that increase in knowledge.

The aim of this second edition is the same – to stimulate further interest in the ecological relationships which determine the distribution of Australian vegetation types, especially among graduate students and fellow botanists around the world.

The chapters were largely completed by the end of 1991; research results published subsequently may not have been included.

A map of all major vegetation types appears on p. xviii of this edition.

ACKNOWLEDGEMENTS

I wish to acknowledge the willingness of all authors to contribute to this second edition, especially those contributing for the first time. As well, I thank my colleague Colin Totterdell for providing a cover photograph from northern Australia that nicely balances the superb portrait of *Eucalyptus pauciflora* from southern Australia used so effectively for the cover of the first edition. Dr John Carnahan kindly allowed the re-use of his simplified map of Australian vegetation from the first edition.

Finally, I record the absence of contributions from G. Singh and O.B. Williams, who co-authored chapters in the first edition. We regret their premature deaths in the period since 1981 and acknowledge their distinctive and major contributions to Australian plant ecology.

R.H. Groves
Montpellier, 1992

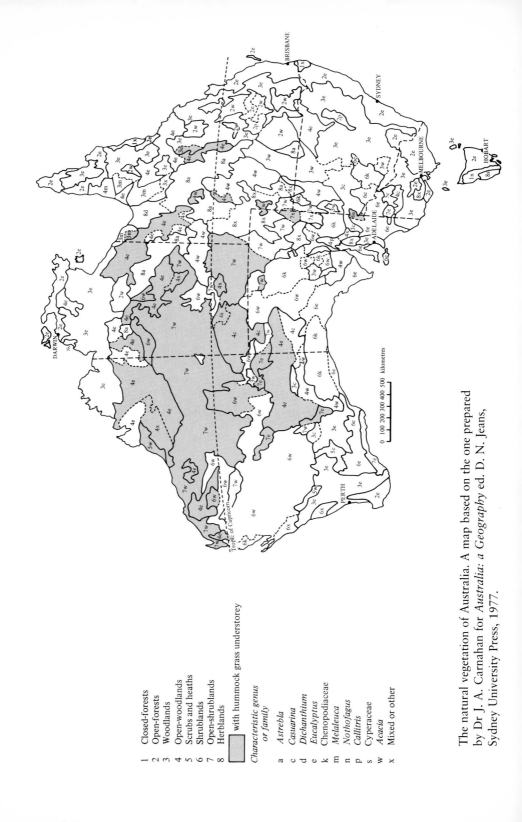

The natural vegetation of Australia. A map based on the one prepared
by Dr J. A. Carnahan for *Australia: a Geography* ed. D. N. Jeans,
Sydney University Press, 1977.

1 Closed-forests
2 Open-forests
3 Woodlands
4 Open-woodlands
5 Scrubs and heaths
6 Shrublands
7 Open-shrublands
8 Herblands

with hummock grass understorey

Characteristic genus
or family

a *Astrebla*
c *Casuarina*
d *Dichanthium*
e *Eucalyptus*
k *Chenopodiaceae*
m *Melaleuca*
n *Nothofagus*
p *Callitris*
s *Cyperaceae*
w *Acacia*
x Mixed or other

PART 1

INTRODUCTION

1

Phytogeography of the Australian region

B.A. BARLOW

THE AUSTRALIAN vegetation has many unique features, both in composition and structure. In composition, it is unique in the sense that a very high proportion of its species is endemic to the continent. It is also unusual in that two large tree and shrub groups, the eucalypts and the phyllodinous acacias, between them dominate almost all the plant associations of the continent, whilst having very limited natural occurrence outside Australia. In structure, it includes vegetation types, such as sclerophyll forest, mallee woodland and hummock grassland, which do not fit easily into global vegetation classifications (Doing, 1981). The deciduous habit is rare, and woody scleromorphic forms are a feature of many communities. In a broad sense, then, the unique character of the Australian vegetation is reflected in plant communities often of unusual structure, dominated by *Eucalyptus* and *Acacia*, and including other genera and species which usually do not occur anywhere else.

Plant form and vegetation structure are ecophysiological consequences of climatic history, and the structural uniqueness of the Australian vegetation can be explained in these terms. The uniqueness in composition of the present-day flora is also a consequence of its evolutionary origins. This uniqueness virtually disappears, however, at higher taxonomic levels, which show the Australian flora to be a typical subset of the world flora. Almost all the angiosperm families in Australia occur widely elsewhere, and conversely, almost all the larger families of the world occur in Australia. The uniqueness of the Australian flora at the generic and specific levels therefore requires explanation in terms of the geophysical and climatic history of the continent, which have provided the isolation and selection pressure under which floristic diversification has occurred as well as the impetus for structural change.

Understanding Australian phytogeography is simplified by the

3

striking diversity of the existing plant communities. As an island continent occupying a latitudinal range from tropical to cool temperate, there are physical environments ranging from maritime to continental, ever-wet to arid, and seasonal to almost aseasonal. There is a diversity of soils, many of them very old, found in a range of habitats from extensive lowlands to localised alpine. Understanding the geophysical history of these environments is the key to understanding the derivation of the plant communities and their relationships to each other.

Landform history

Although the concept of continental displacement dates back to 1858 (Smith, 1982), it has only been generally accepted and applied in phytogeography since about 1960. Whilst its relevance to Australian vegetation history was always very obvious, continental displacement was long thought to raise more biogeographical problems than it solved, and even if it did occur, to have been too early to have had any bearing on angiosperm dispersal. Australian phytogeography was based on the assumption that the earth's geography was fixed, so that Australia's isolated position relative to other continental land masses had not changed. Changes in sea level and/or tectonic movements in the earth's crust were thought to have created land bridges between Australia and other regions. There was resistance to the idea of individual plant migration through long-distance dispersal, and it was assumed that whole plant communities had migrated together. It was further assumed that the angiosperms could not have arisen in Australia, and must therefore have arrived as waves of colonisers from outside.

The challenge to this phytogeographic 'invasion' theory was two-fold. The plant fossil record in Australia did not support it (see below). Secondly, whilst there was no supporting geophysical evidence for the required land bridges, the direct evidence supporting plate tectonics and continental displacement steadily became stronger.

The physical basis and biological consequences of plate tectonics are now embodied in a widely accepted theory of Gondwanan biogeography in which Australia has a pivotal position. At the close of the Cretaceous period, Antarctica, Australia, New Zealand and the components of the present New Guinea were closely assembled at a higher latitude, and had a warm, humid climate. Although the northward separation of Australia from a stationary Antarctica commenced about 125 million years ago (Ma), movement remained slow (4.5 mm/y) until 55 Ma, when the rate increased to 60–70 mm/y (Wellman & McDougall, 1974; Crook & Taylor, 1985). A narrow land connection between Australia and Antarctica may have persisted until about 40 Ma (Kemp, 1978). The Tasman Sea started to open about 80 Ma, with New Zealand and the Lord Howe Rise separating from Australia leaving a clean faulted edge on the Australian side (Ollier, 1986). The Coral Sea also started to open at about this time, with the separation of Papua New Guinea from northeastern Australia. The southern part of western Papua is a continuation of the Queensland part of the Australian plate.

Circumpolar oceanic circulation was initiated with the separation of Australia and Antarctica, thereby reducing heat transport from equator to pole and increasing temperature gradients. Thus at the same time as Australia was moving away from Antarctica the air circulation was becoming cooler and drier, and arid conditions started to overtake Australia about 30 Ma (Bowler, 1982).

In the Miocene, after about 20 million years of comparative geographic isolation, the northward-moving Australian plate collided with the westward-moving volcanic arcs of the Sunda region (Fig. 1.1). The northwestern margin of the Australian plate, as it existed at the time of contact, is now represented by the Outer Banda Arc, extending as a great deformation from Timor to Buru, thence the southeastern and eastern arms of Sulawesi, the Moluccan islands and the Vogelkop of New Guinea (Audley-Charles, 1981, 1987; Audley-Charles, Carter & Milsom, 1972; Audley-Charles, Hurley & Smith, 1981; Audley-Charles, Ballantyne & Hall, 1988). Although there is increasing evidence that certain blocks on the Asian side of the collision boundary are earlier Gondwanan fragments, the Miocene contact ended a long period of isolation of differentiated Australian and Malesian floras (see p. 14).

Land surfaces in the contact zone increased as the convergence continued. It was probably not before latest Miocene or early Pliocene that a land migration route from Celebes to Australia via New Guinea was established (Audley-Charles, 1981), and a strong floristic contact between the Australian and Malesian regions ensued. Prior to this, exchange would have been limited to a filtering of taxa of higher dispersibility (McKenna, 1972). By the late Pliocene, lowered sea levels would have exposed extensive land on both sides of the contact boundary. Whilst the Makassar Strait has long existed as a seaway, and is the most robust part of Wallace's line, Borneo and Celebes would have been almost continuous at its southern end (Audley-

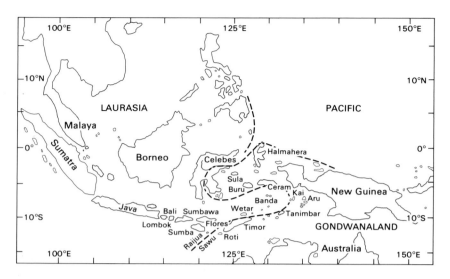

Fig. 1.1. Present-day relationship of the Australian and Sunda plates following the Miocene collision event. (Modified from Audley-Charles, 1981.)

Charles, 1981). By the middle Pleistocene, the Lesser Sunda Islands would have been connected to Asia by continuous land (at least as far as Timor), and would have been separated from mainland Australia by only a short sea distance.

Floristic elements

The first phytogeographical analysis of the Australian flora by Hooker (1860) provided the basis of the invasion theory which persisted for more than a century. Hooker had an excellent general knowledge of the entire world flora, and produced a classical analysis in which the Australian flora was resolved into 'elements' based on taxonomic affinities with the floras of other regions. He identified three elements, and it is not surprising that there is a strong ecological as well as taxonomic distinction between them. These were (1) an autochthonous (Australian) element consisting mainly of endemic taxa occupying temperate open forest, woodland and heath habitats and mainly scleromorphic in character, (2) an Indomalayan element represented in tropical and subtropical rainforest and monsoon habitats and showing taxonomic affinity with plants of similar habitats in the Indo-malayan region, and (3) an Antarctic element represented in temperate rainforest and alpine habitats, characterised by *Nothofagus* and showing taxonomic affinity with plants of New Zealand and temperate South America in particular. Hooker also noted the presence of cosmopolitan plant groups, mostly herbaceous, widely distributed within Australia. Whilst Hooker himself drew no conclusions as to the actual history of the Australian flora, his analysis became the basis of the invasion theory of colonisation of the Australian land mass by separate incursions of different floras, perhaps at different times, over land bridges fortuitously made and broken. For an account of the history and weaknesses of the invasion theory, and its eventual replacement, see Barlow (1981).

Phytogeographic elements now recognised in the Australian flora relate directly to the prevailing concepts of Gondwanan biogeography. The two basic elements are (1) the Gondwanan Element, comprising stocks derived directly from the original Gondwanan flora which was present in Australia at the time of separation from the supercontinent, and (2) the Intrusive Element, comprising plants which have entered Australia subsequent to its separation from Gondwanaland (Barlow, 1981; Nelson, 1981).

The Gondwanan Element comprises two subelements, a Relict one and a derived Autochthonous one (Barlow, 1981; Nelson, 1981). The Relict Subelement includes taxa which are mostly still confined to humid habitats like those presumed to have existed in Gondwanaland. They show strong residual taxonomic affinity with floras of similar habitats in other southern lands, and having apparently shown little evolutionary diversification from the Gondwanan flora, they are presumed to represent the present-day survivors of this stock.

The Autochthonous Subelement developed in response to the onset of climatic events which followed the separation and northward drift of Australia from Antarctica. Of particular importance were the cycles of aridity which increased in frequency and amplitude from the middle of the Tertiary period (Bowler, 1982). This element also developed in response to edaphic evolution, which may have been the catalyst for the widespread evolution of scleromorphy (see below). It shows much evolutionary diversification from its Gondwanan progenitors, and it is the high generic and species endemism of this subelement which imparts the unique character to the Australian flora. It is predominantly temperate and arid-adapted, and is the largest component of the flora.

The Intrusive Element is a composite of several different subelements. Nelson (1981) recognised (1) a Tropical Subelement, comprising Malesian plants of post-Miocene inception, (2) a Cosmopolitan Subelement of widely distributed taxa, especially important in the arid zone (see below), and (3) a Neoaustral Subelement of mainly temperate species of northern hemisphere derivation, thought to be important in the alpine and subalpine zone (see below). Barlow (1981) also recognised an Indogondwanan Subelement, to accommodate plants of ancient Gondwanan derivation which have reached Australia as part of the post-Miocene intrusion. Further analysis suggests, however, that the entire intrusive tropical flora may ultimately be of Gondwanan derivation, and the Tropical and Indogondwanan Subelements may therefore be identical (Barlow & Hyland, 1988; see below).

The phytogeographic elements in the Australian vegetation are summarised in Table 1.1.

Table 1.1. *Phytogeographic composition of the Australian vegetation*

Element	Subelement	Description
1. Gondwanan		Derived directly from the original Gondwanan flora present in Australia
	1a. Relict	Present-day survivors of the humid Gondwanan flora
	1b. Autochthonous	Highly endemic, derived in response to climatic cycles under geographic isolation
2. Intrusive		Reached Australia after separation from Gondwanaland
	2a. Tropical	Malesian plants of post-Miocene inception, although probably ultimately of Gondwanan derivation
	2b. Cosmopolitan	Widely distributed taxa of high dispersibility
	2c. Neoaustral	Temperate species of northern hemisphere derivation

Gondwanan Relict Subelement

The scenario for an Australian flora at the beginning of the Tertiary period was a pan-Australian Gondwanan closed rainforest. Climatic conditions were warm and moist; high rainfall was general in southern Australia and extended through the interior. Temperatures were high in northern and inland Australia (Wopfner, Callen & Harris, 1974) and warm (20–25 °C) in southern Australia. Some ecological zonation probably existed, with limited differentiation between warmer and more temperate habitats. The southern beech *Nothofagus* was widespread, as were plants referable to *Araucaria*, *Podocarpus*, *Dacrydium*, *Anacolosa*, Cupanieae, Myrtaceae, *Nipa*, and several Proteaceae. Retraction of these Gondwanan mesic forests, and replacement by open forests, woodlands and heath, followed from the cycles of aridity which increased in frequency and amplitude from the middle of the Tertiary period (Bowler, 1982), and from associated edaphic evolution.

The uplands of eastern Australia have existed at least since the beginning of the Tertiary period (Ollier, 1982, 1986), and have continuously provided habitats favourable for survival of mesic forests (Barlow & Hyland, 1988). Dissected plateaux, narrow valleys, edaphic mosaics, localised high rainfall and fire-proofing, all associated with the survival of closed forests, have presumably been a continuous feature of the eastern Australian Tertiary–Quaternary environment. The other old uplands in Australia have all been overcome by major climatic cold and arid cycles, and the present day uplands of New Guinea and New Zealand are too young to have become major refugia. The old eastern uplands of the Australian plate may therefore be the only large part of the Australasian/Pacific region where Gondwanan mesic forests have persisted continuously since Cretaceous times. Even here, the rainforests which include the Relict Subelement have a very dissected area.

Component floristic groups of the Australian rainforest vegetation have been identified in detailed analyses by Webb & Tracey (1981), Webb, Tracey & Williams (1984), and Webb, Tracey & Jessup (1986). Numerical analysis of 1316 species in 406 genera from 561 sites identified three major 'ecofloristic' regions. These are (1) temperate and subtropical humid rainforests of southeastern Australia, (2) warm humid rainforests of northeastern Australia, and (3) 'dry' monsoon forests of northern and eastern Australia. These three regions overlap substantially, but a core area can be identified for each.

Outside the core area each floristic entity tends to be represented by outliers, often well within the area of another entity. The outlying occurrences are relictual, being the results of climatic oscillations and sifting of the ecofloristic entities. The core areas correlate in a general way with the optimal growth regimes in Australia calculated by Nix (1982).

These ecofloristic entities are so different in composition that they may not be simple segregates of a single rainforest flora, derived by climatic filtering and attenuation (Webb & Tracey, 1981). Rather, they may reflect old floristic elements differentiated very early from Gondwanan stocks, and with different histories in the Australasian region.

Within the first region, three ecofloristic provinces were recognised (Webb & Tracey, 1981). Two of these are the temperate rainforests of southernmost and southeastern Australia respectively. The third comprises the subtropical rainforests of eastern Australia between approximately 22° and 30°S latitudes. Webb & Tracey argued a separate identity for the latter rainforest province, and in postulating biogeographical relationships treated it as a fourth rainforest component of the Australian vegetation (Fig. 1.2).

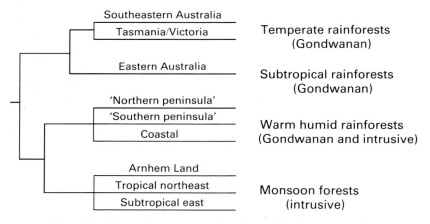

Fig. 1.2. Schematic representation of ecofloristic regions and provinces as proposed by Webb & Tracey (1981).

Of these four humid forest stocks, the temperate and subtropical ones are recognised as direct derivatives, in Australia, of the ancestral Gondwanan flora, and therefore belong in the Relict Subelement. The first is the cool temperate and montane mossy (*Nothofagus*) forest of southeastern Australia and Tasmania. The second is the warmer subtropical vine forest, characterised by *Elaeocarpus*, *Ceratopetalum*, *Doryphora* and *Cinnamomum*, and possibly the extant relict of the 'Cinnamomum flora' of the early Tertiary (Christophel, 1981). These two rainforest stocks may represent the original differentiation of the Gondwanan flora into cool and warm forest formations.

The third ecofloristic region comprises the hot, humid vine forests of Australia's wet tropics. This unit includes many old Gondwanan taxa, but has been greatly modified by later interaction with the Intrusive Malesian flora. Webb & Tracey (1981) recognised three ecofloristic provinces, distinguishing the rainforests of the Cooktown-Atherton Tableland block from, firstly, the seasonal forests which occur mainly in areas of lower relief in northern Cape York Peninsula, and secondly, the specialised forests of sand dune systems on eastern Cape York Peninsula. The Gondwanan Relict Subelement is represented mostly in the first province. The tropical rainforest region is described further under Intrusive Element.

The Relict Subelement in the tropical rainforests of Australia includes a remarkable array of primitive angiosperms. Of the 14 such families recognised by Takhtajan (1969), eight occur in northeastern Australia (Webb & Tracey, 1981). These include two narrowly endemic families (Austrobaileyaceae, Idiospermaceae), and another which is shared only with New Guinea (Eupomatiaceae). Although there are greater concentrations of primitive genera and species in southeast Asia, the concentration of primitive families in the Australasian region is the greatest in the world (Webb & Tracey, 1981). There is no evidence that these families evolved *in situ* in northeastern Australia, and the primitive angiosperms still extant presumably all achieved wide distributions in Gondwanaland in Cretaceous times. The northeastern Australian refugium for the wet tropical component of the Relict Subelement is thus of great antiquity (Barlow & Hyland, 1988).

The fourth region, comprising the monsoon forest flora, can be identified as part of the Intrusive Subelement, and is discussed below.

Gondwanan Autochthonous Subelement

The Tertiary history of the Australian vegetation has been one of differentiation from the original Gondwanan flora, under conditions of increasing geographical isolation until the Miocene contact with the Sunda plate. Climatic changes were progressive until, in the Miocene, conditions comparable with those of the present were established. Decreasing mean temperatures, increasing intensity of atmospheric pressure systems, regional decreasing precipitation and increasing seasonality increased the level of habitat differentiation, and the pan-Australian Gondwanan flora diversified into a spectrum of ecological associations. The Autochthonous Subelement simply comprises those labile components of the original flora which have undergone change to produce typically Australian taxa with high levels of endemism. Conversely, the Relict Subelement simply represents the survivors of the Gondwanan flora which have been more conservative in the evolutionary sense.

The major features of the Autochthonous Subelement were thus established by the middle of the Tertiary period. They include open-forest and woodland formations dominated by *Eucalyptus* and *Acacia*, both of which hardly penetrate the closed-forest associations. They also include shrublands, heath formations, herbfields and grasslands in habitats ranging from tropical to cool temperate and alpine. They are characterised by remarkable diversification in certain families, including Myrtaceae, Proteaceae, Rutaceae, Epacridaceae, Fabaceae, Mimosaceae, Goodeniaceae and Casuarinaceae. Most of the component species are perennial, and often woody.

This flora has been moulded and differentiated by later geophysical and climatic events. Geographical isolation of the temperate floras of eastern and western Australia was initiated by marine incursions into southern Australia which were possibly continuous from late Eocene to mid-Miocene times (Nelson, 1981). The emerged sediments have since remained as dry land, and the limestones of the present Nullarbor Plain, together with the arid dune systems to the north, may have maintained an edaphic and climatic barrier between east and west which has been almost uninterrupted to the present (Crocker & Wood, 1947; Raven & Axelrod, 1974; Nelson, 1981).

This extended isolation is reflected in the high level of specific endemism in the very rich flora of the South West Botanical Province in Western Australia, estimated at 87 per cent by Beard (1969) and modified to 68 per cent by Marchant (1973) and to 75–80 per cent by Hopper (1979). Generic endemism is low, and there are many vicarious species pairs in eastern and western Australia. This is consistent with the imposition of isolation at a time when the major genera of the Autochthonous Subelement were already widely established. Hopper (1979) showed that the greatest species density in the South West Botanical Province occurs in the transitional rainfall zone between the more mesic forest areas and the arid zone, and that a high proportion of the species are local endemics within this zone. He concluded that recurrent climatic fluctuations would have generated greater stresses in the transitional zone and, coupled with the evolution of a mosaic of nutrient-deficient soils, would have produced disruptions in labile population systems ideal for rapid speciation.

Scleromorphy

A striking aspect of the Autochthonous Subelement is its general scleromorphy. Many of its major genera are characterised by relatively small, rigid leaves, by short internodes and by small plant size. Scleromorphy is common in Australian Myrtaceae, Proteaceae, Rutaceae, Epacridaceae, Mimosaceae, Fabaceae and Goodeniaceae. It was long considered that scleromorphy in the Australian flora was an adaptive response to increasing aridity, and in particular to the advent of mediterranean climate. A notable exception was Andrews (1916), who linked scleromorphy with soil properties and suggested that it was an adaptive response to low levels of soil nitrogen and calcium. It is now widely accepted that scleromorphy is an adaptive response to nutrient deficiency, marked among other things by physiological processes leading to reduction in the number of cells formed (Beadle, 1954, 1966, 1968, 1981; Loveless, 1961; Specht, 1972; Johnson & Briggs, 1975, 1981). Beadle (1954, 1968) suggested that, because soil phosphorus levels regulate soil nitrogen fixation, phosphorus status is the basis of scleromorphy as a physiological adaptation.

Conditions favouring the evolution of scleromorphy in the Australian flora probably arose early in the Tertiary period. Heath communities are part of the moist tropical ecosystem, forming a mosaic with closed-forest communities and occurring on deep infertile sandy soils (Specht, 1981). They share families and even genera in common with rainforest. Scleromorphic forms probably differentiated at the margins of rainforest along declining soil fertility gradients (Beadle, 1981; Johnson & Briggs, 1981). The process probably began at least in early Tertiary times, possibly catalysed by climatic change, and has continued through successive differentiations as the climate has been progressively modified. Because of the continuous edaphic processes on so much of the Australian land surface, poor soils are now the rule, and scleromorphic vegetation has become progressively more widespread, resulting in a level of expression not matched elsewhere.

Even the eucalypts have apparently originated at the margins of mesic closed-forests (Pryor & Johnson, 1981; Johnson & Briggs,

1984). The distributions of their less specialised elements, including *Arillastrum* in New Caledonia and others in tropical Australia, suggest a subtropical or tropical origin. A similar pattern is evident in *Melaleuca*, another important and widespread scleromorphic genus of Myrtaceae. It replaces the eucalypts as the dominant tree in seasonally inundated tropical lowlands, and the most primitive section of the genus is primarily tropical in distribution (Barlow, 1988).

Scleromorphy was thus a pre-adaptation to, rather than a direct consequence of, the inception of mediterranean climates in southern Australia (Barlow, 1981). Later in the Tertiary the scleromorphic flora colonised habitats where this climatic regime became established, but it has remained confined to low-nutrient soils even under increasingly dry conditions, not effectively spreading to adjacent higher-nutrient soils. This feature of scleromorphic communities may explain the very limited representation of an Australian floristic element in Malesia (Steenis, 1979).

The scleromorphic heath communities in tropical Australia may therefore in a sense be relictual. The much greater development of such communities in temperate seasonal and arid situations in Australia reflects the later radiation which has occurred as these climatic conditions have overtaken the southern part of the continent.

Intrusive tropical subelement

By the time the contact between the floras of the Australian and Malesian regions was strengthening, through Miocene to Pliocene time, much differentiation had already occurred in the Gondwanan Element in Australia. Seasonally arid areas and open-forest and woodland communities were apparently already widespread (Truswell & Harris, 1982), and mesic forests were probably already somewhat restricted in distribution. The interaction between the existing Australian flora and the intrusive Malesian stocks was strong in New Guinea, which is an integral part of the Australian floristic region (see below). In what is now the Australian mainland, tropical intrusive influences are found in the tropical rainforests of the northeast, in the monsoon vegetation of the north, and secondarily in the arid zone vegetation (see below).

Whilst floristic exchange between the Australian and Sunda plates is now doctrine, there are different interpretations of the nature of the interacting floras. At one extreme it has been viewed as a contact between Laurasian and Gondwanan megafloras (Whitmore, 1981).

At the other extreme it has been argued that the entire tropical flora of the Asian–Australian region is Gondwanan, and the collision event has simply re-established contact between derivatives of the one basic stock (Webb & Tracey, 1981; Webb *et al.*, 1986). The differences, however, are largely semantic, as there is little merit in applying 'Gondwanan' and 'Laurasian' labels too rigorously (Thorne, 1986). There is clear evidence of taxonomic difference between the Asian and Australian tropical floras, and in this sense they may be conveniently described as Laurasian and Gondwanan. There are also obvious phyletic links between the tropical floras of Africa, Asia and Australia, all of which were ultimately derived from the ancestral Gondwanan flora. The world occurrence of tropical rainforest, as illustrated by Richards (1952), is still mostly on Gondwanan land surfaces. Whilst the largest area of apparently Laurasian land surface with tropical forest is in fact the southeast Asian/Malesian area, its close proximity to northward-drifting India about 40 Ma, and its accretion of earlier Gondwanan fragments (Audley-Charles, 1987; Audley-Charles *et al.*, 1988), suggest possible Gondwanan sources of the Malesian tropical flora (Specht, 1981; Barlow & Hyland, 1988). Irrespective of terminology, however, it is clear that two floras, separated by a long period of time and differentiation, were brought into contact as a result of the collision event.

The major area of interaction of the intrusive and Australian rainforest floras is to the east of the contact zone, in New Guinea and northeastern Australia. Most of the New Guinean uplands are very young; in addition to the emplacement of exotic terrains there has been recent vertical uplift along the axis of the island at a rate of *c.* 2 mm/y (Chappell, 1974; Pain & Ollier, 1984). Thus whilst New Guinea is clearly part of the Australian region with respect to floristic history, the extensive ever-wet uplands of the present day did not exist in this form through most of the period of angiosperm history of the region, nor even at the time of the first floristic exchange with Malesia.

Although New Guinea must have initially been occupied by Australian plants, it appears to have been largely over-run by Malesian intrusives, at least in the lowlands (Whitmore, 1981; Hartley, 1986). At the time of the collision, humid tropical conditions prevailed at the northern margin of the Australian plate, and the original Australian (Gondwanan) flora may have been less competitive than the Malesian flora to the west (Hartley, 1986). Hartley's analysis, based on 716 New Guinean genera of ever-wet primary forests, showed that more than three fourths of them may be of direct Malesian origin.

There is a residual ecological differentiation of the major elements in the New Guinean rainforest flora (Whitmore, 1981; Hartley, 1986). Whilst the Malesian component is almost ubiquitous in the lowland ever-wet forests, the Australian element occurs predominantly in lower montane and montane forests, mostly above 900 m elevation. The differentiation is not absolute, and whilst many tree genera of New Guinean rainforests above 1500 m are of southern affinities (Schodde & Calaby, 1972), there are genera of wider ecological amplitude which extend into this zone, and which are often of Malesian affinity (Hartley, 1986). This austral rainforest province is just as clearly evident in New Guinea, however, as it is in northern Queensland, thereby indicating the importance of New Guinea as part of the Australian phytogeographic region.

The New Guinean block has provided the primary route for dispersal of Malesian intrusives to Australia. Of the large number of pantropic and African-Madagascan-Malesian genera in the New Guinean rainforest flora, 79 per cent also extend to Australia (Hartley, 1986). This strong Malesian influence in the rainforest flora of tropical Australia is further indicated by direct analysis of the northern Australian floristic region (Barlow & Hyland, 1988).

The integrated Gondwanan-Intrusive rainforests have been moulded in a dynamic pattern of shifting vegetation mosaics in the Quaternary (Kershaw, 1981; Webb & Tracey, 1981; Barlow & Hyland, 1988). Boundaries between rainforests and open-forests have advanced and retreated, and rainforest areas have been fragmented and coalesced. Vegetation changes at local sites have been frequent, as documented by Kershaw (1978). In the last 120 000 years, a site on the Atherton Tableland has supported tropical rainforest, succeeded by gymnosperm-dominated rainforest with *Araucaria* and sclerophylls, then low mesophyll vine forest similar to the cool subtropical rainforest of southeastern Queensland, then warm subtropical rainforest with increased angiosperm composition, again the *Araucaria*-dominated cooler forest, then sclerophyll woodland with *Eucalyptus* and *Casuarina*, until the present-day forests returned to the area 10 000 years ago.

The shifting vegetation mosaics have left some waifs as evidence of their movement. Many species which are widely distributed in a 'core area' have a limited 'relict' occurrence remote from it, or in a different floristic province (Webb & Tracey, 1981). Thus one consequence of the climatic cycles of the Quaternary has been some weakening of the floristic distinctions between the tropical forest provinces of the region.

Quaternary climatic cycles have regulated recent floristic exchange between New Guinean and Australian rainforests. Whilst the two lands have been broadly connected above sea level through much of Quaternary time, the low-lying Torresian land surface may have been mostly unsuitable for humid forest colonisation. Conditions favouring rainforests in the Quaternary were seldom more favourable than they are at present, and the Torres Strait region may have been an effective ecological barrier to floristic exchange for much of this time (Walker, 1972). Closed-forests may have extended almost continuously along a narrow Torresian isthmus from New Guinea to Queensland 8000 years ago (Nix & Kalma, 1972), and more protracted contact may have occurred around 100 000 years ago when tropical forests had a more durable tenure. During periods of maximum cooling and aridity the mesic forests may have disappeared almost completely, most recently 20 000–17 000 years ago (Nix & Kalma, 1972).

The present rainforests of northern Cape York Peninsula thus represent the limited results of recolonisation, from both New Guinean and Queensland refugia, which probably reached maximum momentum about 8000 years ago. These forests are relatively depauperate, and have low endemism (Barlow & Hyland, 1988). Of 165 rainforest genera which occur in northern Queensland but do not reach northern Cape York Peninsula, about 40 per cent are Gondwanan, reaching their northern range limits in northern Queensland. Almost all of the remainder are further distributed in New Guinea and/or the Malesian region, and their absences from the northern area simply represent disjunctions in their distributions (Fig. 1.3). The northern Peninsula region has thus undergone more pronounced floristic perturbations in the Recent than the areas immediately to the north and south, which are more stable and more refugial in character. Critical systematic studies support the view that Recent colonisation of northern Cape York Peninsula has involved New Guinea as a source area (Whiffin & Hyland, 1986).

The ecofloristic provinces distinguished by Webb & Tracey (1981) thus conform to the vegetation history of the warm humid rainforests. The Cooktown-Atherton Tableland province forms a relatively stable reservoir which provides continuity of rainforests through the extremes of climatic cycles. Most of the Gondwanan Relict Subelement in the tropical forests occurs in this province, integrated with the Intrusive Element. The seasonal forests of the Northern Peninsula province, in contrast, represent opportunistic recolonisation of the core area by components of the Intrusive Element. The coastal province of specialised closed-forest on dune systems is also a differentiate of the Intrusive Element.

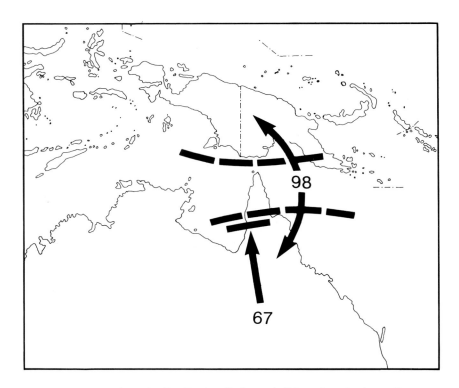

Fig. 1.3. Generic distribution limits and disjunctions in Australian rainforest genera in northeastern Australia (from Barlow, 1990*b*). See text for explanation.

Monsoon vegetation

Monsoon forests occur in tropical areas in which there is a pronounced dry season. They often lack some of the attributes usually associated with tropical rainforests (Webb & Tracey, 1981). The trees may be relatively short, and become increasingly deciduous with increase in length or intensity of the dry season. The canopy is more open, plank buttresses are lost and growth forms such as bottle trees and spinescent shrubs are more conspicuous.

The ecological and floristic affinities of monsoon forests are nevertheless with the rainforests from which they have differentiated (Webb & Tracey, 1981). The monsoon forest flora is distinguished by low species diversity and low endemism. There are no primitive angio-sperms, and conifers are rare. The Australian genera are typically widespread, usually to Malesia, and often to Africa (Steenis, 1979; Whitmore, 1981). The monsoon forest thus represents a dry palaeo-tropical floristic element which may have had an extensive and rela-tively continuous distribution in the past. In Australia it may be restricted to virtually relictual occurrences in local areas, on a broad scale as a consequence of Recent arid maxima, but on a finer scale because of its fire sensitivity. These occurrences extend from Arnhem Land eastwards to Cape York Peninsula and thence southwards to the Upper Hunter region of New South Wales.

In the Malesian region, the two great blocks of tropical rainforest are separated by a broad corridor of seasonally dry climates covering much of the Philippines, Sulawesi, Moluccas, Lesser Sunda and Banda Arc islands, part of Java and part of southern New Guinea (Whit-more, 1981; Fig. 1.4). In this corridor monsoon forest forms a mosaic with rainforest, the latter being confined to gallery and other wetter sites. The Australian occurrence of monsoon forests is an extension of this pattern, and northern Australia and some seasonally drier parts of northeastern Australia are obviously part of the same corridor.

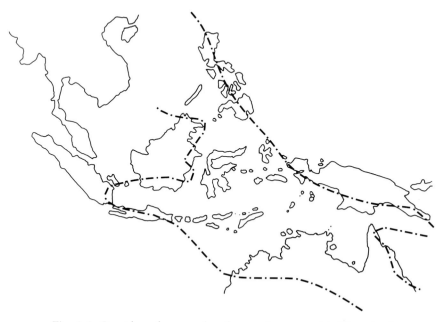

Fig. 1.4. Corridor of seasonal and strongly seasonal habitats in the Malesian–Australian region. (Modified from Whitmore, 1981.)

By the late Pliocene, progressive increase in seasonality of the climate would have resulted in expansion of monsoon forests at the expense of closed mesic forests. Thus at a time when the potential for migration over continuous land in the Australian/Malesian region was greatest, the predominant vegetation in the corridor may have been the monsoon forest type (Barlow & Hyland, 1988). As well as contributing directly to the monsoon forest flora of northern Australia, this intrusive stock has also been significant in the evolution of an Australian arid zone flora (see below).

Webb & Tracey (1981) recognised three ecofloristic provinces within the monsoon forest region, based on floristic composition (Fig. 1.2). These have core areas in Arnhem Land, the tropical northeast and subtropical eastern Australia respectively, and they differ in the level of residual floristic affinity which they have with the warm humid rainforests.

Arid zone vegetation

The Cosmopolitan Subelement of the Intrusive Element recognised by Nelson (1981) is not exclusively represented in the arid zone flora. Its contribution to arid zone plant associations has been more significant, however, than in any other components of the present Australian vegetation.

The arid or eremean zone in Australia is generally defined in terms of the 250 mm isohyet. So defined, it comprises more than one third of the Australian land area. In addition, there are other floristic regions in Australia where seasonal aridity is significant. In monsoonal northern Australia, for example, there is a prolonged winter dry season of up to nine months. Burbidge (1960) recognised three large 'interzones' between the arid zone and the temperate and tropical zones, characterised by seasonal aridity and often by heavy soils, and with a significant representation of arid-adapted plants. In the broad sense, then, more than half the land area of Australia at the present time is continuously or seasonally arid.

Diels (1906) classified the arid zone flora into a palaeotropical group and a cosmopolitan group, mainly in the north, and an Australian group and a littoral group, mainly in the south. Burbidge (1960) seems to have taken up this classification with little change. In her analysis of 363 genera of the arid zone flora, 102 were found to be endemic. Of the remainder, 91 genera were also represented in temperate Australia and 81 in the adjacent tropical lowlands. The cosmopolitan component comprised the other 89 genera, which also occurred throughout all regions of Australia.

These analyses support the view that the arid zone flora is young, having arisen only after extensive arid conditions were established in relatively recent geological time. It was derived by selection from the pre-existing, highly adapted, total Australian flora. Components of tropical lowland affinity and derivation dominate in the northern part of the Eremea (Burbidge, 1960) and components apparently derived from the Autochthonous Subelement dominate in the southern part. The main difficulty has been in the explanation of the cosmopolitan component, including endemic genera in families such as Poaceae, Chenopodiaceae, Brassicaceae, Aizoaceae and Asteraceae, which are well represented in all of the major world deserts. How did these colonisers cross extensive areas of unsuitable and fully occupied territory to become established in the young Australian arid zone?

The explanation is generally based on the fact that these cosmopolitan groups are also well represented in littoral habitats, where salinity and soil type may impose physiological conditions similar to those of deserts. Their colonisation of the deserts may therefore have occurred from coastal habitats, especially in places where the arid zone extends to the coast. Burbidge (1960) gave this hypothesis overriding weight, suggesting that the progenitors of the Australian arid zone vegetation existed on the coastlines from Cretaceous times until the late Tertiary, when the first extensive arid areas were formed.

Because the arid flora is a composite flora derived from adjacent, older plant communities as arid conditions overtook the continent, interest centres mainly on its age. Whilst post-glacial arid cycles were probably the main determinant of the arid zone flora, fossil and climatological evidence suggests that differentiation of eremean plants began in the Tertiary (Crocker & Wood, 1947; Burbidge, 1960). Plate tectonics models suggest that a northward-drifting Australia would have entered the subtropical arid belt in Miocene time, and therefore that extensive aridity would have impinged on the continent from the north 15 Ma (Beard, 1976).

An analysis by Bowler (1982) also suggests that the present expression of aridity in the Australian landscape dates from events which began in Miocene time. It questions, however, the concept that aridity was initiated from the north by Australia's drift into dry subtropical latitudes. Bowler suggested, from palaeoclimatological evidence, that there has been a northerly displacement of weak subtropical anticyclonic pressure systems from winter latitudes near 50° S in the Miocene. Aridity would have therefore overtaken the Australian continent from the south rather than the north, and equable, moist, summer-rainfall conditions would have given way to seasonally dry conditions. The present climatic pattern over Australia was initiated about 2.5 Ma, as the continued northward movement of the pressure systems increased the moisture budget again in southern Australia, leaving the interior with a reduced moisture budget. According to Bowler the present arid zone landforms have a timeframe of 700 000 years but the major wet–dry oscillations of the last 100 000 years have been most significant in determining the landscape.

The palaeobotanical record suggests that open and perhaps arid plant communities have existed in Australia since the Miocene or even earlier. At the Eocene Hale River deposits in central Australia, grass pollen comprises 7 per cent of the total pollen count (Truswell & Harris, 1982), thereby indicating presence of localised grassland formations. In samples of Miocene age grass pollen is as high as 10 per cent and *Casuarina* pollen is sometimes abundant, although cool temperate conditions are generally indicated by the frequent dominance of *Nothofagus* and *Dacrydium*. Lange (1978) recorded from near Woomera a Miocene assemblage of fruits assignable to *Eucalyptus*, *Leptospermum*, *Calothamnus*, *Melaleuca-Callistemon* and *Angophora*. The fossil record for the Miocene, whilst not giving clear evidence of extensive deserts similar to those of today, does indicate the expansion of open-forest vegetation and, given the accepted bias of the fossil record towards wet environments, arid conditions were probably present at least by the Miocene.

Notwithstanding the evidence for the evolution of an eremean flora in Australia over a period of at least 15 million years, there is little evidence that the arid zone has functioned as a major centre of species radiation. In *Acacia* the nearest relatives of adapted arid zone species are mostly found in adjacent temperate areas (Maslin & Hopper, 1982). Many parallel lines of adaptation to aridity thus exist, rather than a single line leading to the explosive radiation of a new arid-adapted species group. A similar situation almost certainly exists in *Eremophila*, Australian Euphorbieae (Hassall, 1982), *Dodonaea* (West, 1982), Gnaphaliinae (Short, 1982), and *Calotis* (Stace, 1982). It is also paralleled by similar evolutionary patterns in vertebrate and invertebrate animal groups (Baverstock, 1982; Greenslade, 1982). The arid zone thus emerges as an area which has been colonised successfully by numerous biotypes selected from adjacent populations growing under more favourable conditions, not as a floristic zone in which internal evolutionary radiation has produced its own characteristic flora.

An exception to this generalisation is the Chenopodiaceae. *Sclerolaena* and *Maireana* have apparently radiated widely in arid habitats. In *Atriplex* all the Australian species appear to be derived from a common specialised immigrant ancestral type, except for one coastal species which may be a separate introduction (Parr-Smith, 1982). Several species clusters have evolved in arid and semi-arid Australia from this ancestral type, all with a distinctive ovule orientation and all with the C_4 dicarboxylic acid metabolism and Kranz anatomy.

The present-day distribution of desert plants (Fig. 1.5) is the result of climatic events of the last few thousand years, because even an established desert flora would be disrupted by periods of extreme aridity. Arid maxima, originally thought to be inversely associated with glacial maxima, are associated with lowered rather than increased temperatures (Galloway & Kemp, 1981). The last glacial/arid maximum in southeastern Australia was about 17 500–16 000 y before present (BP) (Bowler, 1978; Ollier, 1986). It may have been followed by a relatively pluvial phase culminating about 3500 y BP (Gill, 1955), with conditions subsequently moving again towards aridity. Evidence for episodic evolution in the arid zone is found in *Cassia* (Randell, 1970) and *Eremophila* (Ey & Barlow, 1972), in which diploid biotypes occur in or near mountain systems in central and southern Australia which could have been refugia during arid maxima. Subsequent range extensions over the arid plains have been by derived polyploid biotypes, which are now widespread in the interior.

Alpine vegetation

The Neoaustral Subelement of the Intrusive Element recognised by Nelson (1981) was long thought to comprise a major part of the Australian alpine and subalpine vegetation. Recent systematic studies suggest, however, that temperate plant migration from the northern hemisphere during latest Tertiary and Quaternary times has not contributed very greatly to these associations (see below).

Alpine and subalpine habitats in mainland Australia and Tasmania occur over a relatively small area. On the mainland, they are confined to the southeast of the continent above elevations of 1370–1525 m, and their area of 5180 m² is only 0.07 per cent of the mainland (Costin, 1981). In Tasmania, they extend down to 915 m, and their

Fig. 1.5. The level of cover of arid zone vegetation in many parts of central Australia reflects the influence of microtopography and water availability. (Photo: C. J. Totterdell.)

area of 6480 m² is about 10 per cent of the total. In relative terms for Australia, precipitation is high, ranging from 760 to 2540 mm y^{-1}. In the Snowy Mountains, winter snow persists, on average, for 60 days over an area of 1200 km², but for 120 days over only 100 km² (Galloway, 1986).

The alpine areas of southeastern Australia and Tasmania are part of an uplifted plateau of considerable age (Ollier, 1986), and their elevation may be linked with the plate tectonic movements associated with the separation of the Australian and New Zealand blocks. The onset of alpine conditions, however, was probably a Quaternary event determined by macroclimate, and the alpine habitats are probably no older than those of New Zealand and New Guinea. The Australian Alps have experienced only one glaciation, the Last or Margaret Glaciation, which culminated about 19 000 y BP. The alpine habitats are therefore small islands in a lowland area of continental dimensions, of very recent emergence in an ancient landscape with long-adapted vegetation.

Oligo-Miocene wet sclerophyllous heath in Victoria may have been the precursor of alpine heath (Kershaw *et al.*, 1986), and macrofossils suggest a similar situation in Tasmania (Hill & Gibson, 1986). Pollen analysis in Tasmania indicates the presence of cold-adapted shrub genera prior to late Tertiary cooling, with herbaceous genera arriving later, but there was no treeless alpine vegetation before the late Pleistocene (Macphail, 1986).

The more recent record tends to confirm the recent onset of truly alpine conditions. *Colobanthus*, *Gentianella*, *Montia*, *Pimelea* and *Myriophyllum* were all present in western Victoria 20 000 y BP (Kershaw *et al.*, 1986), along with shrubby *Casuarina*, which suggests a mosaic of cold-adapted autochthonous vegetation. Such cold steppe vegetation may have been extensive in southeastern Australia in the last few million years, and may have contributed cyclically to high altitude alpine communities in response to increased precipitation, most recently in the last 12 000 years. Sparse alpine desert with *Plantago* herbfields existed at Kosciusko *c.* 15 700 y BP, with accelerated generic recruitment *c.* 11 700 y BP marking the arrival of low and tall herbfield and feldmark, and a 7-fold increase in plant cover in 1000 years (Martin, 1986). The treeline was still below its present level 7000 y BP. Similarly in Tasmania modern alpine vegetation extended to much lower elevations earlier in the Holocene than at present (Macphail, 1986).

Two processes have been involved in the differentiation of an alpine flora. The newly emergent alpine habitats have been colonised, in part, by plants from the communities of the surrounding lowlands. They have also been colonised by distance dispersal from remote alpine or cool temperate communities. Phytogeographic interpretation therefore depends on establishing the relative significance of the two processes, and for those species apparently derived through distance dispersal, their sources and times of arrival.

There are no genera at all which are endemic to the Kosciusko alpine area (Smith, 1986). Most of the genera (59 per cent) extend beyond the Australasian region, with 19 per cent of them austral and 40 per cent bihemispheric. These data suggest that Gondwanan floristic elements were not of major significance in the origin of the alpine flora, and that distance dispersal of bihemispheric genera was just as important. The raw data conceal, however, a much greater contribution from autochthonous floristic elements than that indicated, because many bihemispheric genera may have had ancient Gondwanan origins (Barlow, 1990*a*).

The converse statistic is that 60 per cent of the Kosciusko alpine genera have either limited or wide distributions in other southern regions. Many occur in adjacent montane areas or lowlands, often because of wide ecological amplitudes of particular species. Thus many of the alpine genera have reached there by expansion from long-established Autochthonous vegetation associations at lower elevations.

Many of the so-called bihemispheric genera now appear to have had ancient southern origins, or to have undergone major radiation in the southern hemisphere. In *Euphrasia*, Barker (1986) has shown that only 3 of the 14 sections occur in the northern hemisphere, and that the genus is probably Gondwanan in origin. Lowland euphrasias were preadapted to colonise alpine habitats, and this has occurred in several different lineages. In the cushion plant genus *Scleranthus*, which has a very large geographical disjunction between Europe and Australasia, West & Garnock-Jones (1986) concluded that the group is paraphyletic or even polyphyletic, and that the Australasian taxa may be more closely related to other austral genera of Caryophyllaceae. Long southern histories are also indicated in the grasses *Poa*, *Festuca*, *Rytidosperma*, *Erythranthera*, *Chionochloa* and *Australopyrum* (Clayton, 1981; Connor & Edgar, 1986); the apiaceous genera *Aciphylla*, *Gingidia*, *Anisotome*, *Lignocarpa* and *Scandia* (Webb, 1986); and in *Ranunculus* (Briggs, 1986).

These examples suggest that the Neoaustral Subelement in the Australian alpine and subalpine vegetation may be over-rated. Gondwanan histories are often fundamental to the genera which have biohistorical connections to the northern hemisphere, and the sources of the alpine species may have been long-established lowland autochthones. In this respect it is notable that the Kosciusko flora is more strongly infiltrated by widespread low altitude genera than the alpine floras of New Guinea or New Zealand (Smith, 1986). This may be because the subalpine vegetation is not dense or closed forest, as in the latter regions, but open subalpine woodland. Alpine plants in Australia may thus have more opportunity to extend down to lower elevations, and this may present them with an important refugial habitat during thermal maxima. Conversely, the open subalpine habitat may provide more opportunity for upward colonisation of the alpine zone. It is noteworthy that a high proportion of the Autochthonous genera in the Kosciusko alpine area are represented there by only a single species (Smith, 1986).

Another component of preadaptation in the lowland flora to the mild alpine conditions might be its long history of response to aridity, because physiological adaptation to aridity may also confer frost tolerance. In *Scleranthus*, only one species in Australia is confined to the alpine zone, whilst the other alpine ones extend to lower elevations, and the ancestral lowland species are arid-adapted (West & Garnock-Jones, 1986). Similarly, *Olearia*, which may represent the ancestral complex of *Celmisia* (Given & Gray, 1986), is widespread in dry lowland habitats in Australia. The richness of the Autochthonous Subelement in the alpine flora may therefore be the result of the facultative character of adaptations to different climates at an earlier time.

For many alpine species of mainland Australia and Tasmania which appear to have their nearest relatives in remote alpine or cool temperate habitats, distance dispersal appears to be a necessary requirement, even allowing for the biogeographic considerations discussed above. Raven (1973) argued a relatively recent origin for the alpine and subalpine flora of New Zealand, with Australia being a significant source. In contrast, Wardle (1978) suggested that several Australian alpine plant groups may have originated from a New Zealand alpine flora which is older (in part) and more diverse than Raven proposed.

Evidence for very recent direct biotic exchange between high altitude floras of Australia and New Zealand is very limited. *Gingidia montana*, widespread in New Zealand, occurs locally on the New England tableland, probably as a result of a recent dispersal event (Webb, 1986). Similarly, *Uncinia sinclairii*, of alpine and non-alpine situations in New Zealand, is probably recently introduced to its one locality in the Australian Alps (Wilson, 1986). The paucity of such examples suggests that distance dispersal between Australia and New Zealand may have had a different pattern over a long time span since the separation of the southern lands (Barlow, 1990a). It may have involved lowland taxa which were later ancestral to new species in the emerging alpine zones. It may have occurred at times when vegetation zones extended to lower elevations, so that alpine and subalpine habitats coalesced into larger target areas. Close vicarious relationships between species which are endemic respectively to alpine Australia and New Zealand could thus be indicators of earlier distance dispersal events. Such patterns are indicated in *Gingidia*, *Aciphylla* and *Aniso-tome* (Webb, 1986), *Scleranthus* (West & Garnock-Jones, 1986) and *Celmisia* (Given & Gray, 1986).

These considerations also apply to dispersal from more remote southern or even northern hemisphere sources, where again it is difficult to distinguish between recent direct transfer to the alpine area and colonisation of the alpine area following an earlier transfer. In the sedge family Cyperaceae recent dispersal from the northern hemisphere has occurred in the 11 species of *Carex* of the alpine areas, but these species also occur at lower elevations in the local region (Wilson, 1986). *Isolepis*, *Uncinia*, *Oreobolus*, *Carpha* and *Schoenus* comprise an ancient Australasian element, also with strong lowland connections (even though *Carpha* and *Oreobolus* are now obligate high altitude genera), but distance dispersal in the southern hemisphere is strongly indicated. In the rush family Juncaceae the three species of *Juncus* in the alpine zone extend well below it (Wilson, 1986), but belong in a widespread subantarctic species group. In *Luzula* the species seem to have more direct affinity with northern hemisphere species groups, but there has been considerable differentiation of endemic species in the alpine zone, and pre-Quaternary dispersal is probably indicated.

The case histories suggest that distance dispersal around the southern latitudes may have been more significant than migration across the equatorial regions from the northern hemisphere. Even for bihemispheric genera with northern origins, an Andean migration route to the southern hemisphere may be more likely than a Malesian one (Smith, 1986).

The Australian Alpine flora, localised in some newly emerged habitats in small islands of uplands in an old continental landscape, is rather ephemeral. Its sites may have coalesced into a larger area of alpine habitat in the recent past, but at other warmer times may have disappeared completely. The subalpine woodlands of the Australian Alps are an important component of the entire plant association, being floristically an extension of the same macroecosystem, and a reservoir for recolonisation of the alpine habitat (Fig. 1.6). In summary, the entire system has been colonised both by infiltration of plants from surrounding communities (Autochthonous Subelement) and by distance dispersal of cool-adapted biotypes (Cosmopolitan and Neoaustral Subelements). The high proportion of autochthones reflects the openness and arid-adaption of the surrounding vegetation. The arrival of widely dispersed elements has occurred through Tertiary and Quaternary time, both to non-alpine and alpine habitats.

Fig. 1.6. *Eucalyptus pauciflora*, the snow gum, defines the boundary between treed subalpine vegetation and treeless alpine vegetation. (Photo: C. J. Totterdell.)

Conclusion

Geomorphological and palaeoclimatic data play an ever-increasing role in theories for the origin, adaptation and differentiation of regional floras. There has always been the need for geophysical and biohistoric theories to be compatible, but the increasing detail of geophysical knowledge gives it more and more the guiding role in such studies. The conformity between the general propositions of southern hemisphere biogeography, based on plate tectonics and associated paleoclimatic inferences, and knowledge of the Australian vegetation, past and present, lends strong support to the phytogeographic description given above. They reveal that the Australian flora of perhaps 25 000 vascular species is a complex integration of relictual and derived, and autochthonous and immigrant components.

References

Andrews, E.C. (1916). The geological history of the Australian flowering plants. *American Journal of Science*, **42**, 171–232.

Audley-Charles, M.G. (1981). Geological history of the region of Wallace's Line. In *Wallace's Line and Plate Tectonics*, ed. T.C. Whitmore, pp. 24–35. Oxford: Clarendon.

Audley-Charles, M.G. (1987). Dispersal of Gondwanaland: relevance to evolution of the angiosperms. In *Biogeographical Evolution of the Malay Archipelago*, ed. T.C. Whitmore, pp. 5–25. Oxford: Clarendon.

Audley-Charles, M.G., Carter, D.J. & Milsom, J.S. (1972). Tectonic development of eastern Indonesia in relation to Gondwanaland dispersal. *Nature Physical Sciences*, **239**, 35–9.

Audley-Charles, M.G., Hurley, A.M. & Smith, A.G. (1981). Continental movements in the Mesozoic and Cenozoic. In *Wallace's Line and Plate Tectonics*, ed. T.C. Whitmore, pp. 9–23. Oxford: Clarendon.

Audley-Charles, M.G., Ballantyne, P.D. & Hall, R. (1988). Mesozoic-Cenozoic rift-drift sequence of Asian fragments from Gondwanaland. *Tectonophysics*, **155**, 317–30.

Barker, W.R. (1986). Biogeography and evolution in *Euphrasia* (Scrophulariaceae), particularly relating to Australasia. In *Flora and Fauna of Alpine Australasia*, ed. B.A. Barlow, pp. 489–510. Melbourne: CSIRO/Brill.

Barlow, B.A. (1981). The Australian flora: its origin and evolution. In *Flora of Australia*, ed. A.S. George, Vol. 1, pp. 25–75. Canberra: Australian Government Publishing Service.

Barlow, B.A. (1988). Patterns of differentiation in tropical species of *Melaleuca*. *Proceedings of the Ecological Society of Australia*, **15**, 239–47.

Barlow, B.A. (1990*a*). The alpine flora: autochthones and peregrines. In *The Scientific Significance of the Australian Alps*, ed. R. Good, pp. 69–78. Canberra: Australian Academy of Science.

Barlow, B.A. (1990*b*). Biogeographical relationships of Australia and Malesia: Loranthaceae as a model. In *The Plant Diversity of Malesia*, eds P. Bass, K. Kalkman & R. Geesink, pp. 273–92. Dordrecht: Kluwer.

Barlow, B.A. & Hyland, B.P.M. (1988). The origins of the flora of Australia's wet tropics. *Proceedings of the Ecological Society of Australia*, **15**, 1–17.

Baverstock, P.R. (1982). Adaptations and evolution of the mammals of arid Australia. In *Evolution of the Flora and Fauna of Arid Australia*, eds W.R. Barker & P.J.M. Greenslade, pp. 175–8. Adelaide: Peacock.

Beadle, N.C.W. (1954). Soil phosphate and the delimitation of plant communities in eastern Australia. *Ecology*, 35, 370–5.

Beadle, N.C.W. (1966). Soil phosphate and its role in molding segments of the Australian flora and vegetation, with special reference to xeromorphy and sclerophylly. *Ecology*, 47, 992–1007.

Beadle, N.C.W. (1968). Some aspects of ecology and physiology of Australian xeromorphic plants. *Australian Journal of Science*, 30, 348–55.

Beadle, N.C.W. (1981). Origins of the Australian angiosperm flora. In *Ecological Biogeography of Australia*, ed. A. Keast, pp. 407–26. The Hague: W. Junk.

Beard, J.S. (1969). Endemism in the Western Australian flora at species level. *Journal of the Royal Society of Western Australia*, 52, 18–20.

Beard, J.S. (1976). The evolution of Australian desert plants. In *Evolution of Desert Biota*, ed. D.W. Goodall, pp. 51–63. Austin: University of Texas Press.

Bowler, J.M. (1978). Quaternary climate and tectonics in the evolution of the Riverina Plain, south-eastern Australia. In *Landform Evolution in Australia*, eds J.L. Davies & M.A.J. Williams, pp. 70–112. Canberra: Australian National University Press.

Bowler, J.M. (1982). Age, origin and landform expression of aridity in Australia. In *Evolution of the Flora and Fauna of Arid Australia*, eds W.R. Barker & P.J.M. Greenslade, pp. 35–45. Adelaide: Peacock.

Briggs, B.G. (1986). Alpine ranunculi of the Kosciusko Plateau: habitat change and hybridization. In *Flora and Fauna of Alpine Australasia*, ed. B.A. Barlow, pp. 401–12. Melbourne: CSIRO/Brill.

Burbidge, N.T. (1960). The phytogeography of the Australian region. *Australian Journal of Botany*, 8, 75–212.

Chappell, J. (1974). Geology of coral terraces, Huon Peninsula, New Guinea: a study of Quaternary tectonic movements and sea-level changes. *Bulletin of the Geological Society of America*, 85, 555–70.

Christophel, D.C. (1981). Tertiary megafossil floras of Australia as indicators of floristic associations and palaeoclimate. In *Ecological Biogeography of Australia*, ed. A. Keast, pp. 377–90. The Hague: W.Junk.

Clayton, W.D. (1981). Evolution and distribution of grasses. *Annals of the Missouri Botanical Garden*, 68, 5–14.

Connor, H.E. & Edgar, E. (1986). Australasian alpine grasses: diversification and specialization. In *Flora and Fauna of Alpine Australasia*, ed. B.A. Barlow, pp 413–34. Melbourne: CSIRO/Brill.

Costin, A.B. (1981). Alpine and sub-alpine vegetation. In *Australian Vegetation*, ed. R.H. Groves, pp. 361–76. Cambridge University Press.

Crocker, R.L. & Wood, J.G. (1947). Some historical influences on the development of South Australian vegetation communities and their bearing on concepts and classification in ecology. *Transactions of the Royal Society of South Australia*, 71, 91–136.

Crook, K.A.W. & Taylor, G.R. (1985). Early opening history of the Southern Ocean between Australia and Antarctica. In *Otway 85: Resources of the Otway Basin: Summary papers and excursion guide*, No. 1. Melbourne: Geological Society of Australia, South Australian and Victorian Divisions.

Diels, L. (1906). *Die Pflanzenwelt von West Australien*. Leipzig: Engelmann.

Doing, H. (1981). Phytogeography of the Australian floristic kingdom. In *Australian Vegetation*, ed. R.H. Groves, pp. 3–25. Cambridge University Press.

Ey, T.M. & Barlow, B.A. (1972). Distribution of chromosome races in the *Eremophila glabra* complex. *Search*, 3, 337–8.

Galloway, R.W. (1986). Australian snowfields past and present. In *Flora and Fauna of Alpine Australasia*, ed. B.A. Barlow, pp. 27–35. Melbourne: CSIRO/Brill.

Galloway, R.W. & Kemp, E.M. (1981). Late Cainozoic environments in Australia. In *Ecological Biogeography of Australia*, ed. A. Keast, pp. 51–80. The Hague: W.Junk.

Gill, E.D. (1955). The Australian arid period. *Australian Journal of Science*, 17, 204–6.

Given, D.R. & Gray, M. (1986). *Celmisia* (Compositae-Astereae) in Australia and New Zealand. In *Flora and Fauna of Alpine Australasia*, ed. B.A. Barlow, pp. 451–70. Melbourne: CSIRO/Brill.

Greenslade, P. (1982). Origins of the Collembolan fauna of arid Australia. In *Evolution of the Flora and Fauna of Arid Australia*, eds W.R. Barker & P.J.M. Greenslade, pp. 267–72. Adelaide: Peacock.

Hartley, T.G. (1986). Floristic relationships of the rainforest flora of New Guinea. *Telopea*, 2, 619–30.

Hassall, D.C. (1982). Historical development in the Australian Euphorbieae (Euphorbiaceae). In *Evolution of the Flora and Fauna of Arid Australia*, eds W.R. Barker & P.J.M. Greenslade, pp. 323–8. Adelaide: Peacock.

Hill, R.S. & Gibson, N. (1986). Macrofossil evidence for the evolution of the alpine and subalpine vegetation of Tasmania. In *Flora and Fauna of Alpine Australasia*, ed. B.A. Barlow, pp. 205–17. Melbourne: CSIRO/Brill.

Hooker, J.D. (1860). Introductory Essay. *Botany of the Antarctic Voyage of H.M. Discovery Ships 'Erebus' and 'Terror' in the Years 1839–1843, III. Flora Tasmaniae*. London: Reeve.

Hopper, S.D. (1979). Biogeographical aspects of speciation in the southwest Australian flora. *Annual Review of Ecology and Systematics*, 10, 399–422.

Johnson, L.A.S. & Briggs, B.G. (1975). On the Proteaceae – the evolution and classification of a southern family. *Botanical Journal of the Linnean Society*, 70, 83–112.

Johnson, L.A.S. & Briggs, B.G. (1981). Three old southern families – Myrtaceae, Proteaceae and Restionaceae. In *Ecological Biogeography of Australia*, ed. A. Keast, pp. 427–69. The Hague: W.Junk.

Johnson, L.A.S. & Briggs, B.G. (1984). Myrtales and Myrtaceae – a phylogenetic analysis. *Annals of the Missouri Botanical Garden*, 71, 700–56.

Kemp, E.M. (1978). Tertiary climatic evolution and vegetation history in the southeast Indian Ocean region. *Palaeogeography, Palaeoclimatology and Palaeoecology*, 24, 169–208.

Kershaw, A.P. (1978). A record of the last interglacial-glacial cycle from north-eastern Queensland, Australia. *Nature*, 272, 159–61.

Kershaw, A.P. (1981). Quaternary vegetation and environments. In *Ecological Biogeography of Australia*, ed. A. Keast, pp. 81–101. The Hague: W.Junk.

Kershaw, A.P., McEwen Mason, J.R., McKenzie, G.M., Strickland, K.M. & Wagstaff, B.E. (1986). Aspects of the development of cold-adapted flora and vegetation in the Cenozoic of southeastern mainland Australia. In *Flora and Fauna of Alpine Australasia*, ed. B.A. Barlow, pp. 147–60. Melbourne: CSIRO/Brill.

Lange, R.T. (1978). Carpological evidence for fossil *Eucalyptus* and other Leptospermae (subfamily Leptospermoideae of Myrtaceae) from a Tertiary deposit in the South Australian arid zone. *Australian Journal of Botany*, **26**, 221–33.

Loveless, A.R. (1961). A nutritional interpretation of sclerophylly based on differences in the chemical composition of sclerophyllous and mesophytic leaves. *Annals of Botany (London), n.s.* **25**, 168–84.

Macphail, M.K. (1986). 'Over the top': pollen-based reconstructions of past alpine floras and vegetation in Tasmania. In *Flora and Fauna of Alpine Australasia*, ed. B.A. Barlow, pp. 173–204. Melbourne: CSIRO/Brill.

Marchant, N.G. (1973). Species diversity in the south-western flora. *Journal of the Royal Society of Western Australia*, **56**, 23–30.

Martin, A.R.H. (1986). Late Glacial and early Holocene vegetation of the alpine zone, Kosciusko National Park. In *Flora and Fauna of Alpine Australasia*, ed. B.A. Barlow, pp. 161–70. Melbourne: CSIRO/Brill.

Maslin, B.R. & Hopper, S.G. (1982). Phytogeography of *Acacia* (Leguminosae: Mimosoideae) in Central Australia. In *Evolution of the Flora and Fauna of Arid Australia*, eds W.R. Barker & P.J.M. Greenslade, pp. 301–15. Adelaide: Peacock.

McKenna, M.C. (1972). Possible biological consequences of plate tectonics. *Bio-Science*, **22**, 519–25.

Nelson, E.C. (1981). Phytogeography of southern Australia. In *Ecological Biogeography of Australia*, ed. A. Keast, pp. 733–59. The Hague: W.Junk.

Nix, H.A. (1982). Environmental determinants of biogeography and evolution in Terra Australis. In *Evolution of the Flora and Fauna of Arid Australia*, eds W.R. Barker & P.J.M. Greenslade, pp. 47–66. Adelaide: Peacock.

Nix, H.A. & Kalma, J.D. (1972). Climate as a dominant control in the biogeography of northern Australia and New Guinea. In *Bridge and Barrier: the Natural and Cultural History of Torres Strait*, ed. D. Walker, pp. 61–91. Canberra: Australian National University.

Ollier, C.D. (1982). The Great Escarpment of eastern Australia: tectonic and geomorphic significance. *Journal of the Geological Society of Australia*, **39**, 431–5.

Ollier, C.D. (1986). The origin of alpine landforms in Australasia. In *Flora and Fauna of Alpine Australasia*, ed. B.A. Barlow, pp. 3–25. Melbourne: CSIRO/Brill.

Pain, C.F. & Ollier C.D. (1984). Drainage patterns and tectonics around Milne Bay, Papua New Guinea. *Revue de Geomorphologie Dynamique*, **32**, 113–20.

Parr-Smith, G.A. (1982). Biogeography and evolution in the shrubby Australian species of *Atriplex* (Chenopodiaceae). In *Evolution of the Flora and Fauna of Arid Australia*, eds W.R. Barker & P.J.M. Greenslade, pp. 291–9. Adelaide: Peacock.

Pryor, L.D. & Johnson L.A.S. (1981). *Eucalyptus*, the universal Australian. In *Ecological Biogeography of Australia*, ed. A. Keast, pp. 499–536. The Hague: W.Junk.

Randell, B.R. (1970). Adaptations in the genetic systems of Australian arid zone *Cassia* species (Leguminosae, Caesalpinioideae). *Australian Journal of Botany*, **18**, 77–97.

Raven, P.H. (1973). Evolution of subalpine and alpine plant groups in New Zealand. *New Zealand Journal of Botany*, **11**, 177–200.

Raven, P.H. & Axelrod, D.I. (1974). Angiosperm biogeography and past continental movements. *Annals of the Missouri Botanical Garden*, **61**, 539–673.

Richards, P.W. (1952). *The Tropical Rain Forest*. Cambridge University Press.

Schodde, R. & Calaby, J.H. (1972). The biogeography of the Australo-Papuan bird and mammal faunas in relation to Torres Strait. In *Bridge and Barrier: the Natural and Cultural History of Torres Strait*, ed. D. Walker, pp. 257–300. Canberra: Australian National University.

Short, P.S. (1982). Breeding systems and distribution patterns of some arid Australian genera of the subtribe Gnaphaliinae (Compositae: Inulae). In *Evolution of the Flora and Fauna of Arid Australia*, eds W.R. Barker & P.J.M. Greenslade, pp. 351–6. Adelaide: Peacock.

Smith, J.M.B. (1982). An introduction to the history of Australasian vegetation. In *A History of Australasian Vegetation*, ed. J.M.B. Smith, pp. 1–31. Sydney: McGraw Hill.

Smith, J.M.B. (1986). Origins of Australasian tropicalpine and alpine floras. In *Flora and Fauna of Alpine Australasia*, ed. B.A. Barlow, pp. 109–28. Melbourne: CSIRO/Brill.

Specht, R.L. (1972). *The Vegetation of South Australia*. Adelaide: Government Printer.

Specht, R.L. (1981). Evolution of the Australian flora: some generalizations. In *Ecological Biogeography of Australia*, ed. A. Keast, pp. 783–805. The Hague: W.Junk.

Stace, H.M. (1982). *Calotis* (Compositae), a Pliocene arid zone genus? In *Evolution of the Flora and Fauna of Arid Australia*, eds W.R. Barker & P.J.M. Greenslade, pp. 357–67. Adelaide: Peacock.

Steenis, C.G.G.J. van (1979). Plant-geography of east Malesia. *Botanical Journal of the Linnean Society*, **79**, 97–178.

Takhtajan, A. (1969). *Flowering Plants: Origin and Dispersal*. Edinburgh: Oliver & Boyd.

Thorne, R.F. (1986). Summary statement. *Telopea*, **2**, 697–704.

Truswell, E.M. & Harris, W.K. (1982). The Cainozoic palaeobotanical record in arid Australia: fossil evidence for the origin of an arid-adapted flora. In *Evolution of the Flora and Fauna of Arid Australia*, eds W.R. Barker & P.J.M. Greenslade, pp. 67–76. Adelaide: Peacock.

Walker, D. (1972). Bridge and barrier. In *Bridge and Barrier: the Natural and*

Cultural History of Torres Strait, ed. D. Walker, pp. 399–405. Canberra: Australian National University.

Wardle, P. (1978). Origin of the New Zealand mountain flora, with special reference to trans-Tasman relationships. *New Zealand Journal of Botany*, **16**, 535–50.

Webb, C.J. (1986). Breeding systems and relationships in *Gingidia* and related Australasian Apiaceae. In *Flora and Fauna of Alpine Australasia*, ed. B.A. Barlow, pp. 383–99. Melbourne: CSIRO/Brill.

Webb, L.J. & Tracey, J.G. (1981). Australian rainforests: pattern and change. In *Ecological Biogeography of Australia*, ed. A. Keast, pp. 605–94. The Hague: W.Junk.

Webb, L.J., Tracey, J.G. & Williams, W.T. (1984). A floristic framework of Australian rainforests. *Australian Journal of Ecology*, **9**, 169–98.

Webb, L.J., Tracey, J.G. & Jessup, L.W. (1986). Recent evidence for autochthony of Australian tropical and subtropical rainforest floristic elements. *Telopea*, **2**, 575–89.

Wellman, P. & McDougall, I. (1974). Cainozoic igneous activity in eastern Australia. *Tectonophysics*, **23**, 49–65.

Wellman, J.G. (1982). Radiation and adaptation of *Dodonaea* (Sapindaceae) in arid Australia. In *Evolution of the Flora and Fauna of Arid Australia*, eds W.R. Barker & P.J.M. Greenslade, pp. 329–33. Adelaide: Peacock.

West, J.G. (1982). Radiation and adaptation of *Dodonaea* (Sapindaceae) in arid Australia. In *Evolution of the Flora and Fauna of Arid Australia*, eds W.R. Barker & P.J.M. Greenslade, pp. 329–33. Adelaide: Peacock.

Whiffin, T. & Hyland B.P.M. (1986). Taxonomic and biogeographic evidence on the relationships of Australian rainforest plants. *Telopea*, **2**, 591–610.

Whitmore, T.C. (1981). Palaeoclimate and vegetation history. In *Wallace's Line and Plate Tectonics*, ed. T.C. Whitmore, pp. 36–42. Oxford: Clarendon.

Wilson, K.L. (1986). Alpine species of Cyperaceae and Juncaceae. In *Flora and Fauna of Alpine Australasia*, ed. B.A. Barlow, pp. 471–88. Melbourne: CSIRO/Brill.

Wopfner, H., Callen, R. & Harris, W.K. (1974). The lower Tertiary Eyre Formation of the southwestern Great Artesian Basin. *Journal of the Geological Society of Australia*, **21**, 17–52.

2

Quaternary vegetation history

J.R. DODSON

Background and major events of the Quaternary

During the Tertiary, Australia drifted northward into climates increasingly dominated by the westerlies and subtropical highs; and there was greater opportunity for dispersal between Asia and Australia, especially as New Guinea increased in size. The Quaternary therefore opened, about 1.8 million years ago (Ma), with a rich flora including elements from ancient Gondwana, primitive angiosperms, especially in rainforests, an indigenous arid flora and a host of scleromorphic species with an abundance of mechanisms to cope with fire and the nutrient-deficient ancient soils (see previous chapter). These same adaptations were useful in highly seasonal and the often perturbed environments which were to mark Quaternary time. The palaeobotanical records provide a reasonable basis to suppose that the Tertiary period closed with the great majority of present-day native plants already distributed within an Australia which was largely forested.

Major Quaternary phenomena which shaped the composition and distribution of the vegetation as we know it today were dramatic climate changes, the migration of people from Asia and the transplantation of European technology and attitudes to exploitation, and lately conservation.

The separation of Australia and South America from Antarctica led to the formation of the Southern Ocean and reduced heat transfer from the equator to the South Pole. Ice-rafted debris shows Antarctica's ice cap developed by mid-Miocene time (Frakes, 1979) and this became a dominating influence on the climate of southern Australia, at least for the last few hundreds of thousands of years. It is thought that the global patterns of more rapid change are related to variations in earth's orbit, the tilt of its axis and procession (Imbrie & Imbrie,

1979), possibly in the sun's energy output (Fröhlich, 1988), although this has not been demonstrated yet (Stuiver *et al.*, 1991), and processes involving ocean turnover (Broecker & Denton, 1990). These patterns have combined to alter the spatial patterns of energy budgets; and hence changes in temperature, strength of seasonality and precipitation resulted. When insolation was reduced, there was a transfer of water to ice caps at high latitude and altitude and lower sea level which resulted in enlarged land areas. Warmer periods, which occupied only about 10 per cent of time in the last 700 000 years, and of which the last 10 000 years is an example, have reduced ice area and relatively high sea levels and hence reduced land area. Whilst the area of ice on the Australian continent was not extensive the effects of climate were large, and more often than not land bridges linked Tasmania and New Guinea to a single land mass 30 per cent larger than present; and the mountains of these with the southeastern Highlands were the only areas to contain significant ice caps (Fig. 2.1).

The global changes are preserved in ocean sediments where carbonate shells of plankton record changing abundance of ^{18}O and ^{16}O, which take place as a result of $H_2^{16}O$ evaporating faster than $H_2^{18}O$ from the oceans to ice caps during colder periods. Figure 2.2 (from Bowen, 1978) shows the saw-tooth pattern of oxygen isotopes over the entire Quaternary. Initially, change was gradual but since about 700 000 years ago the oscillations between extremes have been much greater.

First human influence on this continent is known from about 40 000 years ago (e.g. Allen, 1989; Hiscock, 1990), although some hypotheses place it earlier than this (e.g. Singh & Geissler, 1985; Roberts, Jones & Smith, 1990). Like all people, the Aborigines exploited plant and animal resources and learnt to modify technology and behaviour to make exploitation of the changing resources more efficient. In Australia, use of fire is thought to have been central to landscape management; and whilst cropping was important in New Guinea, it played an insignificant role in Australia right up until European settlement. A large number of mammalian herbivores coexisted with the first human occupants and it may be that human activity, possibly hunting, and climate change caused the extinction of several; although debate continues on the reasons for their loss (e.g. Wright, 1986*a*; Dodson, 1989). It has been argued by Flannery (1990) that removal of the large herbivore biomass resulted in significant vegetation change, perhaps altering the relationship between vegetation and fire in many areas. There is no question, however, that the two centuries of European settlement have caused the greatest

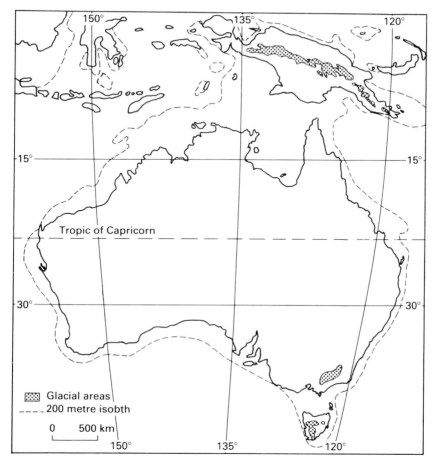

Fig. 2.1. The enlarged Australian land-mass with New Guinea and Tasmania as broad peninsulas, about 18 000 years ago.

upheaval in the composition and structure of plant communities, with some types probably gone forever (Fox, 1990).

In summary, it seems clear that the climatic variations of the Quaternary were more rapid and extreme, especially in aridity and coolness, and when coupled with human impact in the last several tens of millennia have left a significant imprint on the nature and composition of this continent's vegetation. Nothing on this scale had happened in the past.

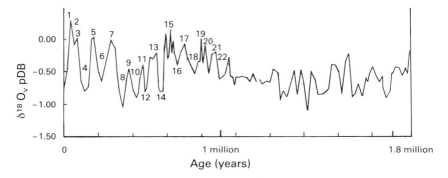

Fig. 2.2. The oxygen isotope record for the entire Quaternary. Peaks represent interglacial conditions, troughs represent lowered ^{16}O in the oceans as ice accumulated during glacials. The numbers refer to oxygen isotope stages.

The data base

Reliable evidence for reconstructing vegetation change comes from well-dated macrofossil assemblages and the pollen analysis of lake, swamp and cave floor sediments. From the spread of such data (Fig. 2.3) it is immediately apparent that large areas of the continent remain unexplored palaeobotanically, particularly the arid and western regions, and that early Quaternary data are scant. Indeed, relatively continuous records covering time since Last Glacial maximum, about 18 000 years ago, are only available from north-eastern Queensland (Kershaw, 1985; Kershaw *et al.*, 1991), north-western New South Wales (Dodson & Wright, 1989), the Southern Tablelands of New South Wales (Singh & Geissler, 1985), central Victoria (Ladd, 1976; Gillespie *et al.*, 1978), western Victoria and adjacent South Australia (Dodson, 1975, 1977; D'Costa *et al.*, 1989; Kershaw *et al.*, 1991; D'Costa & Kershaw, in press) and from northern and western Tasmania (Colhoun, 1977; Sigleo & Colhoun, 1981; Colhoun, van de Geer & Mook, 1982; Macphail & Colhoun, 1985; Colhoun & van de Geer, 1986; van de Geer, Fitzsimmons & Colhoun, 1986; Colhoun, 1988). Only records from Lynch's Crater, Lake George and Darwin Crater extend through at least two glacial maxima and two interglacials, and these are pivotal for understanding the coarse patterns of change, for example with and without the presence of humans.

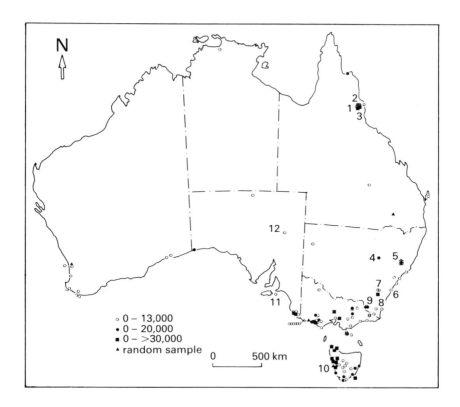

Fig. 2.3. Quaternary vegetation history sites from Australia. Site numbers refer to locations as follows: 1. Lynch's Crater; 2. Lake Barrine; 3. Atherton Tableland; 4. Ulungra Springs; 5. Barrington Tops; 6. Hawkesbury Valley; 7. Breadalbane Basin; 8. Lake George; 9. Burrinjuck Reservoir; 10. Darwin Crater; 11. Lashmar's Lagoon, Kangaroo Island; 12. Lake Frome.

Records of vegetation dynamics for the Holocene, the last 10 000 years, are comparatively abundant, and the Atherton Tablelands (northeastern Queensland), Barrington Plateau (northern New South Wales), Australia's southeastern Highlands, the central plateau of Tasmania and coastal southeastern Australia between Newcastle (New South Wales) and Kangaroo Island (South Australia) have sufficient data to be described as being well known. Clearly, future effort must be directed to other regions; not only to increase the spread of data but to examine more of the vegetation shifts at sensitive ecological boundaries, like those between arid and subhumid regions everywhere, between the monsoon forests and savannas, between rainforest and sclerophyll forests, and in regions where people were abundant; it is these places where much of the vegetation change has taken place.

The long records of change and the early Quaternary

Early Quaternary records of vegetation are few but there are some of fauna which are of assistance in determining conditions at the time. Early to mid-Pleistocene faunas from near Brewarrina, New South Wales (Anderson & Fletcher, 1934) and the Lake Eyre (Stirton, Tedford & Miller, 1961) and Murray-Darling Basins (Firman, 1967; Marshall, 1973) have assemblages showing the occurrence of lung-fish, chelid turtles, crocodiles, cormorants, pelicans and flamingoes. Mammals included rodents, *Sarcophilus*, *Bettongia*, dasyurids, vombatids, potoroids, the extinct large marsupials *Diprotodon*, *Sthenurus*, *Thylacoleo*, *Procoptodon* and *Protemnodon* and the giant ratite bird *Genyornis*. Many of these are preserved because they became bogged in mires and clays, as at Lake Callabonna, South Australia, and Lake Victoria, New South Wales (Hope, 1982). Although the records are poorly dated and probably occupy a huge expanse of time, they clearly show the period was vastly different from today – warmer and with more permanent water in areas now arid. This is not very different from what would be expected from vegetation data of terminal Pliocene age when forest still covered much of Australia (Martin, 1978, 1990). The weight of evidence suggests that the development of open vegetation and restriction of forests which characterises Australia today, were largely due to Quaternary phenomena (see also Chapter 1).

Lynch's Crater, Lake George and Darwin Crater provide the pivotal statements of the broad sweep of late Quaternary vegetation change in Australia because of their length and relatively continuous deposition. To overcome the shortfall in reliable dating for the older parts of the records, the broad patterns of vegetation shifts have been related to the timing of glacials and interglacials from the oxygen isotope record.

Lynch's Crater is today surrounded by complex vine forest and its sediments contain a record of at least two glacial maxima and two interglacials, including the Holocene. The lead-up to the penultimate glacial saw Araucarian vine forest, with sclerophyll taxa prominent, give way to a rich interglacial complex notophyll–mesophyll vine forest. The gymnosperms *Araucaria*, *Podocarpus*, *Dacrydium* and later *Callitris* were present. From 38 000 to 26 000 y BP there was a transitional period, then through glacial maximum and until the Holocene, sclerophyll forest dominated by *Casuarina* and *Eucalyptus* prevailed. Rainforest recovery in the early Holocene almost completely replaced sclerophyll but the rainforest character was different from before. Gymnosperms were reduced to almost insignificant quantities, and *Dacrydium* was extinct in the region (Kershaw, 1985).

Singh & Geissler (1985) argue that the record from Lake George covers about 730 000 years, although some would dispute this (Wright, 1986*b*). Clearly some sections are devoid of pollen or have low counts and the length of some time-breaks remains uncertain. The record clearly shows, however, vegetation assemblages with glacial and interglacial character. The glacial maxima corresponding with oxygen isotope stages 10, 8, 6 and the last one around 18 000 y BP, all show development of grassland and herbfield, with low quantities of sclerophyllous shrubs and trees. The vegetation of interglacial periods was sclerophyllous forest with some cool temperate taxa such as *Cyathea* and *Podocarpus*. *Casuarina* dominated the tree cover of the earlier interglacials but *Eucalyptus* was more prominent in the present one.

Darwin Crater, a meteorite-impact crater about 730 000 years old (Gentner *et al.*, 1973), has some cool temperate rainforest but mostly sclerophyll forest surrounding it today. The first 20 m of a core from the crater contains five major phases of wet lowland temperate forest, regarded as representing interglacials by Colhoun (1988), separated by five communities when the vegetation was herbfield, heathland, woodland or a mosaic of such communities; these are regarded as representing cool to cold climates. The earliest supposed interglacial had *Lagarostrobus* predominant but, at various times, *Nothofagus*, *Casuarina*, *Phyllocladus* and *Eucalyptus* were abundant. Sclerophyll woodland and heathland separated this phase from the next oldest interglacial where *Nothofagus*, *Eucalyptus*, *Phyllocladus*, *Cyathea*, *Casuarina* and Epacridaceae were common. Heathland and herbfield separate the next two interglacial phases where *Lagarostrobus*, with *Phyllocladus*, *Nothofagus*, and *Casuarina*, dominated. The Last Glacial maximum was a herbfield with some heath as indicated by abundant Poaceae, Asteraceae, *Astelia* and Epacridaceae pollen (Colhoun, 1988).

Other records which extend through the Last Glacial maximum confirm that aridity accompanied a 4–8 °C cooling compared to present, and in southeastern Australia a general vegetation cover, including grasses, Asteraceae, chenopods and *Plantago*, and with low tree cover, was everywhere – there is no corollary to this assemblage today, and it included a possibly unknown composite and possibly a now extinct species of *Casuarina* (Dodson, 1977; Singh & Geissler, 1985). Further north, at Ulungra Springs, the vegetation had similar character but was more clearly a semi-arid chenopod shrubland, and the Liguliflorae were prominent (Dodson & Wright, 1989). At this latitude it may have been the case that dryness rather than cooling was important in determining habitat conditions. If so, then for Australia it is evident that glacial maxima were dry and cool (or cold) in the south compared to the warm and relatively wetter interglacials. Over these large patterns would have been superimposed seasonality differences, and possibly changing solar activity, which would have made each glacial and interglacial have its own character. Temperature differences between glacials and interglacials, and across seasons, and effective precipitation would have been important in the south and on mountains, in the response of species to environmental change, but changing effective precipitation would have been of most importance elsewhere. At glacial maxima, lowered sea level increased the degree of continentality, and the alpine communities of the mainland had significantly drier and greater temperature extremes than those of the Central Plateau of Tasmania.

Where well dated, we see each glacial and interglacial flora has its own character. As a glacial maximum approached, the arid core expanded and sclerophyllous forests and rainforests contracted to the coast and well-watered refugia, or where run-on was significant. At the relaxation of aridity, these forests recovered but with different taxonomic composition. Forests never quite reassembled in the same way; presumably because of the distribution of the mosaic of species pools, their composition and the way in which environmental conditions changed. A major gap in the data at present is that for no glacial maximum do we have any record of rainforest; and only from northeastern Queensland is sclerophyll forest seen. We must presume many of the key species were restricted to refugia away from pollen analysis sites, in mountainous areas or along coasts which were over 100 m lower than today. It is possible to envisage that, following each glacial maximum, the reconstitution of the larger areas of rainforest was flavoured by which taxa survived, at least locally, and which were able to migrate to suitable habitat before the rate of sea level rise overwhelmed them.

From these observations we must conclude that many species combinations we observe today are a result of significant past changes, and that the present combinations of species have not necessarily persisted in place over huge expanses of time.

Many rainforest and eucalypt species may have extraordinary geographical ranges because of their genetic plasticity and the lack of competition from many specialised groups. The reason that there are so few animal species restricted entirely to rainforests may be because this was a disastrous strategy in the long term, and those that followed it were eliminated. Both temperate and tropical species which could adapt to more arid environments, even seasonally, were assured of a future in Australia. The absence of some taxa from regions may be explained by slow migration or lack of opportunity because of human activity. The decline of rainforest since the Miocene certainly continued through to the present, but, for example, *Nothofagus cunninghamii* was present in the Snowy Mountains about 30 000 years ago (Caine & Jennings, 1968). Most of its common associates are present and conditions are suitable today (Busby, 1986), but it apparently never recovered from the aridity and cold in the mountains during the Last Glacial maximum. The occurrence of species such as *Eucalyptus pauciflora* in lowland southeastern Australia, or large tracts of land occupied by mallee and heath which have phenologies not in keeping with their prevailing climates (Specht, 1981; Martin, 1989) may be indications of gaps not filled by more specialised groups, and may be a reason why some aliens have been particularly successful, where soil properties are not limiting.

The other large question to be answered, to be considered below, is how much of the vegetation character has been shaped by human activity of the last 40 000 years or so?

The last 13 000 years

We know from a great number of sites that patterns resembling present vegetation didn't begin until about 13 000 years ago, some 5000 years following the Last Glacial maximum. This was apparently because of lingering aridity even though summer–winter temperature differences were smaller than today (COHMAP, 1988). We can document the pattern of forest recovery, and its patchiness, by examining groups of sites from within regions.

It has already been noted that rainforest returned to the Atherton Tableland within the last 10 000 years. From the cluster of sites examined within the region it has been established that the return was not everywhere registered at the same time. It began in the east from

9500 BP and progressed to the most westerly site by 6000 BP (Kershaw, 1970, 1971, 1975, 1976). There were some small differences in composition which, like the forest recovery, reflects a migration in line with the present rainfall gradient. The greatest expression of rainforest occurred during 3600 to 5000 y BP, when Kershaw & Nix (1988) estimate, on the basis of bioclimatic profiles, the precipitation and temperature were 2000 to 2400 mm higher and 2 to 4 °C warmer than present. Since then slightly drier facies of rainforest have developed.

Studies of southern rainforests concerned with vegetation dominated by *Nothofagus moorei* at Barrington Tops (New South Wales) and *N. cunninghamii* in Victoria and Tasmania have been the focus of several studies. At Barrington Tops it has been shown that *N. moorei* was already present at the end of the Pleistocene and its populations expanded to reach maximum Holocene spread between 6500 and 3500 y BP (Dodson, Greenwood & Jones, 1986). In middle to eastern montane Victoria, mixed forests of *N. cunninghamii* and eucalypts developed (Howard & Hope, 1970; Ladd, 1979*a*), and in Tasmania this form and 'pure' *N. cunninghamii* rainforests developed in the west and isolated mountains in the centre and south (Macphail, 1979). A decline in rainforests began around 5000 BP accompanied by an expansion of sclerophyll elements, and grassland in some places, suggesting a response to dryness and possibly a changed fire regime.

Little has been attempted so far to reconstruct changes involving temperate or subtropical rainforests in southern Australia. Jones (1990) concluded that subtropical rainforest on the Illawarra plain has been present for at least 4400 years, with wet sclerophyll restricted to less favourable edaphic and climatic sites. Ladd (1978) also concluded that patches of rainforest (dominated by *Acmena smithii*) in eastern Victoria have been relatively stable for the last 5200 years; however, Dodson & Thom (1992) have concluded that lower Hawkesbury Valley rainforest has undergone considerable destruction and re-establishment from time to time, since 9300 BP, probably as a result of fire.

Tree-lines did not establish their present position in southeastern Australia until after 12 000 years ago. In the Barrington Tops and the Southeastern Highlands of New South Wales and Victoria, herbfields and grasslands were invaded by open eucalypt woodlands from lower altitudes (Ladd, 1979*b,c*). By 11 000 years ago, the tree-line passed 1330 m on the Buffalo Plateau (Binder 1978; Binder & Kershaw, 1978) but in the Kosciusko region did not reach its present position, at about 1750 m, until about 8700 years ago (Raine, 1974; Martin, 1986). In Tasmania, shrublands and scrubs of *Microcachrys tetragona*, *Nothofagus gunnii*, *N. cunninghamii* and *Podocarpus alpina* ascended the mountains from about 11 500 y BP (Macphail, 1979; Markgraf, Bradbury & Busby, 1986). Eucalypt forest replaced grasslands around 11 000 BP (Colhoun *et al.*, 1982) but not until about 8000 BP at Cave Bay (Hope, 1978). Around Mt Field, in Tasmania's wetter regions, *N. cunninghamii* and eucalypt forest developed a tree-line above 1000 m around 7800 BP; but this returned to eucalypt forest after 6000 BP (Macphail, 1979). By way of contrast, the relatively dry Southern Tablelands of New South Wales saw eucalypt forest return at 8000 BP. Lake George and several records from around Breadalbane Basin, some 25 km north, show forest recovery had begun before 9300 BP but was patchy in spread and composition (Singh & Geissler, 1985; Dodson, 1986). Rising temperature was undoubtedly an important factor in the establishment of tree-lines but effective precipitation was also important; however, the significance of any role for fire in delaying the attainment of tree-line position, or indeed its maintenance, remains to be investigated. Galloway (1986) recently argued that a 2 °C warming, possibly as happened in mid-Holocene time, may have been enough to all but eliminate a tree-line in mainland Australia.

In lowland humid mainland Australia, effective precipitation was a more important control on vegetation dynamics. In southwestern Victoria and adjacent South Australia the present woodland dominated by *Casuarina* and *Eucalyptus* replaced an open lightly wooded vegetation with chenopods and grassland steppe elements, from around 12 000 BP (Dodson, 1977, 1979; D'Costa *et al.*, 1989). In the late Holocene, eucalypts gradually became more important relative to *Casuarina* (e.g. Dodson, 1974a; Head, 1988; D'Costa *et al.*, 1989). This may be related to soil development; or associated with fire frequency, since Clark has demonstrated (Singh, Kershaw & Clark, 1981) that *Casuarina* has increased relative to *Eucalyptus* on Kangaroo Island in the relative absence of humans and the occurrence of only occasional fires.

For lowland sites, there is evidence of increased effective moisture between about 7000 and 5000 BP. In Tasmania this shows as an increase in *Pomaderris* (Macphail, 1979), and in southwestern Australia, where *E. diversicolor* increased relative to *E. marginata* (Churchill, 1968); but on the southeastern mainland the evidence is apparently confined to aquatic floras (e.g. Dodson, 1974a,b; Ladd, 1978; Head, 1988). In general, terrestrial vegetation in lowland southeastern Australia did not respond very significantly to changes in climate during the Holocene.

In arid Australia, where responses to effective precipitation change might be expected to be greatest, there are few data. Martin (1973) and Dodson & Wright (1989) described significant shifts in chenopod shrubland and replacement by more mesic vegetation, around 9000 BP and 10 000 BP, respectively, but these are thought to be related to sea level and climate changes at the Pleistocene – Holocene boundary. Singh (1981) attributed changes in abundance of Poaceae and Asteraceae pollen, and minor fluctuations in shrubland, at Lake Frome, as related to possible greater or lesser inland penetration of the summer monsoon or southerly winter-rainfall systems. Luly (1991) argued that the major vegetation change around Lake Tyrrell, in northern Victoria, was the development of mallee and *Callitris* into open *Casuarina* woodland, around 6500 BP. Green *et al.* (1983), Bell, Finlayson & Kershaw (1989), Boyd (1990) and Pearson and Dodson (1993) have all recently demonstrated that spring deposits and stick-nest rat middens in arid and semi-arid Australia, whilst they may have relatively short records, have good potential for studies examining vegetation and environmental stability.

Walker & Singh (1981) noted that Australian vegetation reacted to the world-wide climatic changes of the transition of the Late Pleistocene and Holocene, but ten years later there is still no research to dispute the statement that there is no clearly identifiable date at which the vegetation reactions began. It is clear, however, that once started they had repercussions which lasted millennia. Each case investigated seems to have its own peculiarity and timing of response; because of the proximity of climate, topography, soil and fire regime. The search for underlying trends has revealed universal differences, which is in direct contrast to northwestern Europe and North America where synchronism in vegetation response has been demonstrated (e.g. Wright, 1984).

Human impact and fire

In prehistoric times fire is thought to have been the major anthropogenic effect on vegetation. It was used to clear land for ease of travel, and Jones (1975) coined the term 'fire-stick farming' to describe its use for flushing out game, stimulating 'green-pick', which attracted game, and its function in favouring certain useful plants such as cycads and *Pteridium*; although the extent of the application of fire for this has not been investigated. In addition, fires for warmth, cooking or ceremonial purposes often escaped control. The deliberate use of fire by people must have greatly changed the fire regime, and therefore the nature of vegetation in many areas.

There is no doubt that many species are well-adapted to fire, as evidenced by a myriad of mechanisms, including seeding habits and nutrient cycling and life-cycle strategies, as described elsewhere in this volume. There is direct evidence of fire at least as far back as the Tertiary (Kemp, 1981) and all Quaternary records that have been examined for charcoal show its presence, often in great abundance (Head, 1989).

Even though great strides have been made to quantify charcoal abundance (e.g. Clark, 1982, 1984; Winkler, 1985; Clark, 1988*a*, *b*), interpreting the signal has proved difficult. At Lynch's Crater an increase in charcoal at about 40 000 BP (Kershaw, 1986) is consistent with the archaeological record of first human impact in Australia. At Lake George there is a large increase much earlier than this, perhaps at 130 000 BP, which Singh *et al.* (1981) and Singh & Geissler (1985) suggest may be an anthropogenic signal. Singh *et al.* (1981), however, have shown that charcoal values increased on Kangaroo Island after it was abandoned by humans. At Barrington Tops some sites show increasing charcoal abundance as late as 2000 to 3000 years ago. Recent evidence from New Zealand shows abundant charcoal for at least 15 000 years before human occupation (Dodson, Enright & McLean, 1988).

The only clear signal we have so far is that different quantities of charcoal are associated with different kinds of vegetation; generally less is found where rainforest is abundant. It is not clear if fire hastened the replacement of rainforest with sclerophyll vegetation at Lynch's Crater (Singh *et al.*, 1981) but, certainly, short episodes of rainforest development seemed halted by fire in the Hawkesbury Valley (Dodson & Thom, 1992). Head (1989) concluded that the greatest anthropogenic impact of fire was restricted to marginal areas where vegetation was under stress, for example, in drying climates, where regeneration capacity was low or relative soil erodibility was high. Finer detailed studies of charcoal and vegetation change, preferably from sites with high resolution archaeological data are needed to more directly test the effects of human impact.

There is no doubt about the signal of European impact on the Australian continent. It is marked by the appearance of alien taxa such as *Pinus*, *Plantago lanceolata* and *Echium*, plus all the indicators of clearance. Research in the future is likely to concentrate more on vegetation change and other environmental indicators to provide a more fully rounded picture of the history of human impact in landscape change.

Fine resolution pollen analysis

The palynological examination of sediment samples at greater time resolution than traditionally used is now underway in Australia. The quantitative examination of pollen and spores at time intervals often as short as 1 to 10 years allows the application of a range of modelling and statistical techniques to examine competition and replacement, responses to fire, persistence and to explore other relationships in time. Where the scale is appropriate the technique becomes a powerful ecological tool, of particular use where long-term monitoring, for example of long-lived species, is out of the question.

Green & Dolman (1988) recently summarised some of the ranges of techniques and approaches in data analysis of fine resolution pollen data. Clark (1986) and Clark & Wasson (1986) examined properties of sediments in Burrinjuck Reservoir, southwest of Canberra, and found these could be tied well to historical records. Such corroboration of the palynological with the sedimentary records is important in calibration so that error terms when applied to interpretations of older records are appreciated. For longer time-scales the work by Walker & Chen (1987), Chen (1988), Dodson (1988), and Green *et al.* (1988) provide examples.

Walker & Chen (1987) and Chen (1988) examined the population growth patterns for rainforest trees in the early Holocene at Lake Barrine, in northeastern Queensland. They showed that many secondary and regrowth taxa such as *Trema*, *Macaranga* and *Mallotus* showed characteristic exponential growth patterns and population doubling times of 60 to 120 years, whilst angiosperm canopy trees such as *Elaeocarpus*, *Rapanea* and *Balanops* followed either exponential or logistic growth and had doubling times of 100 to 300 years. The gymnosperm canopy trees *Agathis* and *Podocarpus* had exponential growth but were much slower with doubling times of 200 to 360 years. These data provide a solid basis for understanding vegetation dynamics and for devising management strategies.

Dodson (1988) examined pollen statistics from sediment sections on Barrington Tops in sclerophyll and *Nothofagus moorei* rainforest situations. With a time perspective of millennia, most rainforest and sclerophyll taxa could be shown to exhibit individual responses to change. On time scales of a century or less rapid positive responses and competition after fire occurred in sclerophyll formations. Fire was less common and caused contraction in rainforest, but no competitive effects amongst rainforest species could be identified. Competition between taxa of the two forest formations was not evident, thereby indicating a strong resilience to change under the prevailing environments of the last 2000 years in the region.

It is to be expected that the next review of this kind will contain a larger section devoted to these kinds of studies, especially those examining seral change in forests and the longer term impacts of human activity on vegetation.

References

Allen, J. (1989). When did humans first colonize Australia? *Search*, **20**, 149–54.

Anderson, C. & Fletcher, H.O. (1934). The Cuddie Springs bone bed. *The Australian Museum Magazine*, January 16, 152–8.

Bell, C.J.E., Finlayson, B. & Kershaw, A.P. (1989). Pollen analysis and dynamics of a peat deposit in Carnarvon National Park, central Queensland. *Australian Journal of Ecology*, **14**, 449–56.

Binder, R.M. (1978). Stratigraphy and pollen analysis of a peat deposit, Bunyip Bog, Mt Buffalo, Victoria. *Monash Publications in Geography*, **19**.

Binder, R.M. & Kershaw, A.P. (1978). A late-Quaternary pollen diagram from the southeastern highlands of Australia. *Search*, **9**, 44–5.

Bowen, D.Q. (1978). *Quaternary Geology*. Oxford: Pergamon Press.

Boyd, W.E. (1990). Quaternary pollen analysis in the arid zone of Australia: Dalhousie Springs, Central Australia. *Review of Palaeobotany and Palynology*, **64**, 331–41.

Broecker, W.S. & Denton, G.H. (1990). The role of ocean–atmosphere reorganizations in glacial cycles. *Quaternary Science Reviews*, **9**, 305–41.

Busby, J.R. (1986). A biogeoclimatic analysis of *Nothofagus cunninghamii* (Hook.) Oerst. in southeastern Australia. *Australian Journal of Ecology*, **11**, 1–7.

Caine, N. & Jennings, J.N. (1968). Some blockstreams of the Toolong Range, Kosciusko State Park, New South Wales. *Journal of the Royal Society of New South Wales*, **101**, 93–103.

Chen, Y. (1988). Early Holocene population expansion of some rainforest trees at Lake Barrine basin, Queensland. *Australian Journal of Ecology*, **13**, 225–33.

Churchill, D.M. (1968). The distribution and prehistory of *Eucalyptus diversicolor* F. Muell., *E. marginata* Donn ex Sm., and *E. calophylla* R.Br. in relation to rainfall. *Australian Journal of Botany*, **16**, 125–51.

Clark, J.S. (1988*a*). Particle motion and the theory of stratigraphic charcoal analysis: source area, transport, deposition, and sampling. *Quaternary Research*, **30**, 67–80.

Clark, J.S. (1988*b*). Stratigraphic charcoal analysis on petrographic thin sections: application to fire history in northwestern Minnesota. *Quaternary Research*, **30**, 81–91.

Clark, R.L. (1982). Point count estimation of charcoal in pollen preparations and thin sections of sediment. *Pollen et Spores*, **24**, 523–35.

Clark, R.L. (1984). Effects on charcoal of pollen preparation procedures. *Pollen et Spores*, **26**, 559–76.

Clark, R.L. (1986). Pollen as a chronometer and sediment tracer, Burrinjuck Reservoir, Australia. *Hydrobiologia*, **143**, 63–9.

Clark, R.L. & Wasson, R.J. (1986). Reservoir sediments. In *Limnology in Australia*, eds. P. De Deckker & W.D. Williams, pp. 497–507. Melbourne & Dordrecht: CSIRO & W. Junk.

COHMAP Members (1988). Climatic changes of the last 18,000 years: observations and model simulations. *Science*, **241**, 1043–52.

Colhoun, E.A. (1977). Late Quaternary fan gravels and slope deposits at Rocky Cape, northwestern Tasmania: their palaeoenvironmental significance. *Papers and Proceedings of the Royal Society of Tasmania*, **111**, 13–27.

Colhoun, E.A. (Compiler) (1988). *Cainozoic Vegetation of Tasmania*. Special Paper, Department of Geography, University of Newcastle, Australia.

Colhoun, E.A. & van de Geer, G. (1986). Holocene to middle Last Glaciation vegetation history at Tullabardine Dam, western Tasmania. *Proceedings of the Royal Society of London, Series B*, **229**, 177–207.

Colhoun, E.A., van de Geer, G. & Mook, W.G. (1982). Stratigraphy, pollen analysis and palaeoclimatic interpretation of Pulbeena Swamp, northwestern Tasmania. *Quaternary Research*, **18**, 108–26.

D'Costa, D.M., Edney, P., Kershaw, A.P. & De Deckker, P. (1989). Late Quaternary palaeoecology of Tower Hill, Victoria, Australia. *Journal of Biogeography*, **16**, 461–82.

D'Costa, D.M. & Kershaw, A.P. (in press). A late Pleistocene and Holocene pollen record from Lake Terang, western plains of Victoria, Australia. *Palaeogeography, Palaeoclimatology and Palaeoecology*.

Dodson, J.R. (1974*a*). Vegetation and climatic history near Lake Keilambete, western Victoria. *Australian Journal of Botany*, **22**, 709–17.

Dodson, J.R. (1974*b*). Vegetation history and water fluctuations at Lake Leake, southeastern South Australia. I. 10,000 B.P. to present. *Australian Journal of Botany*, **22**, 719–41.

Dodson, J.R. (1975). Vegetation history and water fluctuations at Lake Leake, southeastern South Australia. II. 50,000 B.P. to 10,000 B.P. *Australian Journal of Botany*, **23**, 815–31.

Dodson, J.R. (1977). Late Quaternary palaeoecology of Wyrie Swamp, southeastern South Australia. *Quaternary Research*, **8**, 97–114.

Dodson, J.R. (1979). Late Pleistocene vegetation and environments near Lake Bullenmerri, western Victoria. *Australian Journal of Ecology*, **4**, 419–27.

Dodson, J.R. (1986). Holocene vegetation and environments near Goulburn, New South Wales. *Australian Journal of Botany*, **34**, 231–49.

Dodson, J.R. (1988). The perspective of pollen records to study response, competition and resilience in vegetation on Barrington Tops, Australia. *Progress in Physical Geography*, **12**, 183–208.

Dodson, J.R. (1989). Late Pleistocene vegetation and environmental shifts in Australia and their bearing on faunal extinctions. *Journal of Archaeological Science*, **16**, 207–17.

Dodson, J.R., Enright, N.J. & McLean, R.F. (1988). A late Quaternary vegetation history for far northern New Zealand. *Journal of Biogeography*, **15**, 647–56.

Dodson, J.R., Greenwood, P.W. & Jones, R.L. (1986). Holocene forest and wetland dynamics at Barrington Tops, New South Wales. *Journal of Biogeography*, **13**, 561–85.

Dodson, J.R. & Thom, B.G. (1992). Holocene vegetation history from the Hawkesbury Valley, New South Wales. *Proceedings of the Linnean Society of New South Wales*, **113**, 121–34.

Dodson, J.R. & Wright, R. (1989). Humid to arid to subhumid vegetation shift on Pilliga Sandstone, Ulungra Springs, New South Wales. *Quaternary Research*, **32**, 182–92.

Firman, J.B. (1967). Stratigraphy of the late Cainozoic deposits in South Australia. *Transactions of the Royal Society of South Australia*, **91**, 165–78.

Flannery, T.F. (1990). Pleistocene faunal loss: implications of the aftershock for Australia's past and future. *Archaeology in Oceania*, **25**, 45–55; 64–7.

Fox, M.D. (1990). Interactions of native and introduced species in new habitats. *Proceedings of the Ecological Society of Australia*, **16**, 141–7.

Frakes, L.A. (1979). *Climates Throughout Geologic Time*. Amsterdam: Elsevier.

Fröhlich, C. (1988). Variability of the solar 'constant'. In *Long and Short Term Variability of Climate*, eds H. Wanner & U. Siegenthaler, pp. 6–17. Berlin: Springer-Verlag.

Galloway, R.W. (1986). Australian snow-fields past and present. In *Flora and Fauna of Alpine Australasia*, ed. B.A. Barlow, pp. 27–36. Melbourne: CSIRO/Brill.

Gentner, W., Kirsten, T., Storzer, D. & Wagner, G.A. (1973). K-Ar and fission track dating of Darwin Crater glass. *Earth and Planetary Science Letters*, **20**, 204–10.

Gillespie, R., Horton, D.R., Ladd, P.G., Macumber, P.G., Rich, T.H. & Wright, R.V.S. (1978). Lancefield Swamp and the extinction of the Australian megafauna. *Science*, **200**, 1044–8.

Green, N., Caldwell, J., Hope, J. & Luly, J. (1983). Pollen from an 1800 year old stick-nest rat (*Leporillis* sp.) midden from Gnalta, western New South Wales. *Quaternary Australasia*, **1**, 31–44.

Green, D.G. & Dolman, G. (1988). Fine resolution pollen analysis. *Journal of Biogeography*, **15**, 685–710.

Green, D.G., Singh, G., Polach, H., Moss, D., Banks, J. & Geissler, E.A. (1988). A fine resolution palaeoecology and palaeoclimatology from southeastern Australia. *Journal of Ecology*, **76**, 790–806.

Head, L. (1988). Holocene vegetation, fire and environmental history for Discovery Bay region, south-western Victoria. *Australian Journal of Ecology*, **13**, 21–49.

Head, L. (1989). Prehistoric Aboriginal impacts on Australian vegetation: an assessment of the evidence. *Australian Geographer*, **20**, 36–46.

Hiscock, P. (1990). How old are the artefacts of Malakananja II? *Archaeology in Oceania*, **25**, 122–5.

Hope, G.S. (1978). The late Pleistocene and Holocene vegetational history of Hunter Island, north-western Tasmania. *Australian Journal of Botany*, **26**, 493–514.

Hope, J.H. (1982). Late Cainozoic vertebrate faunas and the development of aridity in Australia. In *Evolution of the Flora and Fauna of Arid Australia*, eds W.R. Barker & P.J.M. Greenslade, pp. 85–100. Adelaide: Peacock.

Howard, T.M. & Hope, G.S. (1970). The present and past occurrence of beech (*Nothofagus cunninghamii* Oerst.) at Wilson's Promontory, Victoria, Australia. *Proceedings of the Royal Society of Victoria*, **83**, 199–210.

Imbrie, J. & Imbrie, K.P. (1979). *Ice Ages: Solving the Mystery*. London & Basingstoke: Macmillan.

Jones, R. (1975). The Neolithic, Palaeolithic and the hunting gardeners: man & land in the Antipodes. In *Quaternary Studies*, eds R.P. Suggate & M.M. Creswell, pp. 21–34. Wellington: The Royal Society of New Zealand.

Jones, R.L. (1990). Late Holocene vegetational changes on the Illawarra coastal plain, New South Wales, Australia. *Review of Palaeobotany and Palynology*, **65**, 37–46.

Kemp, E.M. (1981). Pre-Quaternary fire in Australia. In *Fire and the Australian Biota*, eds A.M. Gill, R.H. Groves & I.R. Noble, pp. 3–21. Canberra: Australian Academy of Science.

Kershaw, A.P. (1970). A pollen diagram from Lake Euramoo, north-east Queensland, Australia. *New Phytologist*, **69**, 785–805.

Kershaw, A.P. (1971). A pollen diagram from Quincan Crater, north-east Queensland, Australia. *New Phytologist*, **70**, 669–81.

Kershaw, A.P. (1975). Stratigraphy and pollen analysis of Bromfield Swamp, north-eastern Queensland, Australia. *New Phytologist*, **75**, 173–91.

Kershaw, A.P. (1976). A late Pleistocene and Holocene pollen diagram from Lynch's Crater, north-eastern Queensland, Australia. *New Phytologist*, **77**, 468–98.

Kershaw, A.P. (1985). An extended late Quaternary vegetation record from northeastern Queensland and its implications for the seasonal tropics of Australia. *Proceedings of the Ecological Society of Australia*, **13**, 179–89.

Kershaw, A.P. (1986). Climatic change and Aboriginal burning in north-east Australia during the last two glacial/interglacial cycles. *Nature*, **122**, 47–9.

Kershaw, A.P., Baird, J., D'Costa, D.M., Edney, P., Peterson, J.A. & Strickland, K.M. (1991). A comparison of long pollen records from the Atherton and Western Plains volcanic provinces, Australia. In *The Cainozoic of the Australian Region: a reappraisal*, eds M.A.J. Williams, P. De Deckker & A.P. Kershaw, pp. 288–301. Sydney: Geological Society of Australia.

Kershaw, A.P. & Nix, H.A. (1988). Quantitative palaeoclimatic estimates from pollen data using bioclimatic profiles of extant taxa. *Journal of Biogeography*, **15**, 589–602.

Ladd, P.G. (1976). Past and present vegetation of the Lancefield area. *Artefact*, **1**, 113–27.

Ladd, P.G. (1978). Vegetation history at Lake Curlip in lowland eastern Victoria from 5200 B.P. to present. *Australian Journal of Botany*, **26**, 393–414.

Ladd, P.G. (1979*a*). A Holocene vegetation record from the eastern side of Wilson's Promontory, Victoria. *New Phytologist*, **82**, 265–76.

Ladd, P.G. (1979*b*). Past and present vegetation on the Delegate River in the highlands of Victoria. II. Vegetation and climatic history from 12,000 B.P. to present. *Australian Journal of Botany*, **27**, 185–202.

Ladd, P.G. (1979*c*). A short pollen diagram from rainforest in highland Victoria. *Australian Journal of Ecology*, **4**, 229–37.

Luly, J. (1991). A pollen analytical investigation of Holocene palaeoenvironments at Lake Tyrrell, semi-arid north western Victoria, Australia. PhD Thesis, Australian National University.

Macphail, M.K. (1979). Vegetation and climate in southern Tasmania since the last glaciation. *Quaternary Research*, **11**, 306–41.

Macphail, M.K. & Colhoun, E.A. (1985). Late Last Glacial vegetation, climates and fire activity in south-western Tasmania. *Search*, **16**, 43–5.

Markgraf, V., Bradbury, J.P. & Busby, J.R. (1986). Paleoclimates in southwestern Tasmania during the last 13,000 years. *Palaios*, **1**, 368–80.

Marshall, L.G. (1973). Fossil vertebrate faunas from the Lake Victoria region, S.W. New South Wales, Australia. *Memoirs of the National Museum of Victoria*, **34**, 151–71.

Martin, A.R.H. (1986). Late glacial and Holocene alpine pollen diagrams from the Kosciusko National Park, New South Wales, Australia. *Review of Palaeobotany and Palynology*, **47**, 367–409.

Martin, H.A. (1973). Historical ecology of some cave excavations in the Australian Nullarbor. *Australian Journal of Botany*, **21**, 283–316.

Martin, H.A. (1978). Evolution of the Australian flora and vegetation through the Tertiary: evidence from pollen. *Alcheringa*, **2**, 181–202.

Martin, H.A. (1989). Evolution of mallee and its environment. In *Mallee Ecosystems and Their Management*, eds J.C. Noble & R.A. Bradstock, pp. 83–92. Melbourne: CSIRO.

Martin, H.A. (1990). Tertiary climate and phytogeography in southeastern Australia. *Review of Palaeobotany and Palynology*, **65**, 47–55.

Pearson, S. & Dodson, J.R. (1993). Stick nest rat middens as sources of palaeoecological data in Australian deserts. *Quaternary Research*, **39**, 347–54.

Raine, J.I. (1974). Pollen sedimentation in relation to the Quaternary vegetation history of the Snowy Mountains of New South Wales. PhD thesis, Australian National University.

Roberts, R.G., Jones, R. & Smith, M.A. (1990). Thermoluminescence dating of a 50,000 year-old human occupation site in northern Australia. *Nature*, **345**, 153–6.

Sigleo, W.R. & Colhoun, E.A. (1981). A short pollen diagram from Crown Lagoon in the Midlands of Tasmania. *Papers and Proceedings of the Royal Society of Tasmania*, **115**, 181–8.

Singh, G. (1981). Late Quaternary pollen records and seasonal palaeoclimates of Lake Frome, South Australia. *Hydrobiologia*, **82**, 419–30.

Singh, G. & Geissler, E.A. (1985). Late Cainozoic history of vegetation, fire, lake levels and climate at Lake George, New South Wales. *Philosophical Transactions of the Royal Society of London, Series B*, **311**, 379–447.

Singh, G., Kershaw, A.P. & Clark, R.L. (1981). Quaternary vegetation and fire history of Australia. In *Fire and the Australian Biota*, eds A.M. Gill, R.H. Groves & I.R. Noble, pp. 3–21. Canberra: Australian Academy of Science.

Specht, R.L. (1981). Ecophysiological principles determining the biogeography of major vegetation formations in Australia. In *Ecological Biogeography of Australia*, ed. A. Keast, pp. 299–333. The Hague: W. Junk.

Stirton, R.A., Tedford, R.H. & Miller, A.H. (1961). Cenozoic stratigraphy and vertebrate paleontology of the Tirari Desert, South Australia. *Records of the South Australian Museum*, **14**, 19–61.

Stuiver, M., Braziunas, T.F., Becker, B. & Kromer, B. (1991). Climatic, solar, oceanic, and geomagnetic influences on Late-Glacial and Holocene atmospheric $^{14}C/^{12}C$ change. *Quaternary Research*, **35**, 1–24.

van de Geer, G., Fitzsimmons, S.J. & Colhoun, E.A. (1986). Holocene to middle Last Glaciation vegetation history at Newall Creek, western Tasmania. *New Phytologist*, **11**, 549–58.

Walker, D. & Chen, Y. (1987). Palynological light on tropical rainforests. *Quaternary Science Reviews*, **6**, 77–92.

Walker, D. & Singh, G. (1981). Vegetation history. In *Australian Vegetation*, ed. R.H. Groves, pp. 26–43. Cambridge University Press.

Winkler, M.G. (1985). Charcoal analysis for palaeoenvironmental interpretation: a chemical assay. *Quaternary Research*, **23**, 313–24.

Wright, H.E. (1984). Sensitivity and response time of natural systems to climatic change in the late Quaternary. *Quaternary Science Reviews*, **3**, 91–131.

Wright, R.V.S. (1986a). New light on the extinction of the Australian megafauna. *Proceedings of the Linnean Society of New South Wales*, **109**, 1–9.

Wright, R.V.S. (1986b). How old is Zone F at Lake George? *Archaeology in Oceania*, **21**, 138–9.

3

Alien plants

P.W. MICHAEL

> In the early times of a colony, there is comparatively little
> difficulty in distinguishing the colonists from the native species;
> but as the surface of the land becomes artificially disturbed, the
> habits of all its plants are influenced, – the endemic plants are
> driven from their native places, and take refuge in hedgerows,
> ditches and planted copses, and from there associating with the
> introduced plants, are apt to be classed in the same category with
> them; whilst the introduced wander from the cultivated spots and
> eject the native, or taking their places by them, appear like them
> to be truly indigenous. (Hooker, 1860).

ALIEN PLANTS, in the sense used by Watson (1847), are plants
'now more or less established but either presumed or certainly known
to have been originally introduced from other countries'. They are
taken here to include those plants from other countries which are
sufficiently well established to be considered as true constituents of
the Australian flora. They include by far the greater number of plants
naturalised in Australia. In an unpublished list of some 800 weeds
and naturalised plants of New South Wales prepared for a Pan-Pacific
Science Congress by Blakely (1923), about 85 per cent of the species
included are aliens.

The most satisfying definition of naturalised plants is that given by
Thellung (1912), largely following de Candolle (1855), who presented
the first comprehensive and still useful treatment of naturalised plants
in various parts of the world. Thellung's definition is quoted here in
full.

> Nous appelons complètement naturalisée et, par abréviation,
> naturalisée, une espèce qui, n'existant pas dans un pays avant sa
> période historique (au point de vue de l'exploration botanique!),
> venant à y être transportée par l'action volontaire ou inconsciente
> de l'homme ou par une cause inconnue, s'y trouve ensuite avec
> tous les caractères des plantes spontanées indigènes, c'est-à-dire,
> croissant et se multipliant par ses moyens naturels de propagation

(graines, tubercules, bulbilles, drageons, fragments de tiges ou de rhizomes, etc., suivant l'espèce), sans le secours direct de l'homme, se manifestant avec plus ou moins d'abondance et de régularité dans les stations qui lui conviennent, et ayant traversé des séries d'années pendant lesquelles le climat a offert des circonstances exceptionelles.

A species is called completely naturalised or, in short, naturalised, if, not having been present in a country or region before its historic period (from the point of view of botanical exploration) it has been transported there by the intentional or accidental action of man or by a cause unknown and is now found there with all the characteristics of spontaneous indigenous plants, that is to say, increasing and multiplying by natural means (seeds, tubers, bulbs, suckers, fragments of stems or rhizomes etc., according to the species) without the direct help of man, occurring regularly, in greater or less abundance, in situations which suit it, and having gone through series of years during which there have been extreme climatic conditions. (My translation.)

If Thellung's 'pays' is taken to mean both 'country' and 'region', as in my translation, Australian naturalised natives are included. Australian naturalised native species were, before European exploration and settlement, confined to more limited areas. These plants include those which have escaped following their cultivation as ornamentals, for example, *Acacia baileyana*, occurring originally in a limited area of New South Wales, and *Pittosporum undulatum* in coastal regions of southeastern Australia and those which have greatly increased the areas they originally occupied in response to clearing and/or grazing of native communities, for example, *Sclerolaena birchii* and *Chloris truncata*. The eastern Australian tree fern, *Cyathea cooperi*, has become naturalised in the southwest of Western Australia, whilst *Acacia saligna*, native to Western Australia and often used for dune stabilisation in other countries has become naturalised in eastern Australia. Many Australian plants, especially woody ones, have become naturalised, even to the extent of becoming troublesome, outside Australia, for example, in New Zealand, South Africa and Florida and, indeed, some must have shown their propensity to spread very early. Ventenat (1802) who first described *Pittosporum undulatum* thought that it was native to the Canary Islands.

In this chapter, attention is given initially to the difficulties experienced in determining whether Australian plants are alien or not. A brief account of the extent of the alien flora is then presented with lists of the most commonly represented families and selected species characteristic of broad climatic regions. The importance of aliens in

1) S.p ERECT SHRUB [MALVACEAE] good FIBRE.
2) SPANISH NEEDLE [COMP] commer in COMMON
3) [TILIACEAE] FIBRE USED FOR ROPES [CEYLON]

invading natural ecosystems is briefly outlined. The origins, history and spread of certain species are discussed and comparisons and contrasts are made between the distribution of aliens in Australia and in their areas of origin. Brief mention is made of the variation within alien species in Australia and the importance of taxonomic study and knowledge of weedy flora in other parts of the world is stressed. Many of the observations made in this chapter are based on my own unpublished work.

(4) BOMBAY HEMP [LEGUM]

(4) [MALVACEAE] SHRUB CORDAGE.
(5) SPONGE GOURD [CUCURB]
(6) BEAN

Natives or aliens

In Australia, as in other countries, a high proportion of the alien species are weeds of cultivation, pastures, roadsides and waste places. These weedy aliens may be called pioneer species because of their ability to colonise disturbed or denuded land. During the history of land development in Australia relatively few native species have behaved in this way. This is clearly indicated in the list of Blakely (1923).

There are a number of tropical or subtropical species, questionably native to Australia, which have moved southwards into more temperate areas. Some of these may be pan-tropical or pan-subtropical species, for example, *Sida rhombifolia* and *Bidens pilosa*. It is difficult to ascertain whether other species believed to be native to the Indian subcontinent or to the Old World tropics, for example, *Triumfetta rhomboidea*, *Urena lobata*, *Luffa cylindrica*, *Canavalia gladiata*, *Crotalaria mucronata* and *C. retusa*, and *Colubrina asiatica*, are native to northern Australia. Certain species may have been introduced to northern Australia by visitors from southeast Asia before European settlement. Macknight (1976) presented evidence that *Tamarindus indica* was introduced by Makassans in the eighteenth century, for example.

It is, indeed, often difficult to establish whether a particular species is native or alien. As Watson (1859) pointed out, 'the distinction between native and introduced species is absolute and real', but 'there are differences of opinion regarding the evidences in support of either view in reference to individual species'. Plants which were described or renamed early from material collected in Australia must not be assumed, for that reason alone, to be native, for they could have been introduced along with or very soon after European settlement. Robert Brown, in 1802–4, noted a number of alien plants near Sydney, including *Urtica urens*, *Silene anglica*, *Anagallis arvensis*, *Euphorbia*

HERB – URINARY PROBLEMS

peplus and *Poa annua* (Britten, 1906; Maiden, 1916). He also collected the American species, *Gnaphalium pensylvanicum*. The widespread plant *Lepidium africanum*, an introduction from South Africa, was long thought to be the native *Lepidium hyssopifolium* which is, in fact, a much less common species. Another common weed from southern Africa, *Senecio madagascariensis*, was believed for many years to be a form of the native *S. lautus* complex. *Cyclospermum leptophyllum* was recorded by Bentham (1866) under the name *Apium leptophyllum* as occurring presumably native in both Australia and South America, but it seems certain that it is native to subtropical America whence it has spread throughout the subtropical areas of southeast Asia, Australia and elsewhere. On the other hand, Australian native material named *Eryngium rostratum* by Bentham (1866) is different from the Chilean plants originally described as such and should be referred to as *Eryngium ovinum* (Michael, 1987*a*).

Some species are represented by both alien and native forms. A good example is *Oxalis corniculata*, in the broad sense, which includes four or five native forms which warrant specific status, for example, *O. exilis*, as well as perhaps two or three alien forms which are common weeds in gardens and glasshouses. It is likely too that *Cynodon dactylon*, known to have been introduced from India, includes native forms.

Extent of the alien flora

Australia, with an area of 7.7×10^6 km², extends latitudinally from about 10° to 44° S and shows great diversity in climates and soils. Of an estimated total number of species of vascular plants somewhere between 15 000 and 20 000, about 2000 are aliens. In addition, there are a number of alien algae in Australian waters (Australian National Parks & Wildlife Service, 1991), the most important of which is the Japanese brown alga *Undaria pinnatifida* first observed in the early 1980s at the eastern Tasmanian woodchip port, Triabunna, and now spreading in eastern Tasmanian waters where it competes with native species of kelp and obscures the vision of abalone divers (The Mercury, 1992). Accurate estimates of numbers are difficult because of problems in determining taxonomic status and insufficient knowledge, especially of the flora of the 3.0×10^6 km² of land north of the Tropic of Capricorn. Alien plants are often poorly collected, and the native or alien status of some species is, as has been indicated already, difficult to determine.

The following treatment of the vascular flora is based on a circumspect examination especially of the existing Australian floras and of the works of Bailey (1913), Burbidge (1963), Chippendale (1971), Clifford & Ludlow (1972), Aston (1973), Jones & Clemesha (1976), Clifford & Constantine (1980), Baines (1981), Jacobs & Pickard (1981), Sainty & Jacobs (1981, 1988), Jessop (1984), Green (1985), Jacobs & Lapinpuro (1986), Kloot (1986b), Mabberley (1987), Forbes & Ross (1988), Andrews (1990), Hnatiuk (1990) and Simon (1990). Other sources are cited where appropriate.

Ferns and their allies are represented by eight genera and nine species, all of which were introduced as garden plants. They include the horsetail *Equisetum arvense*, well known as a weed in the northern hemisphere, of very restricted occurrence in New South Wales; *Selaginella kraussiana*, a native of southern Africa, in southeastern Australia; *Salvinia molesta*, a floating fern native to Brazil (Fig. 3.1), common in dams and streams along the coast from central New South Wales to northern Queensland and extending into the Northern Territory; and *Pityrogramma austroamericana* and *P. calomelanos*, natives of the Americas, the former naturalised in southeastern Queensland to northeastern New South Wales and the latter in northeastern Queensland and the Northern Territory.

Gymnosperms are represented notably by *Pinus* with a number of species, the most common being *P. radiata*, native of the Monterey Peninsula in California, escaped from plantations in southeastern Australia.

Angiosperms are represented by about 130 families and more than 850 genera, of which more than 70 per cent are alien. About 50 families are represented only by one genus and many genera by only one species. Families represented each by four or more genera, 49 in all, and comprising more than 700 genera are listed below according to the classification of Cronquist (1981). The numbers of genera in each family are given in brackets.

Ranunculaceae (7), Papaveraceae (6), Fumariaceae (4), Moraceae (4), Urticaceae (4), Aizoaceae (13), Cactaceae (7), Chenopodiaceae (9), Amaranthaceae (8), Portulacaceae (4), Caryophyllaceae (21), Polygonaceae (8), Malvaceae (15), Cucurbitaceae (7), Brassicaceae (39), Crassulaceae (6), Rosaceae (21), Mimosaceae (8), Caesalpiniaceae (9), Fabaceae (56), Onagraceae (5), Euphorbiaceae (13), Rhamnaceae (4), Araliaceae (4), Apiaceae (22), Apocynaceae (6), Asclepiadaceae (7), Solanaceae (17), Convolvulaceae

Fig. 3.1. *Salvinia molesta*, native to Brazil, covering a lake in the Botanic Gardens, Townsville. (Photo: P. W. Michael.)

(6), Boraginaceae (12), Verbenaceae (5), Lamiaceae (26), Oleaceae (4), Scrophulariaceae (22), Acanthaceae (6), Pedaliaceae (4), Bignoniaceae (6), Campanulaceae (4), Rubiaceae (11), Caprifoliaceae (4), Asteraceae (114), Araceae (4), Commelinaceae (6), Cyperaceae (7), Poaceae (101), Zingiberaceae (4), Liliaceae (27), Iridaceae (24) and Agavaceae (5).

More than one-third of the families represented in the Australian alien flora have so far been treated in *Flora of Australia*. They are Ulmaceae, Cannabaceae, Moraceae, Urticaceae, Betulaceae and Fagaceae (Vol. 3, 1989); Phytolaccaceae, Nyctaginaceae, Aizoaceae, Cactaceae and Chenopodiaceae (Vol. 4, 1984); Flacourtiaceae, Bixaceae, Cistaceae, Violaceae, Tamaricaceae, Passifloraceae, Cucurbitaceae, Salicaceae, Capparaceae, Brassicaceae, Moringaceae and Resedaceae (Vol. 8, 1982); Haloragaceae, Lythraceae, Thymelaeaceae, Punicaceae, Onagraceae and Combretaceae (Vol. 18, 1990); Aquifoliaceae (Vol. 22, 1984); Melianthaceae, Sapindaceae, Aceraceae, Anacardiaceae and Simaroubaceae (Vol. 25, 1985); Solanaceae (Vol. 29, 1982); Sparganiaceae, Typhaceae, Bromeliaceae, Zingiberaceae, Cannaceae, Pontederiaceae, Haemodoraceae and Liliaceae (Vol. 45, 1987); and Iridaceae, Aloeaceae and Agavaceae (Vol. 46, 1986). I refer the reader to these accounts for further useful background.

There is nothing unusual in the composition of the Australian alien flora, naturalised plants from the above listed families being freely represented in many floras and checklists of plants from other countries.

Of the 2000 alien species, about 35 per cent belong to the three families Fabaceae (about 180 species), Asteraceae (about 230 species) and Poaceae (about 310 species), comprising around 15 per cent, 20 per cent and 25 per cent, respectively of the total number of Australian representatives in these three families.

The proportion of alien species, between 10 and 11 per cent in Australia as a whole, varies considerably from state to state, within states and according to agricultural, ecological and environmental status and history. Proportions vary from around 25 per cent in South Australia and Tasmania to 10 per cent in Western Australia and between 7 and 8 per cent in the Northern Territory; within Victoria from over 30 per cent in the Melbourne and Murray Valley regions to around 17 per cent in the alpine and East Gippsland regions (Beauglehole, 1980); within New South Wales from over 30 per cent in the Riverine Plain (Leigh & Mulham, 1977; Mulham & Jones,

1981), 23 per cent at Mt Tomah in the Blue Mountains (Ingram, 1987), around 20 per cent in the subalpine and alpine zones of the Kosciusko region (Thompson & Gray, 1981), to 16 per cent in rocky outcrops and ranges in the southwest (Norris & Thomas, 1991), 14 per cent in exclosure sites in western areas (Cunningham & Milthorpe, 1981), 12 to 13 per cent in the far southwest of the state near the junction of the Darling and Murray rivers (Fox, 1991) and under 10 per cent in the northwest near Bourke (Moore, 1984); within Western Australia from over 25 per cent in the Perth region to around 7 per cent in the area covered in the Flora of Central Australia, embracing a large part of the arid region.

The proportion of aliens in island floras varies from over 40 per cent on Lord Howe Island (Rodd & Pickard, 1983) and 20 per cent on Kangaroo Island to 9 per cent on Magnetic Island near Townsville (Sandercoe, 1990). Very low proportions have been observed, around 3 per cent for New England National Park in New South Wales (Williams, 1976) and Isla Gorge National Park in Queensland (Everist, Sharpe & Hockings, 1976) and around 2 per cent in the Blackdown Tableland in Queensland (Henderson, 1976) and coastal heath in New South Wales, north of the Macleay River (Williams, 1985). These low proportions may, in part, reflect the interests of the collectors. A footpath flora of parts of Sydney would show as low a proportion of natives. No aliens were recorded by Boyland (1970) in the Simpson Desert National Park.

To indicate briefly the wide range of alien species in the Australian flora, some widespread species belonging to the families Caryophyllaceae, the orders Fabales and Lamiales, and the families Asteraceae and Poaceae, are given here. They are listed according to whether they occur essentially in southern Australia in areas with mediterranean or more temperate climates or in northern Australia in areas with subtropical or tropical climates.

Southern Australia. *Stellaria media, Trifolium subterraneum, Ulex europaeus, Amsinckia* spp., *Echium plantagineum, Cirsium vulgare, Hypochaeris radicata, Phalaris aquatica, Lolium perenne.*

Northern Australia. *Drymaria cordata, Stylosanthes humilis, Mimosa pigra, Stachytarpheta* spp., *Hyptis suaveolens, Tithonia diversifolia, Tridax procumbens, Panicum maximum, Axonopus compressus.*

The area occupied by each of the large number of alien species in Australia varies greatly from the vast areas of southern Australia colonised by *Arctotheca calendula* and *Hordeum* spp. to the scattered occurrences of *Chrysanthemum leucanthemum* in the wetter areas of southeastern Australia; from the large tree, *Cinnamomum camphora*, widespread in the far north coast of New South Wales and the isolated occurrences throughout southern New South Wales of *Ailanthus altissima*, a species which extends its distribution locally by suckering, to the occasional infestations of *Alternanthera philoxeroides* in watercourses in eastern New South Wales; from the dense stands of *Ligustrum* spp. in the Sydney area to the restricted open areas colonised by *Hebenstretia dentata* in and around Newcastle; from the wide areas of *Homeria* spp. in Western Australia to the occasional garden escape, *Freesia refracta* in wetter parts of southern Australia; from the thousands of hectares of *Oxalis pes-caprae* in South Australia and the abundant weeds, *O. corymbosa* and *O. latifolia* in Sydney gardens to the single weedy occurrence of *O. tetraphylla* in an old garden near Sydney, *O. perdicaria* (Fig. 3.2) known only to occupy a few hectares near Grenfell in New South Wales and the tiny isolated occurrences of *O. braziliensis* along the Hunter River.

Fig. 3.2. *Oxalis perdicaria*, native to South America, grown from bulbs collected from the site of its only known Australian occurrence near Grenfell, New South Wales. (Photo: P. W. Michael.)

Invasions of natural ecosystems

In an overall view of Australian vegetation, we are concerned not
only with those alien plants which are now widespread and occupy
vast areas, or which are prominent features of the vegetation of particular regions or which repeatedly occur in similar situations, but
also with those plants termed 'environmental' weeds, the presence of
which is in some ways deleterious to the natural environment and to
human enjoyment of that environment (Groves, 1991). The definition
by Robin (1991) of an environmental weed as any plant that occurs
outside of its natural geographic range, and that is capable of replication without direct human intervention in undisturbed, semi-
disturbed or disturbed indigenous vegetation, includes both native
and alien plants and really makes the study of naturalised plants
environmentally appropriate.

There are obvious difficulties in distinguishing between modified
and natural situations even in Australia where European settlement
began only just over 200 years ago. There are few aliens which have
invaded apparently undisturbed native vegetation in Australia. *Baccharis halimifolia* in uncleared swamps of *Melaleuca quinquenervia*
is one such example (Westman, Panetta & Stanley, 1975). Sometimes
the original disturbance which allowed the ingress of aliens has been
obscured. The presence of a few plants of *Ulex europaeus* in the wild,
almost completely uninvaded heathlands of Rocky Cape National
Park in northwestern Tasmania begs such an explanation.

Alien plants do indeed have profound effects on disturbed natural
plant communities as, for example, in the invasion of the vast Mitchell
grass plains in Queensland by *Acacia nilotica*, in the replacement of
wetland vegetation by pure stands of *Mimosa pigra* in the Northern
Territory and of coastal stands of *Acacia longifolia* var. *sophorae* by
Chrysanthemoides monilifera in southeastern Australia or the displacement of the native flora by *Tamarix aphylla* on watercourses in
arid Australia with consequent adverse effects on the watertable, soil
salinisation and on riverflow and sedimentation regimes. Carr, Robin
& Robinson (1986) have called such invasions 'Australia's greatest
conservation problem' and it is satisfying to point to the recent publications of Groves & Burdon (1986), Adair & Shepherd (1991), Australian National Parks & Wildlife Service (1991) and Kirkpatrick
(1991) in which the problems of plant invasions are clearly stated
and in which attention to research already done is given.

Table 3.1. *Proportion of alien angiosperms from various geographical regions for South Australia and the Cook region, Queensland*

	South Australia	Cook region, Queensland
Total number of species	880	320
Origin		
Eurasian–Mediterranean	64%	10%
American	16%	41%
African and African–Asian (including Old-world tropics)	17%	24%
Asian–Pacific	3%	13%
Pan-tropical	0%	12%

Origins, history and spread

In ascertaining the origin of Australian plants, one has to rely mostly upon the scant information in existing floras and standard references like Willis (1973), Good (1974), Mabberley (1987) and on monographic or detailed treatments of genera in a much-fragmented botanical literature. Be that as it may, the alien Australian flora has originated in all broad geographical regions of the world, the proportions from these regions varying greatly in areas of widely differing climate, as, for example, in South Australia, of Mediterranean aspect, and the Cook region of subtropical Queensland, including Cape York Peninsula and the adjacent Atherton Tableland and tropical rainforests. The proportions of alien angiosperms originating from the broad regions indicated are given in Table 3.1 for South Australia (modified from Kloot (1986*b*, 1987), and for the Cook region (after cautious use of Hnatiuk (1990) and from many other sources).

The African element in South Australia is strongly Mediterranean, in contrast to the strong subtropical or tropical element in northern Queensland. Likewise, the more temperate elements from North and South America are more important in South Australia than in northern Queensland where the tropical American element is of greatest significance.

Examples of aliens, now abundant in Australia, originating from various broad regions are given overleaf:

Europe
 Malva parviflora, Verbascum virgatum, Carduus nutans, Cirsium vulgare, Dactylis glomerata, Poa annua;
Mediterranean regions
 Ranunculus muricatus, Trifolium subterraneum, Parentucellia latifolia, Cynara cardunculus, Bromus rubens, Phalaris minor;
North America
 Solanum elaeagnifolium, Amsinckia spp., *Ambrosia artemisiifolia, Solidago canadensis, Andropogon virginicus, Panicum capillare;*
Tropical America
 Passiflora foetida, Stylosanthes humilis, Jatropha curcas, Hyptis suaveolens, Axonopus compressus, Chloris virgata;
South America
 Nicandra physalodes, Verbena bonariensis, Conyza bonariensis, Xanthium spinosum, Paspalum dilatatum, Nassella trichotoma;
South Africa
 Emex australis, Oxalis pes-caprae, Arctotheca calendula, Melinis (Rhynchelytrum) repens, Sporobolus africanus, Homeria spp.;
Tropical Africa
 Cucumis metulifera, Acacia nilotica, Tamarindus indica, Crotalaria goreensis, Ricinus communis, Sorghum verticilliflorum;
East Asia
 Cinnamomum camphora, Persicaria capitata, Rhus succedanea, Ligustrum lucidum and *L. sinense, Lonicera japonica*

It is important to try and relate the origins of the ancient weeds of cultivation which form a considerable proportion of the alien plants of Australia to the chief primary (essentially neolithic) regions of agriculture as outlined and illustrated by Darlington (1963) following Vavilov (1926, 1935) and Kuptsov (1955), namely the Indian and Indonesian regions expanding to cover southeast Asian and Pacific regions; Ethiopian, near Asiatic, Mediterranean and central Asian regions expanding to cover Europe and Western Asia; Nigerian region expanding to cover a great part of Africa north and south of the tropics; Mexican and Peruvian regions expanding to cover a great part of the Americas and the West Indies and the Chinese region expanding to cover eastern Asia.

Better still, the origin of Australian alien plants may largely be traced to the regions of origin of crop plants as outlined by Darlington (1963) following many authors. These regions include southwest

Asia, the Mediterranean, Europe, Abyssinia, Central Africa, Central Asia, Indo-Burma, southeast Asia, China, Mexico, United States of America, Peru, Chile and Brazil–Paraguay.

In a consideration of the history of some Australian alien species, attention will now be given, where possible, to their places of origin. The early history of many of our alien plants has been documented by: Maiden (1920), McBarron (1955), Everist (1960), Michael (1964, 1972, 1987*b*), Mann (1970), Auld (1977), Mitchell (1978), Kloot (1980, 1983, 1984, 1985, 1986*b*, 1987), and Swarbrick (1986), and the reader is referred to these papers for additional information on some of the species mentioned.

Xanthium spinosum, whose world history is admirably documented by Widder (1923), originated in Chile whence it was transported to Australia on animals or in grain in the first half of the nineteenth century, just as it had been transported to Europe much earlier. It may also have been introduced to Australia as an ornamental. *Xanthium occidentale*, originating in North America and the West Indies, was apparently introduced to Australia with cotton seed. *Xanthium italicum*, native to the United States of America and Mexico, was probably introduced in the same way. A form of *X. orientale* appears to have been introduced to Australia from California. This variable species, which I believe is native to the west coast of the United States of America, extending into Mexico, was probably introduced to Europe by the Spanish quite early; and from southern Europe, where it is now quite common, it appears to have been re-introduced to America to areas around the St Lawrence River. *Xanthium cavanillesii*, known only from two general localities in New South Wales, both associated with very old settlements, must have been introduced soon after European settlement from South America. The two occurrences are of different forms of the species, indicating different times or ways of introduction.

The four most significant alien species of *Echinochloa* present in Australia can be traced to two original centres, the most important being the Indo–Burma region, the source of different forms of *E. colona* and *E. crus-galli* and the obligate weed of rice, *E. oryzoides*. *Echinochloa microstachya* originated in North America. A long-awned form of *E. crus-galli* and *E. oryzoides* appear to have been introduced, along with *E. microstachya* to Australia in rice seed from California. These two former occur in rice-fields in southern Europe and Iraq also.

Hordeum leporinum and *H. glaucum*, both native to the Mediterranean region, must have been introduced quite early (Cocks, Boyce

& Kloot, 1976) and undoubtedly many times from various places in Europe or on the way to Australia. In their discussion of the origin of the *H. murinum* complex in Australia, Cocks *et al.* (1976) neglect to mention the probable importance of South America, especially in view of the early trade between Portugal, Spain and the American colonies and the prominence of both species in the Mediterranean region.

Some of the thistles, notably *Cirsium vulgare* and *Silybum marianum*, natives of Europe and/or the Mediterranean regions, were introduced to Australia in the very early days of settlement, *Cirsium vulgare* as a contaminant of pasture seed and *Silybum marianum*, both as a medicinal plant and as Scotch thistle. Many Australian alien plants were introduced originally as ornamentals and in a study of catalogues of Australian Botanic Gardens and nurseries, as referred to by Michael (1972), numerous species will be found. A number of weeds of great economic importance are included under the names given or their synonyms, for example, *Ageratina adenophora*, *Homeria* spp., *Lantana camara*, *Oxalis pes-caprae*, *Eriocereus martinii*, and *Opuntia* spp. *Amaranthus powellii*, a common weed of cultivation throughout southern Australia, may well have been introduced with *A. hypochondriacus* which was grown as an ornamental in the very early days of the settlement of New South Wales. *Amaranthus powellii* belongs to a group of species native to the Americas, including also *A. hybridus*, *A. quitensis* and *A. retroflexus*, all pioneers especially occuring along riverbanks or in other open areas in their places of origin (Sauer, 1967).

The spread of aliens after introduction to Australia has, as elsewhere, essentially followed settlement. Plants not transported by humans with their animals, crop or pasture seeds, have been spread by wind, water or birds and have colonised disturbed situations. Ornamental plants grown in more and more home gardens have led to more and more foci from which the plants have been able to move into previously uncolonised areas. Intentionally sown pasture species, sometimes, like *Trifolium subterraneum* originally appearing here and there of their own accord, have escaped from cultivation.

In South Australia, where *Oxalis pes-caprae* was introduced before 1850, we may imagine bulbs being grown in all home gardens in the closely settled plains north of Adelaide. With occasional removal and discarding of excess growth from gardens, cultivation and floods along the close streams rising in the Mount Lofty Ranges and traversing the plains, considerable spread was accomplished in the second half of last century. Such favourable combinations of factors do not seem to have occurred over large areas in other southern states.

In contrast, the recent explosion of the American *Parthenium hysterophorus* in Queensland (Haseler, 1976; Armstrong, 1978) was apparently initiated from the sowing in 1960 of a single large area of impure grass seed imported from Texas. This plant was first collected in the Upper Brisbane Valley in 1955, where it still occurs, but it required a massive inoculum to stimulate large-scale spread. The plant now occupies many thousands of square kilometres.

The rapid spread of this plant is exceptional in the history of aliens in Australia. Doley (1977) and Williams & Groves (1980) have predicted a much wider potential distribution. Doing, Biddiscombe & Knedlhans (1969) and Medd & Smith (1978) have predicted further increases in the area occupied by *Carduus nutans* ssp. *nutans* which was first collected in Australia in 1950, almost certainly introduced in seed of *Lolium perenne* from New Zealand. Sindel & Michael (1992) have predicted further expansion of *Senecio madagascariensis*, first collected in the Hunter Valley, near Newcastle, over 70 years ago.

There is little concrete information on the rate of spread of aliens in Australia, mainly because of the lack of interested observers. McVean (1966) calculated that *Chondrilla juncea* spread throughout southern Australia at an average rate of about 24 km per year for 40 years after its initial introduction to Australia. The more recent work in relation to the spread of *Opuntia aurantiaca*, *Parthenium hysterophorus* and *Nassella trichotoma* is summarised in Auld, Menz & Tisdell (1987). Auld & Coote (1980) developed a generalised spread model.

In a period of 20 years, rapid increases in the area occupied by *Andropogon virginicus* in the neighbourhood of Sydney and beyond have been observed. It is now approaching the Victorian border south of Eden. *Ambrosia tenuifolia* from South America, long established around Newcastle, has greatly extended its range between Newcastle and Sydney. The South American *Facelis retusa* originally collected near Newcastle is now extended well beyond Sydney. Macarthur (Select Committee, 1852) first noted *Xanthium spinosum* along the Nepean River in the late 1830s. By the early 1850s it had become a weed of public concern. For many years and probably since the early days of Sydney, *Parietaria judaica* has been firmly established in a relatively small area near the harbour, but in the last few years, with the tremendous disturbance associated with new building and freeway operations, there has been a sudden increase in area occupied by the species. The use of herbicides may also have hastened its spread.

In the last few years there appears to have been an enormous expansion in the area occupied by *Foeniculum vulgare*, once more or less confined to areas close to old settlements. This plant has prompted a recent paper in California (Beatty & Licari, 1992).

Moore (1959) has given an account of changes in vegetation in southeastern Australia following clearing of woodlands, grazing and use of superphosphate and the sowing of *Trifolium subterraneum*. These management factors, leading to increased soil fertility, have encouraged the spread of many weedy aliens, including plants well known as nitrophilous species in Europe, for example, *Cirsium vulgare* and *Onopordum acanthium*, confined in the initial stages of settlement to stock camps and rabbit warrens. Kloot (1986a) has related an ecological history of the alien flora of the cereal areas of South Australia. A comprehensive account of the effects of European settlement on the plants and plant communities in New South Wales has been given by Benson (1991).

Distribution of aliens in Australia and areas of origin

This subject has been reviewed briefly in relation to important Australian pasture weeds by Michael (1970), who differentiated between those aliens originating in Europe or the Mediterranean region, areas with a long history of disturbance due to man's activities, and those originating from the Cape Province of South Africa, a region subjected to a much shorter period of disturbance by man. The former species may have developed much greater adaptability to different environments than the latter, and, accordingly, when introduced to Australia have colonised areas not exactly comparable with the areas of optimum range in Europe.

Echium plantagineum occurs in lower latitudes in Australia than in the northern hemisphere (Moore, 1967). *Carduus nutans* ssp. *nutans*, a biennial with a qualitative vernalisation requirement (Medd, 1977), occurs in areas in Australia with higher winter temperatures and lower relative humidities than in Europe (Doing *et al.*, 1969). It is difficult to relate the distribution of *Avena fatua* and *A. ludoviciana* in Australia to their distribution in Europe and the Mediterranean region. Although *A. ludoviciana* is common in the Mediterranean region, it is of little importance in areas of Australia with mediterranean-type climate, where *A. fatua*, characteristic of the more northern parts of Europe, prevails. In Australia, *A. ludoviciana* is most abundant in the wheat belt of northern New South Wales and southern Queensland where there is appreciable summer rainfall.

On the other hand, species from the Cape Province, for example, *Oxalis pes-caprae*, *Arctotheca calendula* and *Homeria* spp., occupy essentially similar climatic zones in Australia and South Africa. It appears from the data presented by Gray (1976) that *Chrysanthemoides monilifera* ssp. *monilifera* and ssp. *rotundata*, from the southwestern Cape Province and the southeastern coastal areas of South Africa, respectively, each occupy corresponding climatic zones in Australia.

Habitat preferences of plants in Europe may be reflected in their distribution in Australia. A good example is *Trifolium subterraneum* ssp. *yanninicum*, essentially confined in Europe to waterlogged soils in the Balkan peninsula and naturalised only in similar situations in southern Australia (Morley & Katznelson, 1965).

Species originating in the Americas may sometimes show clear distributional relationships, but, more often, the relationships are confused. *Eremocarpus setigerus*, native to California with mediterranean-type climate, is naturalised in Australia only in Western Australia and South Australia in areas with similar climate. *Conyza canadensis* var. *canadensis*, abundant especially in the eastern United States of America and in Canada, occurs only sporadically in coastal and tableland regions of eastern Australia, whilst *C. parva*, widespread in the eastern United States of America, the West Indies, Mexico, Central America and northern South America, occurs only in the coastal, mostly subtropical, parts of eastern Australia.

McMillan (1975) related occurrences of four *Xanthium* spp. in Australia to particular areas in the United States and in South America from which he believes them to have been introduced, essentially on the basis of photoperiodic requirements. All the Australian collections of *X. occidentale* (referred to as *X. chinense*), examined by McMillan represented a type photoperiodically adapted to the southeastern United States and similar to populations occurring on the Mississippi Delta south of New Orleans, Louisiana, situated at about 30° N. In Australia, however, *X. occidentale* extends in great abundance from latitudes below 20° to 33° S. *Xanthium cavanillesii* from near Sydney was shown to be nearly identical with the same species from Buenos Aires, Argentina, at about the same latitude (34° S).

There are many examples of pairs of closely related species which show overlapping distributions in Australia. *Digitaria ciliaris* of subtropical or tropical origin and *D. sanguinalis* of more temperate origin often occur together in central New South Wales but, at lower latitudes in Queensland, *D. ciliaris* occurs to the virtual exclusion of *D. sanguinalis*. The reverse situation applies at higher latitudes in Victoria. The American species *Gnaphalium americanum*, of subtropical and tropical origin, and *G. coarctatum* of more temperate origin, show essentially the same pattern. Relationships of *Gnaphalium* spp. are, however, clouded by inadequate taxonomy.

Historical factors which have greatly influenced the occurrence of aliens in Australia may often obscure ecological relationships. *Echium vulgare*, from Europe, was apparently introduced to Australia much earlier than *E. plantagineum* (Piggin, 1977) and is now confined to a few of the older settled areas in southeastern Australia, where it is often accompanied by the generally much more successful *E. plantagineum*. The introduction and subsequent establishment of *Oxalis perdicaria* in a single small area in New South Wales referred to earlier could be termed an historical accident. The abundance of *Amaranthus quitensis* in the Hunter Valley, New South Wales and its virtual absence elsewhere in Australia (Michael, 1977) can perhaps be attributed to its early introduction there from South America.

Variation in alien species in Australia

In a country where many alien species occur in profusion over wide areas it is to be expected that considerable variation in individual seed-producing species would be evident. Few species, however, have been studied in detail. Perhaps the most well-known work is that concerned with *Trifolium subterraneum* (Aitken & Drake, 1941; Morley, 1961; Morley & Katznelson, 1965), where a great variety of forms differing in vegetative and floral characters and flowering times have been observed. Similar studies have been made on *T. glomeratum* (Woodward & Morley, 1974) where numerous variations have been noted. Much variation in growth habit, flowering times and spikelet characters have been observed in *Avena fatua*, *A. ludoviciana* and *A. barbata* (Whalley & Burfitt, 1972; Paterson, Boyd & Goodchild, 1976). There is little evidence, however, that distinct Australian forms have developed. Rather, it is supposed that the variation in Australian material is the direct result of a wide range of introductions.

Three forms of *Chondrilla juncea*, differing in the shape of rosette leaves, inflorescence morphology and fruit characters, were described by Hull & Groves (1973). One of the forms closely resembles plants collected from southern Italy and it is likely that the other two forms of this apomictic species could be matched with forms occurring in Europe.

Little variation occurs in the major occurrences of the tristylic species *Oxalis pes-caprae* throughout southeastern Australia. These occurrences are of a short-styled pentaploid clone, originally introduced as a garden plant which reproduces only by vegetative means. There are, however, other forms, all tetraploid and with all three style lengths, naturalised especially in Western Australia (Michael, 1964). Where tetraploid forms of different style lengths occur together, viable seed may be produced, encouraging further variation in the species. The freely cross-breeding species, *Lolium perenne, L. multiflorum* and *L. rigidum* are very difficult to distinguish from each other in areas in Australia where any two or all three species have been introduced.

Proper attention to studies of variation in certain broad species groups has led, in recent years, to the definition of a number of distinct species within these groups, for example, in *Solanum nigrum sens. lat.* (Henderson, 1974), *Rubus fruticosus* sp. agg. (Amor & Miles, 1974) and *Amaranthus hybridus sens. lat.* (Michael, 1977).

Conclusions

Many of the studies on Australian alien species have been prompted by their economic significance, both favourable and unfavourable. Successful and desirable species, such as *Trifolium subterraneum, Stylosanthes humilis, Lolium rigidum* and *Chloris gayana* appear to share certain biological attributes with undesirable weeds, such as *Cirsium vulgare, Hyptis suaveolens, Hordeum* spp. and *Xanthium occidentale*. Both groups are able to establish themselves, often forming extensive populations, without man's deliberate action, and tend to be aggressive, competitive and adaptable. Efficient reproduction and the ability to survive under temporarily unfavourable conditions are held in common. We return thus to Thellung's definition of naturalised species given earlier.

There are big gaps in our knowledge of the alien flora as a whole. Problems of identity are foremost. It is hoped that taxonomic treatments like those of *Lupinus* (Gladstones, 1974), *Adonis* (Kloot, 1976), *Hordeum* (Cocks *et al.*, 1976), *Datura* (Haegi, 1976a), *Lycium* (Haegi, 1976b), *Paronychia* (Aston, 1977), *Amaranthus* and *Conyza* (Michael, 1977), *Echium* (Piggin, 1977; *Solanum* Symon, 1981), *Echinochloa* (Michael, 1983), *Cortaderia* (Harradine, 1991), *Stachytarpheta* (Munir, 1992) and others already treated in the *Flora of Australia* will be extended to cover a much wider range of genera.

Work on alien genera is fraught with many difficulties, notably the general paucity of specimens collected in Australia and the difficulties in acquiring collections and literature from other countries. Co-operation with taxonomists in other parts of the world is indispensable.

In view of the interest in methods of predicting the potential spread of species like *Carduus nutans, Parthenium hysterophorus, Senecio madagascariensis* and *Nassella trichotoma* and the recent emphasis on the large number of 'environmental' weeds threatening natural ecosystems in Australia, familiarity with alien and weedy native species established in other parts of the world but not yet in Australia, is desirable. *Nassella trichotoma* (under the name *Stipa trichotoma*) was recorded by Coste (1906) as being completely naturalised on the banks of the River Orba in the Piedmont, Italy and *Parthenium hysterophorus* was recorded by Hauman (1928) as a garden weed in Tucuman, Argentina, at about 27° S.

It is especially important that close attention be paid to the alien and weedy species of subtropical and tropical America and Africa and to those of Asia where certain plants from America and Africa, apparently not yet established in Australia, are prevalent. One such species thought to be a potentially serious invader in Australia and recently discussed in detail by Cruttwell McFadyen (1989) is the American *Chromolaena odorata*. An African species, *Echinochloa stagnina* could become a problem in deep water in the wetlands of northern Australia.

The rapid spread in southern Africa of the toxic South American plant, *Sesbania punicea* (syn. *Daubentonia punicea*) noted by Pienaar (1983) alerts us to the possible dangers of a plant apparently not yet established in Australia, but which, like a number of important Australian weeds, was listed in early Australian botanical garden and nursery catalogues. Groves (1986) highlights the ever-increasing numbers of naturalised plants in Australia. It is best if we learn as much as we can of the likely invaders before they become established.

References

Adair, R. & Shepherd, R. (eds) (1991). Proceedings of a third symposium on the control of environmental weeds, Monash University, 6–7 November 1991. *Plant Protection Quarterly*, 6, 95–153.

Aitken, Y. & Drake, F.R. (1941). Studies of the varieties of subterranean clover. *Proceedings of the Royal Society of Victoria*, 53, 342–93.

Amor, R.L. & Miles, B.A. (1974). Taxonomy and distribution of *Rubus fruticosus* L.agg. (Rosaceae) naturalized in Victoria. *Muelleria*, **3**, 37–62.

Andrews, S.B. (1990). *Ferns of Queensland.* Information Series Q 189008. Brisbane: Queensland Department of Primary Industries.

Armstrong, T.R. (1978). Herbicidal control of *Parthenium hysterophorus.* In *Proceedings of the First Conference of the Council of Australian Weed Science Societies*, pp. 157–64. Melbourne: Council of Australian Weed Science Societies.

Aston, H.I. (1973). *Aquatic Plants of Australia.* Melbourne: Melbourne University Press.

Aston, H.I. (1977). The species of *Paronychia* (Caryophyllaceae) in Victoria. *Muelleria*, **3**, 209–14.

Auld, B.A. (1977). The introduction of *Eupatorium* species to Australia. *Journal of the Australian Institute of Agricultural Science*, **43**, 146–7.

Auld, B.A. & Coote, B.G. (1980). A model of a spreading plant population. *Oikos*, **34**, 287–92.

Auld, B.A., Menz, K.M. & Tisdell, C.A. (1987). *Weed Control Economics.* London: Academic Press.

Australian National Parks and Wildlife Service (1991). *Plant Invasions. The Incidence Of Environmental Weeds in Australia.* Canberra: Australian National Parks & Wildlife Service.

Bailey, F.M. (1913). *Comprehensive Catalogue of Queensland Plants.* Brisbane: Government Printer.

Baines, J.A. (1981). *Australian Plant Genera.* Chipping Norton, New South Wales: Surrey Beatty & Sons.

Beatty, S.W. & Licari, D.L. (1992). Invasion of fennel (*Foeniculum vulgare*) into shrub communities on Santa Cruz Island, California. *Madroño*, **39**, 54–66.

Beauglehole, A.C. (1980). *Victorian Vascular Plant Checklists- 13 Study Area and 24- Grid Distribution.* Portland, Victoria: Western Victorian Field Naturalists' Clubs Association.

Benson, J. (1991). The effect of 200 years of European settlement on the vegetation and flora of New South Wales. *Cunninghamia*, **2**, 343–70.

Bentham, G. (1866). *Flora Australiensis.* Vol. 3. London: Lovell Reeve & Co.

Blakely, W.F. (1923). A census of the weeds and naturalised plants of New South Wales (Abstract). *Proceedings of the Pan-Pacific Science Congress, Australia*, Vol. 1, p. 330. (The list of plants was never published but what appears to be the final draft is held in the Library of the National Herbarium of New South Wales.)

Boyland, D.E. (1970). Ecological and floristic studies in the Simpson Desert National Park, southwestern Queensland. *Proceedings of the Royal Society of Queensland*, **82**, 1–16.

Britten, J. (ed.) (1906). Introduced plants at Sydney, 1802–4. *Journal of Botany (British & Foreign)*, **44**, 234–5.

Burbidge, N.T. (1963). *Dictionary of Australian Plant Genera.* Sydney: Angus & Robertson.

Candolle, A.L.P.P. de (1855). *Geographie Botanique Raisonnée*, Vol. 2. Paris: Librairie de Victor Masson.

Carr, G.W., Robin, J.M. & Robinson, R.W. (1986). Environmental weed invasion of natural ecosystems: Australia's greatest conservation problem. In *Ecology of Biological Invasions: An Australian Perspective*, eds R.H. Groves & J.J. Burdon, p. 150. Canberra: Australian Academy of Science.

Chippendale, G.M. (1971). Checklist of Northern Territory plants. *Proceedings of the Linnean Society of New South Wales*, **96**, 206–67.

Clifford, H.T. & Constantine, J. (1980). *Ferns, Fern Allies and Conifers of Australia*. Brisbane: University of Queensland Press.

Clifford, H.T. & Ludlow, G. (1972). *Keys to the Families and Genera of Queensland Flowering Plants*. Brisbane: University of Queensland Press.

Cocks, P.S., Boyce, K.G. & Kloot, P.M. (1976). The *Hordeum murinum* complex in Australia. *Australian Journal of Botany*, **24**, 651–62.

Coste, H.J. (1906). *Flore Descriptive et Illustrée de la France*, Vol. 3. Paris: Librairie des Sciences et des Arts.

Cronquist, A. (1981). *An Integrated System of Classification of Flowering Plants*. New York: Columbia University Press.

Cruttwell McFadyen, R.E. (1989). Siam weed: a new threat to Australia's north. *Plant Protection Quarterly*, **4**, 3–7.

Cunningham, G.M. & Milthorpe, P.L. (1981). The vascular plants of five exclosure sites in western New South Wales. *Cunninghamia*, **1**, 23–34.

Darlington, C.D. (1963). *Chromosome Botany and the Origins of Cultivated Plants*, 2nd edn. London: George Allen & Unwin.

Doing, H., Biddiscombe, E.F. & Knedlhans, S. (1969). Ecology and distribution of the *Carduus nutans* group (nodding thistles) in Australia. *Vegetatio*, **17**, 313–51.

Doley, D. (1977). Parthenium weed (*Parthenium hysterophorus* L.): gas exchange characteristics as a basis for prediction of its geographical distribution. *Australian Journal of Agricultural Research*, **28**, 449–60.

Everist, S.L. (1960). Strangers within the gates. *Queensland Naturalist*, **16**, 49–60.

Everist, S.L., Sharpe, P.R. & Hockings, F.D. (1976). Plant list – Isla Gorge National Park. *Queensland Naturalist*, **21**, 107–10.

Flora of Australia (1982). Vol. 8. Lecythidales to Batales. Canberra: Australian Government Publishing Service.

Flora of Australia (1982). Vol. 29. Solanaceae. Canberra: Australian Government Publishing Service.

Flora of Australia (1984). Vol. 4. Phytolaccaceae to Chenopodiaceae. Canberra: Australian Government Publishing Service.

Flora of Australia (1984). Vol. 22. Rhizophorales to Celastrales. Canberra: Australian Government Publishing Service.

Flora of Australia (1985). Vol. 25. Melianthaceae to Simaroubaceae. Canberra: Australian Government Publishing Service.

Flora of Australia (1986). Vol. 46. Iridaceae to Dioscoreaceae. Canberra: Australian Government Publishing Service.

Flora of Australia (1987). Vol. 45. Hydatellaceae to Liliaceae. Canberra: Australian Government Publishing Service.

Flora of Australia (1989). Vol. 3. Hamamelidales to Casuarinales. Canberra: Australian Government Publishing Service.

Flora of Australia (1990). Vol. 18. Podostemaceae to Combretaceae. Canberra: Australian Government Publishing Service.

Forbes, S.J. & Ross, J.H. (1988). *A Census of the Vascular Plants of Victoria*, 2nd edn. Melbourne: National Herbarium of Victoria.

Fox, M.D. (1991). The natural vegetation of the Ana Branch – Mildura 1:250 000 map sheet of New South Wales. *Cunninghamia*, 2, 443–98.

Gladstones, J.S. (1974). Lupins of the Mediterranean region and Africa. Technical Bulletin No. 26. Department of Agriculture, Western Australia.

Good, R. (1974). *The Geography of the Flowering Plants*, 4th edn. London: Longman.

Gray, M. (1976). Miscellaneous notes on Australian plants. 2. *Chrysanthemoides* (Compositae). *Contributions from Herbarium Australiense*, 16, 1–5.

Green, J.W. (1985). *Census of the Vascular Plants of Western Australia*. Perth: Western Australian Herbarium, Department of Agriculture.

Groves, R.H. (1986). Plant invasions of Australia: an overview. In *Ecology of Biological Invasions: an Australian Perspective*, eds R.H. Groves & J.J. Burdon, pp. 137–49. Canberra: Australian Academy of Science.

Groves, R.H. (1991). Status of environmental weed control in Australia. *Plant Protection Quarterly*, 6, 95–8.

Groves, R.H. & Burdon, J.J. (eds) (1986). *Ecology of Biological Invasions: An Australian Perspective*. Canberra: Australian Academy of Science.

Haegi, L. (1976*a*). Taxonomic account of *Datura* L. (Solanaceae) in Australia with a note on *Brugmansia* Pers. *Australian Journal of Botany*, 24, 415–35.

Haegi, L. (1976*b*). Taxonomic account of *Lycium* (Solanaceae) in Australia. *Australian Journal of Botany*, 24, 669–79.

Harradine, A.R. (1991). The impact of pampas grasses as weeds in southern Australia. *Plant Protection Quarterly*, 6, 111–15.

Haseler, W.H. (1976). *Parthenium hysterophorus* in Australia. *PANS*, 22, 515–17.

Hauman, L. (1928). Les modifications de la flore Argentine sous l'action de la civilisation. *Mémoires de l'Academie Royale de Belgique Classe des Sciences 2s*, 9, 3–95.

Healy, A.J. (1944). Some additions to the naturalised flora of New Zealand. *Transactions of the Royal Society of New Zealand*, 74, 221–31.

Henderson, R.J.F. (1974). *Solanum nigrum* L. (Solanaceae) and related species in Australia. *Contributions from the Queensland Herbarium*, Number 16.

Henderson, R.J.F. (1976). Plants of Blackdown Tableland. *Queensland Naturalist*, 21, 125–32.

Hnatiuk, R. (1990). *Census of Australian Vascular Plants*. Australian Flora and Fauna Series No. 11. Canberra: Australian Government Publishing Service.

Hooker, J.D. (1860). On some of the naturalized plants of Australia. In *The Botany (of) the Antarctic Voyage*, part III. *Flora Tasmaniae*, vol. 1. London: Lovell Reeve.

Hull, V.J. & Groves, R.H. (1973). Variation in *Chondrilla juncea* L. in south-eastern Australia. *Australian Journal of Botany*, 21, 113–35.

Ingram, C.K. (1987). Checklist of the vascular flora of Mount Tomah. In *The Mount Tomah Book*, 2nd edn, pp. 63–85. Sydney: Mount Tomah Society & Royal Botanic Gardens.

Jacobs, S.W.L. & Lapinpuro, L. (1986). Alterations to the census of New South Wales plants. *Telopea*, **2**, 705–14.

Jacobs, S.W.L. & Pickard, J. (1981). *Plants of New South Wales*. Sydney: National Herbarium of New South Wales, Royal Botanic Gardens.

Jessop, J.P. (ed) (1984). *A List of the Vascular Plants of South Australia*, 2nd edn. Adelaide: Botanic Gardens & State Herbarium.

Jones, D.L. & Clemesha, S.C. (1976). *Australian Ferns and Fern Allies*. Sydney: A.H. & A.W. Reed.

Kirkpatrick, J.B. (ed) (1991). *Tasmanian Native Bush: A Management Handbook*. Hobart: Tasmanian Environment Centre.

Kloot, P.M. (1976). The species of *Adonis* L. naturalized in Australia. *Muelleria*, **3**, 199–207.

Kloot, P.M. (1980). Dr Richard Schomburgk's 'Naturalised Weeds' (1879). *Journal of the Adelaide Botanic Gardens*, **2**, 195–220.

Kloot, P.M. (1983). Early records of alien plants naturalised in South Australia. *Journal of the Adelaide Botanic Gardens*, **6**, 93–131.

Kloot, P.M. (1984). The introduced elements of the flora of southern Australia. *Journal of Biogeography*, **11**, 63–78.

Kloot, P.M. (1985). Plant introductions to South Australia prior to 1840. *Journal of the Adelaide Botanic Gardens*, **7**, 217–31.

Kloot, P.M. (1986*a*). A review of the naturalised alien flora of the cereal areas of South Australia. Technical Paper 12. Department of Agriculture, South Australia

Kloot, P.M. (1986*b*). Checklist of the introduced species naturalised in South Australia. Technical Paper 14. Department of Agriculture, South Australia.

Kloot, P.M. (1987). The naturalised flora of South Australia. 1. The documentation of its development. 2. Its development through time. 3. Its origin, introduction, distribution, growth forms and significance. *Journal of the Adelaide Botanic Gardens*, **10**, 81–90, 91–8, 99–111.

Kuptsov, A.I. (1955). Geographical distribution of cultivated flora and its historical development (Russian). *Geograficheskoe obshchestvo, Izvestiya*, **87**, 220–31.

Leigh, J.H. & Mulham, W.E. (1977). Vascular plants of the Riverine Plain of New South Wales with notes on distribution and pastoral use. *Telopea*, **1**, 225–93.

Mabberley, D.J. (1987). *The Plant-Book*. Cambridge: Cambridge University Press.

McBarron, E.J. (1955). An enumeration of plants in the Albury, Holbrook and Tumbarumba districts of New South Wales. *Contributions from the New South Wales National Herbarium*, **2**, 89–247.

Macknight, C.C. (1976). *The Voyage to Marege*. Melbourne: Melbourne University Press.

McMillan, C. (1975). The *Xanthium strumarium* complexes in Australia. *Australian Journal of Botany*, **23**, 173–92.

McVean, D.N. (1966). Ecology of *Chondrilla juncea* L. in southeastern Australia. *Journal of Ecology*, 54, 345–65.

Maiden, J.H. (1916). Weeds at Sydney in 1802–4. *Agricultural Gazette, New South Wales*, 27, 40.

Maiden, J.H. (1920). *The Weeds of New South Wales*, Part 1. Sydney: Government Printer.

Mann, J. (1970). *Cacti Naturalised in Australia and their Control*. Brisbane: Department of Lands.

Medd, R.W. (1977). Some aspects of the ecology and control of *Carduus nutans* L. (Nodding Thistle) on the Northern Tablelands, New South Wales. PhD thesis, University of New England, NSW.

Medd, R.W. & Smith, R.C.G. (1978). Prediction of the potential distribution of *Carduus nutans* (nodding thistle) in Australia. *Journal of Applied Ecology*, 15, 603–12.

Michael, P.W. (1964). The identity and origin of varieties of *Oxalis pes-caprae* L. naturalized in South Australia. *Transactions of the Royal Society of South Australia*, 88, 167–74.

Michael, P.W. (1970). Weeds of grasslands. In *Australian Grasslands*, ed. R.M. Moore, pp. 349–60. Canberra: Australian National University Press.

Michael, P.W. (1972). The weeds themselves – early history and identification. In *Symposium: The History of Weed Research in Australia. Proceedings of the Weed Society of New South Wales*, 5, 3–18.

Michael, P.W. (1977). Some weedy species of *Amaranthus* (amaranths) and *Conyza/Erigeron* (fleabanes) naturalized in the Asian–Pacific region. *Proceedings of the sixth Asian–Pacific Weed Science Society Conference*, Vol. 1, pp. 87–95.

Michael, P.W. (1983). Taxonomy and distribution of *Echinochloa* species with special reference to their occurrence as weeds of rice. In *Weed Control in Rice*, pp. 291–306. Los Baños: International Rice Research Institute.

Michael, P.W. (1987a). Blue devil – *Eryngium ovinum* A. Cunn. reinstated. *Australian Systematic Botany Newsletter*, 53, 3–4.

Michael, P.W. (1987b). History of the introduction of weeds into Australia. *Proceedings of the 4th Biennial Noxious Plants Conference*, Supplement, pp. 57–61. Sydney: New South Wales Department of Agriculture.

Mitchell, A.S. (1978). An historical overview of exotic and weedy plants in the Northern Territory. *Proceedings of the First Conference of the Council of Australian Weed Science Societies*, pp. 145–53.

Moore, C.W.E. (1984). Annotated checklist of the vascular plants of New South Wales. Technical Memorandum 84/30. Canberra: CSIRO Institute of Biological Resources.

Moore, R.M. (1959). Ecological observations on plant communities grazed by sheep in Australia. In *Biogeography and Ecology in Australia* (Monographiae Biologicae, VIII), eds A. Keast, R.L. Crocker & C.S. Christian, pp. 500–13. The Hague: W. Junk.

Moore, R.M. (1967). The naturalisation of alien plants in Australia. *International Union for Conservation of Nature and Natural Resources Publications*, New Series, Number 9, 82–97.

Morley, F.H.W. (1961). Subterranean clover. *Advances in Agronomy*, **13**, 57–123.

Morley, F.H.W. & Katznelson, J. (1965). Colonization in Australia by *Trifolium subterraneum* L. In *The Genetics of Colonizing Species*, eds H.G. Baker & G.L. Stebbins, pp. 269–82. London: Academic Press.

Mulham, W.E. & Jones, D.E. (1981). Vascular plants of the Riverine Plain of New South Wales – supplementary list. *Telopea*, **2**, 197–213.

Munir, A.A. (1992). A taxonomic revision of the genus *Stachytarpheta* Vahl (Verbenaceae) in Australia. *Journal of the Adelaide Botanic Gardens*, **14**, 133–68.

Norris, E.H. & Thomas, J. (1991). Vegetation on rocky outcrops and ranges in central and southwestern New South Wales. *Cunninghamia*, **2**, 411–41.

Paterson, J.G., Boyd, W.J.R. & Goodchild, N.A. (1976). Vernalization and photoperiod requirement of naturalized *Avena fatua* and *Avena barbata* Pott. ex Link in Western Australia. *Journal of Applied Ecology*, **13**, 265–72.

Pienaar, K.J. (1983). 25. *Sesbania*. In *Plant Invaders. Beautiful, But Dangerous*, ed. C.H. Stirton, pp. 136–9. Capetown: Department of Nature and Environmental Conservation of the Cape Provincial Administration.

Piggin, C.M. (1977). The herbaceous species of *Echium* (Boraginaceae) naturalised in Australia. *Muelleria*, **3**, 215–44.

Robin, J. (1991). Control of environmental weeds. In *Tasmanian Native Bush. A Management Handbook*, ed. J.B. Kirkpatrick, pp. 174–83. Hobart: Tasmanian Environment Centre.

Rodd, A.N. & Pickard, J. (1983). Census of vascular flora of Lord Howe Island. *Cunninghamia*, **1**, 269–80.

Sainty, G.R. & Jacobs, S.W.L. (1981). *Water Plants of New South Wales*. Sydney: Water Resources Commission of New South Wales.

Sainty, G.R. & Jacobs, S.W.L. (1988). *Water Plants in Australia*. Sydney: Sainty & Associates.

Sandercoe, C. (1990). Vegetation of Magnetic Island. Technical Report No. 1, Queensland National Parks & Wildlife Service.

Sauer, J.D. (1967). The grain amaranths and their relatives: a revised taxonomic and geographic survey. *Annals of the Missouri Botanical Garden*, **54**, 103–37.

Select Committee on the Scotch Thistle and Bathurst Burr Report (1852). Sydney: Government Printer.

Simon, B.K. (1990). *A Key to Australian Grasses*. Information Series Q189019. Brisbane: Queensland Department of Primary Industries.

Sindel, B.M. & Michael, P.W. (1992). Spread and potential distribution of *Senecio madagascariensis* Poir. (fireweed) in Australia. *Australian Journal of Ecology*, **17**, 21–6.

Swarbrick, J.T. (1986). History of the lantanas in Australia and origins of the weedy biotypes. *Plant Protection Quarterly*, **1**, 115–21.

Symon, D.E. (1981). A revision of the genus *Solanum* in Australia. *Journal of the Adelaide Botanic Gardens*, **4**, 1–367.

The Mercury (1992). Action is vital on introduced species. Editorial, March 17, p. 8. Hobart.

Thellung, A. (1912). La flore adventice de Montpellier. *Mémoires de la Société Nationale des Sciences Naturelles et Mathematiques de Cherbourg*, **38**, 57–728.

Thompson, J. & Gray, M. (1981). A checklist of the subalpine and alpine species found in the Kosciusko region of New South Wales. *Telopea*, **2**, 299–346.

Vavilov, N.I. (1926). Studies on the Origin of Cultivated Plants. *Trudy po prikladnoi botanike, genetike i selektsii*, 16, No. 2 (Russian with English translation, pp. 139–245).

Vavilov, N.I. (1935). Phytogeographic basis of plant breeding. Tr. by K.S. Chester from original Russian in Vavilov, N.I. (ed.) *Theoretical Bases of Plant Breeding*, Vol. 1. Moscow & Leningrad: State Agricultural Publishing House. (*Chronica Botanica*, **13**, 14–54, 1949/50).

Ventenat, E.P. (1802). Description des plantes nouvelles et peu connues, cultivées dans le jardin de J.M. Cels. 76, t. 76 Paris: de l'Imprimerie de Crapelet.

Watson, H.C. (1847). *Cybele Britannica*, Vol. 1. London: Longman & Co.

Watson, H.C. (1859). *Cybele Britannica*, Vol. 4. London: Longman & Co.

Westman, W.E., Panetta, F.D. & Stanley, T.D. (1975). Ecological studies on reproduction and establishment of the woody weed groundsel bush (*Baccharis halimifolia* L.: Asteraceae). *Australian Journal of Agricultural Research*, **26**, 855–70.

Whalley, R.D.B. & Burfitt, J.M. (1972). Ecotypic variation in *Avena fatua* L., *A. sterilis* L. (*A. ludoviciana*) and *A. barbata* Pott. in New South Wales and Queensland. *Australian Journal of Agricultural Research*, **23**, 799–810.

Widder, F.J. (1923). Die Arten der Gattung *Xanthium*. *Feddes Repertorium* Beihefte Bd. 20.

Williams, J.B. (1976). *The Flora of the New England National Park*. Armidale: Botany Department, University of New England, New South Wales.

Williams, J.B. (1985). *Plants of Coastal Heath Scrub and Swamp – heath in northern New South Wales*. Armidale: Botany Department, University of New England, New South Wales.

Williams, J.D. & Groves, R.H. (1980). The influence of temperature and photoperiod on growth and development of *Parthenium hysterophorus* L. *Weed Research*, **20**, 47–52.

Willis, J.C. (1973). *A Dictionary of Flowering Plants and Ferns*, revised by H.K. Airy Shaw, 1897, 8th edn. London: Cambridge University Press.

Woodward, R.G. & Morley, F.H.W. (1974). Variation in Australian and European collections of *Trifolium glomeratum* L. and the provisional distribution of the species in southern Australia. *Australian Journal of Agricultural Research*, **25**, 73–88.

PART 2

MAJOR VEGETATION TYPES

Preamble

The concept that vegetation types change along a continuum or gradient runs counter to the classification of vegetation types into neat categories. This conflict was apparent in the planning of this book and in the writing of individual chapters, especially those on forests and woodlands and those distinctively Australian scrublands called 'mallees'.

The allocation of chapters in this section has been based on Specht's classification of Australia's major vegetation types (see Table 18.1, p. 531), but no effort has been made to constrain individual authors to the terminology of that classification. Whether a chapter refers to 'tall open-forest' or to the 'wet sclerophyll forest' of earlier writers matters little, provided the individual author conveys to the reader something of the intrinsic structure, floristic composition and ecological relationships of that vegetation type. The authors and the editor have tried to describe the vegetation types in their own ways without adhering to rigid definitions.

The chapters in the next section (Vegetation of Extreme Habitats), on the other hand, are based on readily identifiable regions of Australia, such as the alpine region, and together describe vegetation of a number of types. There is therefore some inevitable overlap between the sections.

I hope that there are sufficient cross-references to help the reader through the fundamental difficulty of viewing vegetation as a continuum when reading about it in separate chapters written by highly individualistic Australian botanists.

R.H. Groves

4

The rainforests of northern Australia

L.J. WEBB & J.G. TRACEY

DESPITE human incursions over the last century, the remaining Australian rainforest vegetation has a unique ecological status, taxonomic interest, and aesthetic appeal that continue to attract increasing attention from scientific specialists and the general community. The original area of rainforest when Europeans arrived was relatively small (less than 1 per cent of the surface of the continent) and its distribution pattern resembled an 'archipelago of habitats' (Herbert, 1967).

The rainforest habitats preserve a remarkable wealth of endemic and, in some areas, primitive biota, as well as exhibiting strong affinities at the generic level with surrounding countries that were continuous with the Australian land mass in Gondwanan time. Although the processes of evolution and community development responsible for the patterns of Australian rainforests are being unravelled only now, evidence already forthcoming indicates a need to revise traditional concepts in Australian phytogeography that previously regarded the floristic elements of the northern rainforests as alien and invasive (Webb & Tracey, 1981).

The earliest descriptions of Australian rainforests were taxonomic, and the northern types were soon recognised as 'of high interest for the genesis of the Australian flora' (translation of Diels, 1906). Ecological studies did not begin until about 60 years ago, at first near Sydney and Melbourne, and expanded northwards and then to the northwest (for references, see Webb & Tracey, 1981).

Except for the scattered pockets of monsoon forest in northern Australia, many of the remaining rainforests along the eastern coast are relatively accessible. In this chapter we shall describe briefly the widely differing types, their distribution and dynamics, and offer some interpretations of them.

Ecological characteristics

Australian rainforests have several striking characteristics that readily set them apart from the rest of Australian vegetation.

Definition

The canopy is closed and the trees densely spaced, in contrast to the open and generally scattered sclerophyllous vegetation that covers most of the forested area of the continent in the moister coastal and subcoastal zones, but mainly in the east.

In tropical and subtropical rainforests there are three or more tree layers, with or without emergents. The tree layers become reduced to two and eventually to one distinct layer at higher latitudes and altitudes, i.e. temperate and montane forest types that are floristically impoverished. Rainforests are also distinguished from other forests by combinations of characteristic life forms, e.g. epiphytes, lianes, certain root and stem structures, certain tree-ferns and palms, absence of annual herbs on the forest floor. Species composition is most complex and the interdependence of the different niches of the forest is most complete under optimal environmental conditions and on relatively large ground surfaces of long stability. Lowering of moisture and temperature, decrease in soil nutrient availability and aeration, and reduction in habitat size and stability are accompanied by lower species diversity. All rainforests, especially the most complex tropical ones, are composed of a mixture of species representing different stages of succession following different kinds of disturbance. These successions ensure the maximum saturation of an area in time as well as in space. Disturbances should be understood as an integral factor of the rainforest environment, and they strongly influence species diversity.

Rainforests, including monsoonal types of drier northern areas, vary in the proportion of evergreen and deciduous species in the canopy, and belong to the category of closed-forests and closed-scrubs, as opposed to open-forests, woodlands and scrubs (Specht, 1970). The closed-forests correspond to the *'forêts denses'*, referred to mainly in the African and French literature (e.g. Aubréville, 1965), and to the *'ombrophilous'* forests of the UNESCO world classification (UNESCO, 1969).

The term rainforest is well entrenched in the literature of Australia (e.g. Diels, 1906; Francis, 1929; Fraser & Vickery, 1938; Baur, 1957; Webb, 1959) and elsewhere (e.g. Schimper, 1903; Richards, 1952;

Odum & Pigeon, 1970; Walter, 1971; Whitmore, 1984) so that there are advantages in retaining it. One of the most cogent botanical reasons is that tropical, subtropical, monsoonal and temperate types of rainforest in Australia resemble formation-types in other countries, with which they have many biological affinities and can be readily compared. This is in spite of the lack of uniformity in terminology for rainforests in Africa, Asia and America noted by Letouzey (1978). Within Australia, seemingly disparate types of rainforest occur throughout a wide geographical range (see Fig. 4.1) stretching in an arc from the northwest to the southeast through some 6000 km, but they all form part of a floristic continuum, especially when other plants besides trees are included.

The spelling 'rainforest' as a single word adopted here follows Baur (1968), and denotes a series of formations that are generally

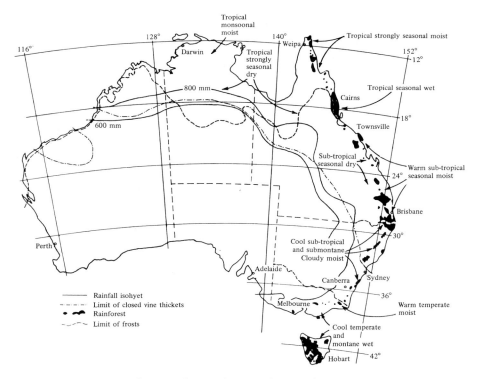

Fig. 4.1. Distribution of Australian rainforests (from Webb & Tracey, 1981). The arrows indicate the approximate centre of distribution ('core area') of each climatic type but the boundaries of each type overlap because of past climatic changes.

independent of other forest types. The spelling also avoids undue emphasis on rain as the sole determining environmental factor (Baur, 1968). It is, however, considered that the term rainforest should be restricted to the general formation-type, and that further subdivisions, equivalent to formations and subformations which can be broadly correlated with climate and soils, are best described as structural types following the nomenclature of Webb (1959, 1968, 1978). Floristic provinces or elements (see below), whose distribution among structural types and their habitats is the result of climatic sifting and historical factors, are most conveniently designated as climatic types in relation to their 'core' areas at the present time. These are assumed, perhaps arguably, to indicate the particular climates under which the floristic elements originally evolved.

To avoid circular reasoning, rainforest is strictly defined to exclude species that do not regenerate within a well-developed or slightly disturbed canopy of closed-forest. A corollary is that such excluded species are usually integral members of other formation-types such as eucalypt forests. This definition excludes, for example, all species of *Eucalyptus*, *Casuarina* and *Melaleuca*; all but one species of *Callitris* (*C. macleayana*); all but a few species of *Acacia* (e.g. *A. bakeri*, *A. fasciculifera* in the dry subtropics; and debatably *A. melanoxylon* in attenuated patches of warm temperate rainforest in the southeast, and *A. mangium*, *A. aulacocarpa*, *A. cincinnata* in the wet tropics of the northeast). Recent taxonomic revisions in Myrtaceae (Wilson & Waterhouse, 1982), however, have emphasised the linkages between evergreen rainforest taxa and the sclerophyll flora. Some evergreen rainforest types are dominated in the canopy by capsular-fruited myrtaceous trees in *Backhousia*, *Ristantia*, *Xanthostemon*, *Stockwellia* and *Allosyncarpia* with endemic or disjunct regional distributions throughout Australian rainforests. The genus *Tristania*, previously regarded as sclerophyll, has been redefined and is now restricted to *Tristania laurina*, a riparian species in rainforests of New South Wales. *Ristantia*, with different species of rainforest trees in northern and central Queensland, and *Tristaniopsis*, with different species in northern Queensland and southern Queensland/NSW rainforest, have been excluded. All are accepted as rainforest trees. Also split from *Tristania* is *Lophostemon*, with brush box (*L. confertus*) intimately associated with subtropical rainforests, swamp mahogany (*L. Suaveolens*) with a widespread and disjunct distribution associated with seasonal swamps, and *Lophostemon grandiflorus* widely scattered throughout the seasonally dry water courses in the wet/dry tropics of northern Australia often associated with species linked with rainforests at generic level. Clearly, the definition of cut-off points for rainforest and sclerophyll vegetation is arbitrary when floristic relationships are considered.

Discontinuity

Another striking feature of Australian rainforests is their small area compared with that of sclerophyll forests and woodlands, and their scattered distribution, especially in drier inland areas of the north where they become restricted to fire-proof niches. Rainforest 'pockets' representing well integrated and distinctive biotic communities may occupy less than a hectare and typically do not exceed a few hectares. Discontinuity and fragmentation are especially characteristic of a rainforest floristic element towards its limits, for instance, monsoon rainforests in permanent soakage pockets in the northwest; warm temperate forests in fire-protected moist gullies in the southeast; warm subtropical forests on cool cloudy summits with increasing latitudes. More continuous distributions occur in more favourable and extensive habitats, e.g. the Cooktown–Ingham massif in northern Queensland (approx. 360 km by 80 km – for a descriptive account of the vegetation of this region see Tracey, 1982), and highland segments with orographic rainfall and frequent clouds along the eastern coast, e.g. the Border Ranges (Macpherson Range) of Queensland and New South Wales, and the scarps and foothills of basaltic plateaux farther southwards in New South Wales. The small-scale maps available tend to disguise, however, the ecological fact that the scattered 'islands' of rainforests have tenuous connections by frequent outliers in gullies, along riparian alluvia, on cloudy mountain peaks and residual basaltic caps, and bouldery outcrops and rocky screes that function as 'fire shadows'. This virtual continuity of rainforest habitats is of great biological significance, as for bird-migration routes (Kikkawa, 1968) and preservation of related taxa of insects (Darlington, 1961; Parsons & Bock, 1977) and other organisms discussed in Keast (1981).

Airphoto patterns strikingly confirm impressions gathered from ground surveys that the distribution and composition of rainforest pockets reflect historical processes of climatic–edaphic–topographic sifting which predated the use of fire by Aborigines, and that many of the rainforest fragments were more continuous in earlier and more favourable climatic periods. Geomorphological and palynological evidence for these inferences, and for the profound influence of past climatic changes on the stability and trend of contemporary rainforest patterns, is discussed later in this chapter.

Segregation

In Australia the dominance in the sclerophyll vegetation of genera not found in the rainforest, e.g. *Eucalyptus* and *Melaleuca*, is extraordinary and so different from the rainforests and 'campos cerrados' of Brazil or the evergreen, moist–dry deciduous forest series of India, where the trees and shrubs are species in genera shared with the adjacent rainforests. This seemingly complete segregation of rainforest and sclerophyll floras in Australia prompted Aubréville (1965) on a visit to northern Queensland to exclaim: 'les flores paraissent issues de deux mondes différents' (the floras seem to be derived from two different worlds). We have, however, good reason to believe that the genera found in the Australian sclerophyll vegetation are derived from the closed evergreen forests which had preceded them. The segregation of the two floras is also incomplete, varying in extent between north and south and east and west. Interspersion of the floras may occur in ecotonal communities, among sclerophyll woodlands on suitable soils in the tropics and subtropics, and in temperate and montane situations. Additionally, a few taxa are common to both well-developed rainforests and genuinely sclerophyll vegetation types such as the scrubs and heaths on nutrient-poor sands, as well as strand and mangrove communities.

The significance of these shared taxa in the phylogenetic development of Australian vegetation communities is not properly understood (Webb & Tracey, 1981), but a list of examples at the level of families includes:

> Capparidaceae, Celastraceae, Combretaceae, Dilleniaceae, Ebenaceae, Euphorbiaceae, Fabaceae, Flindersiaceae, Meliaceae, Mimosaceae, Myrtaceae, Pittosporaceae, Proteaceae, Rhamnaceae, Rubiaceae, Rutaceae, Solanaceae, Sterculiaceae, Verbenaceae.

Each of these families has one or more genera shared between closed- and open-forest communities, and their distributions are presumably the result of sifting and co-adaptation in evolutionary time.

Distribution and area

The main areas of distribution are shown in Figure 4.1, but as already noted, allowance should be made for many small interconnecting 'islands' (see Fig. 4.2) between the larger massifs (for a detailed

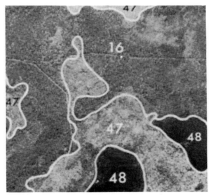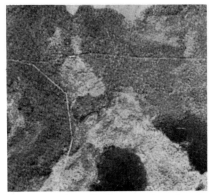

Fig. 4.2. Typical 'islands' of rainforest in sclerophyll vegetation. Land unit 48 is microphyll vine thicket on stony basaltic soils, surrounded by 47 (*Eucalyptus orgadophila* grassy woodland on basaltic cracking clays) and 16 (*E. maculata* open-forest on lateritized sediments) (from Gunn & Nix, 1977, p. 85). (Photo: CSIRO.)

summary of the nature, distribution and status of Australian rain-forests see Anon., 1987). Xeric replace mesic types in 'dry corridors' along the northeastern coast, e.g. Coen–Cooktown, Ingham–Bowen, and Sarina–Gladstone areas in Queensland, as well as in rain-shadows in the lee of coastal ranges. The amount of rainfall is broadly correlated with the presence of coastal mountains which receive orographic rain from the southeastern trade winds and where precipitation is augmented by cloud-drip. Discontinuous patches of rainforest extend away from the coast in northeastern Australia (north of severe frosts at latitude approx. 30° S) to about the 600 mm isohyet, and in northern Australia to about the 800 mm isohyet.

The geographic tropic has little significance in the differentiation of the climatic types. Structural and floristic classifications (see next sections) indicate that tropical and subtropical types extend south-wards to the Illawarra district south of Sydney, e.g. to Minna Murra Falls. Frosts extend north of the tropic in Queensland on the lowlands to about Ingham (latitude 19° S). The Cardwell Range near Ingham, the adjacent mangroves of Hinchinbrook Channel, and the palm swamps just to the north near Tully all contribute to a distinct tropical ecological boundary for many plant and animal taxa.

There are also outliers northwards within the strictly tropical zone of northern Queensland of temperate and submontane floristic elements, and outliers southwards along the Queensland coast of northern monsoonal types.

Correlations of climatic factors and particular rainforest types are therefore not clearcut and are valid only for core areas, outside which there are many extraneous floristic elements as the result of past climatic changes.

It is estimated that the total area of rainforests remaining in coastal eastern Australia is approximately two million hectares, representing between a quarter and a third of the original area. In addition there were several hundred thousand hectares of rainforests (the so-called 'Bottletree Scrubs') interspersed among the country dominated by brigalow (*Acacia harpophylla*) in subcoastal Queensland and northern New South Wales. This totalled approximately five million hectares before clearing, which is now practically complete. The scattered patches and monsoon forests in northern and northwestern Australia average from two to twenty hectares for each stand, but there are many biologically significant fragments less than one hectare. Amalgamation of the larger stands may be imagined to total a few thousand hectares.

Structural classification

An intuitive classification for Australian rainforests was developed by Webb (1959, 1968, 1978) following the use of physiognomic–structural features by earlier workers, notably Beard (1944, 1955). This classification was later elaborated into a methodology by the numerical analysis of data collected in an 'open-ended' *pro forma* (Webb *et al.*, 1970; Webb, Tracey & Williams, 1976). The method is suitable for systematic and large-scale surveys in tropical and subtropical rainforest areas to establish vegetation types and associated environmental types, and does not require taxonomic knowledge.

The intuitive classification provides a useful perspective, from the synecological point of view, of the main kinds of rainforest at local, regional, and continental levels. A field key based on this classification is given in Webb (1978). The major structural types are *vine forests* (tropical and subtropical), *fern forests* and vine-fern forests (submontane and warm temperate), and *mossy forests* and fern-moss forests (montane and temperate). These are further subdivided by periodicity of leaf fall, leaf size, and structural complexity to provide an empirical –inductive classification of types. Specific environmental relationships may then be deduced. The striking contrast in physiognomy and structure between these forests and adjacent sclerophyll vegetation is unique to Australia, so that the use of special life forms to differentiate them is singularly apt. In other countries, however, where rainforest vegetation is a continuum, life forms such as vine, fern or moss are universal within a given climatic zone and such nomenclature is not applicable.

As with any 'pigeon-hole' classification, there are many misfits and overlaps, but the classification is flexible enough to allow prefixes to be omitted and different nodal types to be linked to allow for sites that fall 'somewhere in between' the key units. The environmental relationships of the structural types are summarised in Figure 4.3, and the role of edaphic factors is discussed subsequently.

Floristic classification

Community rank

Unlike structure and physiognomy, which reflect the integrated impact of the physical environment on vegetation, floristic patterns are also the result of phytosociological processes and climatic–edaphic–topographic sifting on a variety of time scales. Definition of structural attributes, although lacking precise cut-off points, does not have the formidable taxonomic difficulties of species identification in complex forests, where in Australia many taxa remained undescribed.

Moreover, structural classification has a built-in constraint, depending on the number of attributes selected, which decides the most useful level in hierachical groupings of sites. The determination of ecological rank in floristic analysis is complicated, however, by historical factors whose influence can only be inferred, and by purposes that often conflict. For instance, as in the present chapter, it may be most convenient to limit classification to trees that form the 'frame of the forest' and presumably carry most of the ecological information (Webb *et al.*, 1967). Floristic classification on the continental scale (see Webb, Tracey & Williams, 1984) has established the present-day patterns of plant distributions and relationships, and provides ample evidence for climatic and edaphic sifting of a much more widespread rainforest vegetation in the past. The floristic patterns correlate well with the structural types of Australian rainforests recognised by Webb (1968).

Floristic classification on a regional scale (740 species × 146 sites) in the wet tropics of Queensland (Williams & Tracey, 1984) also produced strong correlations between floristics and structure at a

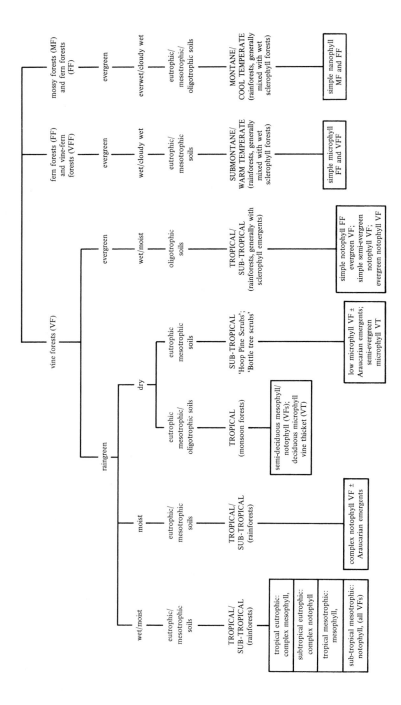

Fig. 4.3. Environmental relationships of the structural types of Australian rainforest vegetation. (After Webb, 1968.)

scale of 1:100 000, and the method used also was able to subdivide the structural types into groups characterised by the presence of endemics or narrowly restricted occurrences of particular species. These results were obtained using only rainforest trees and shrubs.

Nevertheless, there is good evidence that uncommon species of high fidelity may be of diagnostic value and that species of other synusiae (liane, epiphyte, understorey, ground layer) may be useful. Such sophisticated classifications must be consistent in sampling all synusiae, and this is not feasible in general surveys.

Typology in rainforest is determined by need, e.g. forest types aimed at logging operations need to assess volumes of commercial species so that dominants and clusters of favoured timber trees create their own ranking in the typing exercise. These types, however, may be useless as indicators of patterns of occurrence of, say, rare and endangered species.

The floristic elements

Floristic data from systematic spot-listing of tree species, with notes on species of other synusiae, have been accumulated by the authors over a period of twenty years for sites in eastern Australia, and more recently in the Northern Territory. Baur (1965) provided a general floristic classification of rainforests in New South Wales. Other workers have extended more recently the floristic data for Victoria (Howard & Ashton, 1974; Ashton & Frankenberg, 1976; Busby & Bridgewater, 1977; Parsons, Kirpatrick & Carr, 1977; D.M. Cameron in Anon., 1987); Tasmania (Jackson, 1972; Busby & Bridgewater, 1977; Jarman *et al.* in Anon., 1987); New South Wales (Bowden & Turner, 1976; Helman in Anon., 1987; Mills in Anon., 1987; Floyd, 1990); Queensland (Lavarack & Stanton, 1977; Hyland, 1983, 1989; Whiffin & Hyland, 1986; Young & McDonald in Anon., 1987; see also other papers in Anon., 1987); Northern Territory (Russell-Smith, 1991); and northwestern Australia (George & Kenneally, 1975; Beard, 1976; Kabay, George & Kenneally, 1977; Smith, 1977; Kenneally & Beard in Anon., 1987). A recent summary of plant systematics, collections and vegetation in Australia and recommendations for future studies is presented by Kershaw & Whiffin (1989).

Although the data were patchy and incomplete, especially for rainforest pockets in remote areas, a numerical analysis was undertaken by W.T. Williams in 1980. Its results form the basis of discussions here, derived from Webb & Tracey, (1981). Subsequent floristic analysis

(Webb *et al.*, 1984) has not changed the general pattern but has clarified the floristic relationship of the monsoon tropics and dry inland subtropics.

The groupings of sites (8-group level) shown in the dendrogram in Figure 4.4 were produced from Williams' analysis of 561 sites representing 1316 tree species. For an amplified and updated discussion see Webb *et al.* (1984). The distribution of the sites is given in Figure 4.5. Each site-grouping has a broad geographical reality, although there are many areas of overlap, both latitudinally and altitudinally, and one site-grouping is differentiated within the tropical, strongly-seasonal zone by its occupation of nutrient-poor sands. The densest and most characteristic aggregation of sites is interpreted as a core area for a particular floristic element. It is convenient to regard the distribution of the elements at the 3-group level as floristic regions, and at the 8-group level as floristic provinces. It

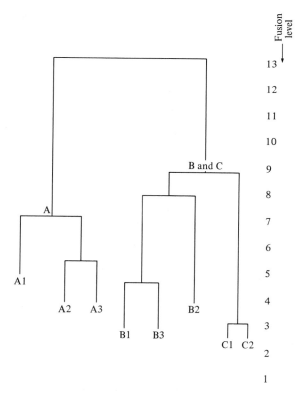

Fig. 4.4. Dendrogram of numerical analysis showing grouping of sites into floristic regions and provinces (elements).

should be noted, however, that these are established from floristic data from specific habitats and community-types, and not, as in classical floristic geography, from the boundaries of combinations of species with approximately similar ranges.

Environmental relationships

Each floristic element or province is defined primarily as a climatic type roughly correlated with its core area. The distribution of the elements, the structural types within them, and their broad environmental relationships are shown in Table 4.1. The meteorological data available from recording stations remote from rainforest patches,

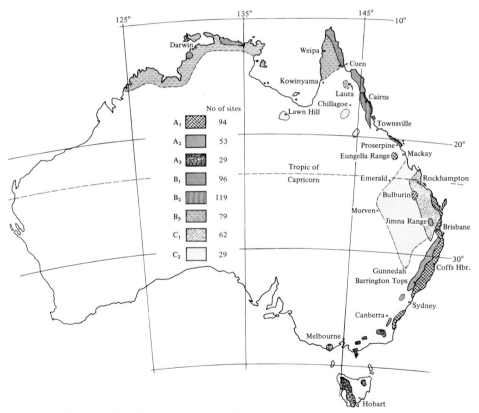

Fig. 4.5. Distribution of sites in floristic regions and provinces of different types of Australian rainforests.

Table 4.1. *Climatic and edaphic features of core areas of the rainforest ecofloristic provinces (species classification) and their equivalent structural types (after Webb et al., 1984)*

Eco-floristic province	Structural type	Climatic type	Edaphic type	'Average' climatic station	Alt. (m)	Mean annual rainfall (mm)	Mean annual raindays	Mean rainfall driest 6 consecutive months	Mean ann. temp. (°C)	Mean minimum temp. coldest month (°C)	Inferred soil mineral nutrient status
A_1	complex notophyll vine forest (CNVF)	warm subtropical (mesotherm/megatherm) aseasonal/ slightly seasonal humid	eutropic	Condong (NSW) 28°19'S, 153°26'E	5	1722	142	550	19.3	5.8	high/medium
	simple notophyll vine forest (SNVF)	ditto	mesotrophic/ oligotrophic	Coffs Harbour (NSW) 30°19'S, 153°07'E	3	1759	147	569	18.3	6.6	medium/low
	SNVF or simple microphyll vine-fern forest (SMVFF)	cool subtropical (mesotherm[1]) aseasonal/ slightly seasonal humid (+ cloud)	mesotrophic/ oligotrophic	Whian Whian (NSW) 28°36'S, 153°23'E	381	2388	147	762	16.6	5.8	medium/low
				Yarras (Mt Seaview) (NSW) 31°23'S, 152°15'E	146	1603	159	499	17.3	3.8	

	Forest type	Climate	Trophic status	Location							
A₂	microphyll fern forest (MFF), merging with microphyll moss forest	subtropical lower montane (microtherm²/mesotherm) aseasonal slightly seasonal humid (+ cloud)	eutrophic/mesotrophic	Wentworth Falls (NSW) 33°42'S, 150°22'E	883	1374	149	406	12.3	1.5	high/medium
				Styx River (NSW) 30°37'S, 152°11'E	1036	1516	130	444	13.1	2.8	
	microphyll vine-fern forest (MVFF)	warm temperate (microtherm/mesotherm) slightly seasonal humid	eutrophic/mesotrophic	Cann River (Vic.) 37°32'S, 149°09'E	88	1004	146	448	14.1	2.0	high/medium
A₃	nanophyll fern or moss forest (NFF, NMF), merging with MVFF in E. Victoria	cool temperate inc. upper montane (microtherm) aseasonal slightly seasonal humid (+ cloud)	eutrophic/mesotrophic/oligotrophic	Waratah (Tas.) 41°26'S, 145°31'E	624	2201	252	820	8.3	1.6	high/medium/low
				Cradle Valley (Tas.) 41°38'S, 145°57'E	915	2774	237	1066	6.7	0.0	

Table 4.1. (*Cont.*)

Eco-floristic province	Structural type	Climatic type	Edaphic type	'Average' climatic station	Alt. (m)	Mean annual rainfall (mm)	Mean annual raindays	Mean rainfall driest 6 consecutive months	Mean ann. temp. (°C)	Mean minimum temp. coldest month (°C)	Inferred soil mineral nutrient status
B₁	semi-deciduous mesophyll and notophyll vine forest (SDMVF, SDNVF)	tropical (megatherm) seasonal humid/subhumid (monsoonal)	eutrophic/mesotrophic/oligotrophic	Darwin (NT) 12°26'S, 130°52'E	29	1594	109	110	27.5	18.9	high/medium/low
				Oenpelli (NT) 12°91'S, 138°03'E	7	1360	92	51	27.8	17.9	
				Iron Range (Q) 12°47'S, 143°18'E	19	2049	202	215	25.4	18.4	
				Moreton (Q) 2°27'S, 142°38'E	39	1362	102	66	26.2	17.0	
B₂	complex mesophyll vine forest (CMVF) at lower altitudes in coastal zone[3]	tropical (megatherm) aseasonal/seasonal humid (± cloud)	eutrophic/mesotrophic	Innisfail (Q) 17°32'S, 145°58'E	40	3644	155	760	23.5	15.1	high/medium
				Cardwell (Q) 18°16'S, 146°02'E	5	2127	122	289	24.1	13.3	
				Kairi (Q) 17°12'S, 145°34'E	715	1260	113	189	20.2	10.8	

	Structural type	Climate	Soil (trophic status)	Locality					Mean temp.	Mean temp.	Fire
B₃	deciduous microphyll vine thicket (DVT)	tropical (megatherm) highly seasonal subhumid (monsoonal)	eutrophic/ mesotrophic/ oligotrophic	Coen (Q) 13°57'S, 143°12'E	193	1146	86	44	25.4	16.7	high/medium/ low
				Kowanyama (Q) 15°28'S, 141°25'E	13	1222	71	36	26.9	14.8	
				Katherine (NT) 14°28'S, 132°16'E	107	952	62	46	27.2	12.9	
C₁	microphyll vine forest with araucarian emergents	subtropical (mesotherm/ megatherm) moderately seasonal humid/ subhumid	eutrophic/ mesotrophic	Brisbane (Q) 27°28'S, 153°02'E	38	1146	123	366	20.7	9.8	high/medium
				Gympie (Q) 26°11'S, 152°40'E	94	1161	117	354	20.4	6.0	
				Kalpower (Q) 24°42'S, 151°18'E	339	905	89	245	18.7	3.7	
C₂	semi-evergreen microphyll vine thicket (SEVT)	subtropical (mesotherm/ megatherm) moderately seasonal subhumid	eutrophic/ mesotrophic	Biloela (Q) 24°24'S, 150°30'E	173	699	75	187	20.7	5.1	high/medium
				Mt Surprise (Q) 18°09'S, 144°19'E	453	799	57	67	23.8	9.5	
				Mitchell (Q) 26°30'S, 147°59'E	335	575	61	188	19.8	3.3	

[1] equivalent to 'warm temperature rainforest' of New South Wales (Baur, 1965).

[2] equivalent to 'cool temperate rainforest' of New South Wales (Baur, 1965).

[3] with subtropical (mesotherm) and marginal lower montane (microtherm) type at higher altitudes (CNVF, SNVF, microphyll vine-fern thicket (MVFT)), and more seasonal subhumid types in 'rain-shadows' and drier subcoastal areas.

(*a*)

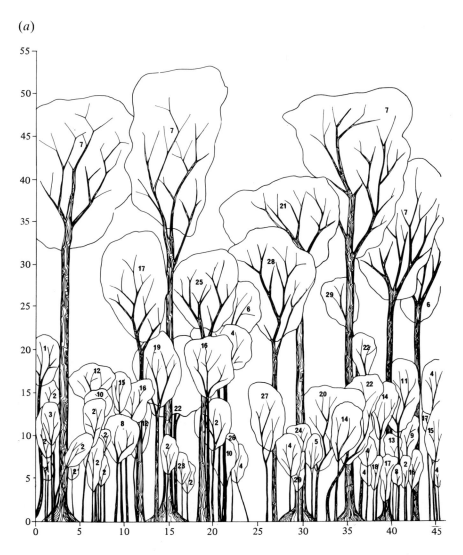

Fig. 4.6. Profile diagrams illustrating selected structural types of rainforest: (*a*) complex notophyll vine forests (Cannabullen, Queensland, map reference Tully 540430). 1. *Myristica muelleri*; 2. *Rockinghamia angustifolia*; 3. *Austromyrtus dallachyana*; 4. *Endiandra sankeyana*; 5. *Euodia haplophylla*; 6. *Xanthophyllum octandrum* 7. *Argyrodendron peralatum*; 8. *Synima cordierorum*; 9. *Doryphora aromatica*; 10. *Sloanea australis*; 11. *Tetrasynandra*

(*b*)

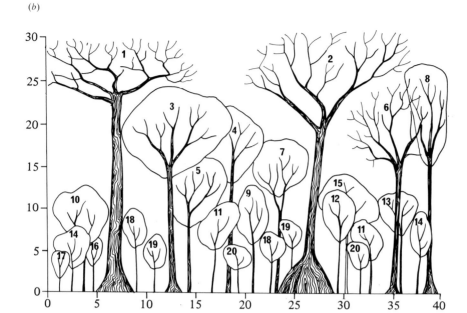

pubescens; 12. *Endiandra monothyra*; 13. *Dysoxylum klanderi*;
14. *Beilschmiedia bancroftii*; 15. *Neisosperma poweri*;
16. *Beilschmiedia tooram*; 17. *Hylandia dockrillii*; 18. *Siphonodon
membranaceus*; 19. *Dysoxylum oppositifolium*; 20. *Apodytes
brachystylis*; 21. *Flindersia acuminata*; 22. *Daphnandra repandula*;
23. *Acronychia vestita*; 24. *Toechima erythrocarpum* 25. *Alangium
villosum* spp. *tomentosum*; 26. *Alphitonia whitei*; 27. *Elaeocarpus
culminicola*; 28. *Cardwellia sublimis*; 29. *Gmelina fasciculiflora*.

(*b*) semi-deciduous mesophyll vine forest (Dowlings hill, near Helenvale,
Queensland, map reference Helenvale 180225). 1. *Bombax ceiba*;
2. *Ficus albipila*; 3. *Paraserianthes toona*; 4. *Vitex acuminata*;
5. *Garuga floribunda*; 6. *Terminalia sericocarpa*; 7. *Canarium
australianum* 8. *Argyrodendron polyandrum*; 9. *Acacia polystachya*;
10. *Cupaniopsis anacardioides*; 11. *Chionanthus ramiflora*;
12. *Harpullia pendula*; 13. *Polyalthia nitidissima*; 14. *Aidia
cochinchinensis*; 15. *Mimusops elengi*; 16. *Antidesma ghaesembilla*;
17. *Wrightia pubescens*; 18. *Aglaia elaegnoidea*; 19. *Glycosmis
pentaphylla*; 20. *Ixora klanderana*. (*c*) simple microphyll vine-fern
thicket (Bellenden Ker, Queensland, map reference Bartle Frere

(c)

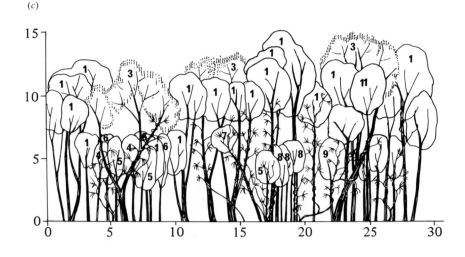

c. 786912). 1. *Cinnamomum propinquum*; 2. *Dracophyllum sayeri*;
3. *Leptospermum wooroonooran*; 4. *Trochocarpa laurina*;
5. *Drimys membranea*; 6. *Rapanea achridifolia*; 7. *Rhodomyrtus
sericea*; 8. *Alyxia orophila*; 9. *Syzigium apodophyllum*;
10. *Planchonella singuliflora*; 11. *Balanops australiana*; 12. *Wilkiea
macooraia*; 13. *Uromyrtus metrosideros*; 14. *Polyscias
bellendenkerensis*; 15. *Flindersia unifoliolata*; 16. *Orites fragrans*;
17. *Laccospadix australasicus*; 18. *Linospadix palmeranus*.

(*d*) simple notophyll vine-fern forest (Mt Spurgeon, Queensland, map
reference Mossman 120720). 1. *Elaeocarpus largiflorens*;
2. *Polyscias murrayi*; 3. *Pullea stutzeri*; 4. *Rhodomyrtus trineura*;
5. *Apodytes brachystylis*; 6. *Opisthiolepis heterophylla*; 7. *Cryptocarya
densiflora*; 8. *Ceratopetalum virchowii*; 9. *Sarcopteryx martyana*;
10. *Cryptocarya corrugata*; 11. *Orania appendiculata*;
12. *Cinnamomum laubatii*; 13. *Prumnopitys ladei*; 14. *Rapanea
achradifolia*; 15. *Carnavonia* sp. ('Mt Haig' BRI 160356);

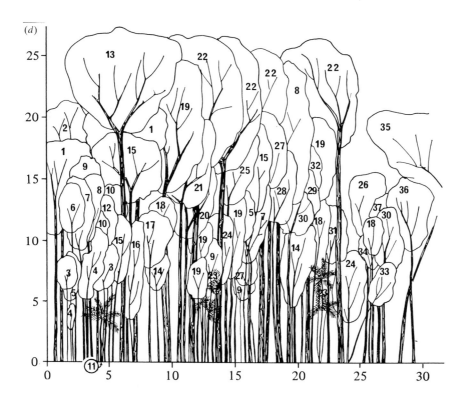

16. *Guettardella tenuiflora*; 17. *Syzygium endophloium*; 18. *Acmena hemilampra* spp. *orophila*; 19. *Sphalmium racemosum*; 20. *Endiandra montana*; 21. *Symplocos cyanocarpa*; 22. *Flindersia pimenteliana*; 23. *Laccospadix australasicus*; 24. *Rhodamnia blairiana*; 25. *Garcinia brassii*; 26. *Syzygium cormiflorum*; 27. *Planchonella obovoidea*; 28. *Cryptocarya angulata*; 29. *Diospyros* cf. *hemicycloides* 30. *Goniothalamus australis*; 31. *Beilschmiedia collina*; 32. *Sphenostemon lobosporus*; 33. *Wilkiea macooraia*; 34. *Aceratium ferrugineum*; 35. *Alphitonia petriei*; 36. *Planchonella macrocarpa*; 37. *Elaeocarpus stellularis*.

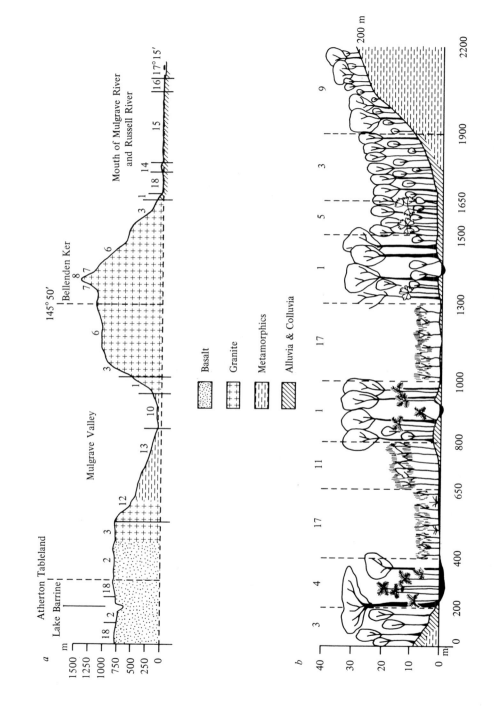

Atherton Tableland

Lake Barrine

145° 50′

Bellenden Ker

Mulgrave Valley

Mouth of Mulgrave River
and Russell River

145°50′

Basalt

Granite

Metamorphics

Alluvia & Colluvia

Fig. 4.7. Topographic sequences at different scales in northern Queensland (Sugarcane Creek, Tully-Mission Beach road, map reference approx. 17°55'S, 146°00'E). Vertical scale is exaggerated as shown. (*a*) transect of 40 km; (*b*) transect of 2200 m. Key to vegetation types: Rainforest: 1. complex mesophyll vineforest on wet lowlands, alluvial and colluvial soils; 2. complex mesophyll vineforest on wet cloudy uplands, basalt; 3. mesophyll vineforest on lowlands and foothills, granite and schist; 4. mesophyll vineforest with dominant feather palms (*Archontophoenix alexandrae*) lowland swamps; 5. mesophyll vineforest with dominant fan palms (*Licuala ramsayi*) lowland swamps; 6. simple notophyll vineforest on cloudy wet uplands, granite; 7. simple microphyll vinefern forest on cloudy wet highlands, granite; 8. simple microphyll vine-fern thicket on exposed cloudy wet highlands, granite. Rainforest with sclerophyll emergents and co-dominants: 9. mesophyll vineforest with *Acacia mangium*, *A. aulacocarpa* emergents on lowlands, schist; 10. mesophyll vineforest with *Eucalyptus grandis*, *Acacia melanoxylon*, *A. aulacocarpa* emergents on foothills, uplands, and highlands, granite. Open sclerophyll forests and woodlands: 11. medium open-forest *Melaleuca quinquenervia*, lowland swamps; 12. medium layered woodland with *Syncarpia glomulifera* and *Eucalyptus intermedia*, upland granite; 13. medium layered woodland with *Eucalyptus tereticornis*, *E. tessellaris* and *E. intermedia*, lowland schists. Vegetation complexes and mosaics: 14. swampy coastal plains complex with *Melaleuca quinquenervia* as main component; 15. saline littoral complex with mangrove forest and scrub as main component; 16. coastal beach ridge and swale complex with medium layered woodland *Eucalyptus tereticornis*, *E. tessellaris*, *E. pellita* as main component; 17. coastal plains (impeded drainage) complex with stunted paperbark forest *Melaleuca viridiflora* as main component. Cleared land: 18. mainly sugarcane on lowlands and mixed farming and grazing on uplands.

(which are, in addition, widely scattered and 'out of phase') do not include important factors such as mountain fogs and cloud-drip, so that the climatic parameters inferred from one or two stations for each province are very approximate.

There is broad agreement between the climatic zones correlated with major structural types (see Fig. 4.3), and the climatic types correlated with the core areas of the different floristic provinces. Within each climatic zone there may be considerable variation in structural type depending on altitude, rainfall, soil fertility, and soil drainage. Habitat diversity is highest for the tropical, seasonally wet province, and least for the tropical, strongly seasonally dry areas. To show the structural types possible within one province would make Table 4.1 too complicated, so that only the optimal type is shown, i.e. for conditions of highest temperature, rainfall, and soil fertility.

A few profile diagrams have been selected to illustrate the range of stratification, canopy height, tree density, and occurrence of life forms such as tree ferns and palms (Fig. 4.6). The topographic sequences selected for two transects in northern Queensland illustrate the diversity of vegetation types within this province (Fig. 4.7).

Soils exert a strict control on structural type within a given climatic zone (Tracey, 1969; Webb, 1969a), although at the extremes of the tropical-monsoonal dry type and the cool-temperate wet type, soil differences tend to be overwhelmed by climate. Soil drainage, whether excessively free to produce droughty sites or seasonally impeded to produce swampy sites, is generally obvious enough. Soil nutrient status is, however, of great importance in differentiating simple evergreen and complex raingreen structural types in the tropical and subtropical moist zones. Nutrient availability may be inferred from the mineral composition of the parent rock, provided that climatic and topographic conditions are not extreme (Webb, 1968).

An idea of the complicated interplay of climatic and edaphic factors correlated with the structural types listed in Table 4.1 may be gained from Figure 4.3.

Biogeographic significance

Structural patterns

The distribution of the main structural types provides a general climatic typology for Australian rainforests, based primarily on latitude and altitude (temperature), and secondarily on rainfall and topography (moisture) as in Table 4.1, and as discussed by Webb (1959, 1968). At the tertiary level, and sometimes even higher, edaphic factors including climatic-edaphic compensation become important.

It is clear, however, that the elucidation of phylogenetic processes that are central to biogeography requires the use of taxa.

Floristic patterns

At the 3-group level (Fig. 4.4) the floristic regions denote hot wet, hot dry and cool wet forests, as in the broad structural classification (Fig. 4.3) of tall mostly evergreen vine forests, low mostly deciduous vine thickets, and tall fern or mossy forests respectively.

The primary separation of the hot forests from the cool forests indicates that they are relatively remote phylogenetically, suggesting relatively long periods of independent evolution. It does not support the traditional view (e.g. Herbert, 1960) that the tropical lowland rainforest of northern Queensland simply becomes attenuated southwards in response to a gradient of decreasing temperature.

The wet and the dry forests of the tropics are also relatively remote phylogenetically, at least at the species level. There is no continuous gradient of impoverishment westwards from the north eastern coast along a classical formation-series of decreasing moisture as idealised, for example, by Beard (1944) and Webb (1968). The humid eastern rainforest massif of northern Queensland ends abruptly. These sharp boundaries are formed by regular fires correlated with rainshadows behind mountains and cutoff points at the limit of penetration of regular cloud cover and drizzle, particularly in the winter months. Fire following disturbances, e.g. cyclones, in the higher rainfall areas and on drier ridges with shallow poorer soils produces a sclerophyll-dominated canopy layer of *Acacia* and *Eucalyptus* with an understorey of rainforest whose development relates to time since the last wildfire. Small pockets of rainforest survive intact in gullies amongst *Eucalyptus/Acacia* communities until rainfall decreases to the degree that these are eliminated by fire. Small pockets of dry deciduous forests and Araucarian vine thickets occur widely amongst the woodlands to the west of the main rainforest massif and even occur in dry enclaves within the wet tropics on granite headlands near the sea.

In contrast, in the subtropics of southern Queensland, there is a more or less continuous gradation westwards from the wet coastal types to the increasingly fragmented dry semi-evergreen types. This interpretation may be relevant to late Quaternary changes described by Kershaw (1988; see Walker & Singh, 1981, for references). Except for

Agathis and *Podocarpus*, these climatic changes practically obliterated the conifer forests that flourished before about 30 000 BP on the Atherton Tableland, northern Queensland.

The subtropical dry floristic region tapers southwards into subcoastal New South Wales, suggesting much earlier extensions southwards of the dry monsoonal climates of the north and northwest of the continent. Ancient and widespread floristic elements in the dry tropics of the Old and New Worlds are recognised as paleotropical or archeotropical (e.g. Schnell, 1970).

Examples of common tree species exclusive to each of the three regions are:

Tropical wet
> *Acmena graveolens, Agathis microstachya, Aglaia argentea, Backhousia bancroftii, Cardwellia sublimis, Caryota rumphiana, Cryptocarya oblata, Darlingia ferruginea, Elaeocarpus bancroftii, E. ferruginiflorus, Endiandra palmerstonii, Flindersia ifflaiana, Hypsophila halleyana, Idiospermum australiense, Maniltoa lenticellata, Neorites kevediana, Normanbya normanbyi, Placospermum coriaceum, Toechima erythrocarpum, Ristantia pachysperma.*

Tropical dry
> *Alstonia constricta, Backhousia angustifolia, Brachychiton australis, B. rupestris, Cochlospermum gregorii, Croton arnhemicus, Dolichandrone heterophylla, Ehretia membranifolia, Erythroxylon australe, Flindersia collina, F. maculosa, Gyrocarpus americanus, Lysiphyllum cunninghamii, Notelaea microcarpa, Planchonella cotinifolia, Strychnos lucida, Terminalia aridicola, T. oblongata, Ventilago viminalis, Wrightia saligna.*

Subtropical and temperate wet
> *Acmena ingens, Anopterus macleayanus, Archontophoenix cunninghamiana, Argyrodendron actinophyllum, Atherosperma moschatum, Ceratopetalum apetalum, Cryptocarya meisnerana, Doryphora sassafras, Elaeocarpus holopetalus, Flindersia bennettiana, Geissois benthamii, Nothofagus cunninghamii, N. moorei, Orites excelsa, Polyosma cunninghamii, Sarcopteryx stipitata, Schizomeria ovata, Sloanea woolsii, Synoum glandulosum, Syzygium moorei.*

It is significant that the fairly clear-cut geographical separation of the three regions occurs only at the species level. At the generic level, the boundaries of the regions (and more so those of the provinces) are blurred (Webb & Tracey, 1981). The distribution among the

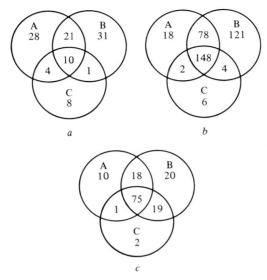

Fig. 4.8. Distribution of (*a*) 103 endemic rainforest genera; (*b*) 442 non-endemic rainforest genera; and (*c*) 145 rainforest families, among the Australian rainforest floristic regions A, B, and C (from Webb & Tracey, 1981).

regions of endemic and non-endemic species of the total of species sampled in the present analysis is given in Figure 4.8.

At the 8-group level (Fig. 4.4) the floristic provinces can be correlated with climatic types as in Table 4.1. Despite the desirability of following a standard international classification (Letouzey, 1978), it is not possible to match many of the Australian rainforest types with those in the UNESCO world classification (UNESCO, 1969).

The floristic elements are no longer neatly segregated geographically, and extend in various directions outside the core areas of the provinces. The extent of overlap by these outliers is shown in Figure 4.5.

It would have been preferable to identify the different floristic elements taxonomically as well as climatically, but this becomes too complicated for other than the simple types, e.g. *Nothofagus-Atherosperma* (cool temperate), *Nothofagus-Ceratopetalum* (warm temperate and submontane moist), *Bombax-Gyrocarpus* (tropical monsoonal dry).

Species vary in their fidelity for a particular floristic element and province. Besides exclusive species which are completely or almost completely confined to one element, there are species with decreasing degrees of fidelity: selective, preferential, indifferent, and accidental (Braun-Blanquet, 1932; Mueller-Dombois & Ellenberg, 1974). Certain of these less exclusive species are highly vagile in relation to seed dispersal mechanisms.

The community-types

The term 'community-type' is adopted as a neutral term and denotes, in this chapter, a characteristic and widespread species group recognisable at a high level in the hierarchy of phytosociological units provided by numerical analysis of site floristic data. One or more community-types characterise a particular core area of a floristic element or province. The number is greatest in the large rainforest massifs which have the greatest habitat diversity, such as the tropical seasonal wet province of northern Queensland. Even in this complex area, however, the number of community-types may be limited to those situations where there is obvious floristic discontinuity, i.e. to distinctly different habitats, such as mountain summits, swamps, and parent rocks within a wide altitudinal zone in the high rainfall belt. It may be argued that the community-type is definable at the point where emphasis changes from discontinuities between stand groups to continuities within them (Dale & Webb, 1975). Unless it is necessary to subdivide them for a particular purpose, differences in frequency of tree species within one extensive and continuously varying phytosociological unit are treated as 'segregates' in the sense of Braun (1950) and Schulz (1960).

The segregates at different places are characterised by variations in frequency of leading species which may become rare, absent, or abundant. Variations that are obviously related to local site differences, generally soils, are regarded as 'lociations'. Where the variations in dominants over a geographical or altitudinal range are not correlated with obvious site differences, they are termed 'faciations' and interpreted as the result of historical changes, such as climatic sifting. This interpretation of a widespread community-type segregating in deterministic or probabilistic ways (cf. Webb, Tracey & Williams, 1972; Dale & Webb, 1975; Diamond, 1975) emphasises the floristic continuity of rainforest patches that are widely scattered, and recognises the dynamic nature of communities. It implies that the boundaries between different floristic regions and provinces are not definitive or fixed, although it may be convenient to regard a concentration of sites of a particular community-type as a core area. It is to be expected that many examples of extraneous floristic elements occur outside core areas. These elements may be interdigitated in different niches in other core areas, or interspersed as community-fragments with other community-types. Both occurences reflect the historical changes discussed below.

A detailed catalogue of community-types and their segregates is outside the scope of this chapter. A floristic framework of Australian rainforests is presented in Webb *et al.* (1984).

Dynamics of the rainforests

The role of refugia

There are many examples of disjunct, narrowly endemic, relict, and vicarious species within one or more provinces, some of which are listed by Webb & Tracey (1981). There are also examples, less readily identified in the absence of a comprehensive framework of community-types for all the provinces, of segregates and fragmentary communities which occur outside the province typified by a particular community-type. As already noted, these outliers, whether individual species or recognisable community fragments, are interpreted most satisfactorily as the result of historical changes.

There are many examples of disjunctions between north and south in eastern Australia, but a much greater number of southern species occur as relicts in the uplands of northern Queensland than do northern species in the south. This implies that southern floristic elements previously extended during cooler periods to the north. There are also outliers in northern New South Wales of temperate flora now common in Tasmania (Howard, 1981).

Relict and fragmentary tropical strongly-seasonal communities also occur scattered farther south along the coast of central Queensland, suggesting earlier extensions of dry seasonal climates southwards. The scattered occurrence of grasslands characterised by relict montane and subalpine species in northern New South Wales (e.g. in the Deervale area, east of the New England scarp) and in southern Queensland (e.g. Emuvale, Bunya Mts) may be interpreted as indicating a previous cool and dry steppe climate (Webb, 1964).

There is also considerable evidence for the relict distribution of organisms other than plants along the discontinuous tract of eastern Australian rainforest which is summarised by various authors in Keast (1981).

The evolution of contemporary rainforest patterns, and the degree of instability implied by them, are conveniently considered on three time scales: geological, prehistorical, and historical (Webb & Tracey, 1981). Historical changes in the strict sense of written history since the arrival of Europeans occupy only two centuries. Unlike other countries, where natural and human disturbances of vegetation occupying many thousands of years are confused, recent human impacts on Australian rainforests are readily identified, but for reasons of space will not be pursued here (see e.g. Seddon & Davis, 1976, for references).

On a geological scale, evidence from modern patterns of plant distribution, fossils, and geomagnetic studies indicates that Australia was once part of the Gondwanaland continent in the Cretaceous (Keast, 1981). Of 436 non-endemic genera of woody rainforest plants listed by Webb & Tracey (1981), substantial percentages are found today in what were parts of Gondwanaland, e.g. Australasia (New Zealand, New Hebrides, New Caledonia, Fiji, Lord Howe Is., Norfolk Is.) and Indo-Malayan (syn. Malesian) (28%); Africa (26%); world tropics (23%); New Caledonia (21%); and India (21%). It now seems reasonable to suppose that what were traditionally regarded as Indo-Malayan immigrant elements by Hooker (1860) and his followers originated mainly *in situ*, and differentiated since Gondwanan times. At least 103 genera of woody rainforest plants are endemic. Of 98 rainforest genera of primitive angiosperms and gymnosperms listed by Specht, Roe & Boughton (1974), more than 73 (allowing for recently described Proteaceae not included) are found in the tropical seasonal wet province, and 42 in the subtropical moist province. Of 14 primitive angiosperm families listed by Takhtajan (1969), eight occur in Australasia. This suggests that these refugia are of great antiquity, dating back to the Cretaceous and possibly earlier for taxa such as certain conifers and cycads, that now have very narrow and presumably relict distributions in northeastern Australia. Of course, the possibility of seed dispersal over moderate and, more rarely, long distances cannot be excluded, as we shall discuss subsequently.

On a prehistorical time scale, there is now direct evidence from palynological studies on the Atherton Tableland, northern Queensland (for details, see Walker & Singh, 1981). On a world scale, additional evidence suggests that the distribution patterns of tropical rainforests at the present day are less than 4000 to 12 000 years old (Synnott, 1977).

Viewed in the perspective of these now well-documented climatic changes, and allowing for the influences of topography, soils, wildfires, and, in historical times, European settlement, the archipelago of rainforest habitats that characterises the eastern and northern Australian landscape becomes an *archipelago of refugia.*

Australian rainforest refugia were classified into four main types by Webb & Tracey (1981): (1) large, relatively wet areas such as lowland gorges and cloudy wet summits and gullies on slopes of coastal mountains, (2) small, relatively dry and fire-proof topographic isolates such as bouldery outcrops, (3) small, often narrow edaphic isolates such as riverine alluvia and perennial spring-fed soaks, and calcareous coastal dune systems, including areas exposed on the continental shelf during post-glacial periods, and (4) topographic– edaphic–climatic isolates in lower rainfall areas such as residual basaltic caps of coastal areas and mountains.

Given the extent of past climatic changes and their interactions on different time scales, and recognising the role of refugia of different types and ages in the ebb and flow of rainforest vegetations, the overlap between the floristic elements, the faciations, and the consistent patchiness of rainforest patterns away from previous refugia become explicable. Further details are presented by several authors for different regions within Australian rainforests in Anon. (1987).

Interspersions

The use of structure (i.e. closed canopy) to separate rainforests from the rest of Australian vegetation loses much of the biogeographical significance of the floristic elements. There is a limited but significant floristic continuity between closed- and open-forests under extreme environmental conditions, notably lowered temperature and rainfall, and low soil fertility.

Three kinds of interspersions may be recognised:

Boundaries and ecotones

These are narrow in the tropics and subtropics of northern Australia, and wide (forming tall open-forests or 'wet sclerophyll forests', see Ashton, 1981) in warm and cool temperate regions of the southeast. Differences in ecotones are related to different fire behaviour under particular seasonal climatic, topographic, and edaphic conditions. Surface fires in late winter–spring, and running crown fires in summer–autumn, characterise the tropical and temperate regions respectively (Webb, 1968).

Where the rainforest communities become fragmented at the limits of their range, the boundaries become involuted, forming a mosaic of ecotones in what was originally a rainforest matrix (see Fig. 4.2).

It should also be noted that certain species belonging to sclerophyll and rainforest genera are exclusive to ecotones, which accordingly deserve consideration as ecological entities in their own right.

Scattered understorey in grassy eucalypt woodland and in Acacia open-forest

In the former case the floristic elements are often deciduous and characterise tropical mixed deciduous woodland on well-drained soils of higher fertility in northern Australia. The species, unlike those of the closed rainforests, are fire tolerant.

There is a remarkable convergence ('epharmosis') of leaf size and shape in certain taxa to narrow linear and often falcate to resemble sclerophylls such as eucalypts, e.g. *Syzygium eucalyptoides* (Webb & Tracey, 1981).

Mixed communities in at least two different ecological situations

The intermingling of floristic elements from closed rainforests and open sclerophyll forests is so intimate that sample plots of 20 m² or less cannot avoid both floristic types. The mixing is so complete that the elements do not appear to represent fragmentary associations, and that species are widespread and characteristic of similar habitats, implying long periods of co-adaptation. They form an exception to the strict definition of rainforest given earlier.

(a) Nutrient-poor sands in coastal areas of northern Australia support vegetation gradients from sclerophyll heath and scrub through layered open forest to vine forest from which sclerophylls are never absent.

(b) Submontane and montane exposed situations on eastern Australian highlands and in Tasmania.

Vagility

The distribution of rainforest patches and the configuration of floristic elements at present depend partly on the suitability of habitat factors and past climatic–edaphic sifting, and partly on the rate and distance of dispersion of diaspores, i.e. on the vagility of the plants.

Dispersal of seeds is least effective by gravity, and increasingly effective by wind, mammals, water, birds and bats. The subject is, however, controversial and the data are meagre, for example, length of time that viable seeds are retained in the digestive tract of birds in relation to distances traversed, migration routes, and habitats visited (see, e.g. Proctor, 1968; Steenis, 1972; McKey, 1975; Regal, 1977; Webb & Tracey, 1981; Whitmore, 1984). In the absence of firm and comprehensive evidence, it is assumed that the efficiency of long-distance dispersal (to be reckoned in hundreds and thousands of kilometres) remains questionable and its operation rare, especially when the sizes of many rainforest seeds (10–20 mm diameter or more) are considered.

It has also been argued that rainforests composed of many biotically interdependent synusiae and involving entire phylogenetic series must migrate communally (Takhtajan, 1969), which further decreases the possibility of transfer of adequate breeding populations of plants over large distances. Perhaps there are also ecological barriers to ecesis in new habitats even when seeds arrive (Regal, 1977). Given the restricted vagility of many rainforest species, and accepting the role of refugia, the interpretation of scattered populations as relict seems more reasonable.

Rainforest as a habitat for birds

Northern Australian rainforests form a closed community in which, as in other humid tropical regions of the world, plants and animals are closely associated, and animals often act specifically as pollinators and seed dispersers. The 'island effect' of rainforest distribution is also paralleled by high fidelity of many faunal elements that have been ecologically trapped as the result of aridity (Kikkawa 1974; Kikkawa, Monteith & Ingram 1981). The following is a summary of circumstantial evidence for these trends as seen in the birds of Australian rainforests.

With increasing latitudes and along mesic-xeric gradients the species richness of land birds generally decreases over the Australian continent (Kikkawa 1968; Kikkawa & Pearse, 1969; Recher, 1971). In the subtropical region of eastern Australia the rainforest birds, compared with birds of semi-arid habitats, are characterised by stronger association of species, relative abundance of tree-nesting frugivores and small clutch size (Kikkawa, 1974). The same tendencies are more pronounced in the tropical rainforests compared with subtropical rainforests within Australia (Kikkawa & Webb, 1967; Kikkawa, 1968) and in the lowland rainforests compared with the montane rainforests in New Guinea (Kikkawa & Williams, 1971a,b). In the subtropical region of Australia the species diversity of birds in the rainforest is not as high as in the surrounding open-forests (wet sclerophyll forests) but in northeastern Australia this relation is reversed. The guilds of birds that contain additional elements in the tropical rainforests are (1) tree-feeding herbivores and omnivores, (2) small predators and (3) ground-feeding insectivores, indicating greater niche volumes of these guilds (Kikkawa, 1988). For example, nectar feeders are almost absent from subtropical rainforests whilst they are abundant in tropical rainforests. In the subtropical region most species of honeyeaters (Meliphagidae) occur in sclerophyll vegetation; only one species occurs regularly in rainforests. On the other hand, most tropical species occur inside the rainforest rather than outside (functional response). In the case of lorikeets (Loriidae) the species abundant in tropical rainforest vegetation (numerical response) are found in sclerophyll vegetation in the subtropical region. Since these birds are residents, both functional and numerical responses to increased resources reflect a sustained supply of nectar and other food sources in tropical rainforests, hence interdependence of nectar-feeders and bird-pollinated plants (e.g. Proteaceae). In the case of rainforest meliphagids most nectar-feeders are facultative in their diet and consume a large amount of insects as well. Their foraging activity is restricted, however, to the rainforest habitat. Fruit-eating birds (mostly pigeons) also show both numerical and functional responses suggesting their increased role as seed dispersers in tropical rainforests. Two species of lowland rainforest fruit-eaters perform regular long-distance migration to New Guinea and adjacent islands reflecting the seasonality of fruiting. Their role in long-distance seed dispersal is therefore potentially greater than that of resident species. Two ground species of lowland rainforests are also migrants from New Guinea and arrive only after the wet season begins locally (Kikkawa, 1976). The guild of foliage-feeding insectivores on the other hand is increased in tropical rainforest when fruit-eaters and ground-feeders meet adverse conditions in the lowlands. The increase is because of an influx of southern migrants escaping the winter conditions of subtropical rainforests. There are at least three species of rainforest flycatchers (Muscicapidae) in this category. Since these birds are associated with rainforest vegetation wherever they occur the patches of rainforests on the route of migration play an important role for their survival. Finally, association of birds with vegetation structure and floristics within tropical rainforests has been demonstrated, showing parallel trends of ecology and biogeography between the northern rainforests and the birds that inhabit them (Webb *et al.*, 1973; Kikkawa, 1982).

Social values and conservation status

For those who have seen something of the vast, life-teeming evergreen and deciduous forests – where they survive – in other tropical countries, a first impression of northern Australia must be surprise and even disbelief that the areas of rainforests are so meagre. Yet, as we have suggested, it must have been like this even at the arrival of humans on this continent sometime between 100 000 and 40 000 years ago. The tropical rainforest flora, whose evolution is rooted in Gondwanaland, has not been extensive since Tertiary times when the sclerophylls became rampant. For the Aboriginal immigrants, who had resource utilisation patterns adapted to a tropical rainforest environment, botany began anew in sclerophyll Australia (Webb, 1973).

The fragmented and restricted distribution of the remaining rainforest communities highlights the need to conserve them. As a rare resource they are precious for a variety of reasons. The patterns of distribution and organisation of separate communities are themselves of great biogeographical and ecological interest, and provide keys to evolutionary processes and niche specialisation. An understanding of these processes is of more than academic interest, and is essential for successful agricultural and forestry practice. For example, the pockets of rainforest serve as 'markers' for particular kinds of biotic environments or habitat types that can be matched elsewhere in the world tropics which lack structural and floristic counterparts of the all-pervasive Australian sclerophyll vegetation. Habitat-matching based on similar structural types was first advocated by Webb (1966) for reciprocal introduction of tree species, as well as of potential pasture species that inhabit forest fringes. More recent refinement of habitat-matching methods using rainforest and monsoon forest sites selected throughout the Indo-Malesian region and northern Australia has further demonstrated its potential for guiding species introductions as well as establishing a 'common ecological language' (Smitinand *et al.*, 1982; Webb, Tracey & Williams, 1985).

Australian aborigines as hunter gatherers used a wide range of leaves, fruits, seeds and tubers from the tropical rainforest and monsoon forest flora in northern Australia. These plants did not enjoy millennia of selection and cultivation as in, say, India and Thailand. Some of the species would undoubtedly qualify after selection as edible fruits, e.g. *Antidesma*, *Elaeagnus*, *Elaeocarpus*, *Parinari*, *Securinega*, *Syzygium*. Some of the Aboriginal bush medicines and poisons

have also been shown by phytochemical analysis to contain physiologically active compounds, e.g. *Alstonia, Barringtonia, Duboisia,
Isotoma, Nicotiana, Stephania, Strychnos, Tylophora* (Webb,
1969*b,c*; Collins *et al.*, 1990).

Despite this considerable knowledge, the Australian rainforests
have yielded very few foodstuffs, spices, medicines, etc., compared
with the forests of Malesia. *Macadamia* (Proteaceae) is the only edible
nut from a native plant to be developed commercially. Some plants
are now widely used in horticulture, for example lilly pillies (*Syzygium*
spp.), native franjipani (*Hymenosporum flavum*), flowering species
of Proteaceae such as silky oak (*Grevillea robusta*), scrub waratah
(*Oreocallis wickhamii*), wheel of fire (*Stenocarpus sinuatus*), and
many ornamental forms of palms, ferns, vines and epiphytes.

Australian rainforests were exploited by Europeans for a variety
of cabinet woods from earliest times. The first red cedar (*Toona
australis*) was exported seven years after the settlement at Sydney in
1788. Continued logging to the present day has degraded practically
all of the accessible rainforests outside of National Parks. Most of
the rainforests that survive in eastern Australia, but to a lesser extent
in northern Australia, are now located in National Parks and State
Forests. Nearly all the tall rainforests that have not been cleared have
been logged. Unlogged areas are mostly on steep upper slopes and
skeletal soils of mountains that support trees of poor form for saw-
milling. If these areas do not contain minerals, they are often gazetted
as National Parks, more or less by default, or remain under the administration of State Forestry authorities as 'protection' (i.e. catchment
protection) forests.

Increased scrutiny of forestry operations has generated considerable
public debate and an unfortunate polarisation between the foresters
and wood-using industries on the one hand and the conservationists
on the other, who reflect the growing international concern to manage
in the most conservative manner the remaining forests (Perlin, 1989)
and especially the rainforests (Hope, 1974). In Australia patches of
rainforest in gullies and gorges, often with spectacular waterfalls,
on isolated mountain tops and fringing rivers and beaches represent
exceptional environments among the dominant sclerophyll vegetation. Traditionally these spectacular landscape features formed the
focus of National Parks gazettal; they were often small areas, their
straight line boundaries bearing no relationship to the surrounding
natural environment. The trend today is towards conservation of the
surviving rainforests along the eastern highlands in larger reserves,
where the rainforests and surrounding vegetation are delineated and

managed as a connected system. In the wet tropics of northern Queensland this decision was taken in conjunction with the official recognition of their worldwide significance, expressed through their nomination for the World Heritage Register. The major omission to date, and thus of major concern, are the rainforests of Cape York Peninsula.

There are limits to this approach, indicated by the thousands of small patches of vine thickets which form an integral part of the vegetation from inland central New South Wales through subcoastal Queensland, Northern Territory and the Kimberley Region of north-west Western Australia. Their long-term survival depends on land management policies related to land clearing, fire, grazing and weed and feral animal control.

Worldwide, a growing concern for the preservation of rainforest began to emerge in the 1970s. It was realised that the lowland rain-forests in many tropical countries have virtually disappeared and many of their species have become extinct. With the acceptance that Australian rainforests, far from being alien and invasive, are derived from the forests present on this continent since Gondwanan times, came the recognition that they are of international significance as ancient and isolated reservoirs of plant and animal taxa, many of them primitive or endemic. This has only recently been promoted in popular media by the scientific community (e.g. Anon., 1986; Webb & Kikkawa, 1990; Adam, 1992).

The temperate rainforests of southwestern Tasmania, important parts of the subtropical and warm temperate rainforests of New South Wales, the rainforests of Queensland's wet tropics, and the monsoon forests of the Northern Territory in Kakadu, have now been listed on the Word Heritage Register, thus having their value recognised globally.

Australian rainforests are vestigial in evolutionary time, a non-renewable resource when subjected to extensive clearing or intensive logging, yet they are one of the most fascinating elements in our moist tropical and subtropical landscapes. They have become increasingly popular as 'destinations' for people seeking recreation and variety. The increased mobility of urban populations along the eastern sea-board and their need for recreation has meant more demand for access to rainforest areas.

The tropical rainforests symbolise the pinnacle of biological com-plexity on earth. The growing popular acceptance of their World Heritage status (since 1988) in the northeastern Australian wet tropics has profound implications for future political priorities in nature protection, and for new ways to identify and interpret intrinsic values of the natural environment. Australia is in a specially favourable position to develop the ethical principles that should govern human relationships to the natural world.

Acknowledgments

The section on rainforest as a habitat for birds was kindly written by Professor Jiro Kikkawa of the Zoology Department, University of Queensland, St Lucia. We thank Dr Reinhild Tracey, of Yungaburra, for her contribution to updating this chapter and typing the manuscript.

References

Adam, P. (1992). *Australian Rainforests*. Oxford University Press.

Anonymous (1986). *Tropical rainforests of north Queensland. Their conservation significance*. Special Australian Heritage Publication Series No. 3. Canberra: Australian Government Publishing Service.

Anonymous (1987). *The rainforest legacy. Australian national rainforest study*, Vol. 1. Special Australian Heritage Publication Series No. 7. Canberra: Australian Government Publishing Service.

Ashton, D.H. (1981). Tall open-forests. In *Australian Vegetation*, ed. R.H. Groves, pp. 121–51. Cambridge University Press.

Ashton, D.H. & Frankenberg, J. (1976). Ecological studies of *Acmena smithii* (Poir.) Merrill & Perry with special reference to Wilson's Promontory. *Australian Journal of Botany*, 24, 453–87.

Aubréville, A. (1965). Principes d'une systématique des formations végétales tropicales. *Adansonia*, 5, 153–96.

Baur, G.N. (1957). Nature and distribution of rainforests in New South Wales. *Australian Journal of Botany*, 5, 190–233.

Baur, G.N. (1965). *Forest types in New South Wales*. Forestry Commission of New South Wales Research Note No. 17.

Baur, G.N. (1968). *The Ecological Basis of Rainforest Management*. Sydney: Forestry Commission of New South Wales.

Beard, J.S. (1944). Climax vegetation in tropical America. *Ecology*, 25, 127–58.

Beard, J.S. (1955). The classification of tropical American vegetation-types. *Ecology*, 36, 89–100.

Beard, J.S. (1976). The monsoon forests of the Admiralty Gulf, Western Australia. *Vegetatio*, 31, 177–92.

Bowden, D.C. & Turner, J.C. (1976). *A preliminary survey of stands of temperate rainforest on Gloucester Tops*. University of Newcastle, NSW, Research Papers in Geography Number 10.

Braun, E.L. (1950). *Deciduous Forests of Eastern North America*. New York: Hafner Publ. Coy.

Braun-Blanquet, J. (1932). *Plant Sociology.* New York & London: McGraw-Hill.

Busby, J.R. & Bridgewater, P.B. (1977). Studies in Victorian vegetation. II. A floristic survey of the vegetation associated with *Nothofagus cunninghamii* (Hook.) Oerst. in Victoria and Tasmania. *Proceedings of the Royal Society of Victoria*, 89, 173–82.

Collins, D.J., Culvenor, C.C.J., Lamberton, J.A., Loder, J.W. & Price, J.R. (1990). *Plants for Medicines. A Chemical and Pharmacological Survey of Plants in the Australian Region.* Melbourne: CSIRO.

Dale, M.B. & Webb, L.J. (1975). Numerical methods for the establishment of associations. *Vegetatio*, 30, 77–87.

Darlington, P.J. (1961). Australian carabid beetles. V. Transition of wet forest faunas from New Guinea to Tasmania. *Psyche. Journal of Entomology*, 68, 1–24.

Diamond, J.M. (1975). Assembly of species communities. In *Ecology and Evolution of Communities*, eds M.L. Cody & J.M. Diamond, pp. 342–444. Cambridge, Massachusetts & London: Belknap Press.

Diels, L. (1906). *Die Pflanzenwelt von West-Australian südlich des Wendekreises.* Leipzig: Wilhelm Engelmann.

Floyd, A.G. (1990). *Australian Rainforest in New South Wales*, Vols. 1 & 2. Sydney: Surrey Beatty & Sons Pty Ltd and National Parks & Wildlife Service, NSW.

Francis, W.D. (1929). *Australian Rain Forest Trees.* Brisbane: Government Printer.

Fraser, L. & Vickery, J.W. (1938). The ecology of the Upper Williams River and Barrington Tops districts. II. The rainforest formations. *Proceedings of the Linnean Society of New South Wales*, 63, 139–84.

George, A.S. & Kenneally, K.F. (1975). Flora. In *A Biological Survey of the Prince Regent River Reserve, North-west Kimberley, Western Australia*, eds J.M. Miles & A.A. Burbidge, pp. 31–68. Wildlife Research Bulletin of Western Australia No. 3.

Gunn, R.H. & Nix, H.A. (1977). Land units of the Fitzroy region, Queensland. CSIRO, *Australia, Land Research Series No. 39.*

Herbert, D.A. (1960). Tropical and sub-tropical rainforest in Australia. *Australian Journal of Science*, 22, 283–90.

Herbert, D.A. (1967). Ecological segregation and Australian phytogeographic elements. *Proceedings of the Royal Society of Queensland*, 78, 110–11.

Hooker, J.D. (1860). Introductory essay. In *The Botany (of) the Antarctic Voyage*, part III. *Flora Tasmaniae*, vol. 1. London: Lovell Reeve.

Hope, R.M. (1974). *Report of the National Estate.* Canberra: Australian Government Publishing Service.

Howard, T.M. (1981). Southern closed-forests. In *Australian Vegetation*, ed. R.H. Groves, pp. 102–20. Cambridge University Press.

Howard, T.M. & Ashton, D.H. (1974). The distribution of *Nothofagus cunninghamii* rainforest. *Proceedings of the Royal Society of Victoria*, 86, 47–75.

Hyland, B.P.M. (1983). A Revision of *Syzygium* and allied genera (Myrtaceae) in Australia. *Australian Journal of Botany*, Supplementary Series No. 9.

Hyland, B.P.M. (1989). A Revision of Lauraceae in Australia (excluding *Cassytha*). *Australian Systematic Botany*, **2**, 1–164.

Jackson, W.D. (1972). Vegetation of the Central Plateau. In *The Lake Country of Tasmania*, ed. M.R. Banks, pp. 61–85. Hobart: Royal Society of Tasmania.

Kabay, E.D., George, A.S. & Kenneally, K.F. (1977). The Drysdale River National Park Environment. In *A Biological Survey of the Drysdale River National Park, North Kimberley, Western Australia*, eds E.D. Kabay & A.A. Burbidge, pp. 13–30. Wildlife Research Bulletin of Western Australia No. 6.

Keast, A. (ed.) (1981). *Ecological Biogeography in Australia*. The Hague: W. Junk.

Kershaw, A.P. (1988). Australasia. In *Vegetation History*, eds B. Huntly & T. Webb, pp. 236–306. Dordrecht: Kluwer.

Kershaw, A.P. & Whiffin, T. (1989). Australia. Plant systematics, collections and vegetation, plus recommendations for the future. In *Floristic Inventory of Tropical Countries*, eds D.G. Campbell & H.D. Hansford, pp. 149–65. New York: New York Botanic Gardens.

Kikkawa, J. (1968). Ecological association of bird species and habitats in eastern Australia: similarity analysis. *Journal of Animal Ecology*, **37**, 143–65.

Kikkawa, J. (1974). Comparison of avian communities between wet and semi-arid habitats of eastern Australia. *Australian Wildlife Research*, **1**, 107–16.

Kikkawa, J. (1976). The birds of Cape York Peninsula. *Sunbird*, **7**, 25–41, 81–106.

Kikkawa, J. (1982). Ecological association of birds and vegetation structure in wet tropical forests of Australia. *Australian Journal of Ecology*, **7**, 325–45.

Kikkawa J. (1988). Bird communities of rainforests. *Acta XIX Congressus Internationalis Ornithologici, Ottawa 1986*, pp. 1338–45.

Kikkawa, J., Monteith, G.B. & Ingram, G. (1981). Cape York Peninsula: the major region of faunal interchange. In Keast (1981), pp. 1695–1742.

Kikkawa, J. & Pearse, K. (1969). Geographical distribution of land birds in Australia – a numerical analysis. *Australian Journal of Zoology*, **17**, 821–40.

Kikkawa, J. & Webb, L.J. (1967). Niche occupation by birds and the structural classification of forest habitats in the wet tropics, north Queensland. *Proceedings of the XIV Congress of the International Union of Forest Research Organizations* (26), pp. 467–82.

Kikkawa, J. & Williams, W.T. (1971*a*). Altitudinal distribution of land birds in New Guinea. *Search*, **2**, 64–5.

Kikkawa, J. & Williams, W.T. (1971*b*). Ecological grouping of species for conservation of land birds in New Guinea. *Search*, **2**, 66–9.

Lavarack, P.S. & Stanton, J.P. (1977). Vegetation of the Jardine River catchment and adjacent coastal areas. *Proceedings of the Royal Society of Queensland*, **88**, 39–48.

Letouzey, R. (1978). Floristic composition and typology. In *Tropical Forest Ecosystems, A State-of-knowledge Report Prepared by UNESCO/UNEP/ FAO*, pp. 91–111. Paris: UNESCO.

McKey, D. (1975). The ecology of co-evolved seed dispersal systems. In *Co-evolution of Animals and Plants*, eds L.E. Gilbert & P.H. Raven, pp. 159–91. Austin: University of Texas Press.

Mueller-Dombois, D. & Ellenberg, H. (1974). *Aims and Methods of Vegetation Ecology*. New York: John Wiley & Sons.

Odum, H.T. & Pigeon, R.F. (1970). *A Tropical Rain Forest*. Springfield, Virginia, USA: Atomic Energy Commission.

Parsons, P.A. & Bock, I.R. (1977). Australian endemic *Drosophila*. I. Tasmania and Victoria, including descriptions of two new species. *Australian Journal of Zoology*, **25**, 249–68.

Parsons, R.F., Kirkpatrick, J.B. & Carr, G.W. (1977). Native vegetation of the Otway region, Victoria. *Proceedings of the Royal Society of Victoria*, **89**, 77–88.

Perlin, J. (1989). *A Forest Journey*. New York: W.W. Norton & Co.

Proctor, V.W. (1968). Long-distance dispersal of seeds by retention in digestive tract of birds. *Science*, **160**, 321–2.

Recher, H.F. (1971). Bird species diversity: a review of the relation between species number and environment. *Proceedings of the Ecological Society of Australia*, **6**, 135–52.

Regal, P.J. (1977). Ecology and evolution of flowering plant dominance. *Science*, **196**, 622–9.

Richards, P.W. (1952). *The Tropical Rain Forest*. Cambridge University Press.

Russell-Smith, J. (1991). Classification, species richness and environmental relations of monsoon rainforest in northern Australia. *Journal of Vegetation Science*, **2**, 259–78.

Schimper, A.F.W. (1903). *Plant Geography upon a Physiological Basis*. Oxford University Press.

Schnell, R. (1970). *Introduction à la Phytogéographie des Pays Tropicaux: Les Problèmes Généraux*. I. Paris: Gauthier-Villars Editeur.

Schulz, J.P. (1960). Ecological studies on rainforest in northern Suriname. Verhandelingen der Koninkluke Nederlandse Akademie van Wetenschappen, AFD. Natuurkunde, 2 sect., **53**, 1–367.

Seddon, G. & Davis, M. (eds) (1976). *Man and Landscape in Australia*. Australian UNESCO Committee for Man and the Biosphere, Publication No. 2, pp. 1–373. Canberra: Australian Government Publishing Service.

Smith, F.G. (1977). Vegetation map of the Drysdale River National Park. In *A Biological Survey of the Drysdale River National Park, North Kimberley, Western Australia*, eds E.D. Kabay & A.A. Burbidge, pp. 31–78. Wildlife Research Bulletin of Western Australia No. 6.

Smitinand, T., Webb, L.J., Santisuk, T. & Tracey, J.G. (1982). A cooperative attempt to compare the habitats of primary forest in Thailand and northern Australia. In *Ecological Basis for Rational Resource Utilization in the Humid Tropics of South East Asia*, ed. Kamis Awang *et al.*, pp. 77–107. Serdang: Universiti Pertanian Malaysia.

Specht, R.L. (1970). Vegetation. In *The Australian Environment*, 4th edn. (rev.), ed. G.W. Leeper, pp. 44–67. Melbourne: CSIRO & Melbourne University Press.

Specht, R.L., Roe, E.M. & Boughton, V.H. (1974). Conservation of major plant communities in Australia and Papua New Guinea. *Australian Journal of Botany, Supplementary Series* No. 7.

Steenis, C.G.G.J. van (1972). *The Mountain Flora of Java*. Leiden: E.J. Brill.

Synnott, T.J. (1977). Monitoring tropical forests: a review with special references to Africa. *MARC Report, Monitoring and Assessment Research Centre, University of London* (1977), No. 5, 1–45. Oxford: University Department of Forestry.

Takhtajan, A. (1969). *Flowering Plants: Origin and Dispersal*. Edinburgh: Oliver & Boyd.

Tracey, J.G. (1969). Edaphic differentiation of some forest types in eastern Australia. I. Soil physical factors. *Journal of Ecology*, 57, 805–16.

Tracey, J.G. (1982). *The Vegetation of the Humid Tropical Region of North Queensland*. Melbourne: CSIRO.

UNESCO (1969). *A Framework for a Classification of World Vegetation*. Paris: UNESCO.

Walker, D. & Singh, G. (1981). Vegetation history. In *Australian Vegetation*, ed. R.H. Groves, pp. 26–43. Cambridge University Press.

Walter, H. (1971). *Ecology of Tropical and Sub-tropical Vegetation*. Edinburgh: Oliver & Boyd.

Webb, L.J. (1959). A physiognomic classification of Australian rainforests. *Journal of Ecology*, 40, 551–70.

Webb, L.J. (1964). An historical interpretation of the grass balds of the Bunya Mountains, South Queensland. *Ecology*, 45, 159–62.

Webb, L.J. (1966). An ecological comparison of forest-fringe grassland habitats in eastern Australia and eastern Brazil. *Proceedings IX International Grassland Congress, Brazil*, pp. 321–30.

Webb, L.J. (1968). Environmental relationships of the structural types of Australian rainforest vegetation. *Ecology*, 49, 296–311.

Webb, L.J. (1969a). Edaphic differentiation of some forest types in eastern Australia. II. Soil chemical factors. *Journal of Ecology*, 57, 817–30.

Webb, L.J. (1969b). The use of plant medicines and poisons by Australian Aborigines. *Mankind*, 7, 137–46.

Webb, L.J. (1969c). Australian plants and chemical research. In *The Last of Lands*, eds L.J. Webb, D. Whitelock & J. Le Gay Brereton, pp. 82–90. Brisbane: The Jacaranda Press.

Webb, L.J. (1973). 'Eat, die, and learn' – the botany of the Australian Aborigines. *Australian Natural History*, 17, 290–5.

Webb, L.J. (1978). A general classification of Australian rainforests. *Australian Plants*, 9, 349–63.

Webb, L.J. & Kikkawa J. (1990). *Australian Tropical Rainforests. Science–Values–Meaning*. Melbourne: CSIRO.

Webb, L.J. & Tracey, J.G. (1981). Australian rainforests: patterns and change. In *Ecological Biogeography in Australia*, ed. A. Keast, pp. 605–94. The Hague: W. Junk.

Webb, L.J., Tracey, J.G., Kikkawa, J. & Williams, W.T. (1973). Techniques for selecting and allocating land for nature conservation in Australia. In

Nature Conservation in the Pacific, eds A.B. Costin & R.H. Groves, pp. 39–52. Canberra: Australian National University Press.

Webb, L.J., Tracey, J.G. & Williams, W.T. (1972). Regeneration and pattern in the sub-tropical rainforest. *Journal of Ecology*, **60**, 675–95.

Webb, L.J., Tracey, J.G. & Williams, W.T. (1976). The value of structural features in tropical forest typology. *Australian Journal of Ecology*, **1**, 3–28.

Webb, L.J., Tracey, J.G. & Williams, W.T. (1984). A floristic framework of Australian rainforests. *Australian Journal of Ecology*, **9**, 169–98.

Webb, L.J., Tracey, J.G. & Williams, W.T. (1985). Australian tropical forests in a southeast Asian context: a numerical method for site comparison. *Proceedings of the Ecological Society of Australia*, **13**, 269–76.

Webb, L.J., Tracey, J.G., Williams, W.T. & Lance, G.N. (1967). Studies in the numerical analysis of complex rainforest communities. II. The problem of species sampling. *Journal of Ecology*, **55**, 525–38.

Webb, L.J., Tracey, J.G., Williams, W.T. & Lance, G.N. (1970). Studies in the numerical analysis of complex rainforest communities. V. A comparison of the properties of floristic and physiognomic-structural data. *Journal of Ecology*, **58**, 203–32.

Whiffin, T. & Hyland, B.P.M. (1986). Taxonomic and biogeographic evidence on the relationships of Australian rainforest plants. *Telopea*, **2**, 591–610.

Whitmore, T.C. (1984). *Tropical Rainforests of the Far East*. Oxford: Clarendon Press.

Williams, W.T. & Tracey, J.G. (1984). Network analysis of North Queensland tropical rainforests. *Australian Journal of Botany*, **32**, 109–16.

Wilson, P.G. & Waterhouse, J.T. (1982). A review of the genus *Tristania* R.Br. (Myrtaceae): a heterogeneous assemblage of five genera. *Australian Journal of Botany*, **30**, 413–46.

5

Southern rainforests

J.R. BUSBY & M.J. BROWN

RAINFORESTS can be distinguished from other vegetation in high-rainfall areas by a (usually) closed tree canopy, generally comprised of more than one (often many) species, and the ability of the canopy species to regenerate in the absence of broad-scale disturbance (e.g. Jarman & Brown, 1983). Shrubby thickets fringing tidal estuaries (e.g. mangroves) or freshwater swamps (e.g. melaleuca thickets) are not considered to be rainforests. Southern rainforests are distinguished from other rainforest types (see Chapter 4) by the absences of significant buttressing at the tree bases and of woody lianes. The epiphytes tend to be predominantly bryophytes and lichens, with occasional ferns but, unlike other rainforests, comprising very few angiosperms.

The southern rainforests are a type of cool temperate rainforest: evergreen closed-canopy forests occurring in areas with high rainfalls, mild summers and cool to cold winters. Rainforests throughout the world largely occur on or near oceans and temperate rainforests are no exception, being largely confined to relatively small, isolated regions around the rim of the Pacific Ocean.

Temperate rainforests are rare: their original extent has been estimated to be *c.* 5 per cent of the amount of remaining tropical rainforests (Mitchell, Hagenstein & Backus, 1991). In the northern hemisphere, cool temperate rainforests are restricted to southeastern China and to the Pacific-northwest of the USA and near-coastal British Columbia, Canada. It is likely that rainforests also once existed in northwestern Europe and may still exist along the Black Sea coast of Turkey and the former Soviet Republic of Georgia (Mitchell *et al.*, 1991). These forests are quite different in origin, species composition and general ecological characteristics from those in the southern hemisphere. The latter occur in far southern South America (predominantly Chile), New Zealand, Australia (especially Tasmania), New Caledonia and at high altitudes in New Guinea. In Australia, cool temperate rainforests occur predominantly in Tasmania, in disjunct areas in south central Victoria and in comparatively small patches along the Great Divide in New South Wales and far southeastern Queensland (Fig. 5.1).

[handwritten annotations: BRUSH BUSH SPIRAEA. UK SOUTHERN BEECH]

Vegetation history

Many of the genera present in the southern rainforests have species distributed in New Zealand, New Caledonia or South America. These include *Eucryphia, Lagarostrobos, Nothofagus, Richea* and *Tasmannia*, among others (Jarman & Brown, 1983). This is also true of genera found in the fossil record, but now extinct in Australia, e.g. *Dacrycarpus* (Oligocene-Miocene: Tas.; present: New Zealand and elsewhere, Wells & Hill, 1989) and *Dacrydium* (Oligocene: Tas.; present: New Zealand and elsewhere, Wells & Hill, 1989). The origins of a number of the genera, particularly the Australian endemics (e.g. *Acradenia, Agastachys, Anodopetalum, Anopterus, Atherosperma, Athrotaxis, Bauera, Cenarrhenes, Diselma, Drymophila, Microstrobos, Monotoca, Notelaea, Prionotes, Prostanthera, Richea, Telopea* and *Tetracarpaea*) are more problematical, but evidence suggests that they are ancient themselves or have been derived from ancient stock with relatively few changes (Jarman & Brown, 1983). The presence of the Southern Hemisphere taxa provides evidence of the origin of these rainforests in the Gondwana supercontinent, where southeastern Australia was linked to South America via New Zealand and West Antarctica (e.g. Hill, 1990).

The last 60 million years (Ma) have seen enormous climatic changes in southeastern Australia, following the breakup of Gondwana. These changes have been associated with the separation of Australia from Antarctica and its northward movement to its present position, for example 13° of latitude in the last 30 Ma. Climate changes because of changing landmass positions, declining sea surface temperatures and development of the Antarctic ice cap had profound effects on the regional vegetation, as indicated in the fossil record (Hill, 1990). The southern cool temperate rainforests are perhaps unique among all Australian vegetation types in that they have occurred continuously in more or less the same regions since the Gondwanan breakup. All other vegetation types, with the possible exception of some of the northern rainforests, have either evolved subsequent to that time or migrated to their present locations from elsewhere on the continent.

The general climate trend between 65 and 2.5 Ma was an overall cooling, particularly at high latitudes. Rainfall patterns also altered

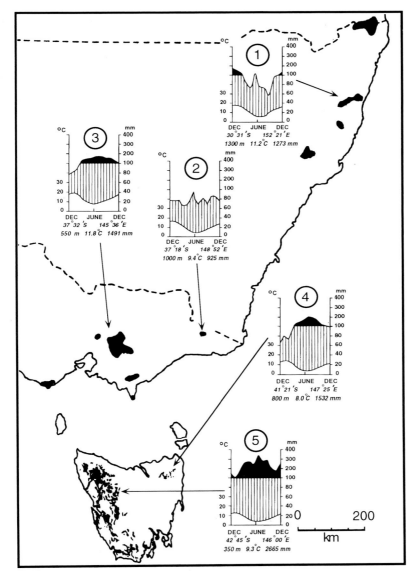

Fig. 5.1 Distribution and climatic characteristics of southern rainforest in Australia. Mainland distributions enclose virtually all stands, along with considerable areas of non-rainforest vegetation; Tasmanian distributions are of 'pure' rainforest, with smaller stands not shown. Climate diagrams indicate trends in mean monthly temperature and precipitation (after Walter & Lieth, 1960), along with latitude, longitude, altitude, mean annual temperature and precipitation, for 'typical' stands.

substantially, with a change from high, year-round rainfall to lower rainfall totals with more seasonality. Initially, rainfall may have been higher in summer but, at some point, changed to being winter-dominated (Bowler, 1982). This deterioration in climate, including increases in the occurrence of frosts and droughts, reduced the areal extent and species diversity within the southern rainforests. Tasmania appears to have been disproportionately affected, most likely due to its island nature. Its rainforests are extremely species-poor, even by comparison with rainforest at similar latitudes in South America and New Zealand (Jarman & Brown, 1983).

Tertiary fossil deposits in Tasmania, however, show evidence of a much greater diversity of species in the Tertiary than are now present (e.g. Hill, 1990). Around 50 Ma many of the taxa had relatives presently well to the north of Tasmania, often in high-altitude equatorial forests (Hill, 1990). Some taxa present today show a very long fossil history. Pollen referable to species of the present-day genera *Lagarostrobos*, *Phyllocladus*, *Nothofagus* and *Tasmannia* from Australasian or South American deposits date back to the Eocene and earlier (Jarman & Brown, 1983). The leaf structures of some of these species have remained virtually unchanged during the Tertiary (Hill, 1990).

During the last 2.5 Ma there has been a series of relatively rapid climatic fluctuations, including massive glaciations at high altitudes and latitudes. It appears that, although its areal extent may have subsequently fluctuated, the southern rainforest in Tasmania was close to its present species diversity prior to this time. This diversity was lower than at any time during the recorded past (Hill, 1990).

Despite the fact that many species have become extinct in southern rainforests, there is evidence to suggest that evolution has occurred, perhaps in response to changing climates. It has been shown, for example, that *Nothofagus cunninghamii* and *N. moorei* may have had a common ancestor in the early Tertiary (Hill, 1983) and subsequently diverged in their morphology and distribution. For example, the northern *N. moorei* has larger leaves than the southern *N. cunninghamii*. Their climatic responses, however, remained very similar in many respects, the one major difference being in seasonality of precipitation (see below). Other Tasmanian rainforest elements such as *Acradenia*, *Anopterus*, *Eucryphia* and *Trochocarpa* also have smaller leaf sizes than their counterparts in more northern warm temperate rainforests; the fossil record indicates that smaller leaf sizes are the derived form (Hill & Read, 1987).

Distribution

The canopy within the southern rainforests is species-poor by comparison with other rainforest types; it seldom comprises more than three or four species and occasionally there is only one. The major genera (focussing only on their southern rainforest species) include *Atherosperma* (1 sp., NSW-Tas.), *Athrotaxis* (3 spp., Tas.), *Ceratopetalum* (1 sp., Qld-NSW.), *Diselma* (1 sp., Tas.), *Doryphora* (1 sp., Qld-NSW), *Eucryphia* (1 sp., NSW; 2 spp., Tas.), *Lagarostrobos* (1 sp., Tas.), *Nothofagus* (1 sp., NSW-S. Qld; 1 sp., Vic.-Tas.; 1 sp., Tas.) and *Phyllocladus* (1 sp., Tas.). In some areas species such as *Richea pandanifolia* may be important. Species prominent in rainforest tend to have restricted distributions in other forest communities (Jarman & Brown, 1983).

At lower elevations, the southern rainforests generally grade into tall eucalypt forests, often through an ecotone of 'mixed' forest, i.e. a eucalypt canopy over a rainforest understorey. The width of this ecotone varies from several metres in the north to several kilometres in the south, depending largely on fire frequency and intensity. The most extensive areas of mixed forest occupy a broad north–south belt along the eastern extremities of Tasmanian rainforest distribution (Jarman & Brown, 1983). At higher elevations, the rainforests grade into subalpine woodland and alpine scrub vegetation. Where there are no clear floristic or physiognomic boundaries between vegetation types, it is necessary to use a minimal height requirement (trees greater than 5 m) to delimit rainforest (Jarman, Kantvilas & Brown, 1991).

Tasmania has the largest area of rainforest in the country, some 765 000 ha, or about 10 per cent of the State (Brown & Hickey, 1990). Rainforest predominates in the west and northwest, with significant patches in the southwest, south and northeast (Brown *et al.*, 1990). Small isolated pockets occur in the east and southeast, chiefly as narrow riparian strips, on steep southeastern-facing slopes in deeply incised lowland areas, or on southern- to eastern-facing slopes on hill tops where cloud-trapping augments precipitation (Neyland, 1990). The most common dominant species is *Nothofagus cunninghamii*, but other species are also prominent in various areas.

Cool temperate rainforest similar to the Tasmanian type occurs in several isolated regions in south-central Victoria: the Otways, Central Highlands, West Gippsland and Wilsons Promontory, with different types occurring on the Errinundra Plateau in the far east of the State (Howard & Ashton, 1973; Busby, 1984, 1986).

Different types again occur along the higher elevations and eastern-facing crests of the Great Escarpment which bounds the coastal lowlands of New South Wales north into southern Queensland. Vegetation dominated by *Eucryphia moorei* occurs in small patches between 400 and 1000 m in southeastern NSW (Beadle, 1981), although the species itself occurs between 180 and 1137 metres (Table 5.1, data from Hill, Read & Busby, 1988). At lower elevations *Doryphora sassafras* is a co-dominant (Beadle, 1981). This species, along with *Ceratopetalum apetalum*, occurs both in southern rainforests and in warmer rainforest types north to around the Queensland border. Significant patches of southern rainforest, dominated by *Nothofagus moorei*, occur in New South Wales on the Barrington Tops, the New England area and in the Border Ranges between New South Wales and Queensland. These forests are confined to high-fertility soils, mostly krasnozems derived from basalt (Beadle, 1981).

Botanical characteristics

The predominant life forms in the southern rainforests comprise trees, shrubs, ferns, bryophytes and lichens (Jarman, Brown & Kantvilas, 1984). Most of the trees and shrubs grow upright, with single or multiple stems, but some species, notably *Anodopetalum biglandulosum*, can produce a dense tangle of interlocking stems. Tree ferns (*Dicksonia antarctica* and *Cyathea australis*) are common on fertile sites (e.g. Beadle, 1981). These not only support epiphytes but also provide suitable germination substrates for the dominant trees and shrubs. Although buttressed tree trunks and woody lianes are virtually absent, scrambling plants (*Trochocarpa* spp., *Aristotelia peduncularis*) and rosette shrubs (*Richea pandanifolia, Dracophyllum milliganii*) contribute to the structural diversity. *Prionotes cerinthoides* (monotypic within the Epacridaceae) and *Fieldia australis* are among the few true angiosperm epiphytes present. There is, however, an abundance of epiphytic ferns (particularly Hymenophyllaceae), bryophytes and lichens. One recently fallen individual stem of the long-lived conifer *Lagarostrobos franklinii* supported 13 vascular plants, 40 bryophytes and 74 lichens (J. Jarman & G. Kantvilas, pers. comm.).

Tree heights seldom exceed 30 m, even under optimal conditions. At high altitudes and on infertile or poorly-drained sites, heights are restricted to 5 m or less. With the exception of the winter-deciduous *Nothofagus gunnii*, all of the trees and shrubs are evergreen. Their

Table 5.1. Selected environmental attributes of southern rainforest elements [means ± standard deviation (top) and range (below)]. Site data compiled by the authors or from Busby (1986) and Hill et al. (1988). Climatic data from BIOCLIM (Busby, 1986; Nix, 1986)

Element	Number of sites	Elevation (m)	Mean annual temperature (°C)	Mean annual precipitation (mm)	Mean temp. wettest quarter (°C)	Precipitation driest month (mm)
Ceratopetalum apetalum (NSW)	55[1]	677 ± 331	14.1 ± 1.4	1411 ± 479	18.7 ± 1.9	62 ± 13
		20–1240	11.5–17.6	694–2805	11.2–22.7	40–93
Doryphora sassafras	65[1]	878 ± 300	12.8 ± 1.4	1261 ± 343	17.8 ± 1.6	62 ± 11
		20–1450	10.2–17.3	694–2081	11.2–20.4	40–93
Eucryphia moorei (NSW)	27	634 ± 246	12.2 ± 1.3	1057 ± 276	16.4 ± 2.6	65 ± 10
		180–1137	9.7–14.7	802–1974	9.8–19.8	51–93
Nothofagus cunninghamii (Tasmania)	191	541 ± 331	8.9 ± 2.0	2120 ± 679	5.4 ± 2.2	104 ± 36
		5–1260	4.7–12.2	930–3523	0.4–9.7	46–188
Nothofagus cunninghamii (Victoria)	124	751 ± 303	10.6 ± 1.6	1469 ± 137	6.8 ± 1.8	74 ± 7
		352[2]–1440	6.2–13.9	971–1679	1.6–10.6	51–85
Nothofagus gunnii (Tasmania)	64	1031 ± 146	6.0 ± 0.9	2596 ± 532	2.2 ± 1.2	124 ± 20
		670–1260	4.3–7.8	1417–3300	0.2–6.3	88–169
Nothofagus moorei (Qld, NSW)	49	1149 ± 247	11.8 ± 1.9	1513 ± 333	16.9 ± 1.2	70 ± 9
		500–1550	8.9–16.1	1113–2373	14.5–20.6	52–83

[1]sites where species co-occurred with southern rainforest species (Hill et al., 1988).
[2]this species occurs down to sea level on the eastern side of Wilsons Promontory (not sampled).

leaves are usually simple microphylls or nanophylls (less than *c*. 75 mm in length and *c*. 2000 mm² in area), although there are a few species with compound leaves and yet others with larger leaves. The leaves are mostly entire or serrated (but not deeply dissected), glabrous or with an indumentum of wax or hair on one surface only, usually the underside (Jarman *et al.*, 1984).

The woody angiosperms are generally hermaphrodites, although genera such as *Atherosperma*, *Nothofagus* and *Tasmannia* and the conifers are monoecious or dioecious. Few of the species have succulent fruits and these appear to be bird-dispersed (e.g. *Tasmannia*, *Cyathodes*, *Phyllocladus*). Some species have winged fruit or seed (e.g. *Nothofagus*, *Eucryphia*, *Anopterus*), but many have no such apparent aids to dispersal (Jarman *et al.*, 1984).

Species composition and vegetation variation

Whilst vascular plants comprise by far the greater proportion of the rainforest biomass, the cryptogams contribute the majority of the plant biodiversity. For example in Tasmania, where the most diverse forests occur, about 70 species of angiosperms may be considered typical of *Nothofagus* forests although, at any one site, only one or two (up to *c*. 20) species may be present. Gymnosperms are represented by 6 species and 48 species of pteridophytes have been recorded (Jarman, Kantvilas & Brown, 1986; Jarman *et al.*, 1991). This contrasts with the 168 species of bryophytes and 217 lichens that have also been recorded. At any one site, a community might contain 2–6 angiosperm, 1 gymnosperm, 10–15 fern, 30–70 bryophyte and 60–100 lichen species (after Jarman *et al.*, 1991). Such sites have a depauperate and mostly adventive vertebrate fauna, but are relatively rich in invertebrate species with, for example, 70–96 species of Coleoptera (Coy, 1990).

Studies of the plant community composition of the southern rainforests have been conducted. In Tasmania, two major 'alliances' have been distinguished: one characterised by *Nothofagus cunninghamii*, the other by *Athrotaxis* species (Busby, 1984, 1987; Jarman *et al.*, 1984, 1991). The *Athrotaxis* Alliance, termed 'open montane forest' by Jarman *et al.* (1984, 1991), is restricted to Tasmania and is found at altitudes above 900 m. It is dominated by *A. cupressoides* and contains floristic elements from the associated treeless vegetation at high altitude (e.g. Kirkpatrick, 1983), as well as some of the more typical rainforest elements of the *N. cunninghamii* Alliance.

The *Nothofagus cunninghamii* Alliance forms the bulk of the rainforest in Tasmania and occurs from sea level to *c.* 1000 m (at or near the treeline). On the fertile and well-drained krasnozems derived from basalt and dolerite, these forests contain tall (30+ m), well-formed trees, with a park-like open or ferny understorey. These forests contain few angiosperms, but are rich in ferns and cryptogams. Such forests have been termed *callidendrous* (= beautiful tree) by Jarman *et al.* (1984, 1991).

On poorer substrates the rainforest trees generally retain their form, but the forest is strongly layered, with an increased abundance and diversity of smaller trees and shrubs. These *thamnic* (= shrubby) forests have a lower diversity of ferns. On the poorest substrates, siliceous quartzites and conglomerates, or on very poorly-drained fertile substrates, the forests are of low stature with poor separation of understorey layers. There is an increased diversity of shrub species and of co-dominant canopy trees, with a concomitant open forest structure. These *implicate* (= tangled) forests are comparatively rich in angiosperms, but have very few ferns. In all, 40 distinct communities have been recognised for the purposes of management and conservation evaluation in Tasmania (Jarman *et al.*, 1991). Profile diagrams of the four major types are shown in Figure 5.2.

In Victoria, the southern rainforests occupy a much smaller area than in Tasmania and exhibit a lower diversity of communities (Busby, 1984). The forests are dominated by *Nothofagus cunninghamii*, except in far eastern Victoria where this species is absent and the dominants are *Atherosperma moschatum*, *Elaeocarpus holopetalus* and *Telopea oreades*. *A. moschatum* is otherwise widespread as a co-dominant with *N. cunninghamii*, or as an understorey species in the wetter eucalypt forests, but is absent from the Otways.

A very unusual and very restricted rainforest variant occurs on the Errinundra Plateau in eastern Victoria. This is dominated by a species of *Podocarpus* which forms a canopy *c.* 12 m high. This is the only occurrence of a conifer in southern rainforest on the Australian mainland, although conifer species are otherwise common in the canopy of temperate rainforests in Tasmania and overseas.

The Victorian rainforest flora is not particularly rich. One survey of 360 sites listed a total of only 110 species of vascular plants, the most common of which included, apart from the trees mentioned above, nine species of fern (including one tree fern and four epiphytic species), one herb, one climber, one small shrub and three tall shrub or small tree species (P.K. Gullan, pers. comm.). The bryophyte flora is rich: Howard & Ashton (1973, Table 1) listed 58 species of mosses

(a) *Callidendrous rainforest*

BEAUTIFUL

(b) *Thamnic (and gallery) rainforest*

SHRUB   *LIKE*

Fig. 5.2. Generalised profile diagrams for the major types of Tasmanian southern rainforest. Terminology follows Jarman *et al.* (1984): (*a*) Callidendrous rainforest; (*b*) Thamnic (and gallery) rainforest;

(c) *Implicate rainforest*

(d) *Open-montane rainforest*

Fig. 5.2 (*continued*) (c) Implicate rainforest; and (d) Open-montane rainforest.

and 47 of liverworts, but it is certain that, as in Tasmania, many other species also occur. The overwhelming majority of the vascular plant species found in these rainforests occur also in other vegetation types, for example, riparian vegetation or wet eucalypt forest.

The southern rainforest communities in Victoria are largely confined to gullies within tall eucalypt forest, in particular *Eucalyptus regnans*, but also *E. delegatensis*, *E. nitens*, *E. viminalis* and *E. pauciflora* (Howard & Ashton, 1973). Eucalypts are commonly present because the rainforest stands are sufficiently small to allow severe fires to penetrate and open up the canopy, thereby facilitating eucalypt regeneration. Occasionally, for example on Mt Donna Buang and high on the Errinundra Plateau, stands occupy mountain slopes, but these are very restricted compared with their Tasmanian equivalents.

At higher elevations, as in parts of Tasmania, the rainforest elements form a tall shrub understorey in subalpine forest, usually dominated by *Eucalyptus delegatensis* or *E. pauciflora*. In wetter areas, the eucalypts drop out and species such as *Leptospermum grandifolium* co-dominate in the rainforest canopy.

In southern New South Wales, *Eucryphia moorei* occurs in pure stands at higher elevations and, lower down, with *Doryphora sassafras*. Although this type shares the same tree ferns with the Victorian and Tasmanian rainforests, it shares none of its more common shrub species such as *Pittosporum undulatum*, *Beyeria lasiocarpa* or *Backhousia myrtifolia* (Beadle, 1981).

Nothofagus moorei reaches heights of 50 m with trunk diameters up to 1.5 m. Individuals are sometimes multi-trunked, frequently with adventitious shoots (Hewson, 1989). The rainforests which it dominates usually have a closed canopy with a lower, usually discontinuous, tree or tall shrub layer with scattered smaller shrubs, ferns (including tree ferns), climbers and epiphytes. There is a generally narrow ecotone to lower-elevation forests, often dominated by *Ceratopetalum apetalum*, *Schizomeria ovata* or *Doryphora sassafras* (Beadle, 1981).

Habitat

The southern rainforests occur sporadically across a wide latitudinal belt between 28° and 44°S (Fig. 5.1). Even so, climatic conditions for the *Nothofagus moorei* and *N. cunninghamii* rainforest types, separated by over 700 km, show significant overlaps (Table 5.1). For example, mean annual temperatures at *N. moorei* rainforest sites in

NSW overlap those for *N. cunninghamii* in Victoria by 5° C (i.e. 69% of the range of the former overlaps 65% of that of the latter). This is despite the fact that the mean elevations differ by *c.* 400 m! There is, however, no overlap in this parameter between *N. moorei* and *N. gunnii*. On the other hand, mean temperatures for the wettest quarter show no overlap at all between *N. moorei* sites and those for the other *Nothofagus* species. This reflects the predominance of summer rainfall in the north and winter rainfall in the south. It is possible that precipitation seasonality has been a major factor influencing the evolution of these species. *Ceratopetalum apetalum* and *Doryphora sassafras* show a pattern similar to *N. moorei*.

Unlike other rainforest types, southern rainforests are subject to frequent freezing temperatures and occasional snow, particularly at higher elevations. Snow can fall even in summer in southwestern Tasmania. For those sites summarised in Table 5.1, mean minimum temperatures for the coldest month are, on average, below zero for *Nothofagus gunnii* ($-1.2\pm0.9°$ C, mean±standard deviation) and *N. moorei* ($-0.1\pm1.7°$ C), and for 24 per cent of *Ceratopetalum apetalum*, 51 per cent of *Doryphora sassafras*, 59 per cent of *Eucryphia moorei* and 13 per cent of Victorian and 26 per cent of Tasmanian *N. cunninghamii* sites. The winter minimum temperatures tend to be milder in the low elevation Tasmanian and Victorian forests; perhaps because of ameliorating effects of cloud cover.

Maximum summer temperatures tend to be mild to warm rather than hot. Mean maxima for the warmest month average $23.0\pm1.6°$ C (max. 26.6) for *Nothofagus moorei*, $25.1\pm1.2°$ C (max. 27.0) for *Eucryphia moorei*, $23.4\pm1.6°$ C (max. 26.8) [Victoria] and $19.0\pm1.9°$ C (max. 23.4) [Tasmania] for *N. cunninghamii* and 16.4 ± 0.9 (max. 18.3) for *N. gunnii*. The similarity between the first three maxima is suggestive of an absolute temperature limit on these geographically separated species. Maxima for the other two species are somewhat higher (29.2° C for *Ceratopetalum apetalum* and 28.0° C for *Doryphora sassafras*, but see note in Table 5.1), which is consistent with their distribution extending into warmer rainforest types.

Mean annual precipitation generally exceeds 1000 mm per year throughout these rainforests although, very occasionally, small stands can occur in sheltered valleys outside this range. Mesoclimate effects can be very important in some areas, for example Howard (1973*a*) noted that clouds often gather on the southern edge of Mt Donna Buang in Victoria and persist for long periods, even in summer. Fog drip and snow are important supplementary sources of moisture in some areas. *Nothofagus* species in Tasmania extend into the wettest region of that State: the western mountains where mean annual precipitation reaches around 3500 mm. This area is the wettest in Australia outside the Queensland wet tropics.

One climate parameter which appears to be particularly critical to these rainforests is the precipitation of the driest month which, in *Nothofagus moorei*, *Eucryphia moorei* and both the Victorian and Tasmanian sites of *N. cunninghamii*, show almost identical lower limits of around 50 mm. The limits for *Ceratopetalum apetalum* and *Doryphora sassafras* are slightly lower, which probably reflects the fact that the driest period is in winter when evapotranspiration could be expected to be lower. All these species, therefore, appear to have very similar minimum requirements. The apparent similarity in the upper range limit in *N. moorei* and the Victorian *N. cunninghamii* is merely a coincidence. The significance of dry season precipitation in the lower-elevation limits of *N. cunninghamii* has been discussed by Busby (1986).

Tasmanian rainforests are found on a range of soil types, including krasnozems formed on basalt and dolerite, other mineral soils, and peats formed on quartzites and other siliceous rocks (Jarman *et al.*, 1991).

The extensive areas of rainforest in northwestern Tasmania occur primarily on gradational soils, often with a peat layer (Richley, 1978). There is a strong effect of soil fertility on both the structure and floristics of the forests. The forests become increasingly complex from the Tertiary basalt-derived loams and clay-loams to the gravelly clay-loams and peats found on the Lower Devonian sedimentary substrates. In southwestern Tasmania, the rainforest soils typically have an organic horizon of (dark) (reddish) brown fibrous peat to 1 m, overlying deeper clay or loamy mineral horizons (Tarvydas, 1978).

The montane rainforests of the Tasmanian Central Highlands are found on dolerite solifluction soils and are best developed on relatively deep, yellowish brown gradational soils, which are often overlain by peat (Pemberton, 1986).

Life-cycle attributes

A rainforest species is defined as one able to perpetuate itself (either vegetatively or from seed) within the relatively undisturbed forest. Species which rely on fire or other catastrophic events for their survival in rainforest are not considered as rainforest species (Jarman & Brown, 1983).

Comparatively little is known of the population dynamics of the major southern rainforest species, i.e. the life cycle stages from seed dispersal through seedling germination and establishment to growth, reproduction and mortality. *Nothofagus cunninghamii* is one species on which some studies have been conducted (e.g. Howard, 1973*a*,*b*, *c*).

The majority of seed is shed in late summer or early autumn. In most years no germination occurs until the following late spring or early summer. Very little viable seed survives over the winter, but this may be because of predation rather than seed death. Seedling establishment may be abundant on logs or freshly-disturbed but compacted soil, but is rare on undisturbed friable soil with litter. The critical factor appears to be dehydration during the early stages of root development (Howard, 1973*a*).

The mature *N. cunninghamii* forest is self-maintaining, through the successful regeneration of seedlings under canopy gaps formed by the death of one or more trees (Howard, 1973*c*). Even though *N. cunninghamii* has a higher light compensation point than other rainforest species, it is lower than more open-forest species such as *Eucalyptus regnans* (Howard, 1973*c*). This explains why *N. cunninghamii* can establish under a *E. regnans* canopy, leading to the formation of a 'mixed' forest, whereas the latter species cannot establish under its own canopy (Howard, 1973*c*).

Observations of adult and seedling *N. cunninghamii* found that growth is confined to one flush in the spring, with an occasional second flush in autumn. Experimental results showed that growth responds to a combination of temperature and day length change, with low soil moisture being an important factor limiting summer growth (Howard, 1973*b*). The last observation is consistent with bioclimatic analyses of the species' distribution, which showed that low-elevation distribution limits in both Tasmania and Victoria were correlated with high summer temperature or low precipitation (Busby, 1986).

The average life of a single leaf on a mature tree is just over three years and the major external cause of shoot death is shading by the canopy above. This natural pruning means that, when trees are close together, the lower branches all grow upward. In isolated trees or those around gaps or on the forest margin, the foliage zone reaches ground level on its exposed sides (Howard, 1973*b*). This, along with moisture levels generally higher than in neighbouring habitats, means that low-intensity fires rarely penetrate well-established rainforest.

In *N. cunninghamii*, dormant buds can develop on burls at the base of the tree and also on the main stem and larger branches. The buds allow the tree to recover from wind or fire damage. Buds high up in the tree may occasionally produce roots, which can reach the ground either down the outside of the trunk or through decayed heartwood. There may be genetic differences in epicormic shoot formation among various populations of the species (Howard, 1973*b*).

Flowering in *N. cunninghamii* appears to require a minimum shoot age of about 25 years, with the seed crop influenced by weather conditions and degree of insect attack. Thus, intervals between fires of less than 25 years could be expected to eliminate the species' capacity to reproduce by seed (Howard, 1973*b*).

Population dynamics

The dominant angiosperms in the southern rainforests are capable of continuous regeneration with self replacement (e.g. Read & Hill, 1988). The coniferous dominants: *Athrotaxis*, *Lagarostrobos* and *Phyllocladus*, each exhibit continuous regeneration in implicate forests or open situations. In mature closed-forests, however, the age structures suggest initial establishment and extended recruitment after catastrophic disturbances such as landslides, floods or wildfire.

Rainforest dynamics

Fire regimes strongly influence the distribution and status of rainforest in the landscape matrix of eucalypt forest and other communities in the high rainfall areas of southeastern Australia. Elaborate feed-back mechanisms between the structural and floristic forms of the vegetation and the nutrient and fire regimes have been detected (e.g. Gilbert, 1959; Jackson, 1968). A mature closed-forest of *Nothofagus cunninghamii*, for example, is reasonably resistant to fire. Once a closed canopy is established, it preserves a humid interior with rapid decomposition of fine fuels. When the trees die, however, they remain as dead stags in the forest, supporting a column of lichens and bryophytes. In severe drying weather, accompanied by strong winds, these 'ladder fuels' act as traps for spot fires and provide ignition and entry points for rainforest fires (A.B. Mount, pers. comm.).

If severe fires do penetrate the forest, the component species regenerate from seed or from coppice and, within 50–60 years, the forest form is regained. If, however, the forest is burnt before crown closure has occurred greater damage results, with many trees and most of the understorey being killed. The more open conditions favour the establishment of dense eucalypt regeneration. More frequent fires progressively favour *Acacia* dominance, until most rainforest species are eliminated (Howard, 1973*a*).

Fire dynamics can therefore be quite complex. For example, a single wildfire in Tasmanian rainforest may stimulate a flush of mosses (e.g. *Funaria*), liverworts (e.g. *Marchantia*), vascular plant species such as *Senecio* spp., *Pteridium esculentum*, *Gahnia grandis* and *Leptospermum* spp., along with the slower seedling and vegetative regeneration of the original rainforest elements. The non-rainforest, fire-sere species generally have attributes such as high essential oil content and rapid production of flammable litter and of phenols which inhibit litter decomposition, which promote the spread of any subsequent fires. Thus there is a positive feedback between these species and fire, with each promoting the other to the detriment of the rainforest species. The complete conversion of areas of implicate rainforest to moorland and scrub by repeated firing over a 60-year period has been documented (Podger, Bird & Brown, 1989). There is also evidence of rainforest invasion into long unburnt moorland in western Tasmania (Podger, 1990) and on grassland, shrublands and eucalypt forest on fertile soils in northeastern Tasmania (Ellis, 1985).

In the absence of repeated fires, the rainforest structure will eventually be re-established. There are many 'pole' stands of even-aged *Nothofagus* forest in areas where rainforest fires have occurred in the recent past. If undisturbed for 400–600 years, the stands begin to break up through deaths of *Nothofagus* and other canopy dominants. The mature forest reaches a dynamic equilibrium in which the rainforest elements regenerate through time by gap-phase replacement.

The 'mixed' forests are maintained in a manner similar to the rainforests, except that the average fire frequency is 80–400 years (Gilbert, 1959). The eucalypt dominants die after about 400 years (see Chapter 6, this volume), since they cannot regenerate under the closed rainforest canopy, and the stand is converted to rainforest. If fires occur before this stage, the eucalypts are regenerated and may also expand into any adjacent areas of burnt rainforest.

Present status

The area of southern rainforests in Australia is approximately one million hectares, with around 75 per cent occurring in Tasmania (Brown & Hickey, 1990). It has been estimated that 15 per cent of the pre-European rainforest vegetation cover in Tasmania has been lost to clearing for agriculture or forest plantations or by inundation for hydro-electric schemes (Hickey, 1990). Very few of the mainland forests have been affected, largely because they occur in steep-sided mountain gullies or high on mountain slopes. The rainforest of the Strzelecki Ranges in Victoria has been considerably reduced, however, by land clearing in the early 1900s (P.K. Gullan, pers. comm.). Virtually all that remains in this area is contained within the Tarra/Bulga National Park.

Most of the forests remaining today are in relatively pristine condition, but there are some extensive and locally severe disturbances because of logging access, mineral exploration and mining, tourism and recreation. Chief among the impacts are localised clearing and direct and indirect impacts of fire, pests and diseases.

Direct human impacts

Past logging has affected some communities more than others. For example, Peterson (1990) estimated that *c.* 85 per cent of the Huon Pine (*Lagarostrobos*) stands had been selectively logged in the past two centuries. In contrast, the various *Nothofagus* forests have not generally been impacted by logging except indirectly, where occurring as understorey to eucalypt forest.

Other direct impacts have involved localised clearing, either for linear infrastructure such as roads or powerlines, or for settlements, mines, etc. Some stands have been inundated by impoundments for hydro-electricity. Although each individual impact has been (generally) relatively minor, the cumulative impacts are becoming increasingly significant.

Fire

Fire is a perennial problem, in that it does not respect the boundaries of national parks or other managed areas. The advent of Europeans has led to a significantly changed fire regime: fires seem to have become less frequent, but more intense and covering larger areas (e.g. Bowman & Brown, 1986). For example, a single wildfire in

Nothofagus-dominated lowlands of northwestern Tasmania in 1982 burnt 25 000 ha, including 15 000 ha of rainforest. Much of the rainforest was burnt in under 12 hours (Barker, 1990).

The changed fire regime could have major conservation implications. For example, approximately 30 per cent of the 49 000 ha of *Athrotaxis selaginoides* forests of Tasmania have been burned in the past 100 years (Brown, 1988). Given that the older trees in these stands are 800–1000+ years old, such a fire impact is clearly not conducive to their long-term survival.

Pests and diseases

Increased access, largely through roading, has resulted in the introduction and spread of pests and diseases. Weed species such as *Digitalis, Leycesteria, Fuchsia* and *Sarothamnus* are major problems on roadsides and heavily used tracks, but are rarely invasive of undisturbed vegetation.

A major disease of *Nothofagus cunninghamii* is myrtle wilt, caused by a lethal pathogen *Chalara australis* (Elliott *et al.*, 1987; Kile & Walker, 1987). The evidence suggests that the fungus is spread by air- or water-borne infection of tree wounds, together with below-ground spread by root contact (Kile, Packham & Elliott, 1989). The fungus occurs in forests remote from human disturbance and surveys have recorded cumulative mortalities averaging 25 per cent (Elliott *et al.*, 1987), with currently-dying trees occurring at a rate of 1.6 per cent of live trees. The gaps created by myrtle wilt appear to form part of the natural dynamics of the forest. Since the gaps are rapidly recolonised by *N. cunninghamii*, there appears to be little long-term change in the nature of the forest (Packham *et al.*, 1990). The expression of the disease is, however, exacerbated greatly by disturbance such as roading, tree blazing or logging, and management practices are required to minimise these effects (Kile *et al.*, 1989).

Another potentially harmful pathogen is the root-rot fungus *Phytophthora cinnamomi*. This fungus has been introduced into Tasmania (Podger & Brown, 1989) and has the capacity to kill many native plant species, including those in rainforest. The fungus is soil-borne and its natural movement is slow, unless the soil is transported by other means. Its infection pattern is concentrated along lines of human access (Podger *et al.*, 1990). Thus earth-moving equipment, bushwalkers' boots and animals are all possible vectors.

There is no evidence of damage by the fungus to undisturbed rainforest. It has been shown, however, to attack many rainforest species in implicate rainforests on oligotrophic soils along roadside margins. It has been extracted from dead and dying rainforest seedlings and resprouts after fire (Podger & Brown, 1989). The microclimate under the undisturbed rainforest canopy is inimical to the spread of the fungus (Podger & Brown, 1989; Podger *et al.*, 1990) but, since there are no known practical control methods, good hygiene and quarantine are required.

Reservation of rainforest

Around 96 per cent of the 765 000 ha of Tasmanian rainforests and most if not all of the mainland rainforests are on crown lands. Some 60 per cent of the Tasmanian forests are in dedicated reserves or in reserves recommended and accepted in principle by government (Brown & Hickey, 1990). Another 15 per cent has been allocated to areas for evaluation of conservation potential. Most of the larger mainland rainforest patches are in nature conservation reserves.

All areas of 'pure' rainforest on the mainland and the outliers in eastern Tasmania are reserved from logging by explicit prescriptions. Many areas of 'mixed forest' (tall eucalypt forest with rainforest understorey) are however available for logging. Some 22 per cent of Tasmania's wet eucalypt forests are in current or proposed reserves. On the mainland, the 'mixed forest' is found in a narrower ecological zone and a much smaller area is involved. The precise boundary between protected rainforest and harvestable eucalypt forest is a matter of debate.

Studies of several rainforest dominants have found that 84 per cent of *Athrotaxis selaginoides* forests, 100 per cent of *A. cupressoides*, 76 per cent of *Lagarostrobos franklinii* and 95 per cent of *Nothofagus gunnii* forests occur in reserves (Jarman *et al.*, 1991). Most of the *N. moorei* stands also occur in reserves.

At the plant community level, it appears likely that all of the communities recognised by Jarman *et al.* (1991) in Tasmania are represented in reserves. Some communities in the outlying patches in eastern Tasmania are, however, poorly reserved (Neyland, 1991). The situation on the mainland is not so well known, but human impacts have been relatively minor. Although some of the vascular plants of rainforests are relatively rare, the vast majority of the species are well represented in reserves, have reasonably extensive distributions outside rainforests in other vegetation types or are otherwise not believed to be at risk from threatening processes associated with people. The available data are inadequate to assess the status of the non-vascular plants, although fewer of these would be likely to have extensive ex-rainforest distributions. Few vertebrates are restricted to temperate rainforests, although one species, *Gymnobelideus leadbeateri* (Leadbeater's Possum), appears restricted to eucalypt forest bordering rainforest in central Victoria (Lindenmayer *et al.*, 1991). The status of invertebrate species is largely unknown.

Global change

The nature and magnitude of any direct effects of climatic change on the structure and composition of the southern rainforests are as yet unknown. The impacts, however, may be dramatic. Busby (1988) showed that, under one climate change scenario, given a number of assumptions, the climate for almost all of Tasmania and central Victoria, the bulk of the present distribution of *Nothofagus cunninghamii*, was predicted to become climatically 'marginal' for that species. Climates for the presently-isolated populations in the Otway Ranges and West Gippsland in Victoria were predicted to become unsuitable.

Any more realistic appraisal must await outputs from detailed regional climate and ecosystem dynamic models, currently under development. In the meantime, considerable information on impacts of past climatic changes can be gleaned from the fossil record and from core samples taken from long-lived *Lagarostrobos franklinii* trees (e.g. Francey, 1990).

Global warming will have obvious indirect impacts on the rainforests through changes in fire regime and disease epidemiology. Changes in either may prove to be adverse for the rainforests.

It is clear that the southern rainforests, although comparatively robust, given their long history in Australia, have been progressively whittled away over geological time, largely by adverse climate change. Now reduced to mere fragments of their former area and diversity, they are vulnerable to inappropriate land management policies and practices, particularly fire and disturbance. The potentially adverse impacts of enhanced-greenhouse climate change imply that management of rainforest areas and their constituent biodiversity needs to be cautious and conservative. The key to their management is to manage the whole landscape of which they are a component. This means management of fire regimes, forestry, vehicle access and all other activities on a landscape basis. Land management must become more integrated over much longer time horizons if these magnificent forests are to continue to evoke feelings of awe and wonder in future generations.

References

Barker, M.J. (1990). Effects of fire on the floristic composition, structure and flammability of rainforest vegetation. *Tasforests*, 2, 117–20.

Beadle, N.C.W. (1981). *The Vegetation of Australia*. Cambridge University Press.

Bowler, J.M. (1982). Aridity in the late Tertiary and Quaternary of Australia. In *Evolution of the Flora and Fauna of Arid Australia*, eds W.R. Barker & P.J.M. Greenslade, pp. 35–45. Adelaide: Peacock.

Bowman, D.M.J.S. & Brown, M.J. (1986). Bushfires in Tasmania: a botanical approach to anthropological questions. *Archaelogy in Oceania*, 21, 166–71.

Brown, M.J. (1988). *Distribution and Conservation of King Billy Pine*. Hobart: Forestry Commission.

Brown, M.J. & Hickey, J.E. (1990). Tasmanian forest – genes or wilderness? *Search*, 21, 86–7.

Brown, M.J., Jarman, J., Grant, J., Corbett, K. & Kantvilas, G. (1990). Tasmanian rainforest communities: their description, environmental relationships and conservation status. In *Tasmanian Rainforest Research*. Tasmanian National Rainforest Conservation Program Report No. 1, pp. 3–11. Hobart: Government Printer.

Busby, J.R. (1984). *Nothofagus cunninghamii* (Southern Beech) *Vegetation in Australia*. Australian Flora and Fauna Series Number 1. Canberra: Australian Government Publishing Service.

Busby, J.R. (1986). A biogeoclimatic analysis of *Nothofagus cunninghamii* (Hook.) Oerst. in southeastern Australia. *Australian Journal of Ecology*, 11, 1–7.

Busby, J.R. (1987). Floristic communities and bioclimates of *Nothofagus cunninghamii* rainforest in southeastern Australia. In *The Rainforest Legacy*. Australian National Rainforests Study, Vol. 1, pp. 23–31. Canberra: Australian Government Publishing Service.

Busby, J.R. (1988). Potential impacts of climate change on Australia's flora and fauna. In *Greenhouse: Planning for Climate Change*, ed. G.I. Pearman, pp. 387–98. Melbourne: CSIRO.

Coy, R. (1990). A survey of invertebrate animals in Tasmanian rainforest. In *Tasmanian Rainforest Research*. Tasmanian National Rainforest Conservation Program Report No. 1, pp. 55–63. Hobart: Government Printer.

Elliott, H.J., Kile, G.A., Candy, S.G. & Ratkowsky, D.A. (1987). The incidence and spatial pattern of *Nothofagus cunninghamii* (Hook.) Oerst. attacked by *Platypus subgranosus* Schedl in Tasmania's cool temperate rainforest. *Australian Journal of Ecology*, 12, 125–38.

Ellis, R.C. (1985). The relationships among eucalypt forest, grassland and rainforest in a highland area in north-eastern Tasmania. *Australian Journal of Ecology*, **10**, 297–314.

Francey, R. (1990). Tasmanian tree rings: a treasure trove for globally significant palaeo-environment reconstructions. *Climate Change Newsletter*, **2**(2), 1–2.

Gilbert, J.M. (1959). Forest succession in the Florentine Valley, Tasmania. *Papers and Proceedings of the Royal Society of Tasmania*, **93**, 129–51.

Hewson, H.J. (1989). Fagaceae. In *Flora of Australia*, 3, 97–100. Canberra: Australian Government Publishing Service.

Hickey, J. (1990). Change in rainforest vegetation in Tasmania. *Tasforests*, **2**, 143–49.

Hill, R.S. (1983). Evolution of *Nothofagus cunninghamii* and its relationship to *N. moorei* as inferred from Tasmanian macrofossils. *Australian Journal of Botany*, **31**, 453–65.

Hill, R.S. (1990). Sixty million years of change in Tasmania's climate and vegetation. *Tasforests*, **2**, 89–98.

Hill, R.S. & Read, J. (1987). Endemism in Tasmanian cool temperate rainforest: alternative hypotheses. *Botanical Journal of the Linnean Society*, **95**, 113–24.

Hill, R.S., Read, J. & Busby, J.R. (1988). The temperature-dependence of photosynthesis of some Australian temperate rainforest trees and its biogeographical significance. *Journal of Biogeography*, **15**, 431–49.

Howard, T.M. (1973*a*). Studies in the ecology of *Nothofagus cunninghamii* Oerst. I. Natural regeneration on the Mt Donna Buang Massif, Victoria. *Australian Journal of Botany*, **21**, 67–78.

Howard, T.M. (1973*b*). Studies in the ecology of *Nothofagus cunninghamii* Oerst. II. Phenology. *Australian Journal of Botany*, **21**, 79–92.

Howard, T.M. (1973*c*). Studies in the ecology of *Nothofagus cunninghamii* Oerst. III. Two limiting factors: light intensity and water stress. *Australian Journal of Botany*, **21**, 93–102.

Howard, T.M. & Ashton, D.H. (1973). The distribution of *Nothofagus cunninghamii* rainforest. *Proceedings of the Royal Society of Victoria*, **86**, 47–75.

Jackson, W.D. (1968). Fire, air, water and earth – an elemental ecology of Tasmania. *Proceedings of the Ecological Society of Australia*, **3**, 9–16.

Jarman, S.J. & Brown, M.J. (1983). A definition of cool temperate rainforest in Tasmania. *Search*, **14**, 81–7.

Jarman, S.J. Brown, M.J. & Kantvilas, G. (1984). *Rainforest in Tasmania*. Hobart: National Parks & Wildlife Service.

Jarman, S.J., Brown, M.J. & Kantvilas, G. (1987). The classification, distribution and conservation status of Tasmanian rainforests. In *The Rainforest Legacy: Australian National Rainforests Study*, Vol. 1: The Nature, Distribution and Status of Rainforest Types, pp. 9–22. Special Australian Heritage Publication Series Number 7(1), Australian Heritage Commission. Canberra: Australian Government Publishing Service.

Jarman, S.J., Kantvilas, G. & Brown, M.J. (1986). The ecology of pteridophytes in Tasmanian cool temperate rainforest. *Fern Gazette*, **13**, 77–86.

Jarman, S.J., Kantvilas, G. & Brown, M.J. (1991). *Floristic and ecological studies in Tasmanian rainforest.* Tasmanian National Rainforest Conservation Program Report No. 3. Hobart: Forestry Commission.

Kile, G.A., Packham, J.M. & Elliott, H.J. (1989). Myrtle wilt and its possible management in association with human disturbance of rainforest in Tasmania. *New Zealand Journal of Forestry Science*, 19, 256–64.

Kile, G.A. & Walker, J. (1987). *Chalara australis* n. sp. a vascular pathogen of *Nothofagus cunninghamii* (Fagaceae) in Australia and its relationship to other *Chalara* species. *Australian Journal of Botany*, 35, 1–32.

Kirkpatrick, J.B. (1983). Treeless plant communities of the Tasmanian high country. *Proceedings of the Ecological Society of Australia*, 12, 61–77.

Lindenmayer, D.B., Nix, H.A., McMahon, J.P., Hutchinson, M.F. & Tanton, M.T. (1991). The conservation of Leadbeater's possum, *Gymnobelideus Leadbeateri* (McCoy): a case study of the use of bioclimatic modelling. *Journal of Biogeography*, 18, 371–83.

Mitchell, A., Hagenstein, R. & Backus, E. (1991). Mapping the world's temperate rain forests. *ARC News*, 13(4), 31–2.

Neyland, M. (1990). Rainforest floristics and boundaries in eastern Tasmania. In *Tasmanian Rainforest Research.* Tasmanian National Rainforest Conservation Program Report No. 1, pp. 13–19. Hobart: Government Printer.

Neyland, M. (1991). *Relict rainforest in eastern Tasmania.* Tasmanian National Rainforest Conservation Program. Report. Hobart: Dept Parks, Wildlife & Heritage.

Nix, H.A. (1986). A biogeographic analysis of Australian elapid snakes. In *Atlas of Elapid Snakes of Australia*, ed. R. Longmore, pp. 4–15. Australian Flora and Fauna Series Number 7. Canberra: Australian Government Publishing Service.

Packham, J., Elliott, H.J. & Brown, M.J. (1990) Myrtle wilt. In *Tasmanian Rainforest Research.* Tasmanian National Rainforests Conservation Program Report No. 1, pp. 45–53. Hobart: Government Printer.

Pemberton, M. (1986). *Land Systems of Tasmania. Region 5 – Central Plateau.* Hobart: Department of Agriculture.

Peterson, M.J. (1990). *Distribution and Conservation of Huon Pine.* Hobart: Forestry Commission.

Podger, F.D. (1990). *Phytophthora*, fire and change in vegetation. *Tasforests*, 2, 125–28.

Podger, F.D., Bird, T. & Brown, M.J. (1989). Human activity, fire and change in the forest at Hogsback Plain, Southern Tasmania. In *Australia's Everchanging Forests*, eds K.I. Frawley & N.M. Sample, pp. 119–40. Canberra: Australian Defence Force Academy.

Podger, F.D. & Brown, M.J. (1989). Vegetation damage caused by *Phytophthora cinnamomi* on disturbed sites in temperate rainforest in western Tasmania. *Australian Journal of Botany*, 37, 443–80.

Podger, F.D., Mummery, D.C., Palzer, C.R. & Brown, M.J. (1990). Bioclimatic analysis of the distribution of damage to native plants in Tasmania by

Phytophthora cinnamomi. Australian Journal of Ecology, **15**, 281–9.

Read, J. & Hill, R.S. (1988). The dynamics of some rainforest associations in Tasmania. *Journal of Ecology*, **76**, 558–84.

Richley, L. (1978). *Land Systems of Tasmania. Region 3—North west*. Hobart: Department of Agriculture.

Tarvydas, R. (1978). Soils. In *Lower Gordon River Scientific Survey*. Hobart: The Hydro-Electric Commission.

Walter, H. & Lieth, H. (1960). Klimadiagramm. *Weltatlas*. Jena: Fisher.

Wells, P.M. & Hill, R.S. (1989). Fossil imbricate-leaved Podocarpaceae from Tertiary sediments in Tasmania. *Australian Systematic Botany*, **2**, 387–423.

6

Tall open-forests

D.H. ASHTON & P.M. ATTIWILL

TALL open-forests include all stands which at maturity are greater than 30 m tall and possess a foliage cover between 30 and 70 per cent of the area (Specht, 1970). Eucalypts are by far the commonest dominants and reach their supreme expression in this formation where heights of *E. regnans* occasionally exceed 100 m. They are unique in that their shaft-like trunks support open crowns with pendant leaves which permit the penetration of up to 40 per cent of daylight. The understoreys are variable, however, in more mesic sites where the canopy exceeds 40–50 m, the understorey is luxuriant with tall mesomorphic shrubs, ferns and tree ferns. These peculiarly Australian forests have been called 'wet sclerophyll forests' (Beadle & Costin, 1952) or 'sclerophyll fern forests' (Webb, 1959). In cooler or less mesic sites, canopy height is less (30–50 m) and the understorey simplified. Thus in Victoria between altitudes of 1000 and 1500 m, the understorey to *E. delegatensis* may range from grassy to the wet sclerophyll fern type. At lower altitudes in somewhat drier sites a layered grass-shrub understorey occurs beneath *E. obliqua*, *E. viminalis* and *E. st. johnii*. In some mesic sites on poor soils a microphyllous-leptophyllous shrub layer of the dry sclerophyll type may occur beneath *E. obliqua* and *E. sieberi* (Ashton, 1976a). Along the Murray River mid-tract, under conditions of optimal flood regime, elite trees of *E. camaldulensis*, 42 m tall, occur above a hydrophytic grassy understorey in what must be regarded as a peculiar variant of this formation. In this chapter, most discussion will centre on the commercially valuable wet sclerophyll type of tall open-forest, rather than on the variants described above.

In many places the mature wet sclerophyll eucalypts emerge above a well-developed rainforest (or closed-forest); see the previous two chapters and Webb & Tracey (1981); Howard (1981). Since the closed-forests are self-perpetuating and since the eucalypts are light-demanding, this forest-in-forest structure is clearly successional. Such stands have been called 'mixed' forests (Gilbert, 1959). The canopies of some of the eucalypts on the extreme southwestern coast of Western Australia bear unusually heavy crowns and this may have led Gardner (1942) to regard these wet sclerophyll forests as a form of temperate rainforest.

Genera other than *Eucalyptus* may form tall open-forests in certain areas. In northern Australia some flood-plain areas are dominated by *Melaleuca leucadendron*. In limited areas of wet flats in Tasmania and the Otway Ranges, forests of *Acacia melanoxylon* may occur over ferns or *Melaleuca ericifolia*. In New South Wales and Queensland, *Lophostemon (Tristania) conferta* may form wet sclerophyll forests or be part of adjacent rainforest.

In Australia, the tall open-forests of the wet sclerophyll type are distributed in a discontinuous arc of high rainfall country from northern and southern Queensland (latitude 17° S, 25° S) to southern Tasmania (latitude 42°S) then across a low rainfall gap of 2100 km from western Victoria to the southwest of Western Australia (latitude 35° S). Over this enormous range, a wide floristic variation occurs, not only in the eucalypt component but also in the understorey species (Tables 6.1, 6.2). A marked floristic discontinuity is evident between southwestern Australia and the eastern coast. In the latter region two trends are evident: firstly, a gradual replacement of species dominance with decreasing latitude; and secondly, the extension of 'southern' species into northern latitudes where compensating altitudinal increases exist.

Three broad groups of wet sclerophyll eucalypt are discernible: those of central New South Wales and southern Queensland (Fig. 6.1), *Eucalyptus cloeziana, E. microcorys, E. pilularis, E. saligna* and *E. grandis*; those of Victoria (Fig. 6.2) and Tasmania (and some in the highlands of northern New South Wales) *Eucalptus regnans, E. viminalis, E. obliqua, E. globulus, E. fastigata, E. delagatensis, E. cypellocarpa, E. dalrympleana* and *E. nitens*; those in southwest Western Australia (Fig. 6.3), *Eucalyptus diversicolor, E. calophylla, E. guilfoylei,* and *E. jacksonii*.

About 11 species of *Eucalyptus* are more or less confined to the wet sclerophyll forests, and another 27 to 30 species occur in them at the wetter ends of their range. It is likely that these variable species, e.g. *E. pilularis, E. obliqua*, are made up of numerous ecotypes which enable them to occur also in dry sclerophyll and grassy forests, woodland or as scattered low trees in heathland.

Of the eucalypts exclusive to wet sclerophyll forests, six are smooth-barked, three are completely fibrous and two are 'half-barks'. These eucalypts are amongst the tallest trees in the world, although

Table 6.1. *Characteristic understorey species of certain wet sclerophyll forests in five regions of Australia*

	Subtropics (southern Queensland and northern NSW)	Warm temperate (central and southern NSW)	Cool temperate (central and southern Victoria)	Cool temperate (southern Tasmania)	Warm temperate (southwest Western Australia)
Tall trees	E. cloeziana E. grandis E. pilularis E. microcorys	E. saligna E. paniculata E. pilularis E. smithii	E. regnans E. viminalis E. obliqua E. cypellocarpa	E. regnans E. obliqua E. ovata E. delegatensis	E. diversicolor E. calophylla E. jacksonii E. guilfoylei
Small trees and shrubs	Syncarpia glomulifera Tristania conferta Casuarina torulosa Callicoma serratifolia Rapanea howittiana Dodonaea triquetra Persoonia attenuata Elaeocarpus reticulatus	Syncarpia glomulifera Acacia binervata Casuarina torulosa Callicoma serratifolia Rapanea howittiana Dodonaea triquetra Persoonia linearis Elaeocarpus reticulatus	Acacia melanoxylon Acacia dealbata Pomaderris aspera Olearia argophylla Bedfordia arborescens Hedycarya angustifolia Zieria arborescens Prostanthera lasianthos	Acacia melanoxylon Acacia dealbata Pomaderris apetala Acacia pentadenia Bedfordia salicina Hedycarya angustifolia Zieria arborescens Phebalium squameum	Casuarina decussata Agonis flexuosa Trymalium spathulatum Acacia urophylla Lasiopetalum floribundum Bossiaea laidlawiana Oxylobium lanceolatum Logania vaginalis

Table 6.1. *Characteristic understorey species of certain wet sclerophyll forests in five regions of Australia*

	Subtropics (southern Queensland and northern NSW)	Warm temperate (central and southern NSW)	Cool temperate (central and southern Victoria)	Cool temperate (southern Tasmania)	Warm temperate (southwest Western Australia)
Pachycauls	*Brachychiton acerifolium*	*Zieria arborescens*	*Correa laurenciana*	*Tasmannia lanceolata*	*Hibbertia amplexicaulis*
	Backhousia myrtifolia	*Tieghemopanax sambucifolius*	*Tieghemopanax sambucifolius*	*Cenarrhenes nitida*	
	Archontophoenix cunninghamiana	*Livistona australis*	*Cyathea australis*	*Dicksonia antarctica*	*Macrozamia riedlei*
	Livistona australis	*Cyathea australis*	*Dicksonia antarctica*		
	Cyathea australis	*Macrozamia communis*			
	Cyathea leichardtiana				
	Lepidozamia peroffskyana				
Lianes	*Marsdenia rostrata*	*Marsdenia rostrata*	*Clematis aristata*	*Clematis aristata*	*Clematis aristata*
	Smilax australis	*Smilax australis*	*Parsonsia brownii*		*Billardiera floribunda*
	Flagellaria indica	*Clematis aristata*	*Pandorea pandorana*		*Hardenbergia comptoniana*

PACHY - STOUT?
THICK FERNS
TREE

	Cissus hypoglauca	Cissus hypoglauca	Billardiera longiflora		Adiantum aethiopicum
	Parsonsia brownii	Parsonsia brownii			Pteridium esculentum
	Pandorea pandorana	Pandorea pandorana			
	Morinda jasminoides	Morinda jasminoides			
	Celastrus australis	Geitonoplesium cymosum			
Ground ferns	Ripogonum discolor				
	Culcita dubia	Culcita dubia	Polystichum proliferum	Polystichum proliferum	
	Blechnum cartilagineum	Blechnum cartilagineum	Blechnum nudum	Blechnum nudum	
	Hypolepis rugosula	Adiantum hispidulum	Blechnum wattsii	Blechnum wattsii	
	Asplenium australasicum	Pteridium esculentum	Histiopteris incisa	Histiopteris incisa	
			Pteridium esculentum	Pteridium esculentum	
			Culcita dubia		
Herbs	Oplismenus aemulus	Oplismenus aemulus	Australina muelleri	Australina pusilla	Hydrocotyle hirta
	Imperata cylindrica	Imperata cylindrica	Hydrocotyle hirta	Hydrocotyle hirta	Veronica calycina
	Viola betonicifolia	Hydrocotyle hirta	Viola hederacea	Viola hederacea	

Table 6.2. *Forest-in-forest communities in eastern Australian rainforest with eucalypt emergents, a late stage in secondary succession*

	Southern Queensland and northern NSW	Southern NSW and eastern Victoria	Central Victoria	Southern Tasmania
Very tall trees (emergents)	*E. microcorys* *E. cloeziana* *E. grandis*	*E. botryoides* *E. maidenii* *E. viminalis* *E. elata*	*E. regnans* *E. viminalis*	*E. regnans* *E. obliqua*
Tall trees rainforest	*Casuarina torulosa* *Syncarpia glomulifera* *Tristania conferta*	*Acacia melanoxylon*	*Acacia melanoxylon* *Nothofagus cunninghamii*	*Nothofagus cunninghamii*
Trees	*Argyrodendron actinophyllum* *Cryptocarya rigida* *Litsea reticulata* *Ackama paniculata* *Callicoma serratifolia* *Cinnamomum oliveri* *Rhodamnia trinervia*	*Acmena smithii* *Eucryphea moorei* *Rapanea howittiana* *Acronychia oblongifolia*	*Atherosperma moschatum* *Pittosporum bicolor* *Hedycarya angustifolia*	*Atherosperma moschatum* *Phyllocladus aspleniifolius* *Pittosporum bicolor* *Anodopetalum biglandulosum*

Shrubs	Citriobatus multiflorus	Elaeocarpus reticulatus	Coprosma quadrifida	Tasmannia lanceolata
	Wilkiea huegeliana	Eupomatia laurina		Coprosma nitida
	Eupomatia laurina			Cenarrhenes nitida
				Dicksonia antarctica
Pachycauls	Archontophoenix cunninghamiana	Livistona australis	Cyathea australis	
	Livistona australis	Cyathea australis	Cyathea cunninghamii	
			Dicksonia antarctica	
Lianes	Cyathea leichardtiana	Cyathea leichardtiana	Clematis aristata	
	Ripogonum discolor	Marsdenia rostrata	Parsonsia brownii	
	Cissus hypoglauca	Celastrus australis		
	Morinda jasminoides	Parsonsia brownii		
	Piper novaehollandiae	Smilax australis		
	Celastrus australis			
	Smilax australis			

Fig. 6.1. Tall open-forest of *Eucalyptus pilularis* near Taree, New South Wales. (Reproduced by kind permission of the Forestry Commission of New South Wales.)

Fig. 6.2. Mature tall open-forest of *Eucalyptus regnans* at Wallaby Creek, Victoria, with a *Pomaderris aspera* understorey.

Fig. 6.3. Mature tall open-forest of *Eucalyptus diversicolor*, near Pemberton, Western Australia, with *Trymalium floribundum* in the understorey.

today only vestiges of the once magnificent forests remain. The tallest *E. regnans* surveyed in Victoria last century was 110 m (Hardy, 1935); today the tallest tree stands at 98.75 m in the Styx Valley, southwestern Tasmania (Australian Newsprint Mills, personal communication). The tallest *E. diversicolor* in Western Australia is 83.3 m. A recently discovered, tall *E. viminalis* in northwestern Tasmania has been authentically measured to 89.9 m (Institute of Foresters of Australia, 1976).

Similar groupings of the understorey components could also be made. Tree-ferns are a conspicuous feature of the eastern coast forests, especially in the south, but are totally absent from Western Australia. Palms, however, are characteristic of the warmer coastal forests in New South Wales and Queensland. In the northern forests, the understorey of tall shrubs and vines includes notophyllous and microphyllous forms such as *Callicoma serratifolia*, *Dodonaea*, *Elaeocarpus reticulatus* and *Brachychiton acerifolium*. 'Small trees' (10 to 40 m) include the myrtaceous *Syncarpia glomulifera* and *Tristania conferta* as well as *Casuarina torulosa*. Palms (*Livistona australis*) and vines are common. In the southeast the understorey has similar leaf-size classes and includes *Pomaderris aspera* and *P. apetala*, the composites, *Olearia argophylla* and *Bedfordia arborescens* and the tree-ferns *Cyathea australis* and *Dicksonia antarctica*. Small trees (20 to 30 m) consist of acacias such as *A. dealbata* and *A. melanoxylon*. In southwest Western Australia the appearance of the undergrowth is astonishingly similar to some of the southeastern forests, except for the absence of tree-ferns and the low diversity of ground ferns. The understorey includes such tall shrubs as *Trymalium spathulatum*, *Lasiopetalum floribundum* and *Bossiaea laidlawiana*. Small trees including *Agonis flexuosa* and *Casuarina decussata* form a second storey. Perhaps the latter species may prove to form a climax stand in the prolonged absence of fire in over-mature *E. jacksonii* forest. No rainforest trees of Malesian or Antarctic affinities now occur in this wet southwestern corner of Australia.

The environment

Climate

The outstanding features of the environment are high, reliable rainfall and shelter from the worst desiccation of the fire-promoting winds. Since wet sclerophyll forest is potentially capable of developing in most rainforest sites under appropriate fire regimes, annual rainfalls may exceed 1500 to 2000 mm and that of the driest month may exceed 50 mm. Wet sclerophyll forest will extend over large areas of mountainous terrain receiving between 1000 and 1500 mm per annum, however. In the drier sites, the annual total may be compensated for by protection in deep gullies or by extremely equable maritime conditions. In areas receiving only 25 to 30 mm in the driest month, rainforest elements are lacking or rare; such areas, which are found in western Victoria and southwestern Australia, are likely to be more fire-prone. Along all the coastal mountains benefits are likely to accrue from the incidence of low cloud and fog-drip. This may add significantly (*c.* 15–20%) to the total precipitation in some mature *E. regnans* forests at altitudes of 600 to 700 m (Brookes, 1949). In addition, the great height of the mature forest may increase orographic rainfall, since a rise in altitude of 100 m may increase annual precipitation by about 150 mm on the Great Dividing Range in central Victoria.

The seasonal distribution of rainfall follows the general latitudinal trend in which winter-dominant rains occur in the southwest and south, uniform distributions occur in the east, and summer-dominant rains occur in the north. Droughts are a recurrent feature of the Australian continent and rainfalls less than half the average total may occur at intervals of 20 to 50 years. When this situation developed in Victoria in 1967, wet sclerophyll forests were killed or severely damaged only in dense stands on rocky ridge-tops and northern slopes, or where streams dried out as water tables fell. In some cases tree and shrub death could have been due to a synergistic combination of droughts and fungal or insect parasites. The most erratic climate is that encountered in the northern range of the wet sclerophyll forests in southern Queensland where the coefficient of variability of annual rainfall is of the order of 40 per cent. The most reliable rainfalls, on the other hand, are in the western half of Tasmania and southwest Western Australia where the coefficient of variability is only 10 per cent (Leeper, 1970).

The temperature regimes of wet sclerophyll forest areas are cool to mild and warm since they encompass all climates from subtropical in southern Queensland to submontane in Victoria, Tasmania and New England in New South Wales. In coastal southern Queensland and in southwest Western Australia, frosts and snowfalls are uncommon, whereas in the cooler and higher areas in southeastern Australia these phenomena may be occasional to frequent. Relative humidities of wet sclerophyll forest areas are controlled by the temperature and the source of the winds. Because their habitats are situated along mountain ranges and plateaux which fringe the oceans, the inflow of maritime air contributes greatly to the amelioration of the climate. The generally brief periods dominated by desiccating winds radiating from the Australian interior constitute a climatic hazard which encourages the spread of fires. Above all else, this factor is of the greatest importance to the survival and perpetuation of these forests.

Fires

There can be no doubt that fire is an integral part of the environment of these wet eucalypt forests. Fire requires dry fuel, oxygen, low humidities and high temperatures for its unimpeded progress (Luke & McArthur, 1978). The high forests provide an enormous potential fuel supply, hence fires are dependent on the combinations of meteorological conditions for which the whole of the so-called wet sclerophyll belt is notorious. The onset of prolonged, erratic, dry summer conditions following dry winters can provide explosive conditions in these areas. The most hazardous conditions arise when slow-moving high-pressure systems direct hot, dry, desert winds outwards to the coast. Temperatures reach and exceed 40 °C, relative humidities fall to between 10 and 20 per cent and winds may gust up to 70 to 80 km/h. In general terms the lower, drier forests ignite most readily and therefore are burnt more frequently than the high forests. Forests in the wettest climates are least likely to dry out and thus least likely to be disturbed by fire; ridges and spurs exposed to dry winds will tend to carry fire more easily and constitute 'fire paths'. In heavily dissected terrain, fires will tend to burn more severely on the windward slopes than on the lee slopes and gullies may be 'jumped' and constitute 'fire shadows' (Webb, 1970). Fires are notoriously capricious in these forests and are under the control of wind, slope and above all the flammability, quantity and compaction of fuel.

In the summer of 1938–9, enormous fires razed 1 380 000 ha of mountain forests in Victoria (in total, 13.5 per cent of the State). Similar holocausts occurred at the turn of the century and before that, in the early gold mining times of 1851. A reasoned review by King (1963) suggested that small fires were more frequent in Aboriginal times, and large damaging fires more characteristic of the period of European occupation.

The outstanding characteristic of the high eucalypt forests is their even-agedness. This, together with the presence of persistent soil charcoal is suggestive of catastrophic fires in the past centuries. The commonest fires in these forests on mainland Australia are the surface fires which consume the ground-stratum of ferns and grasses and kill or ignite the shrub and sensitive small tree strata. Under these conditions, mature trees may survive and regeneration may be promoted in large canopy gaps. Thus a 2-, 3- or 4-aged forest may result, depending on the exact fire regime. An important principle here is that the fire resistance increases with age and the thickness of lower bark on the trunk, and hence destruction of older forests requires more intense fires (see later). Under severe conditions of higher wind speeds, fires sweep up into the tree canopy. The crown fires which result may be free-running under explosive conditions, or dependent on the fuel of the undergrowth (Luke, 1961). As pointed out by Mount (1969), crown fires often totally destroy the forests, but permit dense regeneration to follow. In the wet country, such conditions originally did not recur more than once every one or two hundred years. Since the arrival of Europeans these types of fires have been of a frequency such that even eucalypts have been eliminated from large areas. In cooler climates of Tasmania and certain limited sites in Victoria humus accumulation in the top soil may allow slow smouldering humus fires to persist. These fires girdle giant eucalypt trees as well as rainforest trees (Cremer, 1962) and set in train a whole sequence of recolonisation that may take centuries to complete.

In eucalypt forests the presence of deciduous bark streamers, rough fibrous bark or lichen epiphytes (e.g. *Usnea*), tend to carry fire up to the canopy under some conditions. These materials also serve to disseminate fire ahead of the main front, leading to a common Australian phenomenon of spot-firing (Luke, 1961). A chaotic melee of fires can develop anywhere up to 30 km ahead of the main fire front (Stretton, 1939).

The quality of the fuel is modified by nutrient content (Vines, 1975), so that poor soils supporting markedly sclerophyllous, flammable undergrowth are likely to be sites of severe fires. As Jackson (1968) and Ashton (1976a) have pointed out, the fertility of soils is likely to be depleted in areas of high rainfall subjected to repeated burning. In terms of ecosystem potential it is a 'downward spiral'.

Soils

The soils of the tall open-forest are variable and are derived from a wide spectrum of parent materials. Since they are relatively deep and their infiltration rates are good, their water relations are favourable for the sustenance of large and complex forests. In relatively 'dry' climates in Victoria (1000 mm per annum), fertile krasnozems from basalt support grassy tall open-forest of *Eucalyptus globulus*, *E. obliqua* and *E. viminalis*. On arkoses, podsolic soils support layered forest with an understorey of both shrubs and grasses. Tall open-forests of *E. delegatensis*, with grassy understoreys, also occur in submontane areas (1200–1500 m), where soils are humus-rich brown loams.

Where rainfalls in Victoria are higher (1500 mm or more per annum), krasnozems and related brown, acid, friable soils occur under wet sclerophyll forest on igneous (granites, granodiorites, dacites and basalts) and sedimentary rocks (mudstones and shales). In Tasmania, such forests occur on similar red-brown soils derived from dolerite. Yellow-grey, podzolic, silty clay soils support these forests on arkose in Victoria, and on limestone in Tasmania. A similar pattern of soils and parent materials may be found in wet sclerophyll forests in the subtropics. In southwestern Australia, forests of *E. diversicolor* occur on deep red-earth soils formed where the under-lying gneiss has been exposed by erosion of the lateritic plateau (McArthur & Clifton, 1975). Although low in nutrients such soils are much more fertile than those of the *E. marginata* dry sclerophyll forests on nearby laterite, and the boundary between *E. diversicolor* and *E. marginata* is often sharp and distinct.

In the wet, monsoonal climates of coastal Arnhem Land, layered tall open-forests occur on very poor soils and contain an understorey of tall, rank grasses and broad-leaf shrubs of Indo-Malayan affinities.

The influence of the litter cycle is indicated by the marked concentration of nutrients in the top soil. The litter fall of the wet sclerophyll forests is relatively high by world standards in relation to climate (Bray & Gorham, 1964; Vogt, Grier & Vogt, 1986). In *Eucalyptus regnans* forests at Wallaby Creek in central Victoria it is 8.1 t/ha in mature forest and 6.8 t/ha in pole-stage forests. The *E. regnans* leaf component, which makes up about half of the total, is greatest in half-grown spar-stage forests 40–50 years old. Polglase & Attiwill (1992) measured litterfall over a chronosequence of *E. regnans* forests near Toolangi, Victoria, from 5 years to 250 years old. Leaf litterfall was almost constant (2.6 t/ha) from age 5 years indicating that the site is more or less fully occupied by that time; total litterfall increased from 4.9 t/ha at age 5 years to 8.6 t/ha at age 250 years. The amount of leaf material dropped annually varies with the climate: it is greatest in subtropical *E. pilularis* forests and least in the mediterranean-type climate of southwest Western Australia.

Table 6.3. *Annual litter fall (t/ha) and its nutrient content (kg/ha) in tall open-forests (developed from O'Connell & Menage, 1982)*

	Litter weight	N	P	K	Ca
E. diversicolor[1]					
Leaf	3.54	32.1	0.7	10.6	54.1
Total	9.45	58.1	1.9	31.7	122.3
E. regnans[2]					
Leaf	4.02	43.5	1.4	5.7	34.8
Total	7.76	57.6	1.9	7.5	48.8
E. pilularis[3]					
Leaf	6.50	41.1	1.4	8.3	24.9
E. grandis[4]					
Leaf	4.22	34.5	1.3	7.2	30.8
Total	9.60	66.7	2.7	14.9	64.2

[1]Mature forest: O'Connell & Menage (1982).
[2]Mature forest: Ashton (1975*b*).
[3]Webb, Tracey, Williams & Lance (1969).
[4]27-year-old plantation: Turner & Lambert (1983).

The return of nutrients to the top soil in leaf fall is relatively large and similar in the *E. pilularis* and *E. regnans* (both in the subgenus Monocalyptus) forests. However, the return of calcium by the smooth-barked *E. grandis* plantation (even as young as 27 years) and *E. diversicolor* (both in the subgenus Symphyomyrtus) forest is much greater than in *E. regnans* and *E. pilularis*. This greater return of calcium in the litterfall of Symphyomyrtus species follows the greater uptake of calcium (and also for some other elements, e.g. manganese) by these species (Lambert, 1981; Attiwill & Leeper, 1987; Lambert & Turner, 1991).

The well-developed understorey of tall open-forests is of great importance in nutrient cycling. For a number of tall open-forests reviewed by Baker & Attiwill (1981) the understorey contains only 5–20 per cent of the weight but 20–35 per cent of the nitrogen in the above-ground stand. Ashton (1975*b*) showed that *Pomaderris aspera* contributes 28 per cent of the weight of leaf litterfall but 56 per cent of the calcium in leaf litterfall in *E. regnans* forest. Potassium values are likely to be underestimated because of pre-fall leaching. In general, the whole of the nutrient return is somewhat underestimated because of the persistent and sometimes substantial attacks on the canopy foliage by insects. Much of this cycling would thus occur by means of their frass.

The nitrogen levels of these forests are most likely to have been enhanced by the nitrogen-fixing plants. Acacias, which are almost universally associated with eucalypts in Australia, could be important factors, especially in the wake of fires. Thus in *Eucalyptus regnans* forests *Acacia dealbata*, *A. melanoxylon* and *A. obliquinervia* are common, in *E. pilularis* forest *A. binervata* and *A. silvestris*, and in *E. diversicolor* forest *A. pentadenia* and *A. urophylla*. In addition, *Casuarina* may be involved in this contribution, notably *C. torulosa* in the northern forests and *C. decussata* in those of southwestern Australia. The southwestern *E. diversicolor* forests also include well-developed understoreys of *Bossiaea laidlawiana*. Despite the apparent importance of these species to the nitrogen economy of tall open-forest, their contribution remains mostly unquantified. Adams & Attiwill (1984a,b) estimate that *Acacia* may fix 20 kg/ha/year of nitrogen over the first 5 years after fire in regenerating *E. regnans* forest. N_2-fixation by *Bossiaea laidlawiana* is of the same order: 8–14 kg/ha/year (Grove & Malajczuk, 1991; O'Connell & Grove, 1991). Polglase, Attiwill & Adams (1992) concluded that nitrogen fixation and rapid turnover mean that nitrogen is never limiting to growth of *E. regnans* forest.

The status of phosphorus is low, as is usual for most Australian soils (Chapter 7, Gill, 1981), yet rarely does it appear to be so low as to be a primary nutrient deficiency. Reports of responses of tall open-forests to added nutrients are few and too variable to be conclusive. O'Connell & Grove (1991) cite studies of growth reponses in 5-year-old *E. diversicolor* forest regenerating after clear-felling and slash-burning; Attiwill (1991a) found no net growth response to heavy applications of fertilizer in *E. regnans* regeneration.

The soils of the wet sclerophyll forests are acid, pH values ranging from 4 to 6. The general availability and potential of nutrient supply is greater than in the dry sclerophyll forest (Beadle, 1962; McColl, 1966; Ashton, 1976a). Undoubtedly, the rate of turnover of litter and dead roots is a major factor contributing to fertility. Ellis (1969, 1971) showed in Victoria that the respiratory activity in soils under wet sclerophyll forest was also greater than that under dry sclerophyll forest.

On mainland Australia, even in the montane zone, temperatures are warm enough to allow the active decomposition of litter, and mull humus is the rule. In the east and southeast, the forest floor is vigorously cultivated into the top soil by foraging lyre-birds (*Menura novaehollandiae*). At Beenak in central Victoria, 20 per cent of the bare floor of a pole-stage *E. regnans* forest is turned over each year by lyre-birds (D.H. Ashton & O. Bassett, unpublished data). In mature *E. regnans* forest at Wallaby Creek in central Victoria, the weight of the forest floor amounts to 22 t/ha, of which the leaf material component is very little more than the current year's fall. In Tasmania, however, cooler maximum temperatures result in the accumulation of fibrous and amorphous humus on the soil surface, which may amount to 160 t/ha (Frankcombe, 1966). Although the nutrient return of the higher-quality wet sclerophyll forests is about equal to the average conifer forest in North America (Lutz & Chandler, 1947), the rate of turnover of the relatively nutrient-rich leaf material in the former is very rapid, because of decomposition by fungi and bacteria and through disintegration by soil fauna. The abundant moisture, good aeration of soil and the rapid flux of nutrients via the litter cycle is probably responsible for a rapid utilisation of energy and very high growth rates.

In fact, the growth rate of *E. regnans* forest regenerating after major bushfire ranks among the highest productivity in the world (Attiwill, 1991a). The response of *E. regnans* to this major disturbance is typical of more fertile sites (Vitousek *et al.*, 1982; Polglase, Attiwill & Adams, 1986) where high rates of nitrogen fixation, rapid uptake of nitrogen by plants and immobilisation of nitrogen by microorganisms are key factors conveying resilience (and the conservation of nutrients, Boerner, 1982) to forest ecosystems. The capacity for rapid uptake depends both on quick re-establishment after disturbance and on rapid growth rates of the regeneration. It is interesting to note that Lockett & Candy (1984) found that the benefits to early growth after clear-felling and slashburning are greater in tall open-forest than in drier forest.

Whereas in most forests of the world the dominant form of inorganic nitrogen is ammonium ion, tall open-forests, at least in eastern Australia, are characterised by the mineralisation of nitrogen proceeding to nitrification (Adams & Attiwill, 1982, 1986; Richards *et al.*, 1985). This capacity to nitrify is related to the balance between nitrogen capital, the rate of nitrogen turnover, and the demand for nitrogen by heterotrophic decomposers (Adams & Attiwill, 1986; Polglase, Attiwill & Adams, 1992). Despite this production of readily leached nitrate ion, the cycling of nitrogen is tight; even after major disturbance in which nitrate production is enhanced, the mineralisation–immobilisation balance returns to pre-disturbance levels within two years (Weston & Attiwill, 1990).

Insect attack and herbivory

In *E. regnans* forest, some two-thirds of young leaves are chewed to some extent (by larvae of moths, beetles and sawflies and by adult beetles, weevils, leaf miners and gall insects; Ashton, 1975*d*). Buds, flowers and fruits are eaten or damaged by gall-forming and other insects, and by parrots and cockatoos.

On average, it seems that herbivory accounts for a relatively small proportion of gross primary production in most eucalypt forests, including tall open-forest (Attiwill, 1979; Ohmart, Stewart & Thomas, 1983; Lamb, 1985). In some years, however, forests have been severely defoliated. The Christmas beetle (*Chrysoptharta bimaculata*, Chrysomelidae) attacks young regeneration of *E. regnans* and *E. delegatensis* in Tasmania (Greaves, 1966) and *E. regnans* in the Central Highlands of Victoria. Forty-year-old forest in central Victoria has been almost completely defoliated by a stick insect (or phasmatid) *Didymuria violescens* (Neumann *et al.*, 1977).

In the mid-1980s *E. regnans* forest near Tanjil Bren, Victoria was almost completely defoliated by psyllids (*Cardiaspina bilobata*). Adams & Atkinson (1991) suggest that this outbreak may be related to the unusual and strongly nitrifying soil at Tanjil Bren (with a concentration ratio NO_3^--N: $NH_4^+-N = 4.5$, most probably resulting from the very large numbers of acacias in the understorey). The leaves of these trees were richer in glutamine and a hundred times more heavily infested with psyllids than were leaves from similar-aged *E. regnans* forest at Britannia Creek where the soil was ammonifying (NO_3^--N: $NH_4^+=0.7$). This proposed nutritional basis for herbivory is a promising line of research; clearly, there is a need to improve our knowledge both of the magnitude of herbivory and of the factors which predispose forest trees to attack. This is especially important for a number of the fast-growing trees of the tall open-forest, since these are the species most preferred in commercial plantations, both in Australia and overseas.

Growth and development

The forest form

Some of the fastest growing eucalypts are found in the wet sclerophyll forests. For trees in their juvenile state, annual height growth increments of 2 m during the first 10 years are not uncommon. In *Eucalyptus regnans* forests, half of the final mature height is achieved in the first 25 to 35 years (Jackson, 1968; Ashton, 1976*b*). The greatest standing biomass may be achieved between 75 and 100 years, although the culmination of the annual increment may be achieved earlier in the spar-stage between 40 and 60 years. The wet sclerophyll forests characteristically commence as even-aged stands following intense fires. Mineral soil is exposed and the nutrient status and biological level stimulated by the heat treatment imposed on the organic horizon by the fires. Under these initially non-competitive conditions, light, water and nutrients are adequate for prodigious growth. Seeds of the understorey species are often stored in the soil, although some species may regenerate vegetatively from organs at or below ground level, or be dispersed into the site from outside sources. The acacias grow very fast and may, in some cases, temporarily outgrow the eucalypts. This may be an advantage in some mountain sites where radiation frosts in hollows can be lethal. At Noojee, Victoria 2.4 × 10^6 seedlings per hectare of both *E. regnans* and *A. dealbata* occurred immediately after the 1939 fires (Beetham, 1950). The thinning out of wet sclerophyll eucalypts with age is very rapid. In *E. regnans* forests, a marked change in physiognomy of the stand takes place after 25 to 40 years when most of the early suppressed individuals finally die. During this period, tree numbers may diminish from nearly 2000/ha at 16 years to 380/ha at 40 years. The rate of decrease is reduced thereafter and the number stabilises at about 40 to 80/ha at maturity after 150 years. Similar kinds of changes are found in other wet sclerophyll forests, although the denser crowns of *E. microcorys* and *E. calophylla* usually have more or less horizontally disposed foliage and their capacity to survive low light under canopies is much greater (Jacobs, 1955). Suppression of weaker trees is likely to be considerably delayed if root fusion with more vigorous trees has taken place (Ashton, 1975*c*).

The generalisation that trees from wet sites tend to develop root systems at an initially slower rate than those from dry sites (Toumey & Korstian, 1947) seems to hold for those species of eucalypt studied (Zimmer & Grose, 1958; Ashton, 1975*c*). Seedlings of *E. regnans* are relatively slow to develop both the juvenile shoot and root in the first few months, but thereafter growth is accelerated and continues unabated for several years. The vigorous growth in the dense pole-stages results in marked apical dominance and conical crowns confer on the stand a conifer-like appearance. Continued shedding of branches leads to the development of a clear bole. The mature crown becomes open and infilled with secondary branch systems (Jacobs, 1955). Not until late maturity does die-back of the apical portions persist and lead to stag-headedness. Epicormic growth on the trunks descends to lower heights in very old trees (300–400 years), especially if such trees have been damaged by fires (Jacobs, 1955).

Results of a study of the water relations of tall eucalypts by Connor, Legge & Turner (1977) have revealed that moisture tensions at the tops of trees are developed in accordance with a static hydraulic head of about −0.1 bar/m. Such water potential in mature *E. regnans* may reach −8 bars at 60 m height in winter. On hot afternoons the water potentials at the summits of mature and pole stage trees may be similar at about −17 to −20 bars. The gradient, however, is very much steeper in the young dense forests (−0.45 bars/m), compared with that of mature forests (−0.27 bars/m).

Not surprisingly, shoot growth at the tops of tall mature trees is very slow, and amounts to only about 10 cm per year. The leaves are consequently small and possess thick cuticles. In *E. regnans*, the uppermost leaves of mature trees 75 m high are about 8 to 9 cm^2 in area and 8 to 10 cm long. Spar-sized trees 50 m tall develop leaves 20 to 40 cm^2 in area, 20 cm long and with cuticles 12 to 15 μm thick. Saplings 5 m tall may have leaf areas of 50 to 60 cm^2 and cuticle thicknesses of only 7 to 9 μm. The average diameter of vessels in secondary xylem of roots is much larger than that of stems. For instance, twigs and roots 1.5 mm in diameter have vessels averaging 44 and 74 μm in diameter respectively, whereas those from similar organs 1 cm in diameter are 72 and 155 μm. N. Legge (personal communication) has recorded vessels nearly 1 mm in diameter in long ropy roots of *E. regnans*. Such features are likely to facilitate water movement. The ability of these species to utilise the water supply and sustain rapid growth in their early decades must in part be due to their capacity to develop sufficiently low tensions. The long growing season of 7 to 9 months (Ashton, 1975*d*; Cremer, 1975) of *E. regnans* ensures the maintenance of large leaf areas in the vigorous early stages of the development of the stand. Hopkins (1964) suggested, from the results of seedling experiments, that the greater growth of *E. regnans* in fertile topsoils is a self-imposed factor restricting this species to sites with a sustained water supply.

The catchments of wet sclerophyll forests near large cities are frequently utilised for their water yield. In central Victoria, *E. regnans* covers at least half of the area of the water supply catchments for Greater Melbourne. Following the destruction of many of these forests by bushfires, especially those of 1939, young dense forests have grown up. In the succeeding decades the harvest of water has steadily declined, irrespective of the trends of annual precipitation. This is thought to be because of the increased water use by the dense young stands (Langford, 1974). The extent of the reductions of water yield in four catchments ranges from 31 to 13 per cent of that yielded in the pre-fire period and is strongly correlated with the area occupied by young dense regeneration of *E. regnans* (62 to 20%). The explanation of reduced transpiration with increasing age of *E. regnans* forest appears to lie principally with changes in sapwood area (Dunn & Connor, 1991). The precise reasons limiting height growth and the maintenance of sapwood in tall trees presumably reside in the relationships between cell expansion and water potential. Or is the limitation to size structurally based (Attiwill, 1991*b*)? As a structural limit is reached, the crown begins to fail and to break; tissues are then exposed to fungal attack so that the vigour of the tree would decline. The general question, however – what is the limit to the size of a tree – is still to be answered.

For the *E. regnans* forests which are the water catchments of the City of Melbourne, however, once the 1939 regrowth has developed beyond the phase of maximum leaf area index (and of maximum overstorey sapwood area), an improvement of water yield may be anticipated. O'Shaughnessy & Jayasuriya (1987) estimate that, if the present policy of no logging in the catchments is maintained, the eventual increase in streamflow will be equivalent to 30 per cent of the current annual consumption of water. Such an estimate assumes, of course, that the forests will be protected from the seemingly inevitable cycle of disturbance by bushfire, followed by regeneration.

Regeneration

The failure of eucalypts to regenerate in mature forest in the absence of fire is primarily related to low light levels and the morphological changes in leaf orientation and anatomy. Young seedlings of *E. regnans* with horizontal, dorsiventral foliage have low light compensation points and may persist under low light intensities provided that there is no fungal or insect attack or browsing by herbivores. With the development of iso-bilateral foliage in the second and third years, the light compensation point increases sharply (Ashton & Turner, 1979) and the demand for high light intensities to sustain growth rates increases greatly.

Some species, especially those with horizontally-oriented leaves (such as *Eucalyptus calophylla* and perhaps *E. microcorys*), are relatively persistent under moderate amounts of shade, and survive damage because of their effective lignotuber development. Under high light intensities, eucalypts recover from many such attacks with remarkable tenacity and vigour. The absence of the lignotuber in many of the wet sclerophyll eucalypts diminishes their chances of persistence. In forests of *E. regnans*, *E. obliqua*, *E. delegatensis* and many others, the harvesting of seed by insects is rife. Ants have been implicated (Jacobs, 1955; Cunningham, 1960*a,b*; Grose, 1963) as have also lygaeid bugs (Grose, 1960, 1963; Cremer, 1965*a*). In the *E. regnans* forests at Wallaby Creek (37°24'S; 145°15'E), more than 60 per cent of seed in the annual seed fall of *E. regnans* may be removed, which in the warmer months involved nearly all the full, fertile seed (Ashton, 1979). In addition, growth of eucalypt seedlings in the mature forest soils is frequently relatively poor. In *E. pilularis* forest, Florence & Crocker (1962) ascribed this to a microbiological antagonistic effect. Evans, Cartwright & White (1967) reported a phytotoxin (nectroline) from some root-surface isolates of the fungus *Cylindrocarpon destructens*. This fungus also occurs in *E. regnans* top soils and may be involved in poor growth of seedlings in such soils if conditions are kept cool and moist. However, subsequent detailed experiments by Ashton and Kelliher (in press) have failed to establish a clear, causative correlation in glasshouse, growth cabinet or field plots even though *in vitro* necrosis of young seedling root-tips can be induced by exposure to Seitz-filtered culture solutions of this fungus (Willis & Ashton, 1978). Earlier work suggested that the proximity of old root systems may have an inhibitory effect on young seedlings (Ashton, 1962), although definitive proof of allelopathic processes is notoriously difficult. Air-drying of soil followed by re-wetting can result in an enormous increase in seedling growth (Ashton & Kelliher, in press). Much of this effect is related to the near total sterilisation of the soil and the release of substantial amounts of inorganic nitrogen and of readily available phosphorus.

The importance of this phenomenon may lie in the necessity for massive disturbance of all strata of the forest, leading to the solarisation of the bared soil in large gaps. Under these conditions of massive disturbance, vigorous regeneration of *E. regnans* may establish and overcome competition from understorey species. The importance of solarisation in agriculture and forestry has been clearly described by Warcup (1983).

The removal of all these problems by fire is dramatic. Inhibition is replaced by stimulation. This 'ash-bed' effect has been ascribed variously to the enhancement of phosphorus availability (Attiwill, 1962; Loneragan & Loneragan, 1964), to an increased surge of nitrogen availability (Hatch, 1960), and to the death of micro-organisms and the consequent release of their nutrients, and to the replacement of the normal microbial flora by an entirely different one over the period of increased fertility (Renbuss, Chilvers & Pryor, 1972). All of these effects have been intensively studied in soil from *E. regnans* forest (Chambers & Attiwill, 1993*a*, *b*); the stimulatory effects of the ash-bed appear to be primarily because of greatly increased availability of phosphorus and nitrogen after heating. Increases in available manganese because of heating are temporary (a few weeks), but these increases are sufficient to be toxic to plants which establish immediately after fire.

One of the significant features of the wet sclerophyll eucalypts is the virtual absence of hard, dormant seed. Usually they are virtually the only members of the forest association not represented as soil-stored seed (Gilbert, 1959; Cunningham, 1960*b*; Cremer, 1965*b*; Floyd, 1976; Ashton, 1979). The recent seed fall may be found occasionally in the uppermost 1 cm of soil but it is likely that such temporarily buried seed would be destroyed by the heat of the fire. The storage of eucalypt seed is in the capsules of the canopies. The persistence of capsules for three or more years usually ensures that at least the crop of one good flowering season is present. In *E. regnans* forest, a 20-fold difference in flowering and seed-set may occur from a good to a bad season. After a peak flowering year many millions of seeds are stored per hectare of canopy, and released over several years or after fires, apparently in one large fall (Cremer, 1965*a*; Christensen, 1971; Ashton, 1975*a*).

The ecology of the wet sclerophyll forests therefore hangs on an extraordinarily fine thread. It seems almost inconceivable that the small capsules of these species – often 1 cm or less long – can protect seed sufficiently long in the holocaust of raging crown fires. Mount (1969) suggested that, because of the peculiar phenology of many of the eucalypts, the moist capsules are situated well below the current foliage. The heat therefore would tend to be swept upwards and away. Results of preliminary research by Webb (1966) indicated that seeds within large green *E. globulus* fruits were protected for about 9 min from a lethal rise in temperature when such capsules were subjected to intense red heat at 440 °C in an electric furnace. Results of experiments by Gloury (1978) suggest that when green capsules of *E. regnans* were subjected to uniform heat in an oven the germination capacity of the enclosed seed was diminished below that of seed exposed nakedly under the same conditions. Thus, these attempts to gauge the 'protectiveness' of capsules of different ages of various species indicate the complexity of the problem. Ashton (1985) and Judd & Ashton (1991) concluded that, since the survival time of *E. regnans* seed in capsules exposed to high temperatures is so short, the exposure of capsules to very high temperatures during a crown fire must also be very limited, of the order of 10–30 seconds (Ashton, 1985).

Succession

Because of the vicissitudes of fire, a forest may be partially damaged or totally razed. Repetitive burning within the primary non-flowering period of an obligate seed-regenerating species can result in its complete elimination from a site. The variability of vegetation structure and floristics in burnt forests is therefore great, and the boundaries of the various communities are often dramatically sharp. Throughout the tall open-forest, even-aged stands from pole to mature form occur, each being related to a specific fire history. However, partially damaged forest may show two or even three distinct age-classes, the older ones being invariably butt-scarred. This characteristic damage is due to litter accumulation on uphill sides of trunks and to the chimney-flue effect of flame vortices in the lee of the trunk bases (Gill, 1975). Surface fires in the ground stratum and the shrub understorey may leave the eucalypts relatively undamaged yet initiate even-aged strata of undergrowth seedlings or coppice shoots. In southern Tasmania where the humus layers burn above the mineral soil, the entire forest community may be destroyed (Cremer, 1965a), the colonising species arriving by wind and bird dispersal. In such instances it has been estimated by Cremer (1965a) that eucalypt seed sources need to be within 200 m for replacement of this dominant life form to occur. In the half-barked *E. regnans* stands, the butt-bark appears to be sufficiently thick to protect the tree from heat, if surface fires of low intensity occur at intervals greater than about 25 years. Therefore, within this age limit, the whole forest is even-aged, but beyond this period the understorey may be either contemporaneous or younger than the overstorey.

If rainforest species are present in the general vicinity, they may be expected to invade the maturing understorey of the wet sclerophyll forest within 100 years, and completely replace it in the succeeding century. Within the following two centuries the eucalypts will disappear as their life span is reached. The higher the rainfall, the less likely is this sequence likely to be interrupted by fire. The Tasmanian maritime environment is less conducive to destructive fires than the more continentally-influenced mainland areas in Victoria. Large areas of rainforest of *Nothofagus cunninghamii* and *Atherosperma moschatum* are therefore rare in Victoria, although, as shown by Howard (1973), the potential for such to occur is relatively great (see Chapter 5 and Howard, 1981). In Victoria, where rainforest species are remote from the burnt stands, successions in the understorey types of the wet sclerophyll forests proceed along different lines. There is a tendency for shrubs and small trees to reach their life span after specific intervals. In *E. regnans* forests a light-crowned understorey of the composite, *Cassinia aculeata*, begins to senesce after 25 to 30 years. It is then replaced by *Pomaderris aspera*, *Olearia argophylla* and *Bedfordia arborescens*, a common trio of understorey dominants in this forest type in central Victoria. After about 100 years, *P. aspera* becomes senescent and the tree ferns and ground-stratum of rosette ferns (*Polystichum proliferum*) or tussock grass (*Poa ensiformis*) increases in abundance. This may inhibit the regeneration of the understorey shrubs and trees unless fire occurs. Only those shrubs with vegetative reproduction (e.g. layering and root suckering) persist. Cunningham & Cremer (1965) suggested that certain seeds may be stored in the soil for one or two centuries or more. Fires at intervals longer than 100 years do not result in the renaissance of an understorey of *Pomaderris apetala*.

The big forests of both *E. diversicolor* and *E. regnans* possessed structurally and floristically variable understoreys at the time of discovery by Europeans. Persistent folk-lore relates both to the riding of a horse through the virgin forests and to the hacking of a path for progress (Select Committee, 1876; Galbraith, 1937; Underwood, 1973). Such variation can be interpreted in terms of persistent, patchy, fire regimes.

The first communities to arise after fires are derived from soil-borne seeds, from fire-resistant capsules and from rapidly dispersed disseminules. Bryophytes, such as *Funaria hygrometrica* and *Ceraton purpureus* or *Marchantia polymorpha*, may become particularly abundant in severely burnt soil in the first two years after fire but their lawn-like mats and cushions are eventually overcome by more vigorous and taller herbaceous and woody regrowth (Cremer & Mount, 1965). During the first few years, the communities are often dominated by so-called fire weeds. These are usually composites, both native and introduced. In warmer climates, weeds from other families are also involved, such as the introduced *Phytolacca octandra* and *Lantana camara* (Floyd, 1966, 1976), and wiry lianes (Van Loon, 1969).

In southeastern Australia the invasion of rainforest species may be sufficiently delayed by fires or long dispersal distances so that large tracts of wet sclerophyll forest continue to persist. The climatic climax to much of the tall eucalypt forest along the east coast is derived from the so-called Indo-Malayan element (Hooker, 1860) which forms rainforests of the 'notophyll vine' and 'microphyll vine' forest types (Webb, 1959). The climatic climax of the southern areas and the wet cool plateaux to the north are of the so-called Antarctic element which forms 'microphyll mossy' and 'microphyll fern' forests, epitomised by *Nothofagus*. In Victoria, variations of the temperate rainforest are climax to a whole range of wet sclerophyll forests from near sea level to about 1450 m (Howard & Ashton, 1973).

The northernmost distribution of tall open-forests occur on the Atherton Plateau along the western edge of the area of notophyll vine forest (Tracey, 1982). In some areas on rhyolite, a zone of wet sclerophyll forest up to 1 km wide occurs adjacent to simple notophyll vine forest. The eucalypt forest is dominated by *E. grandis* with an understorey of microphyllous shrubs (*Phebalium* sp., *Rhodomyrtus* sp. and *Acacia melanoxylon*) with a ground stratum of sedges (*Gahnia sieberiana*) and ferns (*Dicranopteris*). Such forest is reminiscent of some wet sclerophyll forests on Wilson's Promontory, Victoria. In other sites on the Atherton Plateau, the tall open-forest is represented by a very narrow ecotone 20–100 m wide between notophyll vine forest and grassy open-forest dominated by *E. intermedia* and *E. acmenioides*. Unwin (Unwin, Stocker & Sanderson, 1985; Unwin, 1989) suggested that repeated fire has caused the attrition of the rainforest boundary and permitted the establishment of *E. grandis* in a narrow band where fire is either less intense or less frequent. The invasion of the ecotone by rainforest species appears to be rapid in this area and a distinctive wet sclerophyll understorey is not developed. Tracey (1982) describes a series of tall open-forests whose understoreys are related to fires on relatively oligotrophic soils. Further south, on the Macpherson Ranges, notophyll vine forest has advanced and stranded mature *E. microcorys* 100–200 m inside the present rainforest boundary. In central New South Wales (Watagen

Mountains) a tall open-forest of *E. paniculata*, *E. acmenioides*, *E. deanii* and *E. pilularis* with a layered, grassy understorey abuts onto simple notophyll vine forest. Recent surface fires which burnt up to the rainforest have resulted in the vegetative regrowth of many of the rainforest species such as *Synoum glandulosum* and *Rapanea howittiana* from lignotubers or root suckers. Such coppice resistance is described by Unwin *et al.* (1985) on rainforest margins on the Atherton Plateau. *Acmena smithii* is relatively abundant on the rainforest margins in this region (Burges & Johnston, 1953). At Wilson's Promontory and in East Gippsland, this species exhibits a remarkable resistance to single fire events because of its lignotuberous development (Ashton & Frankenberg, 1976; Melick & Ashton, 1991). In southwestern Australia, such potential climaxes appear to have been eliminated by climatic changes in the geological past (Cookson, 1954; Crocker, 1959).

The limit of rainforest distribution in the complete cessation of fire is difficult to determine. Certainly, many of their members are more resistant to drought (Williams, 1978) than has hitherto been thought. Howard (1970), for example, was able to establish *N. cunninghamii* in dry sclerophyll forest in a ridge-top site, Mt Donna Buang, and in gullies in the Dandenong Ranges, central Victoria, where it is not native. The limits in drier country appear to be obscured by fire. What exactly is climax again becomes an academic question, the solution of which requires experimentation and long-term observations.

In *E. regnans* forests, two small trees of the family Asteraceae possess lignotubers. Frequently these species, *Olearia argophylla* and *Bedfordia arborescens*, coppice from this organ and rapidly dominate the ensuing understorey. In Victoria, two rainforest trees, *Nothofagus cunninghamii* and *Acmena smithii*, are also endowed with lignotuberous burls from which recovery is possible after moderately intense fires (Howard, 1973; Ashton & Frankenberg, 1976). Repeated surface or crown fires, however, are likely to eliminate these species if the heartwood of the original trunk is incompletely callused, since the stumps may be ignited and the living buds killed from within by smouldering fires.

Almost all of the understorey species are represented in the store of soil seed and are stimulated to germinate by the passage of the fire. In *E. regnans* forests, this particularly applies to *Pomaderris aspera*, *P. apetala*, *Acacia dealbata*, *A. melanoxylon*, *Prostanthera lasianthos* and *Cassinia aculeata*. Dense understoreys of these species may result, their composition depending on the longevity of seed and the original species composition. In the mature *E. regnans* forest at Wallaby Creek, Victoria, 40 years after a surface fire *Pomaderris aspera* is 10 to 15 times more abundant than *Prostanthera lasianthos*; yet in regeneration on large cleared plots the proportion of these two species is about parity. That fires do not always result in the exact replication of the original understorey was noted by the exploiters of the great forest of South Gippsland at the turn of the century (Committee of the South Gippsland Pioneers' Association, 1920).

Since European settlement, magnificent eucalypt forests have been reduced to scrub by top fires at intervals of less than 15 to 20 years. This has been discussed by Gilbert (1959) and Jackson (1965), who both point out that there is a selection for the more fire-tolerant species in the process. *Pomaderris aspera* will likewise be removed if surface fires recur at intervals of less than five to ten years. The ability of the older eucalypts to withstand the heat of an undergrowth fire is almost certainly related to the greater thickness of the bark in the butt region. Many wet sclerophyll eucalypts, particularly *E. regnans*, have very thin gum bark until the spar stage when a thicker persistent subfibrous bark develops and gradually extends up to a height of 10 m by maturity. In general, bark thickness increases with age and diameter of the trunk (Gill & Ashton, 1968; Vines, 1968) so that younger trees will be more likely to be damaged or killed by fire than older. Many of the 'true' wet sclerophyll eucalypts develop a 'half-bark' condition at maturity similar to that of *E. regnans*. This, and the isolation of the high canopy from the understorey, are probably factors permitting a wide expression of understorey types in mature stands.

Rainforest components in mixed forest may persist vegetatively, although fire-sensitive eucalypts are killed. Thus, in Victoria, coppice *Nothofagus cunninghamii* may be found under even-aged pole-stage *E. delegatensis*, *E. nitens* and *E. regnans* on Mt Donna Buang and Mt Baw Baw, and *Acmena smithii* under *E. obliqua* and *E. muellerana* at Wilson's Promontory (see Chapter 5 and Howard, 1981).

In areas of moderate altitude (100–800 m) in central Victoria where mean annual rainfalls are 1100–1250 mm, a mixture of *Pomaderris aspera*, *Olearia argophylla* and *Bedfordia arborescens* tends to form a relatively stable understorey for 100 years or more, provided there is no easily available seed source from rainforest patches in gullies. Under this rainfall regime, fire frequencies are likely to maintain this condition.

The question of the effect of acacias on the whole nitrogen economy of the wet sclerophyll forests has often been verbally expressed by many colleagues. Detailed research on this problem is long overdue. The problem is that most studies of nitrogen fixation have been in drier forests where acacia numbers are much lower than the thousands to millions per hectare which might germinate after fire in tall open-forest. For example, rates of nitrogen fixation for a number of legumes in jarrah (*E. marginata*) forest are mostly less than 1 kg/ha/year (Hansen & Pate, 1987); Hansen, Stoneman & Bell (1988) estimate that under optimal conditions, these rates could increase to 5-10 kg/ha/year. In managed *E. regnans* forest, about 4 kg/ha/year is required to balance the losses of nitrogen by clear-felling and regeneration burning on a rotation of 100 years. If the rate of nitrogen fixation over the first five years, when acacias are dominant, is maintained at 20 kg/ha/year (Adams & Attiwill, 1984*b*), then the required rate over the next 95 years is only 3.2 kg/ha/year.

There seems to be little doubt that the intensive regeneration of legumes after fire is of benefit to tall open-forest. Investigations by Shirrefs (1977) in stands of 1939 origin at Toolangi, Victoria, suggest that the nitrogen level of one-year-old foliage of *E. regnans* in the presence of *A. dealbata* is enhanced 50 per cent (significant at $P = 7\%$), compared with those stands in which it is absent. If this work is substantiated with other sites and species combinations, an important effect of fire in Australian ecosystems may be established. One important effect of soil which cannot be overlooked in the ecology of these forests is the so-called 'ash-bed' effect. This has been shown by many workers (e.g. Pryor, 1963) to cause a two- to threefold increase in tree growth for 2-3 years following fire. It is undoubtedly important in allowing the eucalypt species to exploit their superior growth potential.

Evolution of the tall open-forest or wet sclerophyll type

It seems paradoxical that the forests which produce vast amounts of fuel and generate the most fires, consist of fire-sensitive species. That they need to be burnt at some time in their seed-bearing life is axiomatic. The reason why these eucalypts frequently bear thin bark, lack lignotubers and coppice poorly may be related to the channelling of their energy resources towards the development of a superlative juvenile height growth rate.

At the moment it appears that the sheer flammability of the overstorey may be the secret of the success of the forest regeneration, a philosophy expressed by Mutch (1970). The thermal death point of tissues must be expressed in terms of both temperature and time of exposure. The protection offered by various tissues is chiefly concerned with delaying the penetration of heat to the vital areas. The flammability of the eucalypt crowns therefore may be expected to create conditions of explosive heat for a very short time. Under this regime, one may expect capsules to protect seed for this critically short period. If this is the explanation, the achievement is almost a miracle of timing. In one sense, the tolerance of the high forests to severe fire once every one, two or three centuries is because of their low resistance to it.

It would indeed be intriguing to know the state of the wet sclerophyll component in Pleistocene times before the arrival of humans. As pointed out by Jackson (1968), the incidence of fire was probably greater in the more arid post-Pleistocene times. Successional trends have been accelerated since the control of firing was imposed by forestry authorities 30–40 years ago. A similar successional picture is provided by the analogous forest fire regimes in the Pacific West Coast of North America. One would assume the fires in the drier times would have been initiated by lightning, aided in some areas by vulcanism or the exothermic reactions of cracked and desiccating peats. Possibly, fires were a much rarer event and the wet sclerophyll forests relatively scarce ecotones between rainforests and more drought-resistant eucalypt forests or woodlands. According to Webb (1970) the wet sclerophyll forests in the east, south and southwest of Australia are the special product of intense fires on a grand scale. In the northern areas, surface fires of lower intensity cause the gradual attrition of the rainforest margins. The boundaries in the former areas therefore are likely to be blurred; in the latter areas they are often unbelievably sharp.

The evidence of fire in Tertiary times in Australia has been deduced from fusain in lignites (Kemp, 1980). More recently, the identification of microscopic charcoal in Pleistocene deposits has revealed a long history of fire, albeit at relatively low levels of intensity. From about 40 000 years BP in the Lake George deposits near Canberra (Singh, Kershaw & Clark, 1980), there is a dramatic and sustained increase in charcoal frequency. This time coincides with the known earliest records of Aborigines in Australia. This is no mere coincidence, since these peoples were known to light fires, either accidentally in the course of their fire-stick culture (Mulvaney, 1969), or deliberately in their management and hunting of game. Some early writers at least were impressed with the skill with which they fired the country. Given the erratic unpredictable weather (even to modern meteorologists!)

in southern regions of the continent, it must be assumed that fires were an integral part of much of the country from this time.

Since the arrival of Europeans and the drive for pastures and minerals, the severity and frequency of large fires has been a normal consequence of most long periods of dry weather. Huge areas of high forest lie in a ruin of secondary scrub and testify to this blatant mismanagement. Pollen grains of *Eucalyptus* and other genera typical of wet sclerophyll forests have been traced back well into the last glaciation period (130 000 year BP) in northwestern Tasmania (E.A. Colhoun, personal communication). The assemblages of the pollen record over this time-span in both this area and Lake George, indicate oscillations compatible with wide climatic fluctuations involving both temperature and precipitation. Any decrease in precipitation is likely to result in a tendency for greater fire frequency and the shrinkage of rainforest areas to more sheltered niches. With the arrival of humans such shrinkages of rainforest need not necessarily be the result of climatic change, since fire frequency would be greater.

It seems likely that such eucalypts assert their growth supremacy only on good soils or on those whose fertility was sufficiently stimulated by fire. Only on the very poor quartzose soils in high rainfall areas of western Tasmania are both the complex rainforest and the sclerophyllous *E. nitida* forests of relatively poor stature.

It is possible that these tall eucalypts initially occupied extremely limited niches such as may be found on the slopes of rocky gullies or gorges where open conditions would have provided relief from competition. Certainly, fires would have allowed such fast-growing species to exploit their colonising potential and greatly expand their range.

Whatever the palaeo-fire regime in the wetter parts of Australia, it seems likely that it has been a major factor shaping the evolution of fast-growing, light-demanding eucalypts in close proximity to rainforest elements. Whether the arrival of humans on the continent has accelerated the rate of evolution of the wet sclerophyll eucalypt species as a consequence of increased firing is a question to be pondered upon, but probably never answered adequately.

References

Adams, M.A. & Atkinson, P.I. (1991). Nitrogen supply and insect herbivory in eucalypts: the role of nitrogen assimilation and transport processes. In

Productivity in Perspective: Third Australian Forest Soils and Nutrition Conference, ed. P.J. Ryan, pp. 239–41. Sydney: Forestry Commission of New South Wales.

Adams, M.A. and Attiwill, P.M. (1982). Nitrogen mineralization and nitrate reduction in forests. *Soil Biology and Biochemistry*, **14**, 197–202.

Adams, M.A. & Attiwill, P.M. (1984*a*). Role of *Acacia* spp. in nutrient balance and cycling in regenerating *Eucalyptus regnans* forests. I. Temporal changes in biomass and nutrient content. *Australian Journal of Botany*, **32**, 205–15.

Adams, M.A. & Attiwill, P.M. (1984*b*). Role of *Acacia* spp. in nutrient balance and cycling in regenerating *Eucalyptus regnans* forests. II. Field studies of acetylene reduction. *Australian Journal of Botany*, **32**, 217–23.

Adams, M.A. & Attiwill, P.M. (1986). Nutrient cycling and nitrogen mineralization in eucalypt forests of south-eastern Australia. II. Indices of nitrogen mineralization. *Plant and Soil*, **92**, 341–62.

Ashton, D.H. (1962). Some aspects of root competition in *E. regnans*. In *Proceedings of the Third General Conference Institute of Foresters of Australia*. Melbourne: Institute of Foresters of Australia.

Ashton, D.H. (1975*a*). The flowering behaviour of *Eucalyptus regnans* F. Muell. *Australian Journal of Botany*, **23**, 399–411.

Ashton, D.H. (1975*b*). Studies on the litter of *Eucalyptus regnans* F. Muell. forests. *Australian Journal of Botany*, **23**, 413–33.

Ashton, D.H. (1975*c*). The root and shoot development of *Eucalyptus regnans* F. Muell. *Australian Journal of Botany*, **23**, 867–87.

Ashton, D.H. (1975*d*). The seasonal growth of *Eucalyptus regnans* F. Muell. *Australian Journal of Botany*, **23**, 239–52.

Ashton, D.H. (1976*a*). The development of even-aged stands in *Eucalyptus regnans* F. Muell. in central Victoria. *Australian Journal of Botany*, **24**, 397–414.

Ashton, D.H. (1976*b*). Phosphorus in forest ecosystems at Beenak, Victoria. *Journal of Ecology*, **64**, 171–86.

Ashton, D.H. (1979). Seed harvesting by ants in forests of *Eucalyptus regnans* F. Muell. in central Victoria. *Australian Journal of Ecology*, **4**, 265–77.

Ashton, D.H. (1985). Viability of seeds of *Eucalyptus obliqua* and *Leptospermum juniperinum* from capsules subjected to a crown fire. *Australian Forestry*, **49**, 28–35.

Ashton, D.H. & Frankenberg, J. (1976). Ecological studies of *Acmena smithii* (Poir). Merrill and Perry with special reference to Wilson's Promontory. *Australian Journal of Botany*, **24**, 453–87.

Ashton, D.H. & Kelliher, K.J. The desiccation phenomenon in *Eucalyptus regnans* forest soil and its effect on seedling growth. *Australian Journal of Botany*, (submitted).

Ashton, D.H. & Turner, J.S. (1979). Studies on the light compensation point of *Eucalyptus regnans* F. Muell. *Australian Journal of Botany*, **27**, 589–607.

Attiwill, P.M. (1962). The effect of heat pre-treatment of soil on the growth of *E. obliqua* seedlings. In *Proceedings of the Third General Conference Institute of Foresters of Australia*. Melbourne: Institute of Foresters of Australia.

Attiwill, P.M. (1979). Nutrient cycling in a *Eucalyptus obliqua* L'Herit. forest. III. Growth, biomass and net primary production. *Australian Journal of Botany*, **27**, 439–58.

Attiwill, P.M. (1985). Effects of fire on forest ecosystems. In *Research for Forest Management*, eds J.J. Landsberg & W. Parsons, pp. 249–68. Melbourne: CSIRO.

Attiwill, P.M. (1991*a*). Productivity of *Eucalyptus regnans* forest regenerating after bushfire. In *Symposium on Intensive Forestry: The Role of Eucalypts, IUFRO P2. 02–01 Productivity of Eucalypts*, ed. A.P.G Schonau, pp. 494–504. Pretoria: Southern African Institute of Forestry.

Attiwill, P.M. (1991*b*). The disturbance of forested watersheds. In *Ecosystem Experiments*, eds H.A. Mooney, E. Medina, D.W. Schindler, E-D. Schulze & B.H. Walker, pp. 193–213. Chichester: John Wiley & Sons.

Attiwill, P.M. & Leeper, G.W. (1987). *Forest Soils and Nutrient Cycles*. Melbourne University Press.

Baker, T.G. & Attiwill, P.M. (1981). Nitrogen in Australian eucalypt forests. In *Proceedings, Australian Forest Nutrition Workshop: Productivity in Perpetuity*, pp. 159–72. Melbourne: CSIRO.

Beadle, N.C.W. (1962). Soil phosphate and the delimitation of plant communities in eastern Australia. *Ecology*, **43**, 281–8.

Beadle, N.C.W. & Costin, A.B. (1952). Ecological classification and nomenclature. *Proceedings of the Linnean Society of New South Wales*, **77**, 61–82.

Beetham, A.H. (1950). Aspects of forest practice in the regenerated areas of the Upper Latrobe Valley. Dip. For. thesis, Forests Commission of Victoria, Creswick.

Boerner, R.E.J. (1982). Fire and nutrient cycling in temperate ecosystems. *BioScience*, **32**, 187–92.

Bray, J.R. & Gorham, E. (1964). Litter production in forests of the world. *Advances in Ecological Research*, **2**, 101–52.

Brookes, J.D. (1949). The relation of vegetation cover to water yield in Victorian mountain watersheds. MSc thesis, University of Melbourne.

Burges, A. & Johnston, R.D. (1953). The structure of a New South Wales subtropical forest. *Journal of Ecology*, **41**, 72–83.

Chambers, D. & Attiwill, P.M. (1993*a*). The ashbed effect I. Chemical changes in soil after heat or partial sterilization treatment. *Australian Journal of Botany*, submitted.

Chambers, D. & Attiwill, P.M. (1993*b*). The ashbed effect II. Physical and microbiological changes in soil after heat or partial sterilization treatment. *Australian Journal of Botany*, submitted.

Christensen, P.E. (1971). Stimulation of seedfall in Karri. *Australian Forestry*, **35**, 182–90.

Committee of the South Gippsland Pioneers' Association (eds). (1920). *Land of the Lyrebird*. Melbourne: Gordon & Gotch.

Connor, D.J., Legge, N.J. & Turner, N.C. (1977). Water relations of mountain ash (*Eucalyptus regnans* F. Muell.) forests. *Australian Journal of Plant Physiology*, **4**, 753–62.

Cookson, I. (1954). The occurrence of an older Tertiary microflora in Western Australia. *Australian Journal of Science*, 17, 37–8.

Cremer, K.W. (1962). The effects of fire on eucalypts reserved for seeding. *Australian Forestry*, 26, 129–54.

Cremer, K.W. (1965*a*). Effects of fire on seed shed from *Eucalyptus regnans*. *Australian Forestry*, 29, 251–62.

Cremer, K.W. (1965*b*). Emergence of *Eucalyptus regnans* seedlings from buried seed. *Australian Forestry*, 29, 119–24.

Cremer, K.W. (1975). Temperature and other climatic influences on shoot development and growth of *Eucalyptus regnans*. *Australian Journal of Botany*, 23, 27–44.

Cremer, K.W. & Mount, A.B. (1965). Early stages of plant succession following the complete felling and burning of *Eucalyptus regnans* forest in the Florentine Valley, Tasmania. *Australian Journal of Botany*, 13, 303–22.

Crocker, R.L. (1959). Past climatic fluctuations and their influence upon Australian vegetation. In *Biogeography and Ecology in Australia* (Monographiae Biologicae VIII), eds A. Keast, R.L. Crocker & C.S. Christian, pp. 283–90. The Hague: W. Junk.

Cunningham, T.M. (1960*a*). The natural regeneration of *Eucalyptus regnans*. *School of Forestry, Univ. of Melbourne, Bulletin*, No. 1.

Cunningham, T.M. (1960*b*). Seed and seedling survival of *Eucalyptus regnans* and the natural regeneration of second-growth stands. *Appita*, 13, 124–31.

Cunningham, T.M. & Cremer, K.W. (1965). Control of the understorey in wet eucalypt forest. *Australian Forestry*, 29, 4–14.

Dunn, G.M. & Connor, D.J. (1991). Transpiration and water yield in mountain ash forest. Report, Faculty of Agriculture and Forestry, University of Melbourne.

Ellis, R.C. (1969). The respiration of the soil beneath some eucalypt forest stands as related to the productivity of the stands. *Australian Journal of Soil Research*, 7, 349–58.

Ellis, R.C. (1971). Growth of eucalypt seedlings on four different soils. *Australian Forestry*, 35, 107–18.

Evans, G., Cartwright, J.B. & White, N.H. (1967). The production of a phytotoxin, nectroline, by some root surface isolates of *Cylindrocarpon radicicola* wr. *Plant and Soil*, 26, 253–60.

Florence, R.G. & Crocker, R.L. (1962). Analysis of blackbutt (*E. pilularis* Sm.) seedling growth in a blackbutt forest soil. *Ecology*, 43, 670–9.

Floyd, A.G. (1966). Effect of fire upon weed seeds in the wet sclerophyll forests of northern New South Wales. *Australian Journal of Botany*, 14, 243–56.

Floyd, A.G. (1976). Effect of burning on regeneration from seeds in wet sclerophyll forest. *Australian Forestry*, 39, 210–20.

Frankcombe, D.W. (1966). The regeneration burn. *Appita*, 19, 127–32.

Galbraith, A.V. (1937). *Mountain Ash* (Eucalyptus regnans F. Muell.). *A General Treatise on its Silviculture, Management and Utilization*. Melbourne: Government Printer.

Gardner, C.A. (1942). The vegetation of Western Australia with special reference to the climate and soils. *Journal of the Royal Society of Western Australia*, **28**, 11–87.

Gilbert, J.M. (1959). Forest succession in the Florentine Valley, Tasmania. *Papers and Proceedings of the Royal Society of Tasmania*, **93**, 129–51.

Gill, A.M. (1975). Fire and the Australian flora. *Australian Forestry*, **38**, 4–25.

Gill, A.M. (1981). Patterns and processes in open-forests of *Eucalyptus* in Southern Australia. In *Australian Vegetation*, ed. R.H. Groves, pp. 152–76. Cambridge University Press.

Gill, A.M. & Ashton, D.H. (1968). The role of bark type in relative tolerance to fire of three central Victorian eucalypts. *Australian Journal of Botany*, **16**, 491–8.

Gloury, S.J. (1978). Aspects of fire resistance in three species of *Eucalyptus* L'Herit. with special reference to *E. camaldulensis* Denh. BSc (Hons) thesis, University of Melbourne.

Greaves, R. (1966). Insect defoliation of eucalypt regrowth in the Florentine Valley, Tasmania. *Appita*, **19**, 119–26.

Grose, R.J. (1960). Effective seed supply for the natural regeneration of *Eucalyptus delagatensis* R.T. Baker, syn. *Eucalyptus gigantea* Hook.f. *Appita*, **13**, 141–8.

Grose, R.J. (1963). The silviculture of *Eucalyptus delegatensis*. 1. Germination and seed dormancy. *School of Forestry, Univ. of Melbourne, Bulletin* No. 2

Grove, T.S. & Malajczuk, N. (1991). Nodule production and nitrogen fixation (acetylene production) by an understorey legume (*Bossiaea laidlawiana*) in eucalypt forest. *Journal of Ecology* (in press).

Hansen, A.P. & Pate, J.S. (1987). Comparative growth and symbiotic performance of seedlings of *Acacia* spp. in defined pot culture or as natural understorey components of a eucalypt forest ecosystem in S.W. Australia. *Journal of Experimental Botany*, **38**, 13–25.

Hansen, A.P., Stoneman, G. & Bell, D.T. (1988). Potential inputs of nitrogen by seeder legumes to the jarrah forest ecosystem. *Australian Forestry*, **51**, 226–31.

Hardy, A.D. (1935). Australia's great trees. *Victorian Naturalist*, **51**, 231–41.

Hatch, A.B. (1960). Ash bed effects in Western Australian forest soils. *Forests Department of Western Australia Bulletin* No. 64

Hooker, J.D. (1860). *The Botany (of) the Antarctic Voyage*, Part III. *Flora Tasmaniae*, Vol. 1. London: Lovell Reeve.

Hopkins, E.R. (1964). Water availability in mixed species *Eucalyptus* forest. PhD thesis, University of Melbourne.

Howard, T.M. (1970). The ecology of *Nothofagus cunninghamii*. PhD thesis, University of Melbourne.

Howard, T.M. (1973). Studies in the ecology of *Nothofagus cunninghamii*. I. Natural regeneration on the Mt Donna Buang massif, Victoria. *Australian Journal of Botany*, **21**, 67–78.

Howard, T.M. (1981). Southern closed-forests. In *Australian Vegetation*, ed. R.H. Groves, pp. 102–20. Cambridge University Press.

Howard, T.M. & Ashton, D.H. (1973). The distribution of *Nothofagus cunninghamii* rainforest. *Proceedings of the Royal Society of Victoria*, **86**, 47–76.

Institute of Foresters of Australia. (1976). Newsletter no. 17, p. 1.

Jackson, W.D. (1965). Vegetation. In *Atlas of Tasmania*, ed. J.L. Davies, pp. 30–5. Hobart: Lands & Survey Department.

Jackson, W.D. (1968). Fire, air, water and earth – an elemental ecology of Tasmania. *Proceedings of the Ecological Society of Australia*, **3**, 9–16.

Jacobs, M.R. (1955). *Growth Habits of the Eucalypts*. Canberra: Commonwealth Government Printer.

Judd, T.S. & Ashton, D.H. (1991). Fruit clustering in the Myrtaceae: seed survival in capsules subjected to experimental heating. *Australian Journal of Botany*, **39**, 241–5.

Kemp, E. (1980). Evolution of the Australian biota in relation to fire. In *Fire and the Australian Biota*, eds A.M. Gill, R.H. Groves & I.R. Noble, pp. 3–21. Canberra: Australian Academy of Science.

King, A.R. (1963). *The Influence of Colonisation on the Forests and the Prevalence of Bush Fires in Australia*. Melbourne: CSIRO, Australia, Division of Applied Chemistry (mimeographed report).

Lamb, D. (1985). The influence of insects on nutrient cycling in eucalypt forests: a beneficial role? *Australian Journal of Ecology*, **10**, 1–5.

Lambert, M.J. (1981). *Inorganic constituents in wood and bark of New South Wales forest tree species*. Forestry Commission of New South Wales Research Note No. 45.

Lambert, M.J. & Turner, J. (1991). Nutrient cycling in *Eucalyptus grandis* forests on the New South Wales north coast. In *Productivity in Perspective: Third Australian Forest Soils and Nutrition Conference*, ed. P.J. Ryan, pp. 198–9. Sydney: Forestry Commission of NSW.

Langford, K.J. (1974). *Change in Yield of Water Following a Bushfire in a Forest of Eucalyptus regnans*. Melbourne & Metropolitan Board of Works Departmental Report No. 31.

Leeper, G.W. (1970). Climates. In *The Australian Environment*, 4th edn (rev.), ed. G.W. Leeper, pp. 12–20. Melbourne: CSIRO & Melbourne University Press.

Lockett, E.J. & Candy, S.G. (1984). Growth of eucalypt regeneration established with and without slash burns in Tasmania. *Australian Forestry*, **47**, 119–25.

Loneragan, O.W. & Loneragan, J.F. (1964). Ashbed and nutrients in growth of karri seedlings. *Journal of the Royal Society of Western Australia*, **47**, 74–80.

Luke, R.H. (1961). *Bush Fire Control in Australia*. Melbourne, Sydney & London: Hodder & Stoughton.

Luke, R.H. & McArthur, A.G. (1978). *Bushfires in Australia*. Canberra: Australian Government Publishing Service.

Lutz, H.J. & Chandler, R.F. Jr. (1947). *Forest Soils*. New York: John Wiley & Sons Inc.

McArthur, W.M. & Clifton, A.L. (1975). *Forestry and Agriculture in Relation to Soils in the Pemberton Area of Western Australia*. Soils and Land Use Series No. 54. Melbourne: CSIRO.

McColl, J.B. (1966). Soil-plant relationships in eucalypt forest. *Ecology*, 50, 355–62.

Melick, D.R. & Ashton, D.H. (1991). The effects of disturbance on warm temperate rainforests in southeastern Australia. *Australian Journal of Botany*, 39, 1–30.

Mount, A.B. (1969). Eucalypt ecology as related to fire. *Proceedings of the 9th Tall Timbers Fire Ecology Conference*, 75–108.

Mulvaney, D.J. (1969). *The Prehistory of the Aborigines*. London: Thames & Hudson.

Mutch, R.W. (1970). Wildland fires and ecosystems – a hypothesis. *Ecology*, 51, 1046–51.

Neumann, F.G., Harris, J.A. & Wood, C.H. (1977). The phasmatid problem in mountain ash forests of the Central Highlands of Victoria. *Forests Commission of Victoria Bulletin* No. 25.

O'Connell, A.M. & Grove, T.S. (1991). Processes contributing to the nutritional resilience or vulnerability of jarrah and karri forests in Western Australia. In *Productivity in Perspective: Third Australian Forest Soils and Nutrition Conference*, ed. P.J. Ryan, pp. 180–97. Sydney: Forestry Commission of NSW.

O'Connell, A.M. & Menage, P.M.A. (1982). Litter fall and nutrient cycling in karri (*Eucalyptus diversicolor* F. Muell.) forest in relation to stand age. *Australian Journal of Ecology*, 7, 49–62.

Ohmart, C.P., Stewart, L.G. & Thomas, J.R. (1983). Leaf consumption by insects in three *Eucalyptus* forest types in southeastern Australia and their role in short term nutrient cycling. *Oecologia*, 59, 329–30.

O'Shaughnessy, P.J. & Jayasuriya, M.D.A. (1987). Managing the ash type forests for water production in Victoria. In *Forest Management in Australia: Proceedings 1987 Conference Institute of Foresters, Australia*, pp. 437–63. Perth: Institute of Foresters, Australia.

Polglase, P.J. & Attiwill, P.M. (1992). Nitrogen and phosphorus cycling in relation to stand age of *Eucalyptus regnans* F. Muell. I. Return from plant to soil in litterfall. *Plant and Soil*, 142, 151–66.

Polgase, P.J., Attiwill, P.M. & Adams, M.A. (1986). Immobilization of soil nitrogen following wildfire in two eucalypt forests of south-eastern Australia. *Oecologia Plantarum*, 7, 261–72.

Polglase, P.J., Attiwill, P.M. & Adams, M.A. (1992). Nitrogen and phosphorus cycling in relation to stand age of *Eucalyptus regnans* F. Muell. II. N mineralization and nitrification. *Plant and Soil*, 142, 167–76.

Pryor, L.D. (1963). Ashbed response as a key to plantation establishment on poor sites. *Australian Forestry*, 27, 48–51.

Renbuss, M.A., Chilvers, G.A. & Pryor, L.D. (1972). Microbiology of an ashbed. *Proceedings of the Linnean Society of New South Wales*, 97, 302–10.

Richards, B.N., Smith, J.E.N., White, G.J. & Charley, J.L. (1985). Mineralization of soil nitrogen in three forest communities from the New England region of New South Wales. *Australian Journal of Ecology*, 10, 429–41.

Select Committee, (1876). *Report on Water Reserves, Plenty Ranges*. Victoria Parliamentary Papers Sll.

Shirrefs, P.V. (1977). Interaction between *Eucalyptus regnans* F. Muell. and *Acacia dealbata* Link. BSc (Hons.) thesis, University of Melbourne.

Singh, G., Kershaw, A.P. & Clark, R. (1980). Late Quaternary vegetation and fire history in Australia. In *Fire and the Australian Biota*, eds A.M. Gill, R.H. Groves & I.R. Noble, pp. 23–54. Canberra: Australian Academy of Science.

Specht, R.L. (1970). Vegetation. In *The Australian Environment*, 4th edn (rev.), ed. G.W. Leeper, pp. 44–67. Melbourne: CSIRO & Melbourne University Press.

Stretton, L.E.B. (1939). *Report to the Royal Commission to Enquire into the Causes and the Measures taken to Prevent the Bushfires of January, 1939 and to Protect Life and Property.* Melbourne: Government Printer.

Toumey, J.W. & Korstian, C.F. (1947). *Foundations of Silviculture upon an Ecological Basis.* New York: John Wiley & Sons Inc.

Tracey, J.G. (1982). *The Vegetation of the Humid Tropical Region of North Queensland.* Melbourne: CSIRO.

Turner, J. & Lambert, M.J. (1983). Nutrient cycling within a 27-year-old *Eucalyptus grandis* plantation in New South Wales. *Forest Ecology and Management*, 6, 155–68.

Underwood, R.J. (1973). *Natural Fire Production in the Karri* (E. diversicolor F. Muell.) *forests.* Forests Department of Western Australia Research Paper No. 41.

Unwin, G.N. (1989). Structure and composition of the abrupt rainforest boundary in the Herberton Highland, north Queensland. *Australian Journal of Botany*, 37, 413–28.

Unwin, G.N., Stocker, G.C. & Sanderson, K.D. (1985). Fire and the forest ecotone in the Herberton Highland, north Queensland. *Proceedings of the Ecological Society of Australia*, 13, 215–24.

Van Loon, A.P. (1969). *Prescribed Burning in Blackbutt Forests.* Forests Commission of New South Wales Research Notes No. 23.

Vines, R.G. (1968). Heat transfer through the bark and the resistance of trees to fire. *Australian Journal of Botany*, 16, 499–514.

Vines, R.G. (1975). Bushfire research in CSIRO. *Search*, 6, 73–8.

Vitousek, P.M., Gosz, J.R., Melillo, J.M. & Reiners, W.A. (1982). A comparative analysis of potential nitrification and nitrate mobility in forest ecosystems: a field experiment. *Ecological Monographs*, 52, 155–77.

Vogt, K.A., Grier, C.C. & Vogt, D.J. (1986). Production, turnover and nutrient dynamics of above- and belowground detritus of world forests. *Advances in Ecological Research*, 15, 303–77.

Warcup, J.H. (1983). Effect of fire and sun-baking on the soil microflora and seedling growth in forest soils. In *Soils: An Australian Viewpoint*, Division of Soils, CSIRO, pp. 735–40. Melbourne & London: CSIRO & Academic Press.

Webb, L.J. (1959). A physiognomic classification of Australian rainforests. *Journal of Ecology*, 47, 551–70.

Webb, L.J. (1970). Fire environments in eastern Australia. In *Proceedings 2nd Fire Ecology Symposium, Monash University & Forests Commission of Victoria.*

Webb, L.J. & Tracey, J.G. (1981). The rainforests of northern Australia. In *Australian Vegetation*, ed. R.H. Groves, pp. 67–101. Cambridge University Press.

Webb, L.J., Tracey, J.G., Williams, W.T. & Lance. G.N. (1969). Pattern of mineral return in leaf litter of 3 sub-tropical Australian forests. *Australian Forestry*, 33, 99–110.

Webb, R.N. (1966). The protection of eucalypt seed from fire by their capsules. BSc (Hons.) thesis, University of Melbourne.

Weston, C.J. & Attiwill, P.M. (1990). Effects of fire and harvesting on nitrogen transformations and ionic mobility in soils of *Eucalyptus regnans* forests of south-eastern Australia. *Oecologia*, 83, 20–6.

Williams, R.J. (1978). The comparative strategies of water use in selected xeromorphic and mesomorphic species at Lilly Pilly Gully, Wilson's Promontory, Victoria. BSc (Hons) thesis, University of Melbourne.

Willis, R.J. & Ashton, D.H. (1978). The possible role of soil lipids in the regeneration problem of *Eucalyptus regnans* F. Muell. In *Proceedings of the Open Forum on Ecological Research, Macquarie University, Bulletin of the Ecological Society of Australia*, 8, 11.

Zimmer, W.J. & Grose, R.J. (1958). Root systems and root/shoot ratios of some Victorian eucalypts. *Australian Forestry*, 22, 13–18.

7

Patterns and processes in open-forests of Eucalyptus *in southern Australia*

A.M. GILL

THE MOST species-rich genera of Australia, *Eucalyptus* and tralian flora but in the 200 year history of post-European settlement communities. *Eucalyptus* was novel to the early students of the Australian flora but in the 200 year post history of European settlement of the continent, approximately 700 species have been recognised (Brooker & Kleinig, 1990). *Acacia* (*sens.lat.*) was not a new genus to science when European settlement began near Sydney in 1788 but approximately 1000 new Australian species have been discovered subsequently (M.D. Tindale, personal communication).

Eucalyptus and *Acacia* may be regarded as genera typifying the Australian woody flora. Both are widespread; both have high levels of specific endemism, and both are conspicuous dominants of major plant formations. In more xeric areas, *Acacia* is often the community dominant (Chapter 9 and Johnson & Burrows, 1981), whilst in more mesic areas *Eucalyptus* is dominant (Chapter 6 and Ashton, 1981; Chapter 8 and Gillison & Walker, 1981). The latter dominance reaches an extreme in the open-forests of *Eucalyptus* in southern Australia but *Acacia* retains a strong representation in the understorey.

In this chapter, these uniquely Australian forests will be described in relation to patterns of distribution and processes of community operation. Firstly, however, I shall outline the development of the concept of 'open-forest' and define the forests I shall consider below.

The 'open-forest' concept

Diels (1906) began the difficult journey into the classification of Australia's eucalypt forests; the journey continues. Diels (1906) described the eucalypt forests of southwestern Australia as 'Sklerophyllen-

Wald' (sclerophyll forest) apparently using the adjective 'sclerophyll' to define leaf textures found in the understorey plants rather than those of the eucalypt canopy (Specht, 1970; cf. Beadle & Costin, 1952; Williams 1955; Webb, 1959).

A quarter of a century after Diels, Prescott (1931) introduced the term 'wet sclerophyll forest' for the forests of better site-qualities, particularly in Western Australia and Victoria (dominated by *Eucalyptus diversicolor* and *E. regnans* respectively). Another South Australian author, Wood (1937), later identified the two subdivisions of the sclerophyll forests as 'wet' and 'dry'; the 'wet' forests occurred in areas with more than 100 cm annual rainfall and contained tree-ferns, whilst the 'dry' forests were in areas with less than 100 cm rainfall and contained many sclerophyllous shrubs.

Beadle & Costin (1952) reviewed the classification of Australian plant communities and although they did not attempt to delimit forest communities by rainfall they retained the use of the terms 'wet' and 'dry' to describe these forests; they noted the occurrence of meso-morphic shrubs in the 'wet' sclerophyll forest and xeromorphic shrubs in the 'dry'. In addition, they included a 'swamp' sclerophyll forest with an herbaceous understorey of helophytes. The latter term was not taken up generally and did not appear in the subsequent classification of Williams (1955).

A major change in nomenclature, along with refined definitions, came with Specht (1970) who prepared his new system in conjunction with a number of other Australian ecologists. 'Sclerophyll forest' became 'open-forest' which was defined as a community dominated by trees and which had a canopy cover of between 30 and 70 per cent. Three major subcategories were then distinguished according to the height of the dominants: 'tall open-forest' was greater than 30 m tall and equivalent to the forest described as 'wet sclerophyll' by earlier authors; 'open-forest' was 10 to 30 m tall and equivalent both to the earlier 'dry sclerophyll forest' and to the layered open-forest of Williams (1955); 'low open-forest' was used to describe forests 5 to 10 m tall and largely composed of brigalow (*Acacia harpophylla*) as described in Chapter 9 (Johnson & Burrows, 1981).

In this chapter, the subcategory of forests called 'open-forests' by Specht (1970) will be considered but without his 'layered open-forest' phase. Basically, the forests considered will be the shrubby and grassy phases of the open-forest subformation, a grouping which roughly corresponds to the 'dry sclerophyll forests' of earlier authors. Confusion can occur in nomenclature because the formation-level 'open-forest' and the subformation 'open-forest', are indistinguishable by name. For the rest of the review 'open-forest' refers to the subformation *not* the formation and does not include the 'layered open-forest' phase of the subformation, or the tall open-forest of the previous chapter.

Patterns of distribution in southern Australia

Broadly speaking, we can define three major zones of occurrence of the open-forests in southern Australia. One is in coastal and subcoastal southeastern Australia from Adelaide to Brisbane; the second is in Tasmania; and the third is in the southwestern corner of Western Australia. In each zone, the wetter limits are marked by tall open forests; the drier, more fertile limits by grassy woodlands; the poorer-soil margins tend to abut shrubby woodlands; and sedges and other wetland plants may become prominent on poorly drained margins. Below, three regions have been chosen for closer scrutiny. In each case the open-forest is associated with a variety of other communities and sufficient is known of the factors controlling distribution to warrant a brief review. The regions to be considered are: the Sydney Basin of New South Wales; the Eastern Highlands of Victoria; and the jarrah forest region in the vicinity of Perth, Western Australia. In general, open-forest occurs where the mean annual rainfall is greater than 600 mm.

Sydney Basin

The Sydney Basin is a geologically defined region of Permian and Triassic shales and sandstones (Packham, 1962). In the Blue Mountains to the west of Sydney, the topography is of extensive, gently undulating plateaux dissected by gorges and valleys with steep and colourful cliffs. The horizontally bedded Hawkesbury sandstone often forms the plateau surface. It is underlain by the Narrabeen sandstones and shales but overlain by the Wianamatta shales. Together these three units comprise the Triassic sequence of the region. Study of the vegetation of these units led Beadle (1954, 1962) to suggest that the mosaic of vegetation formations and subformations in eastern Australia was controlled by soil phosphate levels, which, in turn, were a reflection of the phosphate status of the parent materials from which the soils developed.

Beadle (1962) found that where soil phosphate levels were very low in the Sydney region, such as on the Hawkesbury sandstones or on truncated lateritic profiles, 'xeromorphic woodlands and scrubs' occurred (Table 7.1). Where phosphorus levels were higher, however, as in complete lateritic profiles, 'dry sclerophyll' or 'open-forest' was found (Table 7.1). Even higher phosphate levels occurred on the Narrabeen shales and tall open-forests occurred there (Table 7.1), such as are described in the previous chapter. Rainforests were found on the most phosphorus-rich soils of the Narrabeen shales and sandstones (Table 7.1).

In passing, the reader's attention is drawn to the fact that Australian soils are extremely infertile. Phosphorus, particularly, occurs in very low amounts and the addition of superphosphate from the 1880s onwards has had an important effect on raising agricultural yields (Donald & Prescott, 1975). Both the relative stability and antiquity of landscapes and millions of years of leaching perhaps explain the inherent poverty of Australian soils (Donald & Prescott, 1975).

Beadle's (1962) results were for communities on soils of unimpeded drainage. In the region of his study, however, areas of impeded drainage occur as swamps (Davis, 1936) and these have their own distinct flora. Even within the open-forest, local seepages are found where ferns and sedges may be locally abundant.

Other variations in the vegetation of the region may be because of soil depth. Davis (1936), for example, distinguished various stages of primary succession. Some of this variation is not only evident in large patches of seral communities but is also reflected in variations in understorey composition in the open-forest.

Thus, there are wide differences in the vegetation of the region

Table 7.1. *Mean values for surface-soil phosphorus according to substrate and vegetation (data from Beadle, 1962); 'depauperate rainforest' is omitted*

	Hawkesbury sandstone	Laterite	Wianamatta shales	Narrabeen shales and sandstones
'Xeromorphic woodland/ scrub'	37	37	—	—
Open-forest	0	83	—	—
Tall open-forest	—	—	139	180–201
'Rainforest'	—	—	—	430

associated with soil phosphate levels, soil drainage, and soil depth (moisture status?). A statistical survey of the causes of this wide variation was conducted by Burrough, Brown & Morris (1977) across an area of Hawkesbury sandstone about 110 km SSW of Sydney (between Fitzroy Falls and Barren Grounds (34°41′S; 150°44′E)). These communities ranged from sedgelands to heath to woodland to forest. Although underlain by the one geologic substratum (cf. Beadle, 1962), Burrough *et al.* (1977) noted that the distribution of heathlands and sedgelands appeared to be under the control of soil drainage as described by Pidgeon (1937, 1938, 1940) and Davis (1936, 1941), but they devoted most of their attention to plant communities of the well-drained and gentle middle to upper slopes of the landscape.

Across the survey region of Burrough *et al.* (1977), vegetation height and species composition were strongly correlated with mean annual rainfall, thereby suggesting that the availability of soil moisture was an important discriminant. Local variations in this vegetation pattern were interpreted in terms of local variations in effective rainfall or in soil drainage. It was not clear to them whether or not minor variations in vegetation were because of soil texture and porosity differences or to soil fertility. If the latter were the case, Burrough *et al.* (1977) thought that a combination of nutrients was probably involved rather than any particular one.

These studies in the Sydney Basin suggest that the open-forests of the region occur in a habitat with well-drained soils, moderate phosphorus content (for Australia) and moderate rainfall. Areas of poorer drainage may be occupied by sedgelands or heaths. Relatively rich soils with the most favourable soil-moisture conditions will have the largest number of rainforest species (Beadle, 1962), and the well-drained soils of the lowest fertility support 'xeromorphic woodland and scrub'.

Eastern Highlands of Victoria

As their name suggests, the Eastern Highlands are a geomorphic unit of the Victorian landscape. They consist of hills, mountains and, at the highest elevations, of plateaux remnants called 'high plains'. The Highlands are complex structurally and consist mainly of Palaeozoic sedimentary, igneous and metamorphic rocks (Hills, 1955).

In addition to the topographic and geologic complexity, the area has variable temperature and rainfall regimes. Temperatures drop with increasing altitude, which ranges from about 200 m to 2000 m.

Rainfall varies from about 650 mm to more than 1500 mm and heavy snowfalls may occur in the higher country.

The environmental complexity of the region is reflected in the vegetation. To appreciate this complexity, some examples of the effect of aspect will be given later, but here the broad changes with altitude will be described. Because precipitation tends to increase with increasing altitude, the cline described tends to reflect the gradient in both precipitation and temperature. The example chosen is from Rowe (1967); another example, but from southeastern New South Wales, is that of Keith and Sanders (1990).

At the highest elevations are a variety of alpine and sub-alpine communities – fjaeldmark, herbfield, grassland, bog and woodland (see Chapter 16 and Costin, 1981). Tall open-forests and woodlands occur in the next elevational bracket (750–1400 m), the woodlands tending to occur on the drier sites. The open-forests occur at lower elevations and give way to woodlands at still lower elevations. The open-forest here is associated with a 650 to 900 mm annual rainfall on well-drained soils.

Two examples of the importance of aspect are given by Ashton (1976*a,b*). The first is at the dry end of the open-forest spectrum and occurs just outside the Eastern Highlands at Mt Piper (37°12′S; 145°00′E) in central Victoria; rainfall is about 600 mm per year. The second example occurs at the wetter end of the open-forest spectrum and occurs at Beenak, 65 km east of Melbourne; annual rainfall is on average about 1400 mm per year. Below, each of these examples is considered briefly.

The study area at Mt Piper (Ashton, 1976*a*) is a conical hill rising about 230 m above the general surroundings. The form of this landscape feature is because of the indurated quartz capping of this generally sandstone structure. Soils are skeletal. The vegetation is open-forest and woodland. On deeper soils and protected aspects the open-forest is well developed, but on the rockier sites, exposed to greater insolation, the vegetation approaches woodland form. Whilst the contrast is not great between woodland and forest, the example suggests that aspect (insolation and moisture status) and soil depth affect the distribution of these vegetation types across what is really a continuum of types.

The second example is one in which contrasts are very strong. The Beenak study site (Ashton, 1976*b*) is a valley with two contrasted aspects with 15 to 20° slopes to the north and south. On small alluvial flats along the valley are closed-forests of *Nothofagus, Atherosperma* and *Acacia melanoxylon* up to 30 m in height, of the type described by Howard (1981) and in Chapter 5. The south-facing aspect is dominated by tall open-forest up to 66 m in height and the north-facing aspect is covered by open-forest (although it reaches 45 m in height in places, which is greater than Specht (1970) considered for this type). Underlying the whole area is granite and the soils are generally deep. Apart from the differences in insolation and presumably moisture status of the soil, Ashton (1976*b*) found that the taller eucalypt forests occur on soils with higher fertility and that the shorter eucalypt forests are on phosphorus-deficient soils. These results, and similar results of Howard (1973), support Beadle's (1962) contention that soil phosphorus is an important variable associated with the delimitation of tall open-forests and open-forests as well as that of closed-forest and 'open-forest' (at the formation level).

Southwest Western Australia

This region has considerable interest since it is a vegetational and floristic outlier. The forest formations are widely separated from those in eastern Australia and the flora has 75 per cent regional endemism at the species level (Wilson, 1973). Endemism is greatest in heathlands (Wilson, 1973) but is a feature of the forests as well (as we shall see later). Apart from forests and heathlands, the region supports areas of 'shrubby open-scrub' and 'grassy open-scrub, (Specht, 1970), previously known as 'mallee' and described in Chapter 10 and Parsons (1981).

Physiographically (Mulcahy, 1973), the area is dominated by the Archaean shield which forms a low plateau delineated by the Darling scarp on the west and separated from the sea by a coastal plain of Quaternary sediments. To the south, the shield slopes gently to the sea and is covered with a discontinuous and thin veneer of Tertiary and Recent sediments. Areas of the shield are often mantled by laterites.

Tall open-forests of *E. diversicolor* occur in the far southwestern corner of the region where high rainfall (> 750 mm per year) and substrates such as gneiss, granite, limestone, laterite and sand occur (Churchill, 1968). Similarly, a variety of soils support open-forests of *E. marginata* which continue into areas as far east as the 380 mm rainfall isohyet (Churchill, 1968) where the vegetation becomes a woodland. On the southern coastal plain, where soils are derived from limestone, an open-forest of *E. gomphocephala* occurs (Wilson, 1973).

Considerable local variation is obscured by regional descriptions which can give a false impression of the nature and complexity of landscape variation. To illustrate, some of this variety on the coastal plain just north of Perth is described. Hopkins (1960) found that *E. gomphocephala* forest occurred on shallow soils overlying limestone whilst mixed forests of *E. marginata* and *E. calophylla* were found where coffee rock or yellow sand was less than 2 m from the surface. *Eucalyptus* forests were absent from the deep sands and swamps also found in the area.

On the plateau, the forest canopy varies in composition along topographic, climatic and edaphic gradients (Havel, 1975*a*). *Eucalyptus marginata* is dominant on the nutritionally poor gravels of lateritic origin but it does not occur in the loamy soils of relatively high fertility, either in the dry northeast or the moister southwest (Havel, 1975*b*). It is replaced by *E. wandoo* woodlands in the northeast, probably because of inadequate soil-moisture storage, and in the southwest by *E. diversicolor* (tall open-forest) which has a greater growth potential on soils of high fertility (Havel, 1975*b*). Many local patterns occur within the forest according to local variation in topographic and edaphic conditions and have been described for the northern part of the range of *E. marginata* by Havel (1975*a,b*).

Local and regional patterns of distribution of dominant species

The eucalypt dominance of the open-forests of this chapter is almost complete. *Eucalyptus* consists of a number of informally described subgenera (Pryor & Johnson, 1971), three of which are most common in northern Australia and two of which, *viz*. Monocalyptus and Symphyomyrtus, are particularly common in the forested regions of southern Australia (Gill, Belbin & Chippendale, 1985). Ecologically significant differences in subgenera, such as differences in leaf chemistry, root morphology and seedling growth rates, are becoming apparent (Noble, 1989).

In eastern Australia, eucalypt species (and sometimes the closely-related *Angophora*) of open-forests usually grow in association (Fig. 7.1). This is borne out by vegetation surveys in which forest types may be known by the mixture of dominant eucalypt species, e.g. *E. pilularis–E. maculata* (Baur, 1962), *E. sieberana–E. gummifera–E. piperita* (Beadle & Costin, 1952), *E. fastigata–E. cypellocarpa* (Austin, 1978) and *E. obliqua–E. radiata* (Gill & Ashton, 1971). Although this is common it is not ubiquitous, e.g. *E. pauciflora* may

Fig. 7.1. Mixed-species forest of *Eucalyptus* on the southern coast of New South Wales (*E. maculata*, *E. paniculata* and *E. pilularis*).

form pure stands in forests abutting sub-alpine woodland of the same species. In Western Australia the pure stands seem to be common with the major types being dominated by *E. marginata* or *E. gomphocephala*. In New South Wales, results of surveys reported by Baur (1962, 1965) and Costin (1954) suggest that more than half the 'dry sclerophyll forests' required two or more eucalypt species to define them. Such data suggest high proportions for two species in many stands, whilst those designated by one eucalypt species only may have associates, but in smaller proportions.

Pryor (1959) emphasised the importance of breeding barriers between cohabiting species of *Eucalyptus* if the association is to persist. He found that genetic barriers occurred between subgeneric groups of the genus but that the great majority of species within any subgroup could interbreed freely. This suggested, then, that species in association should belong to different subgenera. Using data of Baur (1962, 1965) and Costin (1954) for species associations in 'dry sclerophyll forests', I found that whilst many associations involved species from different subgenera, about 40 per cent had species associates from the same subgenus. Austin, Cunningham and Wood (1983) tested 'Pryor's Rule' statistically and found that, as Pryor had stated, combinations of species from different subgenera do not occur at random. An example where species from the same subgenus occur widely in Victoria is the *E. obliqua–E. radiata* association (both Monocalyptus, Pryor & Johnson, 1971). Investigating this in the Hume Range (37°26′S; 145°08′E), Gill (1966) found that flowering of these species was separated in time, a mechanism which effectively prevented interbreeding. Observations of the same species association elsewhere in Victoria showed that flowering times could coincide but that successful cross pollination was rare or absent (D.H. Ashton, personal communication). Thus Pryor's suggestion that barriers to breeding were necessary was upheld, but that these could be ecological as well as genetic was suggested. An alternative explanation of the association of species in the same subgenus is that interbreeding could occur but that any hybrid seedling formed would be unsuccessful in establishment. Evidence for such a possibility comes from the observation that hybrids typically occur where the habitat is disturbed (Pryor & Johnson, 1971), thereby implying that any habitat patchiness, favouring parental progeny, has been upset.

Another precondition for cohabitation is that tolerances of species overlap in such a way that the habitat is suited to both. That tolerances overlap is perhaps obvious but the way in which they overlap is not; an array of variables is involved. Part of this array of overlapping factors has been shown by the arrangement of species in relation to: temperature and moisture (Morland, 1959; Rundle, 1977); altitude and radiation (Austin, 1978); nutrients and drainage classes (Gibbons & Downes, 1964); and geological substrate, radiation, rainfall and temperature (Austin, Nicholls & Margules, 1990). Overlapping tolerances are reflected in the many combinations of occurrences that a group of species may form in a particular region (Costin, 1954).

Demonstration of the site differences between cohabiting species which is evident from their regional distributions has proved to be very difficult experimentally. Thus, whilst Gill & Ashton (1971) found differences in the proportions of species in different height categories within forest stands in central Victoria, with *E. radiata* more 'tolerant' (in the forestry sense, Baker, 1950) than *E. obliqua*, experimental demonstration of this in the glasshouse was not achieved, although responses to shade, levels of mineral nutrition and drought were examined. The wide variability in seedling progeny, the difficulties of simulating the drastic natural selection of the field, and the fact that most regeneration appeared to be associated with post-fire environments, may explain this result. Other explanations could be that the factors examined were not critical, or that natural fine-grained variation in the field was not duplicated. Perhaps biotic rather than physical factors were more important in some cases.

Biotic explanations for cohabitation of eucalypts have been sought by Burdon & Chilvers (1974) and Morrow (1976). In both cases, leaf damage by invertebrate herbivores or fungi was found to be significant and host specificity was evident among certain groups of the damaging agents. Thus, theoretically, it was possible that compensatory host responses to waves of different host-specific predators could account for the cohabitations observed. Morrow (1976) found that in the absence of insects, one species, *E. stellulata*, had a higher growth rate than the dominant species, *E. pauciflora*. In the natural state, however, where insects were prevalent, *E. stellulata* had only reached half the height of *E. pauciflora* in 35 years. It appeared therefore that the greater intrinsic growth rate of *E. stellulata* in the subalpine environment of these experiments was balanced by the greater tolerance of *E. pauciflora* to insect attack. Indeed, the ability to cope with insect attack seemed to have led to the dominance of *E. pauciflora*.

Eucalypts are considered to have a remarkable sensitivity to habitat (Pryor, 1959) and this is often associated with ecotypic differentiation. Both these observations suggest that eucalypts have high genetic flexibility in time and space. Although gene transport through seeds is for short distances only, e.g. approximately equal to tree height (Grose, 1960), that through pollen may take place over longer distances (Christensen, 1971). With compatible ecotypes (along a cline) or species (of the same taxonomic subgenus) pollen transfer and natural selection can provide the means for high genetic sensitivity to environmental change (e.g. Barber & Jackson, 1957).

National patterns in understorey floristics

If a comparison is made between the floristic lists for open-forests in Australia (from Havel's, 1975*a*,*b*, survey for Western Australia and personal communication; Specht & Perry, 1948; for South Australia; Ratkowsky & Ratkowsky, 1976, for Tasmania; and Gray & McKee, 1969, for the Australian Capital Territory) several points emerge:

   (i) there are numbers of genera common to both east and west;
  (ii) there are numerous disjunctions in species distributions;
 (iii) there are many species and genera endemic to southwest Western Australia, particularly.

These points are discussed briefly below.

    Some genera which are common to both east and west, apart from the ubiquitous dominance of *Eucalyptus*, are shrubs of *Acacia*, *Daviesia*, *Hakea*, *Hibbertia*, *Leptospermum*, *Leucopogon* and *Pultenaea*. All these are large genera having more than 40 species per genus in Australia (Burbidge, 1963). Common herbaceous genera (all rhizomatous), viz. *Dianella*, *Lepidosperma* and *Lomandra*, and the ferns *Adiantum* and *Pteridium*, are found on each of the above species lists. Green (1964), in his comparison of east and west, implicitly included vegetation formations other than forest in his comparisons. He suggested that there are 'several hundred' species common to east and west.

    Green (1964) looked for large discontinuities (*c.* 1200 km) in species distributions across southern Australia. After eliminating many species on the basis of habitat (e.g. littoral niches) he demonstrated the phenomenon for 35 species. Noting that many of these species had very small seeds (e.g. orchids) he emphasised the possible role of long-distance dispersal in explaining the east-to-west disjunction.

    Not only are there obvious species disjunctions between eastern and western Australia, there are disjunctions in eastern Australia as well. The distribution of *Eucalyptus obliqua*, which is found across

Bass Strait in Tasmania as well as on the mainland, is a good example of a disjunction in the range of a dominant species. Among understorey species are some *Acacia* species, such as *A. genistifolia*, *A. verticillata* and *A. dealbata*, all of which occur in Tasmania as well as on the mainland (Gill & Ashton, 1971; Ratkowsky & Ratkowsky, 1976; Purdie, 1977*a*). Other, less obvious disjunctions than that imposed by Bass Strait occur, such as that posed by the Hunter River Valley in New South Wales and that between western Victoria and the Mt Lofty Ranges in South Australia but these have not been studied carefully from a floristic point of view. That they occur between populations of *E. obliqua* from open-forests has been documented by Gill (1966).

Wood & Williams (1960) pointed out that there are similarities in genera but differences between species in open-forests of three areas of eastern Australia, viz. South Australia, New South Wales and the Australian Capital Territory. They called these 'vicarious' species. Green (1964) suggested that the examples of Wood & Williams are better seen as species representative of common genera rather than as 'closely related allopatric species' (part of the definition of vicarious species of Cain, 1944). Unlike Wood & Williams' (1960) examples, Green's (1964) examples are for vicariism between southwestern and southeastern Australia and comprise 47 species pairs.

Not only may the differences and similarities between eastern and western Australia (or differences associated with disjunctions in eastern Australia) be seen in terms of disjunction in species distributions or vicariism, they may also be seen in the extent of local endemism. In comparisons of the species lists referred to at the start of this section, the isolation of the southwestern forests was reflected in the more than 20 endemic genera found there. Endemism to the local region at the generic level was absent or very low in the eastern States.

Disjunctions may be viewed as continuous populations broken by climatic, topographic or edaphic barriers bridged by long-distance dispersal of propagules, or be seen as insurmountable barriers to dispersal requiring a geological rather than a behavioural explanation for their presence. Green (1964), in noting the high proportion of minute-seeded orchids in his discontinuous species ranges between eastern and western Australia, could not dismiss the possibility that long-distance dispersal provided an explanation. For vicarious species, and the highly endemic genera of Western Australia, however, long periods of isolation are implied and geological explanations are required. Parsons (1969) suggested that lowered sea levels in the Pleistocene, with the consequent exposure of coastal lowland now under the sea, may have provided the conditions whereby continuous populations of species could occur across southern Australia (and to Tasmania). Whether or not the time period that has elapsed since then is sufficient to account for the high endemism of the southwest flora remains unresolved.

Biological processes in open-forest

Nutrient cycling

Because Australian soils are so poor in plant nutrients, nutrient cycling is an important topic to a description of open-forests (Attiwill & Leeper, 1987).

The importance of differences in soil phosphorus levels to vegetation distribution has already been emphasised together with the general poverty of this element in Australian soils. In agriculture, many other deficiencies occur, and the country has become renowned for its use of trace elements (Donald & Prescott, 1975). Whilst deficiency of trace elements has been marked, the only study of growth of native species in relation to them points to a possible role of a toxicity (manganese) in affecting eucalypt distribution (Winterhalder, 1963). However, some studies of cycling of trace elements in native forests have begun (Rogers & Westman, 1977; Westman & Rogers, 1977; Lee & Correll, 1978; O'Connell, Grove & Dimmock, 1978).

Emerging from studies of mineral composition of litter are correlations between the concentrations of certain elements. Lee & Correll (1978) found that concentrations of nitrogen, phosphorus, zinc, iron and copper were correlated; O'Connell *et al.* (1978), similarly, found that zinc levels were correlated with those of nitrogen, phosphorus and potassium. Such correlations may reflect functional differences in the plant, as suggested by Garten (1978), who found that nitrogen, phosphorus, iron, copper and sulphur were correlated and formed a set related to nucleic acid and protein metabolism. He also found that concentrations of calcium, magnesium and manganese were related and could be attributed to a 'structural-photosynthetic' set: Lee & Correll (1978) found such a set also in their analyses of eucalypt litter. Garten's (1978) correlations were found between elemental concentrations for 110 North American species of wide ecological amplitude, whilst those of the Australian studies were for litter concentrations within a single forest type. Such groupings of elements may help in the construction of generalisations about the cycling of nutrients in the forest. However, many factors other than their concentrations and affiliations in plant tissue are necessary for the understanding of nutrient cycling – factors such as rainfall accessions, foliar leaching, soil chemistry, mineralisation of litter and dead roots, availability of nutrients, internal recycling and rock weathering. Below, only some of these topics are reviewed, and even then just briefly.

The cycling of elements in the forest varies in rate and quantity according to the nature of the element and its carrier. Thus, elements such as potassium are particularly mobile and are leached from canopies by rain. Other elements may be carried in leaf or twig materials, or as animal excreta. Some elements may be circulated internally within the plant rather than externally through the soil.

Leaching of plant canopies removes a great diversity of materials including both organic and inorganic substances (Tukey, 1970). In *Eucalyptus* forest, Guthrie, Attiwill & Leuning (1978) found foliage leaching of elements to be in the order Na > K > Ca > Mg, whilst Rowe & Hagel (1974) found that leaching from litter was in the order of K > (Ca, Mg) > Na. Among the anions, Rowe & Hagel (1974) found that leaching from litter was in the order of $HCO_3^- >$ $SO_4^{2-} > Cl^-$.

Amounts of materials leached will depend not only on the mobility of the element but also on the amount in the tissue being leached. Guthrie *et al.* (1978) found that, on an annual basis, leaching accounted for 66 per cent Na, 56 per cent K, 33 per cent Mg and 14 per cent Ca of the total present, with the balance being circulated through the litter. Of particular importance to the cycling of some elements, especially nitrogen and phosphorus, is the withdrawal of nutrients prior to leaf fall. This is an internal process of rapid recycling. For phosphorus this may amount to 70–80 per cent in eucalypts (Ashton, 1976*b*; Attiwill, Guthrie & Leuning, 1978; O'Connell *et al.*, 1978). Similarly, O'Connell *et al.* (1978) reported that nitrogen withdrawal may amount to about 60–70 per cent. Westman & Rogers (1977) found, however, in a forest with relatively abundant nitrogen, that nitrogen withdrawal was inconsequential in eucalypt leaves. Perhaps the demand by young leaves affects the rate of withdrawal from older leaves?

Rapid movement of nutrients from the plant canopy to the soil may take place through insect frass and the excreta of vertebrates. Although the importance of some insect feeders has already been mentioned in another context, that of the jarrah leaf-miner, *Perthida glyphopa* (Wallace, 1970), has not. Extensive leaf damage by such insects highlights the importance and role of insect feeders to nutrient cycling. The frass from such feeding is readily decomposed (Springett, 1978), so that, in addition to canopy nutrients being moved prematurely compared to those in an intact crown, the nutrients are also released to the soil more rapidly than for intact leaves.

About 2 to 4 t/ha litter falls from open-forest canopies each year (Lee & Correll, 1978) and is responsible for a large proportion of the movement of calcium and magnesium from the canopy (Guthrie *et al.* 1978). Most of the material, especially the leaves, falls from the eucalypts in summer (e.g. Attiwill, Guthrie & Leuning, 1978). Because of the correlations noted earlier (Lee & Correll, 1978), it might be expected that manganese would behave, in this respect, similarly to calcium and magnesium. Indeed, manganese concentrations parallel those of calcium and magnesium as the leaf matures (Westman & Rogers, 1977; O'Connell *et al.*, 1978), thereby supporting this idea. Interestingly, eucalypts of the informal subgenus Symphyomyrtus tend to have greater manganese concentrations in bark than species of the informal subgenus Monocalyptus (Attiwill & Leeper, 1987).

The rate of return of the less mobile elements to the soil from litter may be estimated from decomposition rates. Gill (1964) estimated that decomposition of eucalypt leaves took from two to three years, and that for twigs took between four and eight years, but decomposition of large branches and logs takes many more years. Rates of decay may be described using the decomposition 'constant' which is the ratio of the average annual accession of materials to the forest floor to the amount of material accumulated there: in southeastern Australian open-forests, the values are typically between 0.2 and 0.4 (Walker, 1981). Such rates of decay, and thereby of mineralisation, are slow compared with rates achieved by burning. However, burning has other consequences as well.

Fire releases elements from organic matter very rapidly but may volatilise elements such as nitrogen together with some phosphorus and sulphur. Some movement of nutrients may also occur in smoke and in local redistributions of ash by wind and water. If the fire is of high intensity and the crown is consumed, release of elements at the relatively high concentrations in the foliage occurs also. Death of trees may be associated with particular intensities, and stores of nutrients in trunk, branches and bark may then become subject to decay processes. The subject is a complicated one as the effects of fire on nutrient cycling depend not only on the characteristics of the plant species and the volumes of material present, but also on the intensities, frequencies and seasons of fire occurrence.

Nitrogen losses through volatilisation during fire may be restored through the activities of symbiotic nitrogen-fixing organisms. Nitrogen-fixing species are common in open-forests and include leguminous shrubs (such as *Acacia*, *Daviesia* and *Pultenaea*), non-leguminous shrubs (such as *Casuarina*) and cycads (such as *Macrozamia*). Halliday & Pate (1976) found that *Macrozamia riedlei* contributed about 19 kg/ha of nitrogen to a coastal community in Western Australia – perhaps an unusually large amount for open forests (Attiwill & Leeper, 1987). *Macrozamia* plants are very fire-resistant and may be expected to fix nitrogen after fire occurs. Their input may vary according to any root damage during fire and according to population changes. For some *Acacia* species, at least, dramatic changes in population (and nitrogen fixation?) may occur according to the characteristics of fires to which they are subjected (Shea, McCormick & Portlock, 1979). Some *Casuarina* species, such as *C. littoralis* in southeastern Australia, seem to be responsive to particular disturbance regimes also.

Fire

Fires are a feature of open-forests. They occur in, and contribute to, one of the most severe fire situations in the world. In open-forest regions large fires may occur once every three to five years (Cheney, 1976). Fuels accumulate rapidly and can reach levels of 15 t/ha after 10 years, 22 t/ha after 20 years and 27 t/ha after 30 years (Wallace, 1966); a quasi-equilibrium is reached in the accumulation of fine-fuel when quantities may reach 10–30 t/ha. These fuels may have very low moisture contents during summer and autumn when hot dry winds can also fan any fire that starts. Fires of high intensity under these conditions can be extremely dangerous to people in the area, as shown by the fires in Victoria and South Australia in 1983 (Keeves & Douglas, 1983; Rawson, Billing & Duncan, 1983).

The effects of fires on the vegetation vary according to sequences of fire events rather than to one event only. Particular sequences may be known as particular 'fire regimes' and consist of the variables of fire intensity, fire frequency and season of occurrence (Gill, 1975). Fire intensities can vary across three or four orders of magnitude (Gill & Moore, 1990), frequencies in open-forests may vary from once every five years to 50 or more, and they can occur in particular sites in the open-forest zone at any time of year if ignited.

Although a single plant community may be affected by the same fire regime, the responses of individual plants and individual species may vary. Seedlings may be killed by a single fire but mature plants of the same species may survive many. Some species have individuals which are all killed by a single fire but the species persists because regeneration occurs from seed stored on the plant or in the soil. Even amongst species of the same genus in the same forest, variations occur: examples may be found in the genus *Acacia* where one species may have fire-sensitive individuals and another species have resistant individuals because of buds buried in the soil.

Most, if not all, the eucalypts of the open-forest have fire-resistant individuals (e.g. Gill & Ashton, 1968). They also have limited seed stores in the canopies of the trees which may be released in greater quantities than usual at the time of fire. Some shrub genera, such as *Hakea*, *Banksia* and *Casuarina*, may have species (Fig. 7.2) which store seed in the canopy and release it at the time of fire also. Some species, like the mistletoes, may have no storage in the canopy and

(a)   (b)

Fig. 7.2. *Banksia serrata* 'cones': (*a*) unburnt; (*b*) burnt. Note that the remnant floral parts surrounding the woody fruits (left) have been burnt by the fire (right) and that the paired valves of the fruits reflex once the resin holding them together has been melted by the fire, thereby allowing their seeds to disperse.

none in the ground and rely on invasion from unburned areas for their persistence when individuals are killed by fire.

The validity of classical theories of secondary succession, including that after individual fires, is being increasingly questioned for Australian plant communities (Purdie & Slatyer, 1976; Noble & Slatyer, 1977). After fire in open-forest, Purdie (1977*a,b*) found that all species present before experimental fires were again present afterwards. She concluded that Egler's (1954) relay floristic model of succession was the most appropriate one to describe this situation (Purdie & Slatyer, 1976). The same conclusion could be drawn from the pioneering studies of Jarrett & Petrie (1929) in Victoria: these authors noted that vegetative regeneration was marked in their open-forests and that 'succession is confined usually to a sere of socies in the lower layers of the forest'. I take this to mean that changes in species abundances occur with time rather than there being changes in numbers of species present. Changes in species' abundances (population sizes) are far more evident than changes in species compositions in open forests.

In South Australian studies, Cochrane, Burnard & Philpott (1962) identified four main 'seral stages' in open-forest. The first was typified by the presence of the liverwort *Marchantia polymorpha*, a lack of species dominance and low ground cover (a stage lasting 18 months); from two to five years after fire the composite *Ixodia achilleoides* was dominant; after six years, dominance changed to that of the slower-growing shrubs such as *Leptospermum myrsinoides*; whilst from seven to ten years, and persisting for at least 20 years, dominance of shrubs of *Acacia myrtifolia* and *Pultenaea daphnoides*, 1 to 2 m tall, was apparent.

Growth and seed dispersal: selected aspects

Eucalypts have small seeds which germinate readily. Seedlings of the open-forest species soon develop a woody swelling in the axils of the cotyledons and first few leaves. These organs, known as 'lignotuber' (Kerr, 1925), are sites of regeneration after stresses, such as insect attack, shade, disease or fire, have been removed. As the plant grows and develops into a tree the lignotuber may be overgrown in some species but the presence of buds beneath the bark may aid survival after stress (Fig. 7.3).

Shoot growth of intact trees occurs either as a unimodal peak in summer (Gill, 1964) or as a bimodal form with peaks in autumn and spring (Specht & Brouwer, 1975). Leaf fall, in southern areas at

Fig. 7.3. Epicormic growth one year after a bushfire in a mixed-species forest of *Eucalyptus* at Canberra, Australian Capital Territory.

least, appears to be associated with shoot growth (Jacobs, 1955; Gill, 1964), although other factors such as drought (Pook, Costin & Moore, 1966), heavy flowering, insect attack and temperature changes (Jacobs, 1955) may affect the process. Whether or not shoot growth has a direct effect on litter fall or merely coincides with temperature (the factor of importance in leaf fall suggested by Attiwill *et al.*, 1978) or a more complex weather set (Gentilli, 1989) is uncertain.

Inflorescence buds of most eucalypts form in leaf axils (Carr & Carr, 1959) as the leaves unfold but flowers may take two years to appear. The actual season of flowering may vary widely among the species although leafing phenology appears to be more uniform. A common pattern is that two years after flowering, capsule dehiscence begins and may continue over a two-year period; many species are unstudied. Seeds are heavily predated by ants (Jacobs, 1955) but massive seed falls stimulated by fire may satiate them (O'Dowd & Gill, 1984).

Ants have been implicated in seed dispersal as well as harvesting. Berg (1975) indicated that many Australian species have seeds with 'elaiosomes' which are particularly attractive to ants. Many of the species with these structures are the shrubs of open-forests. Ants pick up the seeds by the elaiosome, transport it to the nest, remove the elaiosome and discard the seed either to the nest surface or within the nest itself. Diverse genera and families of plants are involved but most species are Australian endemics. Berg (1975) suggested that there may be interacting factors which have led to this abundance of seeds possessing elaiosomes and these factors may include fire and seed predation.

Utilisation

The open-forests of southern Australia are found close to the major centres of population. Although this is convenient for the transport of wood products it also results in demand for other uses, such as conservation of water and plant species, and for recreation.

When untouched eucalypt forests were first exploited, it was found that more than 50 per cent of the volume of commercial species was so defective that it could not be used (Curtin, 1970). In the poorer forests where intensive management for wood production has been uneconomic, this situation has led to a tendency to exploit rather than to manage the resource. In better forests, clearfelling and intensive management have been possible because of the higher profit from wood products such as sawlogs. With recent emphasis on wood chips as a product, some foresters contend that harvesting for such materials will cause an upgrading of these forests for sawlog production (Ovington & Thistlethwaite, 1976). Eucalypts comprise 94 per cent wood volume of native forests (Jacobs, 1970). In open-forests the more important species are *E. marginata* in Western Australia, and *E. obliqua*, *E. maculata* and *E. sieberi* in eastern Australia (Jacobs, 1970). All these species are used for sawlogs but in some areas they are also used for woodchips.

Woodchipping for export has become an important practice in New South Wales, Tasmania and Western Australia. In Western Australia, *E. marginata* and *E. calophylla* are the basis of the industry: in Tasmania, *E. viminalis*, *E. obliqua* and *E. amygdalina* are its backbone; whilst in New South Wales, *E. sieberi* and *E. gummifera* are the principal species (Heyligers, 1975). For further detail, the reader is referred to published reports of enquiries into the woodchip industry (Cromer *et al.*, 1975; Gilbert, 1976; Heyligers, 1975, 1977; Senate Standing Committee on Science and the Environment, 1977) and forest industries in general (Australian Conservation Foundation, 1974; 'FORWOOD', 1974; Routley & Routley, 1974).

Major problems in the management of open-forests concern the use of fire (Attiwill, 1975); the control of the root-rotting disease caused by *Phytophthora* species, especially in Western Australia (Ovington & Thistlewaite, 1976); and the conservation of species (Attiwill, 1975). Both Attiwill (1975) and the Australian Conservation Foundation (1974) stress the need for an integrated view of the forest as a resource: wood production should be only one of a number of equally important goals.

From a species conservation point of view, many alliances of the open-forest are not conserved at all (Specht, Roe & Boughton, 1974) despite the observation of Tyndale-Biscoe & Calaby (1975) that the '*Eucalyptus* forests of southeastern Australia and Tasmania support a rich and varied fauna of mammals and birds and together form the single most important refuge for wildlife in Australia'. The high endemism of the southwestern Australian forests and the threat to their persistence because of *Phytophthora* and bauxite mining makes conservation an important and urgent need in that region where five alliances remain unconserved (Specht *et al.*, 1974).

At the time of writing, a major round of Australian Government enquiries regarding forests was being completed; two draft reports had appeared (Resource Assessment Commission, 1991; Ecologically

Sustainable Working Groups, 1991) which drew attention to the incomplete conservation of open-forests, the overcutting of commercial forests, the need for effective management of biodiversity, and the possible impacts of fire on gaseous emissions and, thereby, the enhanced greenhouse effect.

Conclusions

Eucalyptus forests are an important and conspicuous element of the Australian landscape. Classification and nomenclature of these forests still poses problems which require resolution. In this chapter the subcategory of the formation-level 'open-forests', also known as 'open-forest', has been considered in its range of occurrence in southeastern Australia (including Tasmania) as well as in the southwest. These forests occur in areas of moderate rainfall and temperature where the phosphorus status of the soils is also moderate in the general spectrum of low values for Australian soils. The eucalypts dominating these forests often occur in associations and appear sensitive to slight changes in habitat. Distributions in some cases are disjunct as is the case with many understorey species. Disjunctions may be recognised at the levels of subspecies, species or genus and could be because of behavioural or historical-geological reasons. Because of the poor soils on which they grow, mineral cycling is an important topic and efficient withdrawal of phosphorus from leaves before leaf fall is a feature of these evergreen forests. Fire may strongly affect the cycling process. Plants respond to single fires in different ways and, within a single community subject to a series of fires, many mechanisms for species persistence may be found. Fires tend to occur naturally in summer and autumn in litter dropped from the trees. Leaf fall from the eucalypts may be a response to leaf growth or other internally controlled mechanisms but may be influenced as well by the external environment. The forests are used for wood products, recreation, and water supply as well as the conservation of a unique biota.

Acknowledgments

I would like to thank Dr J.J. Burdon and Mr J.J. Havel for their criticisms of the draft manuscript for the first edition of this book.

References

Ashton, D.H. (1976a). The vegetation of Mount Piper, central Victoria: a study of a continuum. *Journal of Ecology*, **64**, 463–83.

Ashton, D.H. (1976b). Phosphorus in forest ecosystems at Beenak, Victoria. *Journal of Ecology*, **64**, 171–86.

Ashton, D.H. (1981). Tall open-forests. In *Australian Vegetation*, ed. R.H. Groves, pp. 121–51. Cambridge University Press.

Attiwill, P.M., Guthrie, H.B. & Leuning, R. (1978). Nutrient cycling in a *Eucalyptus obliqua* (L'Hérit.) forest. I. Litter production and nutrient return. *Australian Journal of Botany*, **26**, 79–91.

Attiwill, P.M. & Leeper, G.W. (1987). *Forest Soils and Nutrient Cycles*. Melbourne University Press.
Eucalyptus obliqua (L'Hérit.) forest. I. Litter production and nutrient return. *Australian Journal of Botany*, **26**, 79–91.

Austin, M.P. (1978). Vegetation. In *Land Use on the South Coast of New South Wales: A Study in Methods of Acquiring and Using Information to Analyse Regional Land Use Options*, eds M.P. Austin & K.D. Cocks, pp. 44–67. Melbourne: CSIRO.

Austin, M.P., Cunningham, R.B. & Wood, J.T. (1983). The subgeneric composition of eucalypt forest stands in a region of south-eastern Australia. *Australian Journal of Botany*, **31**, 63–71.

Austin, M.P., Nicholls, A.O. & Margules, C.R. (1990). Measurement of the realised qualitative niche: environmental niches of five *Eucalyptus* species. *Ecological Monographs*, **60**, 161–77.

Australian Conservation Foundation (1974). Multiple use on forest land presently used for commercial wood production. *Search*, **5**, 438–43.

Baker, F.S. (1950). *Principles of Silviculture*. New York: McGraw-Hill.

Barber, H.N. & Jackson, W.D. (1957). Natural selection in action in *Eucalyptus*. *Nature*, **179**, 1267–9.

Baur, G.N. (1962). *Forest Vegetation in North-eastern New South Wales*. Forest Commission of New South Wales Research Note Number 8.

Baur, G.N. (1965). *Forest Types in New South Wales*. Forest Commission of New South Wales Research Note Number 17.

Beadle, N.C.W. (1954). Soil phosphate and the delimitation of plant communities in eastern Australia. *Ecology*, **35**, 370–5.

Beadle, N.C.W. (1962). Soil phosphate and the delimitation of plant communities in eastern Australia. II. *Ecology*, **43**, 281–8.

Beadle, N.C.W. & Costin, A.B. (1952). Ecological classification and nomenclature. *Proceedings of the Linnean Society of New South Wales*, **77**, 61–82.

Berg, R.Y. (1975). Myrmecochorous plants in Australia and their dispersal by ants. *Australian Journal of Botany*, **23**, 475–508.

Brooker, M.I.H. & Kleinig, D.A. (1990). *Field Guide to the Eucalypts: 2. Southwestern and Southern Australia*. Melbourne: Inkata.

Burbidge, N.T. (1963). *Dictionary of Australian Plant Genera*. Sydney: Angus & Robertson.

Burdon, J.J. & Chilvers, G.A. (1974). Fungal and insect parasites contributing to niche differentiation in mixed species stands of eucalypt saplings. *Australian Journal of Botany*, **22**, 103–14.

Burrough, P.A., Brown, L. & Morris, E.C. (1977). Variations in vegetation and soil pattern across the Hawkesbury Sandstone plateau from Barren Grounds to Fitzroy Falls, New South Wales. *Australian Journal of Ecology*, **2**, 137–59.

Cain, S.A. (1944). *Foundations of Plant Geography*. New York: Harper.

Carr, D.J. & Carr, S.G.M. (1959). Developmental morphology of the floral organs of *Eucalyptus*. I. The inflorescence. *Australian Journal of Botany*, **7**, 109–41.

Cheney, N.P. (1976). Bushfire disasters in Australia, 1945–1975. *Australian Forestry*, **39**, 245–68.

Christensen, P.E. (1971). The purple-crowned lorikeet and eucalypt pollination. *Australian Forestry*, **35**, 263–70.

Churchill, D.M. (1968). The distribution and prehistory of *Eucalyptus diversicolor* F. Muell., *E. marginata* Donn. ex Sm. and *E. calophylla* R.Br. in relation to rainfall. *Australian Journal of Botany*, **16**, 125–51.

Cochrane, G.R., Burnard, S. & Philpott, J.M. (1962). Land use and forest fires in the Mount Lofty Ranges, South Australia. *Australian Geographer*, **8**, 143–60.

Costin, A.B. (1954). *A Study of the Ecosystems of the Monaro Region of New South Wales*. Sydney: Government Printer.

Costin, A.B. (1981). Alpine and sub-alpine vegetation. In *Australian Vegetation*, ed. R.H. Groves, pp. 361–76. Cambridge University Press.

Cromer, D.A.N., Eldershaw, V.J., Lamb, I.D., McArthur, A.G., Wesney, D. & Girdlestone, J.N. (1975). *Economic and Environmental Aspects of the Export Hardwood Woodchip Industry*, vols. I & II. Canberra: Australian Government Publishing Service.

Curtin, R.A. (1970). Increasing the productivity of eucalypt forests in New South Wales. *Australian Forestry*, **34**, 97–106.

Davis, C. (1936). Plant ecology of the Bulli district. Part I: Stratigraphy, physiography and climate; general distribution of plant communities and interpretation. *Proceedings of the Linnean Society of New South Wales*, **61**, 285–97.

Davis, C. (1941). Plant ecology of the Bulli district. II. Plant communities of the plateau and scarp. *Proceedings of the Linnean Society of New South Wales*, **66**, 1–19.

Diels, L. (1906). *Die Vegetation der Erde 7: Pflanzenwelt von West-Australien*. Leipzig: Engelmann.

Donald, C.M. & Prescott, J.A. (1975). Trace elements in Australian crops and pasture production, 1924–1974. In *Trace elements in Soil–Plant–Animal Systems*, eds D.J.D. Nicholas & A.R. Egan, pp. 7–39. London: Academic Press.

Ecologically Sustainable Working Groups (1991). *Draft Report – Ecologically*

Sustainable Forest Use. Canberra: Australian Government Publishing Service.

Egler, F.E. (1954). Vegetation science concepts. I. Initial floristic composition, a factor in old field vegetation development. *Vegetatio* **4**, 412–17.

'FORWOOD' (1974). *A Series of Reports prepared for the 'Forestry and Wood-Based Industries Development Conference', April 1974.* Canberra: Australian Government Publishing Service.

Garten, G.T. (1978). Multivariate perspectives on the ecology of plant mineral elemental composition. *American Naturalist,* **112**, 533–44.

Gentilli, J. (1989). Climate of the jarrah forest. In *The Jarrah Forest,* eds B. Dell, J.J. Havel & N. Malajczuk, pp. 23–40. Dordrecht: Kluwer.

Gibbons, F.R. & Downes, R.G. (1964). *A Study of the Land in South-western Victoria.* Soil Conservation Authority of Victoria Technical Communication Number 3.

Gilbert, J.M. (ed.) (1976). *Woodchip Symposium Papers, 47th ANZAAS Congress.* Hobart: Tasmanian Forests Commission.

Gill, A.M. (1964). Soil-vegetation relationships near Kinglake West, Victoria. MSc thesis, University of Melbourne.

Gill, A.M. (1966). The ecology of mixed species of *Eucalyptus* in central Victoria, Australia. PhD thesis, University of Melbourne.

Gill, A.M. (1975). Fire and the Australian flora: a review. *Australian Forestry,* **38**, 1–25.

Gill, A.M. & Ashton, D.H. (1968). The role of bark type in relative tolerance to fire of three central Victorian eucalypts. *Australian Journal of Botany,* **16**, 491–8.

Gill, A.M. & Ashton, D.H. (1971). The vegetation and environment of a multi-aged eucalypt forest near Kinglake West, Victoria, Australia. *Proceedings of the Royal Society of Victoria,* **84**, 159–72.

Gill, A.M., Belbin, L. & Chippendale, G.M. (1985). *Phytogeography of Eucalyptus in Australia.* Bureau of Flora and Fauna, Australian Flora and Fauna Series Number 3. Australian Government Printer, Canberra.

Gill, A.M. & Moore, P.H.R. (1990). Fire intensities in eucalypt forests of southeastern Australia. In *International Conference on Forest Fire Research, Coimbra, Portugal.* Paper B.24.

Gillison, A.N. & Walker, J. (1981). Woodlands. In *Australian Vegetation,* ed. R.H. Groves, pp. 177–97. Cambridge University Press.

Gray, M. & McKee, H.S. (1969). *A list of Vascular Plants Occurring on Black Mountain and environs,* Canberra, ACT. CSIRO Australia, Division of Plant Industry Technical Paper Number 26.

Green, J.W. (1964). Discontinuous and presumed vicarious plant species in southern Australia. *Journal of the Royal Society of Western Australia,* **47**, 25–32.

Grose, R.J. (1960). Effective seed supply for the natural regeneration of *Eucalyptus delegatensis* R.T. Baker, syn. *Eucalyptus gigantea* Hook. f. *Journal of the Australian Pulp and Paper Industry Technical Association,* **13**, 141–7.

Guthrie, H.B., Attiwill, P.M. & Leuning, R. (1978). Nutrient cycling in a *Eucalyptus obliqua* (L'Hérit.) forest. II. A study in a small catchment. *Australian Journal of Botany,* **26**, 189–201.

Halliday, J. & Pate, J.S. (1976). Symbiotic nitrogen fixation by coralloid roots of the cycad *Macrozamia riedlei*: physiological characteristics and ecological significances. *Australian Journal of Plant Physiology*, 3, 349–58.

Havel, J.J. (1975a). *Site-vegetation Mapping in the Northern Jarrah Forest (Darling Range). 1. Definition of Site-vegetation Types.* Western Australian Forests Department Bulletin Number 86.

Havel, J.J. (1975b). *Site-vegetation Mapping in the Northern Jarrah Forest (Darling Range). 2. Location and Mapping of Site-vegetation Types.* Western Australian Forests Department Bulletin Number 87.

Heyligers, P.C. (1975). Biological and ecological aspects related to the forestry operations of the export woodchip industry. In *Economic and Environmental Aspects of the Export Hardwood Woodchip Industry*, ed. D.A.N. Cromer *et al.*, Vol. II, pp. 77–119. Canberra: Australian Government Publishing Service.

Heyligers, P.C. (1977). *The Natural History of the Tasmanian, Manjimup and Eden-Bombala Woodchip Export Concession Areas.* Australian Government Department of Environment, Housing & Community Development Studies Bureau Report Number 22.

Hills, E.S. (1955). Physiography and geology. In *Introducing Victoria*, ed. G.W. Leeper, pp. 25–39. Melbourne: ANZAAS.

Hopkins, E.R. (1960). *The Fertilizer Factor in* Pinus pinaster *Ait. Plantations on Sandy Soils of the Swan Coastal Plain, Western Australia.* Western Australian Forests Department Bulletin Number 68.

Howard, T.M. (1973). Studies in the ecology of *Nothofagus cunninghamii* Oerst. 1. Natural regeneration on the Mt Donna Buang massif, Victoria. *Australian Journal of Botany*, 21, 67–78.

Howard, T.M. (1981). Southern closed-forests. In *Australian Vegetation*, ed. R.H. Groves, pp. 102–20. Cambridge University Press.

Jacobs, M.R. (1955). *Growth Habits of the Eucalypts.* Canberra: Commonwealth Government Printer.

Jacobs, M.R. (1970). The forest as a crop. In *The Australian Environment*, 4th edn (rev.), ed. G.W. Leeper, pp. 120–30. Melbourne: CSIRO & Melbourne University Press.

Jarrett, P.H. & Petrie, A.H.K. (1929). The vegetation of Black's Spur region. II. Pyric succession. *Journal of Ecology*, 17, 249–81.

Johnson, R.W. & Burrows, W.H. (1981). *Acacia* open-forests, woodlands and shrublands. In *Australian Vegetation*, ed. R.H. Groves, pp. 198–226. Cambridge University Press.

Keeves, A. & Douglas, D.R. (1983). Forest fires in South Australia on 16 February 1983 and consequent forest management aims. *Australian Forestry*, 46, 148–62.

Keith, D.A. & Sanders, J.M. (1990). Vegetation of the Eden Region, southeastern Australia: species composition, diversity and structure. *Journal of Vegetation Science*, 1, 203–32.

Kerr, L.R. (1925). The lignotubers of eucalypt seedlings. *Proceedings of the Royal Society of Victoria*, 27, 79–97.

Lee, K.E. & Correll, R.L. (1978). Litter fall and its relationship to nutrient cycling in a South Australian dry sclerophyll forest. *Australian Journal of Ecology*, **3**, 243–52.

Morland, R.T. (1959). Erosion survey of the Hume catchment area. VI. Vegetation (cont.). *Journal of the Soil Conservation Service of New South Wales*, **15**, 176–86.

Morrow, P.A. (1976). The significance of phytophagous insects in the *Eucalyptus* forests of Australia. In *The Role of Arthropods in Forest Ecosystems*, ed. W.J. Mattson, pp. 19–29. New York: Springer.

Mulcahy, M.J. (1973). Landforms and soils of southwestern Australia. *Journal of the Royal Society of Western Australia*, **56**, 16–22.

Noble, I.R. (1989). Ecological traits of the *Eucalyptus* l'Hérit. subgenera *Monocalyptus* and *Symphyomyrtus*. *Australian Journal of Botany*, **37**, 207–24.

Noble, I.R. & Slatyer, R.O. (1977). Post-fire succession of plants in Mediterranean ecosystems. In *Proceedings of the Symposium on the Environmental Consequences of Fire and Fuel Management in Mediterranean Ecosystems*, eds H.A. Mooney & C.E. Conrad, pp. 27–36. U.S. Forest Service General Technical Report WO-3.

O'Connell, A.M., Grove, T.S. & Dimmock, G.M. (1978). Nutrients in the litter on jarrah forest soils. *Australian Journal of Ecology*, **3**, 253–60.

O'Dowd, D.J. & Gill, A.M. (1984). Predator satiation and site alteration: mass reproduction of alpine ash (*Eucalyptus delegatensis*) in southeastern Australia. *Ecology*, **65**, 1052–66.

Ovington, J.D. & Thistlethwaite, R.J. (1976). The woodchip industry; environmental effects of cutting and regeneration practices. *Search*, **7**, 383–92.

Packham, G.H. (1962). An outline of the geology of New South Wales. In *A Goodly Heritage. ANZAAS Jubilee, Science in New South Wales*, ed. A.P.P. Elkin, pp. 24–35. Sydney: ANZAAS.

Parsons, R.F. (1969). Distribution and palaeogeography of two mallee species of *Eucalyptus* in southern Australia. *Australian Journal of Botany*, **17**, 323–30.

Parsons, R.F. (1981). *Eucalyptus* scrubs and shrublands. In *Australian Vegetation*, ed. R.H. Groves, pp. 227–52. Cambridge University Press.

Pidgeon, I.M. (1937). The ecology of the Central Coastal area of New South Wales. I. The environment and general features of the vegetation. *Proceedings of the Linnean Society of New South Wales*, **62**, 315–40.

Pidgeon, I.M. (1938). The ecology of the Central Coastal area of New South Wales. II. Plant succession on Hawkesbury Sandstone. *Proceedings of the Linnean Society of New South Wales*, **63**, 1–26.

Pidgeon, I.M. (1940). The ecology of the Central Coastal area of N.S.W. III. Types of primary succession. *Proceedings of the Linnean Society of New South Wales*, **65**, 221–49.

Pook, E.W., Costin, A.B. & Moore, C.W.E. (1966). Water stress in native vegetation during the drought of 1965. *Australian Journal of Botany*, **14**, 257–67.

Prescott, J.A. (1931). *The Soils of Australia in Relation to Vegetation and*

Climate. Council for Scientific & Industrial Research, Australia, Bulletin Number 52.

Pryor, L.D. (1959). Species distribution and association in *Eucalyptus*. In *Biogeography and Ecology in Australia*, eds A. Keast, R.L. Crocker & C.S. Christian, pp. 461–71 (Monographiae Biologicae VIII). The Hague: W. Junk.

Pryor, L.D. & Johnson, L.A.S. (1971). *A Classification of the Eucalypts*. Canberra: Australian National University.

Purdie, R.W. (1977*a*). Early stages of regeneration after burning in dry sclerophyll vegetation. I. Regeneration of the understorey by vegetative means. *Australian Journal of Botany*, **25**, 21–34.

Purdie, R.W. (1977*b*). Early stages of regeneration after burning in dry sclerophyll vegetation. II. Regeneration by seed germination. *Australian Journal of Botany*, **25**, 35–46.

Purdie, R.W. & Slatyer, R.O. (1976). Vegetation succession after fire in sclerophyll woodland communities in south-eastern Australia. *Australian Journal of Ecology*, **1**, 223–36.

Ratkowsky, D.A. & Ratkowsky, A.V. (1976). Changes in the abundance of the vascular plants of the Mount Wellington Range, Tasmania, following a severe fire. *Papers and Proceedings of the Royal Society of Tasmania*, **110**, 63–90.

Rawson, R.P., Billing, P.R. & Duncan, S.F. (1983). The 1982–83 forest fires in Victoria. *Australian Forestry*, **46**, 163–72.

Resource Assessment Commission (1991). *Forest and Timber Enquiry Draft Report. Volumes 1 and 2*. Canberra: Australian Government Publishing Service.

Rogers, R.W. & Westman, W.E. (1977). Seasonal nutrient dynamics of litter in a sub-tropical eucalypt forest, North Stradbroke Island. *Australian Journal of Botany*, **35**, 47–58.

Routley, R. & Routley, V. (1974). *The Fight for the Forests, The Takeover of Australian Forests for Pines, Woodchips and Intensive Forestry*. Canberra: Australian National University Research School of Social Sciences.

Rowe, R.K. (1967). *A Study of the Land in the Victorian Catchment of Lake Hume*. Soil Conservation Authority of Victoria Technical Communication Number 5.

Rowe, R.K. & Hagel, V. (1974). Leaching of plant nutrient ions from burned forest litter. *Australian Forestry*, **36**, 154–63.

Rundle, A.S. (1977). *A Study of the Land in the Catchment of Lake Eildon*. Soil Conservation Authority of Victoria Technical Communication Number 11.

Senate Standing Committee on Science and the Environment (1977). *Woodchips and the Environment*. Canberra: Australian Government Publishing Service

Shea, S.R., McCormick, J. & Portlock, C.C. (1979). The effect of fires on regeneration of leguminous species in the northern jarrah (*Eucalyptus marginata* Sm.) forest of Western Australia. *Australian Journal of Ecology*, **4**, 195–205.

Specht, R.L. (1970). Vegetation. In *The Australian Environment*, 4th edn (rev.), ed. G.W. Leeper, pp. 44–67. Melbourne: CSIRO & Melbourne University Press.

Specht, R.L. & Brouwer, Y.M. (1975). Seasonal shoot growth of *Eucalyptus* spp. in the Brisbane area of Queensland (with notes on shoot growth and litter fall in other areas of Australia). *Australian Journal of Botany*, 23, 459–74.

Specht, R.L. & Perry, R.A. (1948). Plant ecology of part of the Mount Lofty Ranges. *Transactions of the Royal Society of South Australia*, 72, 91–132.

Specht, R.L., Roe, E.M. & Boughton, V.H. (ed.) (1974). Conservation of major plant communities in Australia and Papua New Guinea. *Australian Journal of Botany, Supplementary Series* No. 7.

Springett, B.P. (1978). On the ecological role of insects in Australian eucalypt forests. *Australian Journal of Ecology*, 3, 129–39.

Tukey, H.B. (1970). The leaching of substances from plants. *Annual Review of Plant Physiology*, 21, 305–24.

Tyndale-Biscoe, C.H. & Calaby, J.H. (1975). Eucalypt forest as a refuge for wildlife. *Australian Forestry*, 38, 117–33.

Walker, J. (1981). Fuel dynamics in Australian vegetation. In *Fire and the Australian Biota*, eds A.M. Gill, R.H. Groves & I.R. Noble, pp. 101–27. Canberra: Australian Academy of Science.

Wallace, M.M.H. (1970). The biology of the jarrah leaf miner. *Perthida glyphopa* (Lepidoptera). *Australian Journal of Zoology*, 186, 91–104.

Wallace, W.R. (1966). Fire in the jarrah forest environment. *Journal of the Royal Society of Western Australia*, 49, 33–44.

Webb, L.J. (1959). A physiognomic classification of Australian rainforests. *Journal of Ecology*, 47, 551–70.

Westman, W.E. & Rogers, R.W. (1977). Nutrient stocks in a sub-tropical eucalypt forest, North Stradbroke Island. *Australian Journal of Ecology*, 2, 447–60.

Williams, R.J. (1955). Vegetation regions. In *Atlas of Australian Resources*, 1st Series. Canberra: Department of National Development.

Wilson, P.G. (1973). The vegetation of Western Australia. In *Western Australian Year Book*, Vol. 12, pp. 55–62. Perth: Government Printer.

Winterhalder, E.K. (1963). Differential resistance of two species of *Eucalyptus* to toxic soil manganese levels. *Australian Journal of Science*, 25, 363–4.

Wood, J.G. (1937). *Vegetation of South Australia*. Adelaide: Government Printer.

Wood, J.G. & Williams, R.J. (1960). Vegetation. In *The Australian Environment*, 3rd edn (rev.) pp. 67–84. Melbourne: CSIRO & Melbourne University Press.

8

Woodlands

A.N. GILLISON

AUSTRALIAN vegetation is unique in many respects as a result of its long isolation from other parts of the Gondwanan platform, the low nutrient status of soils, severe fire climate, and low density of humans and browsing fauna. These evolutionary aspects are perhaps nowhere better reflected than in woodlands which cover approximately 25 per cent or 1.94×10^6 km^2 of the continent. Typical Australian woodland elements are also common on many of the off-shore islands and extend into southern Papua New Guinea, with outliers in New Caledonia and Timor. The majority are important grazing lands (*sensu* Moore & Perry, 1970; Moore, 1993) as they are often associated with extensive grass or graminoid formations. Many thousands of square kilometres have been cleared since European settlement (i.e. since the 1780s) to make way for pasture and cultivation (Carnahan, 1990; Moore, 1993). Much of their present distribution reflects patterns of land use and most of the semi-intact woodlands are today under considerable grazing pressure.

The aim of this chapter is to provide a descriptive framework for Australian woodlands which provides a greater level of flexibility and information on structural variability than exists with more traditional classificatory approaches. The content and format follow closely the first edition of this work (Gillison & Walker, 1981). In this volume, a number of vegetation types described by other authors fall within the definition of woodland as applied in this chapter (e.g. *Acacia* woodlands in Chapter 9 and Johnson & Burrows (1981); mallee in Chapter 10 and Parsons (1981); subalpine woodland in Chapter 16 and Costin (1981); and open-forests in Chapter 7 and Gill, 1981*a*). To maintain brevity, and to avoid duplication, only those key woodland examples not discussed by these authors will be presented in detail here.

Definition of woodland

In this chapter, 'woodland' refers to a structural plant formation usually with a graminoid component, dominated by perennial woody plants over 2 m tall which do not have their crowns touching. In broad terms, 'woodland' is included within the open-forest and open-scrub formation classes of Fosberg (1970) and the IUCN/UNESCO units of woodlands and wooded savannas (IUCN, 1973).

In Australia, related vegetation types are included in Specht (1970) and Beard & Webb (1974) classifications of 0.1–70 per cent projective foliage cover classes of woody vegetation over 2 m tall (see also Williams, 1955; Gillison, 1970, 1976, 1983; Moore & Perry, 1970; Walker & Gillison, 1982; Walker & Hopkins, 1984; Johnson & Tothill, 1985; Carnahan, 1976, 1990).

Since most woodlands in Australia are, at best, semi-intact, they may bear little resemblance to their structure prior to European settlement. They are described as they exist today excluding only those subjected to recent gross disturbance apart from areas grazed and annually burned for pastoral reasons.

Because of their structural similarity, especially in the upper strata, the terms 'savanna' and 'woodland' are often interchangeable. This aspect is discussed more fully in a treatment of Australian savannas by Walker & Gillison (1982) where the woodland types described here are referred to as savanna woodland types, 'savanna' referring to the essentially graminoid (grass-like) component. For a more complete discussion on the relative merits of the terms 'savanna' and 'woodland' in Australia see Gillison (1983), Johnson & Tothill (1985) and Gillison (1993).

In Australia, as in many other countries, the widespread and conflicting use of the terms 'tree' and 'shrub' has confused the classification of vegetation types: the commonly used definition of 'shrub' as a low multi-stemmed woody plant 'of smaller structure than a tree' (Carpenter, 1938), has little application in many parts of Australia where multi-stemmed (sympodial) woody plants may exceed 20 m in height (*cf.* an upper limit of 8 m in Specht, Roe & Boughton, 1974) and small single-stemmed (monopodial) woody plants 3 m tall are frequent. Further, some species exhibit both forms depending on the environment, e.g. *Eucalyptus microcarpa* and *Acacia aneura* (see also the term 'clumpwood' proposed by Johnson & Lacey (1984) for certain sympodial woodlands). For these reasons, woodlands that are normally described according to height and projective foliage cover alone, are further described here according to monopodial or sympodial forms (see also Gillison & Walker, 1981).

A descriptive framework for Australian woodlands

The main objectives in establishing a common framework for the description of Australian woodland are to facilitate:

1. the recognition of various static and dynamic woodland types on a continental basis;
2. the correlation of their distributional patterns with major bioclimatic or environmental variables;
3. a systematic basis for the comparison of Australian woodlands with similar vegetation types in other countries.

Despite previous class definitions, woodlands exhibit considerable variation along complex environmental gradients. This inherent variability is further compounded by a variety of past and present land use practices. In the face of such complexity, ecologists have retreated to a simplistic traditional descriptive classification with a strong floristic bias based on the dominant growth form or bark type; e.g. mallee, box, ironbark, bloodwood, paperbark, etc. (Williams, 1955; Moore & Perry, 1970). The use of a classification system based on structural attributes (i.e. Beard & Webb, 1974; Specht *et al.*, 1974; Carnahan, 1976, 1990; Mott & Tothill, 1984) has merely translated the traditional physiognomic woodland classification into a different format often with misleading interpretations, e.g. 'poplar box woodland' – a classification unit that encapsulates six of the structural categories described by Specht (1970) (see Fig. 8.1). The advent of computerised geographic information systems (GIS) and powerful desktop microcomputers has now made feasible the spatial modelling of individual taxa along defined environmental gradients. For management purposes, a GIS thus avoids the necessity to erect simplistic classifications which, by their very nature, exclude important management data. A functional basis for describing vegetation which is amenable to GIS usage has been outlined by Gillison (1981, 1988) using some Australian woodland examples. But for general descriptive purposes, there is, nevertheless, a residual need for a simple descriptive classification method which indicates the degree of variation in the values of descriptive parameters. In developing a descriptive framework, this chapter has addressed this need as well as the desirability of remaining compatible with vegetation structure as used in other chapters.

Woodland types

Four conceptual approaches are used to describe Australian woodland types:

1. The term woodland has already been defined here on a structural basis. Use of the height/cover values differs significantly, however, from the discrete classes of Specht *et al.* (1974). Because the continuum concept is closer to reality than an over-simplified tabular approach, the structure of each woodland type is described using

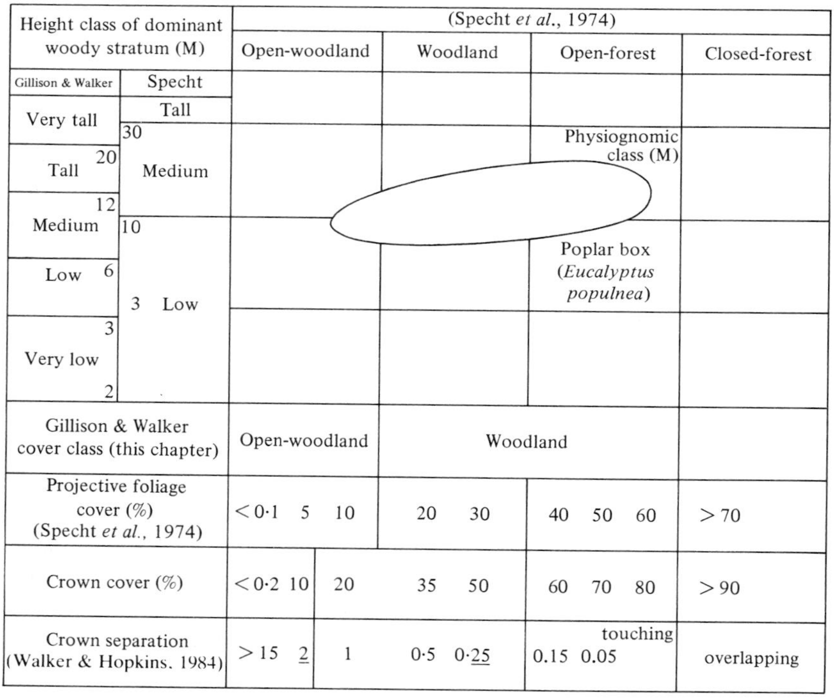

Height class of dominant woody stratum (M)		(Specht *et al.*, 1974)			
		Open-woodland	Woodland	Open-forest	Closed-forest
Gillison & Walker	Specht				
Very tall	Tall				
	30			Physiognomic class (M)	
Tall 20	Medium				
12					
Medium 10				Poplar box (*Eucalyptus populnea*)	
Low 6					
3	Low				
3					
Very low					
2					
Gillison & Walker cover class (this chapter)		Open-woodland	Woodland		
Projective foliage cover (%) (Specht *et al.*, 1974)		<0·1 5 10	20 30	40 50 60	>70
Crown cover (%)		<0·2 10 20	35 50	60 70 80	>90
Crown separation (Walker & Hopkins, 1984)		>15 2 1	0·5 0·25	touching 0.15 0.05	overlapping

Fig. 8.1. Variation in woodland structure as indicated by a structuregram that uses a combination of height and foliage projective cover. The class boundaries shown are those of Specht *et al.* (1974), the values preferred by Gillison & Walker (1981) and used in this chapter being shown for comparison.

an ordinative method developed by Walker & Gillison (1982) for classifying Australian savannas. Values for height and cover for sites within a woodland type are plotted as continuous variable on a height × cover table, and a line drawn around the range of values obtained (here termed a structuregram). The resulting cloud allows the reader to see at a glance the known structural range of a recognised woodland type. In the example given in Figure 8.1 the widespread *Eucalyptus populnea* woodland type exhibits cover values which vary from 6 to 60 per cent and height values from 8 to 20 m. In this way it is possible to retain the use of an otherwise artificial tabular system as a framework for the interpretation and display of real data.

2. For descriptive purposes, it is necessary to subdivide height and cover continua into categories which reflect tallness and openness. Unfortunately, insufficient data exist to define either height or cover classes. Any decision will therefore be largely empirical, and for woodlands, experience suggests modal values or cut off points different from those suggested by Specht (1981).

   The application of both crown and projective foliage cover to delineate cover classes is discussed by Walker & Hopkins (1984). Although classes based on crown separation are preferred, projective foliage cover (pfc) is applied to maintain compatibility with other chapters. 'Open woodlands' are arbitrarily defined as having a projected foliage cover of 0.1–5 per cent (i.e. crowns separated by 2–30 crown diameters) and 'woodlands' have values between 5 and 60 per cent (i.e. less than 2 crown diameters apart to crowns touching).

   The following height classes are similarly applied to Australian woodlands: very tall > 20 m; tall 12–20 m; medium, 6–12 m; low, 3–6 m; and very low, 2–3 m). The relationship between these classes and the Specht *et al.* (1974) classes is shown in Fig. 8.1.

3. The traditional pseudo-floristic/physiognomic use of the dominant growth form and bark type has been retained where possible, for example, box and paperbark woodlands.

4. As discussed above, it appears useful to describe the primary branching character of woody plants using terms other than tree and shrub, e.g 'monopodial' and 'sympodial'. For example, monopodial box-barked (*Eucalyptus microcarpa*) woodlands of central eastern Australia can be distinguished from the sympodial (mallee) form of the same box-species in harsher environments to the south and west. For mulga (*Acacia aneura*), there is change from a mono-

podial to a sympodial structure along a steep mesic to arid continental gradient.

This classification of Australian woodlands purposely ignores a primary division based on seasonal leaf fall (*cf.* Fosberg, 1970) as the majority of woodlands are evergreen. The only partly or wholly deciduous woodlands that occur in northern Australia are dealt with here as an 'anomalous' type. In the extreme south of the continent, there are minor outliers of deciduous woodlands or 'shrublands' of *Nothofagus gunnii* in parts of the central plateau of Tasmania. These extremes merely serve to emphasise the inappropriateness of tabular classifications in dealing with important aspects of vegetation dynamics. The attributes which combine to give a classification of woodland type are shown in Table 8.1.

Woodland distribution and bioclimate

On a continental scale there is a close correspondence between broad woodland structure and bioclimate as defined by indices which reflect plant growth responses to varying combinations of water, light and temperature. Bioclimate is therefore used here as a deterministic framework for woodland types. Whilst, at this scale, edaphic and geomorphological factors appear to be subsidiary determinants to

Table 8.1. *Classification of Australian woodland types*

Woodland type	Structural class	Cover class	(i) woodland 10–60% pfc
			(ii) open woodland 0.1–10% pfc
		Height class	(i) very tall >20 m
			(ii) tall 12–20 m
			(iii) medium 6–12 m
			(iv) low 3–6 m
			(v) very low 2–3 m
	Physiognomic class	Stem class	(i) monopodial (M)
			(ii) sympodial (S)
		Other (bark, leaves, etc.)	

Example: Low (M) paperbark woodland *(Melaleuca viridiflora/M. nervosa).*

bioclimate, a significant exception may be structurally depauperate woodlands on extremely low nutrient soils, especially in climatically buffered maritime regimes, e.g. some of the older subcoastal dune landscapes in temperate and tropical latitudes.

In Walker & Gillison (1982), a set of bioclimatic Provinces for the Australian region was recognised on the basis of plant growth indices together with a 'seasonality' coefficient which expressed variability of annual available water (CV per cent). The general bioclimatic models have been described by Nix (1976) and the specific indices used were calculated by H.A. Nix using several unpublished routines. Further detail on the construction of bioclimatic Provinces is available in Walker & Gillison (1982) and Gillison (1983).

The distribution of the bioclimatic Provinces follows the provisional scheme of Gillison (1983) (Fig. 8.2) with the derivation of the coding detailed in Table 8.2. Each Province is coded according to its region, sub-region and seasonality type, e.g. Province 3 is coded

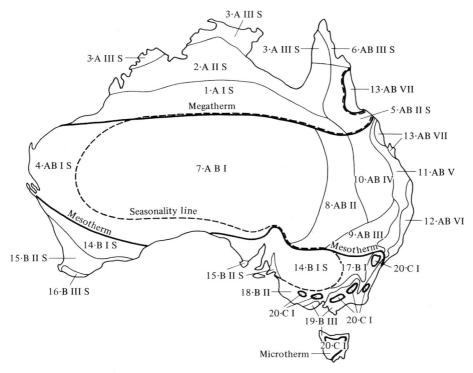

Fig. 8.2. Distribution of bioclimatic provinces in Australia. (See Table 8.2 for details.)

Table 8.2. *Bioclimatic provinces for Australia. See Walker and Gillison (1982) for further details*

Province No.	Code	Bioclimatic region	Bioclimatic sub-region	Thermal optima°C
1	AIS	Megatherm	I	28
2	AIIS		II	28
3	AIIIs		III	28
4	ABIS	Megatherm–	I	28/19
5	ABIIS	Mesotherm	II	28/19
6	ABIIIS		III	28/19
7	ABI	Megatherm–	I	28/19
8	ABII	Mesotherm	II	28/19
9	ABIII		III	28/19
10	ABIV		IV	28/19
11	ABV		V	28/19
12	ABVI		VI	28/19
13	ABVII		VII	28/19
14	BIS	Mesotherm	I	19
15	BIIS		II	19
16	BIIIS		III	19
17	BI	Mesotherm	I	19
18	BII		II	19
19	BIII		III	19
20	CI	Microtherm	I	12

as AIIIS which indicates it belongs to the megatherm region (A), in subregion III and is seasonal (S).

The plant growth index value chosen as a bioclimatic boundary was selected in some cases on the basis of a coincidence with known major vegetation boundaries, e.g. desert-open woodland boundary in northern Australia. To avoid circularity it is emphasised that the bioclimatic Provinces are for descriptive purposes only to indicate the important plant growth conditions experienced by major woodland types.

Broad floristics of Australian woodlands

Woodlands range in species richness of perennial flowering plants from as low as three species per hectare in some low open-woodlands in the interior to 80 species per hectare in some low nutrient, sub-coastal woodlands (Gillison, 1993). Their floristic composition is unique as the majority of dominant woody plants are composed of sclerophyllous taxa from within the Myrtaceae (*Eucalyptus, Melaleuca*) and Fabaceae (*Acacia*). Of the 530 eucalypt taxa currently described, approximately 80 per cent occur in woodlands. Most comprise the relatively small 'subgenera' Blakella, Corymbia, Eudesmia, Idiogenes and the largest 'sub-genus' Symphyomyrtus (after the informal classification of Pryor & Johnson, 1971). The 'sub-genus' Monocalyptus is more-or-less restricted to the more dense forest formations of the coastal and upland regions of the east, although some species occur in highland woodland formations (e.g. *E. coccifera, E. niphophila, E. pauciflora, E. stellulata*).

As with *Eucalyptus*, most *Acacia* spp. are included in woodland formations. *Acacia* tends to replace *Eucalyptus* in northern areas where annual rainfall is < 600 mm, except where there is extra run-on water (see Pryor, 1959, quoted in Pedley, 1978).

Apart from the floristically and structurally dominant *Acacia* and *Eucalyptus*, the following subsidiary woody perennial taxa characterise woodlands: Anacardiaceae (*Buchanania*), Bombacaceae (*Adansonia, Bombax*), Casuarinaceae (*Allocasuarina, Casuarina*), Cochlospermaceae (*Cochlospermum*), Combretaceae (*Terminalia*), Cycadaceae (*Bowenia, Cycas, Macrozamia*), Euphorbiaceae (*Excoecaria, Petalostigma*), Fabaceae (*Albizia, Lysiphyllum* (syn. *Bauhinia*), *Cassia, Erythrophleum*), Malvaceae (*Abelmoschus, Hibiscus*), Myrtaceae (*Melaleuca*), Palmae (*Livistona*), Proteaceae (*Banksia, Grevillea, Hakea, Persoonia*), Rhamnaceae (*Alphitonia, Ventilago*), Rubiaceae (*Gardenia*), Rutaceae (*Atalaya, Eremocitrus, Eremophila, Flindersia, Geijera*), Sapindaceae (*Dodonaea, Heterodendrum*), Sapotaceae (*Planchonia*), Sterculiaceae (*Brachychiton*), and Xanthorrhoeaceae (*Xanthorrhoea*). Whereas some lianes are present in the wetter margins (Apocynaceae, Asclepiadaceae, Vitaceae) they are typically absent from Australian woodland formations. Many of these families and genera intergrade with woodland systems of Papua New Guinea and it is likely many would have occupied the formerly exposed Sahul shelf, a former Pleistocene land bridge linking northern Australia and the present island of Papua New Guinea (Chapter 2, Walker & Singh, 1981). Other than this, the continent with greatest floristic affinity in woodland taxa is Africa where convergent structural development is evident, particularly in the Fabaceae (*Acacia*) and in the Combretaceae and Sterculiaceae. It is of interest that in the Bombacaceae, the baobab *Adansonia*, a characteristically African genus, occurs in the far northwest of Australia and may indicate an extant Gondwanan link (*cf.* Armstrong, 1977) (Fig. 8.3).

Fig. 8.3. Baobab woodland (*Adansonia gregorii*) on a sandstone scree slope in the Newcastle Range, northwestern Australia. (Photo: A.N. Gillison.)

Adaptive traits in Australian woodlands

Although floristic assemblages in Australian woodlands tend to be discrete, certain genera (e.g. *Alphitonia, Dolichandrone, Flindersia, Livistona, Ventilago*) indicate links with a closed seasonal woody vegetation which may have been far more extensive than at present, rather like the so-called 'monsoon' forest in northern Australia which has presumably fluctuated with the arid cycles (the most recent of which took place probably between 10 000 and 15 000 years ago) *cf.* Gillison (1983), Russell-Smith (1991). It is likely that the Quaternary climatic 'norm' was considerably wetter than at present, so that woodlands now exist in a relatively arid phase in a condition that has been further modified by humans in more recent times through changes in fire regimes and the introduction of exotic grazing and browsing animals. Present woodlands therefore reflect a confounded aspect of these elements, both past and present, but there are some overall aspects that strongly suggest adaptation to environmental pressures, of which fire and water stress are paramount.

Many of the life forms and leaf features indicate a xerophilous trend.

Leaf xeromorphy is well developed (i.e. thickened cuticles, increased glaucousness, rolled margins, dense indumentum, high stomate density, high specific gravity and volatile oils, and so on). Woodland eucalypt leaves are commonly pendulous, isobilateral sclerophylls with a high oil content, and there is a unique development of phyllodineous acacias. (Convergence to phyllode-like leaves is also evident in the geographically widespread *Melaleuca leucadendron* alliance). These features, together with solid leaf types, such as *Hakea*, *Callitris* (Fig. 8.4) and the phyllocladous Casuarinaceae characterise many Australian woodland plants. Reduction in leaf size and internode distance is correlated with increased water stress and is conspicuous in rosette forms as in *Cycas media* (Fig. 8.5) and *Pandanus* spp. In the more seasonal regions, especially on heavy cracking clays and/or limestone soils, there is pronounced deciduousness characterised by dominant *Lysiphyllum* (formerly *Bauhinia*) and *Terminalia* in northern Australia. Increases in the amount of volatile oils (Rutaceae, Myrtaceae) and laticiferous tissue (Apocynaceae, Euphorbiaceae, Moraceae) appear to be associated with adaptation to increasing water stress. Xylem water potentials in excess of 7.0 Pa have been recorded by the author in some perennial woody myrtaceous species (Gillison, unpublished observations) in semi-arid woodlands.

Fig. 8.4. Solid-leaved *Callitris intratropica* (dark crown) among eucalypts in a Northern Territory woodland. (Photo: A.N. Gillison.)

Fig. 8.5. Rosette-crowned *Cyas media* understorey showing post-fire foliar regeneration in a woodland in Province 3. (Photo: A.N. Gillison.)

The distribution of 'functional' (i.e. chlorophyllous) leaf tissue is strongly correlated with rainfall seasonality gradients. The proportion of woody species with a green stem subrhytidome increases with seasonality to a point where the photosynthetic apparatus of some species, e.g. *Apophyllum anomalum* may be almost entirely composed of green stems as in some desert green stem succulents. The pattern of different acid decarboxylation pathways in woodland graminoids indicates a marked increase in the proportion of C_4 species towards the warmer and more seasonal woodlands, with C_3 being more evident in the more mesotherm and microtherm areas of the continent (Johnson & Tothill, 1985; Gillison, 1993). The proportion of succulent terrestrial plant species with crassulacean acid metabolism (CAM) pathways also increases towards the more seasonal, megatherm areas and on solodic soils, e.g. on some of the outwash plains of the Gulf of Carpentaria and in seasonally inundated woodlands. CAM epiphytes also occur in megatherm seasonal woodlands and open-forests towards the wetter margins (*Cymbidium, Dendrobium, Dischidia, Hoya*).

Resistance to wildfire is well developed, especially in *Eucalyptus*, where below-ground lignotuberous and sometimes rhizomatous organs or root suckers allow for rapid vegetative regeneration after fire (Jacobs, 1955; Carrodus & Blake, 1970; Lacey, 1974). In some eucalypt species, above-ground epicormic shoots and specific bark types also assist in recovery from fire. Many other woodland genera have below- and above-ground adaptations which ensure adequate fire survival (Gill, 1980, 1981*a*,*b*). Acacias as a whole appear better equipped to withstand extremes of water stress rather than fire, unlike the eucalypts. Mulga (*Acacia aneura*) is capable of withstanding a pressure of −120 bars (Slatyer, 1962) and brigalow (*A. harpophylla*) can withstand −50 bars (Connor & Tunstall, 1968). Australian acacias are distinguished from their African counterparts by the very low incidence of thorns, which may indicate an historical lack of association with an African-type browsing 'savanna' fauna.

Description of selected woodland types

Very tall woodland

Physiognomic class: (M) woollybutt-stringy bark (*Eucalyptus miniata, E. tetrodonta*)

Provinces 2, 3. As indicated in the accompanying diagram (Fig. 8.6), the type structure is variable and changes from short regenerative 'pole' woodland with monopodial stems (Fig. 8.7) to tall woodland that becomes increasingly dense towards Province No. 3.

Understorey tall shrubs (*Acacia* spp., *Gardenia megasperma, Grevillea pteridifolia*) have a crown cover of 20 per cent, and graminoids occur often with a proliferation of small shrubs, *Acacia* spp., *Bossiaea bossiaeoides* and *Petalostigma* spp. This association is usually found on deeply-weathered red sandy soils where, presumably, it is able to tap water at depth. Leaf size varies from microphyll to notophyll (*sensu* Webb, 1959) for *Eucalyptus miniata* and *E. tetrodonta*. In this relatively low nutrient environment with annual ground fires, a conspicuous divergence of life forms is evident in both eucalypts. In *E. miniata*, the one species may occur as two life forms; a tall seed-producing tree out of fire reach, and a low, ligneous rosette propagated by vegetative shoots at or below ground-level in response to fire. *Eucalyptus tetrodonta* is capable of producing extensive root suckers − to date, the only documented case within *Eucalyptus*. This facultative characteristic is evident in many other plants in this bioclimatic Province (see also Lacey, 1974) and may represent an evolution-

Projective foliage cover (pfc)

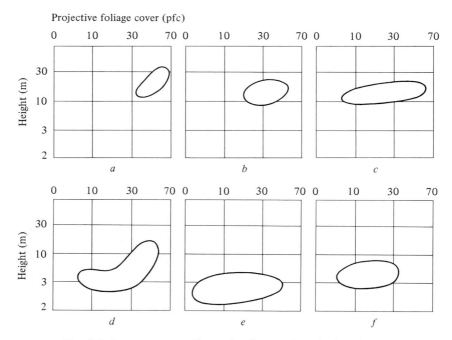

Fig. 8.6. Structuregram of woodland types described in this chapter: (a) Very tall woodland, (M) woollybutt - stringybark (*Eucalyptus miniata, E. tetrodonta*); (b) Tall woodland, (M) grey box (*E. moluccana*); (c) Medium height woodland, (M) poplar box (*E. populnea*); (d) Low woodland, (M) paperbark (*Melaleuca viridiflora, M. nervosa); (e)* Very low woodland, (M) silverleaf box and snappy gum (*E. pruinosa, E. brevifolia*); (f) Low (deciduous) woodland, (M) deciduous *Lysiphyllum, Terminalia*.

ary response to a relatively high fire frequency and low nutrient status. The remarkable capacity of plants in this area to produce underground storage and reproductive tissue is similar in some ways to the geoxylic suffrutices described by White (1977) for parts of central Africa.

On Melville Island in northwestern Australia, the understoreys commonly contain tall, broadleaved shrubs of *Gardenia, Grevillea* and *Owenia* which are replaced in mesic sites by *Acacia, Alphitonia, Cycas* and *Phyllanthus*. In the west of Province 3, *Cycas media* and the palm *Livistona humilis* are frequent understorey elements.

Fig. 8.7. Regenerating very tall woodland of woollybutt (*Eucalyptus miniata*) in northern Australia. Note monopodial form. (Photo: A.N. Gillison.)

Tall woodland

Physiognomic class (M) grey box (*Eucalyptus moluccana*)

Provinces 12, 13, (6). The grey box woodlands form a closely knit taxonomic assemblage comprised of a number of species but mostly of *E. microcarpa* and *E. moluccana*. The former tends to occupy the semi-arid regions of eastern Australia and the latter the wetter zones to the east. As indicated in the structuregram (Fig. 8.6), *E. moluccana* extends into tall open-forest and sometimes closed-forest categories but is commonly tall woodland (Fig. 8.8). The best form of this woodland occurs as pure stands 17 m tall with 50 per cent foliage cover in Province 12, as *E. moluccana* subsp. *queenslandica*. In the more seasonal Province 6 a shorter, more open woodland of subsp. *crassifolia* is common (14 m tall with 45 per cent pfc).

Grey box woodlands become more dense to the south (Province 13) with subsp. *pedicellata* and subsp. *moluccana*. The species is commonly contiguous with ironbark communities that tend to be upslope (*E. crebra*, *E. drepanophylla*, *E. sideroxylon*) or else with

Fig. 8.8. Tall woodland of gum-topped box (*Eucalyptus moluccana*) on solodic soils in Province 6. (Photo: A.N. Gillison.)

blue gums (*E. tereticornis*) downslope on poorly drained flats. In Province 13 it is commonly associated with spotted gum (*E. maculata*). In the northern Provinces 6, 12 and 14 soils tend to be agronomically poor solidics, whereas in the southern Province 13 grey box woodlands are common on brown earths of high agronomic value and this has led to the elimination of many fine stands in favour of cultivated crops and pastures.

Understorey components and strata are both variable. In the north (Provinces 6, 12, 14) mid to tall grasses are common (*Dichanthium* spp., *Heteropogon contortus*, *Imperata cylindrica* and *Themeda triandra* (formerly *T. australis*)) and shrubs of *Acacia* spp. are plentiful. In Province 13 the woodlands have an understorey shrub layer of *Acacia*, *Melaleuca* and *Pultenaea* spp., with graminoids *Dianella*, *Lomandra*, and short tussock grasses *Poa* and *Imperata*.

Within its broad geographic range, genetically fixed growth forms correspond closely with variation in climate. Gillison (1976) found that, when grown in a controlled environment, seedlings from northern Provinces were monopodial, and large-leaved, with small lignotubers, whereas seedlings from the southern Province 13 were shorter, hemi-sympodial, with large lignotubers and relatively small leaves. At a remote southern locality at Bungonia, southwest of Sydney, on low-nutrient soil, Gillison, Lacey & Bennett (1980) subsequently found that *Eucalyptus moluccana* subsp. *moluccana* was capable of reproducing vegetatively by means of rhizo-stolons. This feature, combined with the overall capacity of grey box to coppice following stem damage is indicative of the range of survival modes in a number of species within the genus *Eucalyptus*.

Other tall woodlands in eastern Australia include the ironbarks, *Eucalyptus crebra*, *E. drepanophylla*, *E. melanophloia* and *E. sideroxylon* and the bloodwoods *E. intermedia* and *E. polycarpa*.

In far northern Australia in Provinces 2 and 3, the woollybutt-stringy bark associations also occur as tall woodlands.

Woodlands of medium height

Physiognomic class: (M) poplar box (*Eucalyptus populnea*)

Provinces 9, 10, 11, 12 (18). In Australia, woodlands of medium height are the most widespread. In the example given, the geographic range is over 10° of latitude and, as with grey box, the poplar box woodlands appear to be similarly differentiated in growth form relative to climate, ranging from monopodial forms in the north (Province 14) to hemi-sympodial forms in the south (Province 10) (Fig. 8.9).

Fig. 8.9. Poplar box (*Eucalyptus populnea*) tall woodland in Province 11, showing hemi-sympodial tendency of southern populations of this species. (Photo: A.N. Gillison.)

Poplar box is typical in form of many other eucalypt box-barked species (so-called because of the bark resemblance to turkey box, *Buxus sempervirens*) of the northern hemisphere that form woodlands of medium height, in particular, *E. microcarpa*, *E. microneura*, *E. normantonensis* and *E. tectifica*. The structure of poplar box communities varies with soil type, the red and brown solodics being the most important. The community usually has a simple, two-layered, tree-grass structure, on heavy clays in Province 11, often with *E. orgadophila*, whereas on red earths, extensive shrub associates are common, particularly where grazing has been intensive (the semi-arid shrub woodlands of Moore & Perry, 1970).

Non-eucalypt low trees associated with poplar box on red earths are *Acacia excelsa*, *A. oswaldii*, *Albizia basaltica*, *Flindersia dissosperma*, *Geijera parviflora*, *Heterodendrum oleifolium*, *Lysiphyllum carronii* and *Ventilago viminalis*. On sandy soils, a number of other *Acacia* spp. exist, often with *Callitris columellaris*, and on heavier clays, isolated pockets of *Casuarina cristata* and *C. luehmannii* are common. The eucalypts *E. cambageana*, *E. orgadophila* and *E. thozetiana* are also common on these heavy soil types together with brigalow (*Acacia harpophylla*).

The grasses *Aristida contorta* and *A. jerichoensis* are common throughout poplar box woodlands, whilst in the northern areas *A. inaequiglumis* and *A. pruinosa* are also frequent. Other grass species in the central and southern Provinces are *Bothriochloa decipiens*, *Chloris acicularis* and *C. truncata*, with *Neurachne mitchelliana* and *Stipa variabilis*.

Other tree species that make up medium-height woodlands are too numerous to list here. In northern Australia, the bloodwoods *Eucalyptus dichromophloia*, *E. nesophila*, *E. polycarpa* and *E. porrecta* are typical tree species together with the ironbarks *E. crebra*, *E. drepanophylla*, *E. jensenii*, *E. melanophloia* and *E. shirleyi*. The gum-barked eucalypts *E. alba*, *E. brevifolia*, *E. confertiflora* and *E. papuana* are also conspicuous elements. In the central eastern and southern Provinces west of the Great Dividing Range in eastern Australia, box and ironbark woodlands are common. Apart from poplar box, *E. microcarpa*, *E. odorata* and *E. pilligaensis* are notable dominants in southern box woodlands, although in Province 15 these tend to give way to mallee formations. In Western Australia, *E. ptychocarpa* and *E. salmonophloia* are typical of eucalypt species that form medium to low woodlands on sandy soils.

Low woodland

Physiognomic class: (M) (*Melaleuca viridiflora* and *M. nervosa* complex) (Fig. 8.10)
Provinces 2, 6 (12). Low woodlands are most common in the semi-arid, highly seasonal bioclimatic Provinces in northern Australia, but also occur along the eastern fringe of Queensland, especially in Province 6, in swampy (low-nutrient) localities. A variety of eucalypt species may occur in low woodlands (*E. brevifolia*, *E. dichromophloia*, *E. jensenii*, *E. papuana*, *E. pruinosa*) but perhaps the most unique and physiognomically significant is paperbark woodland. This is characteristically a woodland type dominated by monopodial species of *Melaleuca*, of which the two most common are *M. nervosa* and *M. viridiflora*. The unique feature of this woodland is the laminated paper-thin sections of bark and often mesophyll-sized coriaceous, phyllode-like leaves. The most widespread paperbark woodlands are in the floodplain areas of the Gulf of Carpentaria, although they

Fig. 8.10. Low woodland of the paperbark *Melaleuca nervosa* in Province 2, northern Australia. (Photo: A.N. Gillison.)

occur throughout the megatherm seasonal region of both Australia and Papua New Guinea. Paperbarks, or 'tea-trees' as they are sometimes known, may be dwarfed or twisted, usually about 3–7 m tall but may assume 'pole' forest proportions in backing swamps, often up to 18–20 m tall.

The *M. viridiflora* community is usually two-layered with a short to mid-height (10–15 cm) ground layer of graminoids. Although both *M. viridiflora* and *M. nervosa* occur mainly on solodic plains, they may also occur on quartzite ridges or black soil flats, associated variously with the bloodwoods, *E. dichromophloia* and *E. polycarpa* and particularly deciduous species of *Terminalia* (*T. aridicola*, *T. oblongata*, *T. platyphylla*) and the deciduous *Lysiphyllum* (syn. *Bauhinia*) *cunninghamii*. *Melaleuca tamarascina* sometimes occurs on solodic flats either as a co-dominant with *M. viridiflora* or else as a pure stand. Other species of *Melaleuca* (*M. argentea*, *M. cajuputi* and *M. leucodendron*, *sensu lato*) are scattered throughout the region usually in mixed stands with other woody genera (*Terminalia*, *Petalostigma*, *Alphitonia* and eucalypts).

The low-nutrient site conditions that support paperbark communities usually carry only very mild fires as a consequence of a low fuel load. Because of their unique bark, the melaleucas are also fire-tolerant. Together with the capacity of paperbarks to establish profusely from seed, these factors combine to make the communities highly effective colonisers and competitors in woody plant communities. This capability has created an endemic 'woody weed' problem in parts of northern Australia (Fig. 8.11) and an exotic 'woody weed' problem in Florida in the United States of America (*M. quinquenervia*).

Woodlands also occur in the subalpine regions of Australia where the relatively harsh environment is associated with a reduction in tree height. Such woodlands consist commonly of the snow-gums *Eucalyptus niphophila* and *E. pauciflora* described in Chapter 16 (Costin, 1981). (Fig. 8.12).

Very low woodland

Physiognomic class: (M) silver-leaf box and snappy gum (*Eucalyptus pruinosa-E. brevifolia*)

Provinces 2, 3. These woodlands are characterised by dwarf, mostly monopodial trees 2–4 m tall that vary locally from 2 to 60 per cent cover. Crowns are usually dense and rounded and the communities are typically two-layered, tree-graminoid associations. Such

Fig. 8.11. Paperbark or tea tree (*Melaleuca viridiflora*) regeneration as a woody weed in Province 6, northeastern Queensland. (Photo: A.N. Gillison.)

Fig. 8.12. Low alpine woodland of *Eucalyptus niphophila* in the Australian alps showing almost complete monopodial form. (Photo: A.N. Gillison.)

woodlands occur in harsh environments and exist mainly in Province 2, which is characterised by periods of extreme water stress.

In such an environment that usually includes low-nutrient, shallow soils, dwarf woodland trees tend to be evergreen and relatively fire- and drought-resistant. Trees that occupy this type commonly range into low woodland. In Province 2, dwarf woodland eucalypts are of two leaf types; either small (microphyll) and shiny (*E. brevifolia* and *E. jensenii*) or large (approaching mesophyll size) and glaucous as in *E. pruinosa*.

The distribution of snappy gum (*E. brevifolia*) dwarf woodland ranges from the Ord River in northwestern Australia, through to the Leichhardt River that flows into the Gulf of Carpentaria and commonly occurs on sandstone or quartzite ridges. Silver-leaf box (*E. pruinosa*) tends to favour heavier soils under these extreme seasonal conditions and occupies much the same geographic range.

The ironbark *E. jensenii* is restricted more to northwestern Australia on shallow stony slopes and ridges (Fig. 8.13). Grass species commonly associated with this woodland type are the hummock

Fig. 8.13. Very low woodland of ironbark (*Eucalyptus jensenii*) in the Kimberley region of northwestern Australia. (Photo: A.N. Gillison.)

'spinifex' grasses (*Triodia, Plectrachne*) and the low grasses *Eragrostis* and *Chrysopogon*. These woodlands represent the poorer pasture associations in northern Australia.

Low (deciduous) woodland

Physiognomic class: (M) deciduous *Lysiphyllum, Terminalia* Provinces 1, 2, 3

Throughout the megatherm seasonal region, run-on basins with seasonally-inundated cracking clays are characterised by a unique vegetation type. It is a deciduous variant of the low, evergreen eucalypt woodlands. Here the non-eucalypt genera *Lysiphyllum* (Caesalpinaceae) and *Terminalia* (Combretaceae) predominate (Fig. 8.14).

Stands vary floristically from mixtures to pure stands of each genus, sometimes associated with semi-deciduous eucalypts (*E. alba, E. foelscheana* and *E. bleeseri*). During the post-wet season, these localities support a simple two-layered community of trees and mid-height grasses with the tree crowns dense and green. Under conditions of

Fig. 8.14. Low deciduous *Lysiphyllum, Terminalia* woodland in the Gulf country of northeastern Australia. (Photo: N.H. Speck.)

increasing water-stress towards the latter part of the year, the grasses are usually fired and the trees become either totally or partially deciduous.

A peculiar phenomenon associated with these deciduous woodlands is the precocious growth flush and flowering that takes place about two months before the onset of the monsoonal rains. This feature has not yet been fully explained, but is presumably associated with the highly predictable rainy season where a plant in an otherwise deciduous community can be assumed to have a competitive advantage if it is in full or partial flush before the soil-water store is replenished.

Lysiphyllum cunninghamii, Terminalia aridicola, T. ferdinandiana, T. oblongata and *T. platyphylla* are among the most common species. Other non-eucalypts, particularly in subcoastal areas are *Dolichandrone, Erythrophleum, Ventilago* and *Excoecaria*. Associated grass genera are *Aristida, Astrebla, Chrysopogon, Eragrostis, Sorghum* and *Themeda*.

The type is not restricted to northern Australia and is widespread in the megatherm region of Papua where genera of Malesian affinity (*Albizia, Antidesma, Bombax, Cordia, Desmodium* and *Timonius*) co-exist with eucalypts and terminalias. Detailed descriptions for deciduous woodlands in Australia are generally lacking, although there are limited descriptions in Specht (1958), Perry & Lazarides (1964), Speck (1965), and Story (1970). Minor outliers of sympodial *Nothofagus gunnii* (deciduous beech) woodlands mixed with gymnospermous *Athrotaxus selaginoides* and *Phyllocladus aspleniifolius* occur in parts of the central plateau of Tasmania usually above 1000 m elevation on exposed scree slopes (Fig. 8.15). Despite their restricted distribution they are biogeographically significant as they share close structural and floristic ties with more extensive Gondwanan counterparts in the southern extremities of South America.

Conclusions

In this descriptive and very limited treatment, the variable nature and unique physiognomic character of Australian woodlands has been emphasised. For reasons of brevity important dynamic aspects of woodlands have not been discussed. The effects of disturbance on woodland systems have received attention in more recent literature particularly in relation to grazing by domestic animals. Herbage and shrub response to woodland thinning have been documented by

Fig. 8.15. Low deciduous beech (*Nothofagus gunnii*) woodland outlier in the microtherm Cradle Mountain area of Tasmania. (Photo: A.N. Gillison.)

Walker, Moore & Robertson (1972), and Beale (1973), and responses to fire have been reviewed by Leigh & Noble (1981), Stocker & Mott (1981), Gill (1981*b*), Lacey, Walker & Noble (1981), Gillison (1983), Mott & Tothill (1984), Harrington *et al.* (1984) and Johnson & Tothill (1985). An important aspect of woodland dynamics is the control of unwanted shrubs for grazing purposes (Moore & Walker, 1972; Hodgkinson *et al.*, 1984). Woody plant responses and problems associated with overgrazing in the poplar box (*E. populnea*) woodlands are outlined in Tunstall *et al.* (1981) and have been reviewed more recently by Moore (1993).

Our present understanding of woodland dynamics is far from complete and more work in this area should be carried out in the future. The descriptive framework outlined here may provide one useful focus for such research.

References

Armstrong, P. (1977). Baobabs: remnants of Gondwanaland? *New Scientist*, 73, 212–13.

Beale, I.F. (1973). Tree density effects on yields of herbage and tree components in south west Queensland mulga (*Acacia aneura* F. Muell.) scrub. *Tropical Grasslands*, 7, 135–42.

Beard, J.S. & Webb, M. (1974). *Vegetation Survey of Western Australia: Great Sandy Desert – Part 1, Aims, Objectives and Methods*. Perth: University of Western Australia Press.

Carnahan, J.A. (1976). Natural vegetation. In *Atlas of Australian Resources*, Second Series. Canberra: Department of National Resources.

Carnahan, J.A. (1990). *Vegetation*. In *Atlas of Australian Resources*, Third Series, 6. J.A. Carnahan (Compiler), Australian Land Surveying and Land Information Group, Canberra: Dept of Administrative Services.

Carpenter, J. (1938). *An Ecological Glossary*. New York: Hafner (1962 reprint).

Carrodus, B.B. & Blake, T.J. (1970). Studies on the lignotubers of *Eucalyptus obliqua* L'Herit. I. The nature of the lignotuber. *New Phytologist*, **69**, 1069–72.

Connor, D.J. & Tunstall, B.R. (1968). Tissue water relations for brigalow and mulga. *Australian Journal of Botany*, **16**, 487–90.

Costin, A.B. (1981). Alpine and sub-alpine vegetation. In *Australian Vegetation*, ed. R.H. Groves, pp. 361–76. Cambridge University Press.

Fosberg, F.R. (1970). A classification of vegetation for general purposes. In *Guide to the Check Sheet for IBP Areas (IBP Handbook No. 4)*, ed. G.F. Peterken, Appendix 1, pp. 73–120. Oxford & Edinburgh: Blackwell Scientific Publications.

Gill, A.M. (1980). Fire adaptive traits of vascular plants. In *Fire Regimes and Ecosystem Properties*, eds H.A. Mooney, J.M. Bonnicksen, N.L. Christensen, J.E. Lotan & W.A. Reiners, pp. 208–30. Washington, DC: USDA Forest Science General Technical Report.

Gill, A.M. (1981*a*). Patterns and processes in open-forests of *Eucalyptus* in southern Australia. In *Australian Vegetation*, ed. R.H. Groves, pp. 152–76. Cambridge University Press.

Gill, A.M. (1981*b*). Adaptive responses of Australian vascular plant species to fires. In *Fire and the Australian Biota*, eds A.M. Gill, R.H. Groves & I.R. Noble, pp. 243–71. Canberra: Australian Academy of Science.

Gillison, A.N. (1970). Dynamics of biologically-induced grassland-forest transitions in Papua New Guinea. MSc thesis, Australian National University.

Gillison, A.N. (1976). Taxonomy and autecology of the Grey Box *Eucalyptus moluccana* Roxb. (*s. lat.*). PhD thesis, Australian National University.

Gillison, A.N. (1981). Towards a functional vegetation classification. In *Vegetation Classification in Australia*, eds A.N. Gillison & D.J. Anderson, pp. 30–41. Canberra: CSIRO & Australian National University Press.

Gillison, A.N. (1983). Tropical savannas of Australia and the Southwest Pacific. In *Tropical Savannas*, ed. F. Bourlière, *Ecosystems of the World*, Vol. 13, pp. 183–243. Amsterdam: Elsevier Scientific Publishing Company.

Gillison, A.N. (1988). *A Plant Functional Proforma for Dynamic Vegetation Studies and Natural Resource Surveys*. Technical Memorandum 88/3, CSIRO Division of Water Resources, Canberra.

Gillison, A.N., (1993*a*). Overview of the Grasslands of Oceania. In Natural Temperate Grasslands: Eurasia, Africa, South America and Oceania, ed. R.T. Coupland. In *Ecosystems of the World*, ed. D.W. Goodall, p. 303–13. Amsterdam: Elsevier Scientific Publishing Company.

Gillison, A.N., (1993*b*). Grasslands of the South-West Pacific. In Natural Temperate Grasslands: Eurasia, Africa, South America and Oceania, ed. R.T. Coupland. In *Ecosystems of the World*, ed. D.W. Goodall, p. 435–69. Amsterdam: Elsevier Scientific Publishing Company.

Gillison, A.N., Lacey, C.J. & Bennett, R.H. (1980). Rhizo-stolons in *Eucalyptus*. *Australian Journal of Botany*, **28**, 229–304.

Gillison, A.N. & Walker, J. (1981). Woodlands. In *Australian Vegetation*, ed. R.H. Groves, pp. 177–97. Cambridge University Press.

Harrington, G.N., Mills, M.D., Pressland, A.J. & Hodgkinson, K.C. (1984). Semi-arid woodlands. In *Management of Australia's Rangelands*, eds G.N. Harrington, A.D. Wilson & M.D. Young, pp. 189–208. Melbourne: CSIRO.

Hodgkinson, K.C., Harrington, G.N., Griffin, G.F., Noble, J.C. & Young, M.D. (1984). Management of vegetation with fire. In *Management of Australia's Rangelands*, eds G.N. Harrington, A.D. Wilson & M.D. Young, pp. 141–56. Melbourne: CSIRO.

IUCN (1973). *A Working System for Classification of World Vegetation*. IUCN Occasional Paper No. 6. Morges: International Union for Conservation of Nature and Natural Resources.

Jacobs, M.R. (1955). *Growth Habits of the Eucalypts*. Canberra: Government Printer.

Johnson, R.W. & Burrows, W.H. (1981). *Acacia* open-forests, woodlands and shrubland. In *Australian Vegetation*, ed. R.H. Groves, pp. 198–226. Cambridge University Press.

Johnson, R.D. & Tothill, J.C. (1985). Definition and broad geographic outline of savanna lands. In *Ecology and Management of the World's Savannas*, eds J.C. Tothill & J.J. Mott, pp. 1–13. Canberra: Australian Academy of Science, in association with C.A.B., Farnham Royal, Bucks.

Johnson, R.D. & Lacey, C.J. (1984). A proposal for the classification of tree dominated vegetation in Australia. *Australian Journal of Botany*, **32**, 529–49.

Lacey, C.J. (1974). Rhizomes in tropical eucalypts and their role in recovery from fire damage. *Australian Journal of Botany*, **22**, 29–38.

Lacey, C.J., Walker, J. & Noble, I.R. (1981). Fire in Australian savannas. In *The Ecology of Tropical Savannas*, eds B.J. Huntley & B.H. Walker, pp. 246–72. Berlin: Springer-Verlag.

Leigh, J.H. & Noble, J.C. (1981). In *Fire and the Australian Biota*, eds A.M. Gill, R.H. Groves & I.R. Noble, pp. 471–95. Canberra: Australian Academy of Science.

Moore, R.M. (1993). Grasslands of Australia. In *Ecosystems of the World* Vol. 8B *Natural Temperate Grasslands: Eurasia. Africa, South America and Oceania*, ed. R.T. Coupland, pp. 315–60. Amsterdam: Elsevier Scientific Publishing Company.

Moore, R.M. & Perry, R.A. (1970). Vegetation. In *Australian Grasslands*, ed. R.M. Moore, pp. 59–73. Canberra: Australian National University Press.

Moore, R.M. & Walker, J. (1972). *Eucalyptus populnea* shrub woodlands. Control of regenerating trees and shrubs. *Australian Journal of Experimental Agriculture and Animal Husbandry*, **12**, 437–40.

Mott, J.J. & Tothill, J.C. (1984). Tropical and Sub-Tropical woodlands. In *Management of Australia's Rangelands*, eds G.N. Harrington, A.D. Wilson & M.D. Young, pp. 255–70. Melbourne: CSIRO.

Nix, H.A. (1976). Environmental control of breeding, part-breeding dispersal and migration of birds in the Australian region. In *Proceedings of the 16th International Ornithological Congress*, pp. 272–305. Canberra: Australian Academy of Science.

Nix, H.A. (1983). Climate of tropical savannas. In *Ecosystems of the World*, Vol. 13 *Tropical Savannas*, ed. F. Bourlière, pp. 37–62. Amsterdam: Elsevier Scientific Publishing Company.

Parsons, R.F. (1981). *Eucalyptus* shrubs and shrublands. In *Australian Vegetation*, ed. R.H. Groves, pp. 227–52, Cambridge University Press.

Pedley, L. (1978). A revision of *Acacia* Mill in Queensland. *Austrobaileya*, **1**, 75–337.

Perry, R.A. & Lazarides, M. (1964). Vegetation of the Leichhardt-Gilbert area. CSIRO, Australia, Land Research Series No. 11, pp. 152–91.

Pryor, L.D. & Johnson, L.A.S. (1971). *A Classification of the Eucalypts*. Canberra: Australian National University.

Russell-Smith, J. (1991). Classification, species richness, and environmental relations of monsoon rain-forest vegetation in the Northern Territory, Australia. *Journal of Vegetation Science*, **2**, 259–78.

Slatyer, R.O. (1962). Internal water balance of *Acacia aneura* F. Muell. in relation to environmental conditions. *Arid Zone Research*, **16**, 137–46.

Specht, R.L. (1958). The climate, geology, soils and plant ecology of the northern portion of Arnhem Land. In *Records of the American–Australian Scientific Expedition to Arnhem Land*, Vol. 3, *Botany and Plant Ecology*, eds R.L. Specht & C.P. Mountford, pp. 333–414. Melbourne University Press.

Specht, R.L. (1970). Vegetation. In *The Australian Environment*, (4th edn rev.), ed. G.W. Leeper, pp. 46–67. Melbourne: CSIRO & Melbourne University Press.

Specht, R.L. (1981). The use of foliage projective cover. In *Vegetation Classification in the Australian Region*, eds A.N. Gillison & D.J. Anderson, pp. 10–21. Canberra: CSIRO & Australian National University Press.

Specht, R.L., Roe, E.M. & Boughton, V.H. (1974). Conservation of major plant communities in Australia and Papua New Guinea. *Australian Journal of Botany*, Supplementary Series, No. 7.

Speck, N.H. (1965). Vegetation and pastures of the Tipperary area. CSIRO, Australia, Land Research Series No. 13, pp. 81–98.

Stocker, G.C. & Mott, J.J. (1981). Fire in the tropical forests and woodlands of northern Australia. In *Fire and the Australian Biota*, eds A.M. Gill, R.H. Groves & I.R. Noble, pp. 425–39. Canberra: Australian Academy of Science.

Story, R. (1970). Vegetation of the Mitchell-Normanby area. CSIRO, Australia, Land Research Series No. 26, pp. 75–88.

Tunstall, B.R., Torsell, B.W.R., Walker, J., Moore, R.M. Robertson, J.A. & Goodwin, W.F. (1981). Vegetation changes in a poplar box (*Eucalyptus populnea*) woodland. Effects of tree killing and domestic livestock. *Australian Rangelands Journal*, 3, 123–32.

Walker, J. & Gillison, A.N. (1982). Australian savannas. In *The Ecology of Tropical Savannas*, eds B.J. Huntley & B.H. Walker, pp. 5–24. Berlin: Springer-Verlag.

Walker, J. & Hopkins, M. (1984). Vegetation. In *Australian Soil and Land Survey Handbook*, 1, eds R.C. McDonald, R.F. Isbell & J.G. Speight, pp. 44–67. Melbourne: Inkata.

Walker, J., Moore, R.M. & Robertson, J.A. (1972). Herbage response of tree and shrub thinning in *Eucalyptus populnea* shrub woodlands. *Australian Journal of Agricultural Research*, 23, 405–10.

Walker, D. & Singh, G. (1981). Vegetation history. In *Australian Vegetation*, ed. R.H. Groves, pp. 26–43. Cambridge University Press.

Webb, L.J. (1959). A physiognomic classification of Australian rainforests. *Journal of Ecology*, 47, 551–70.

White, F. (1977). The underground forests of Africa. In *Tropical Botany*, eds D.J. Mabberley & Chang Kiaw Lan, pp. 57–71. *The Gardens Bulletin Singapore*, Vol. XXIX.

Williams, R.J. (1955). Vegetation regions. In *Atlas of Australia*, 1st Series. Canberra: Department of National Development.

9

Acacia *open-forests, woodlands and shrublands*

R.W. JOHNSON & W.H. BURROWS

ACACIA communities, together with the eucalypts, dominate the Australian landscape. Whilst *Eucalyptus* open-forests and woodlands are characteristic of semi-arid and more mesic coastal Australia, *Acacia* shrublands and low woodlands largely replace *Eucalyptus* communities in drier regions of the continent. They become dominant in southern areas when the annual rainfall is winter-incident and less than 250 mm. In northern Australia, they predominate in areas receiving less than 350 mm. Within the central and western arid zone, particularly in the northwest and in drier desert areas, *Acacia* often forms only a sparse shrub layer over *Triodia* grassland. In these communities, scattered mallee eucalypts may occur, although *Eucalyptus* is more frequently restricted to moister habitats along streams and around depressions. Structurally, the *Acacia* communities of central Australia are either shrublands or open-shrublands, although some of the important species, such as mulga (*A. aneura*), form woodlands on more favourable sites.

Acacia communities, however, also extend into more mesic areas in northeastern Australia, where they usually form single dominant open-forests and woodlands interspersed with areas of grassland and semi-arid eucalypt woodlands. In central Queensland, open-forests of brigalow (*A. harpophylla*) reach almost to the coast where they receive more than 900 mm a year. Much less significant areas of *Acacia* open-forests and woodlands are found in temperate southern Australia.

Even in *Eucalyptus* open-forests and woodlands, species of *Acacia* may be common understorey trees and shrubs, particularly on sandy and texture-contrast soils. A few species, such as *A. fasciculifera, A. hylonoma* and *A. bakeri*, occur in rainforests, usually where soils are of low fertility. Tussock grasslands, such as the Mitchell grass (*Astrebla*) and blue grass (*Dichanthium*) downs of Queensland, become associated with scattered trees and shrubs of a number of *Acacia* species.

Stebbins (1972) pointed out that shrubs do not constitute either a taxonomic or evolutionary category. The same taxonomic unit may exist as a tree in one place and as a shrub in another. Also it is often difficult to differentiate low trees from tall shrubs (Moore, 1973). These comments are particularly relevant to many of the dominant *Acacia* spp. which exist as tall trees in wetter and stunted shrubs in drier parts of their range.

These dominant *Acacia* spp., like most other species of *Acacia* in Australia, are phyllodineous and lack spines. This provides the vegetation with a character vastly different from the *Acacia* communities of the Afro-Asian and American continents. The morphological distinctiveness of the endemic Australian acacias caused Pedley (1986) to recommend most species be transferred to a different genus *Racosperma*.

Acacia aneura and *A. harpophylla*, the most widespread species of the shrublands and the open-forests and woodlands respectively, show a number of adaptations which enable them to survive and flourish under low and variable rainfall and high evaporation. Both possess phyllodes which exhibit extreme resistance to desiccation (Connor & Tunstall, 1968). Despite adverse water conditions associated with soils of heavy texture, which are often very saline, *A. harpophylla* is able to grow and form forests of high standing biomass (Moore, Russell & Coaldrake, 1967).

Acacia aneura has terete to flattened phyllodes which are vertically rather than horizontally aligned, thereby aiding water redistribution and minimising heat absorption. The phyllodes have a hairy, scurfy and resinous covering (Francis, 1925) and are sclerophyllous and drought resistant by means of dormancy. The plants aestivate when drought occurs, and resume growth within four days after water again becomes available (Slatyer, 1961). They also have the capacity to channel water down their phyllodes and stems so that rainfall is concentrated at the base of the trunk (Slatyer, 1965; Pressland, 1973). By contrast, the amount of rainfall reaching the soil surface as stem flow is usually small in *A. harpophylla* communities, although it is thought leaf drip is important in water redistribution (Russell, Moore & Coaldrake, 1967).

Spatial patterning is also one of the characteristics of the adaptation of these species to their environment. In central and western Australia and in the more arid parts of Queensland and New South Wales, *A. aneura* frequently occurs in groves that receive run-off water from the sparsely vegetated intergrove areas. The groves are usually positioned along the contour in a discontinuous fashion. About 50 per cent of the soils supporting *A. harpophylla* possess gilgai microrelief. Here, trees are confined mainly to elevated areas so that the depressions, which prevent general run-off, accumulate surface moisture locally yet grow little perennial vegetation.

Acacia communities occupy land mainly used for pastoral production. Semi-arid and arid conditions combined with highly unreliable rainfall impose constraints on productivity and make management difficult. This is further exacerbated by the presence of a standing biomass which is predominantly of little use to humans or their animals, though some species such as *A. aneura* are important drought reserves and are lopped to feed starving sheep. The *Acacia* dominants, in addition, greatly limit the abundance and growth of edible grasses and forbs. In higher rainfall areas *A. harpophylla* communities have been cleared for cultivation and sown pasture, whilst thinning of stands of *A. aneura* in the wetter parts of its range improves native pastures. In the most arid areas, these lands remain unoccupied except for localised use in abnormally wet seasons.

Descriptive accounts of *Acacia* communities covering much of northern Australia have appeared in numerous CSIRO Land Research Series reports. These, together with Queensland Department of Primary Industries' Land Use Study reports and the explanatory notes accompanying the vegetation survey of Western Australia by Beard, have provided much of the background information for this chapter. Other important sources used in compiling the second edition of this review were Aplin (1980), Beadle (1981), Beard (1990) and Wilson *et al.* (1990) and all these are gratefully acknowledged.

Acacia open-forests and woodlands of northeastern Australia

A number of species of *Acacia* of similar growth-type form the almost unspecific canopy layer of open-forests and woodlands in semi-arid regions of northeastern Australia. They occupy two distinct situations in the landscape. One is the shallow, mainly coarse-textured, infertile acid soils on scarp retreats and tablelands. Such soils represent the upper catenary sequence of the dissected Tertiary land surface. The other situation is the fertile, deep, usually alkaline clay soils on undulating to lowland plains associated mainly with the erosional–depositional mid- to lower-slopes of the catenary sequence.

Acacia spp. from different sections of the genus favour these different situations. The most successful species on the Tertiary lateritic surface and exposed underlying sandstones are members of the Section Juliflorae, which have flowers in spikes, whilst the deep clays are dominated by members of the Microneurae group of the Section Plurinerves. The latter group has flowers in heads and phyllodes with crowded longitudinal nerves.

The most widespread species on laterites and sandstone are lancewood (*A. shirleyi*) and bendee (*A. catenulata*) and on the heavy clay soils *A. harpophylla* and gidgee (*A. cambagei*) are characteristic. On very rare occasions, members of different groups form narrow ecotones but usually they are separated by *Eucalyptus* woodlands which occupy intermediate habitats.

Acacia communities are found in areas receiving up to 900 mm rainfall a year, but they are most extensive in rainfall zones receiving less than 750 mm a year. In undisturbed situations in more mesic areas they form open-forest communities, but with decreasing rainfall the canopy becomes more open and the trees become shorter to form low woodlands. Some species, such as *A. cambagei* and *A. catenulata*, penetrate the central drier zone of *Acacia* shrublands where they also form shrubby communities. In contrast *A. aneura*, the most abundant *Acacia* of the arid shrublands, forms extensive open-forests and woodlands in less arid areas of its range.

Although *Acacia* spp. form an almost unispecific upper canopy layer, a few species of *Eucalyptus* enter the canopy or occur as emergents. They are rarely common, however, and are usually found in marginal situations. *Casuarina cristata* alone forms mixed stands with *A. harpophylla* over wide areas of southern Queensland. Occasionally ecotonal areas of mixed *Acacia* spp. of the same group may be found. Even in these zones different species of *Acacia* generally prefer rather distinct habitats.

Whilst the canopy species of each group are from distinct sections of *Acacia*, the associated species also tend to be from taxonomically distinct groups. In mesic areas, woody understorey species in lowland clay situations have affinities with the Australian rainforest element, whilst on shallow soils on tablelands affinities lie with the Eremean flora (Doing, 1981). The grassy ground flora also indicates different origins, with the Chlorideae and Sporoboleae relatively more prominent in the former group and the Aristideae and Andropogoneae in the latter.

On each of these edaphic–physiographic situations, species within each group tend to form a replacement series controlled by moisture availability. *Acacia harpophylla* is replaced, with increasing aridity, by *A. cambagei* on fine-textured soils and a similar but less marked transition occurs on shallow acid soils with *A. shirleyi* being replaced by *A. petraea* (Pedley, 1978).

The present distribution of the major species is controlled primarily by climate and secondarily by soil type within climatic zones, but past history has also had an influence. Cycles of mesic and arid conditions have left isolated post-climaxes of *Acacia* communities in favourable sites well outside their normal climatic range. *Acacia harpophylla*, a species unsuited to spread by long-range dispersal, occurs in widely separated areas receiving as little as 300 mm annual rainfall. Pedley (1973) recorded refugia of *A. aneura* communities which dominate the arid interior in areas where rainfall exceeds 500 mm.

Acacia *open-forests on shallow coarse-textured acid soils*

The most widespread associations are dominated by *A. shirleyi* and *A. catenulata*, although a number of other *Acacia* spp. of the Section Juliflorae (such as *A. sparsiflora*, *A. burrowii*, *A. petraea* and *A. rhodoxylon*) form structurally similar communities.

All the species are associated with the eroded residual Tertiary lateritic surfaces and sandstones, often lateritised, which have been exposed following removal of the Tertiary surface. Soils are usually very acid to acid, occasionally neutral and are of three major types.

Lateritic red and yellow earths
These are generally shallow soils overlying massive or concretionary laterite. They usually occupy gently sloping situations often on the crests of eroded tablelands.

Skeletal soils
They are usually uniform coarse-textured soils developed on lateritised sandstone. They are mainly sands or sandy loams, occasionally light clays, and occur on moderately to steeply dissected slopes often on scarp retreats.

Lateritic podsolic soils
These are generally shallow soils developed on quartzose sandstones capped by laterite. They are less widespread than the previous type and occur on less steep slopes.

The soils are generally infertile with low levels of available nitrogen, phosphorus and carbon; calcium and magnesium may be limiting for growth in some soils.

Lancewood (*Acacia shirleyi*)

Lancewood is the name given to a group of species each of which form unispecific canopy layers of slender, closely-packed trees on residual tablelands and scarps on skeletal soils. Of these, *A. shirleyi* is the most widespread. It occupies these habitats in an arc stretching from northwest of the Darling Downs (31°03′S; 145°45′E) in Queensland to the base of Cape York Peninsula and across the Barkly Tableland into the Northern Territory. Outliers occur as far west as the Victoria River catchment. It is most widespread in the 500 to 750 mm rainfall belt. In wetter areas it forms open-forests up to 15 m in height. In drier parts of its range, low and low-open woodlands predominate.

A wide variety of *Eucalyptus* spp. occur as canopy associates and as emergents although they are rarely abundant. The most prevalent species are members of the bloodwood group, including *E. trachyphloia*, *E. dolichocarpa*, *E. erythrophloia* and *E. citriodora*, and of the ironbark group, including *E. drepanophylla*, *E. decorticans*, *E. melanophloia* and *E. crebra*. Some members of the box group, such as *E. exserta* and *E. thozetiana*, also occur.

Most communities lack a well defined understorey tree or shrub stratum, although scattered small trees and shrubs may occur. Species composition varies with latitude and moisture. In tropical areas *Macropteranthes kekwickii*, *Petalostigma banksii* and *Erythrophleum chlorostachys* may be found. In semi-arid Queensland *Erythroxylum australe*, *Petalostigma pubescens* and, particularly in areas of deeper soil, *Alphitonia excelsa* occur. Ground vegetation is very sparse and comprises mainly wiry grasses of low quality for grazing. *Aristida* spp. are prominent with *A. pruinosa* frequent in tropical and *A. caput-medusae* in subtropical communities. Other widespread grasses include *Cymbopogon*, *Schizachyrium*, and *Cleistochloa subjuncea*. *Heteropogon contortus* and *Themeda triandra* may occur in wetter areas; in more xeric areas are found *Triodia* and *Plectrachne pungens*.

Bendee (*Acacia catenulata*)

This species forms open-forest communities which resemble, structurally and floristically, the lancewood communities. The trees are slender and of even height, up to 13 m in mesic areas, with a mid-storey tree and shrub layer which is usually very sparse but may be conspicuous. It extends onto slightly deeper and more loamy soils than *A. shirleyi* and in these situations the woody understorey layer becomes more prominent. It grows in the 150 to 700 mm rainfall belt, occurring in isolated pockets from south of Charters Towers to the western Darling Downs and the Grey Range, west of Thargomindah.

Associated canopy trees are not common and are mainly other *Acacia* spp., such as *A. shirleyi*, and in drier areas, *A. petraea*, *A. aneura* and *A. aprepta*, together with *Eucalyptus* spp. similar to those occurring in the lancewood communities. *Eucalyptus exserta* is a widespread associate and because the soils may be deeper and more loamy than in the latter community, the boxes such as *E. populnea* and *E. cambageana* are more prominent.

Athough the mid-storey layer is usually very sparse, small-leaved myrtaceous and rutaceous shrubs such as *Lysicarpus angustifolius*, *Micromyrtus* sp., *Phebalium glandulosum*, *Thryptomene hexandra* and *Baeckia jucunda* may form a heath-like understorey. *Eremophila mitchellii* and, in drier western areas, *E. latrobei* and *Senna* spp. are also found.

The ground flora is sparse, being composed mainly of arid scrub grasses such as *Aristida caput-medusae* and other *Aristida* spp., *Thyridolepis mitchelliana*, *Digitaria* spp., and *Paspalidium* spp. The fern *Cheilanthes sieberi* is a frequent associate. *Sida* spp. are among the most abundant forbs and in drier areas *Sclerolaena* spp. become more prominent.

Acacia *open forests and woodlands of deep fine-textured alkaline soils*

Acacia harpophylla and *A. cambagei* dominate the most widely occurring communities in this group. Blackwood (*A. argyrodendron*), boree (*A. tephrina*) and Georgina gidgee (*A. georginae*) communities are less widespread but because all five species occupy similar habitats, pairs of species may form mixed communities where their ranges overlap. A number of other species, all of the Microneurae group of the Section Plurinerves, form structurally and floristically similar communities which are important locally. These include yarran (*A. omalophylla* and *A. melvillei*), womal (*A. maranoensis*) and bowyakka (*A. microsperma*).

In contrast to the previous group these species are usually associated with the mid- to lower-catenary land surfaces within the Tertiary weathered zone. They are most abundant on soils developed on argillaceous sedimentary and basic volcanic rocks exposed following removal of the Tertiary surface and on lateritised Tertiary sediments. They also occur on more recent fine-textured alluvial soils. The soils are usually alkaline, occasionally neutral or even acid, with carbonate concretions and often gypsum at depth, although many of the soils become strongly acid at depth. They can be conveniently divided into three main types.

Clay soils

These are usually deep, grey to red-brown in colour and are by far the most widespread type. Three major sub groups occur depending on the origin of the parent material. Deep cracking clays, often gilgaied, are heavy and often saline soils, developed on pre-Tertiary sediments which have been altered through the lateritisation of the Tertiary land surface or are located on transported Tertiary-weathered zone material. Gilgais vary from incipient to large depressions to 2 m in depth, the latter usually associated with the deeper soil profiles (Isbell, 1962). The second subgroup are the sedentary clay soils which are of moderate depth and occur on argillaceous Permian, Jurassic and Cretaceous sedimentary rocks and basic volcanics which were exposed following the complete removal of the Tertiary land surface. They are mainly medium- to heavy-clays and gilgais are rare. Alluvial clay soils are the third subgroup and are calcareous cracking and non-cracking clays to clay loams occupying the flood plains of major rivers and streams. Gilgais are infrequent.

Texture-contrast soils

These are usually deep soils with a sandy or loamy surface, less than 40 cm thick, on a strongly alkaline clay subsoil. Where these occur the communities often contain various *Eucalyptus* spp.

Red and yellow earths

The soil is invariably loamy at the surface, becoming clayey with depth. They are usually alluvial or aeolian in origin and occur mainly in the more arid areas.

The soils, particularly the clays are relatively fertile though the available phosphate content is low, except in the alluvial clays and the sedentary soils derived from basalt (Isbell, 1962).

Brigalow (*Acacia harpophylla*)

These communities represent the most mesic extreme of this group of associations, generally being found on clay soils on gently undulating lowlands and plains. They extend from about Charters Towers in the north to south of the Queensland–New South Wales border, occupying an area of more than 60 000 km². *Acacia harpophylla* assumes prominence usually as an emergent over a semi-evergreen vine thicket in coastal and subcoastal areas receiving up to 900 mm rainfall a year. The most extensive and richest development of the brigalow communities occurs in the 750 to 500 mm annual rainfall belt in areas receiving two or preferably four months of effective winter rainfall (Farmer *et al.*, 1947). In this region, *A. harpophylla* dominates open-forest communities usually 10 to 20 m high (Fig. 9.1), but with decreasing annual rainfall the trees become lower in stature and less dense. In areas receiving less than 500 mm rainfall, woodlands and low open-woodlands are the most common structural form. In these drier areas *A. harpophylla* communities are found in pockets receiving surface run-on moisture. These outliers extend as far west as the Grey Range, west of Quilpie, where annual rainfall is less than 350 mm.

Throughout its range, *A. harpophylla* commonly occurs as a shrub, often forming dense shrublands. In most circumstances these are the result of major disturbance, such as unsuccessful clearing of open forest communities or fire (Johnson, 1964).

In northern areas, *A. harpophylla* typically occurs as the unispecific canopy of a low-layered open-forest. In subtropical areas *Casuarina cristata* is frequently co-dominant and the canopy varies from all *C. cristata* to all *A. harpophylla*. A number of *Eucalyptus* spp. may also

Fig. 9.1. Open-forest of *Acacia harpophylla* in southeastern Queensland. (Photo: R.W. Johnson.)

be present as canopy trees and emergents. There is less variety of species than in the former group on the shallow acid soils. Members of the box group are the most prominent. The eucalypts also often occur in broad ecotonal communities and their individual presence reflects the habitat. North of Taroom and Injune, *E. cambageana* forms mixed communities with brigalow on red-brown texture-contrast soils, with *Eremophila mitchellii* and *Carissa ovata* frequent as mid-storey species. On alluvial clay soils along flood plains of major rivers, *Eucalyptus coolibah* is a common associate with *Terminalia oblongata* the characteristic understorey tree in tropical areas. *Eucalyptus populnea* forms mixed communities on texture-contrast soils, particularly in subtropical areas with *Eremophila mitchellii* the characteristic mid-storey species. Other associated *Eucalyptus* spp. include *E. thozetiana*, *E. pilligaensis*, *E. orgadophila*, *E. brownii*, *E. crebra* and *E. ochrophloia*. In addition to other *Acacia* and *Eucalyptus* spp., the most common canopy associates are *Brachychiton* spp., *Lysiphyllum* spp., *Cadellia pentastylis*, and *Ventilago viminalis*.

A number of floristic associations based on the composition of the mid-storey low tree and shrub layer can be recognised (Johnson, 1964), their distribution depending on habitat and climatic factors. On uniform clay soils in subtropical areas, *Geijera parviflora* is the characteristic understorey species; in tropical areas *Terminalia oblongata* is predominant. *Eremophila mitchellii* characterises the understorey on texture-contrast soils, although in drier western areas it replaces both *T. oblongata* and *G. parviflora* on more clayey soils. Other frequent mid-storey species include *Eremophila deserti*, *Alectryon diversifolius*, *Carissa ovata*, *Eremocitrus glauca* and *Rhagodia spinescens*. Lianes such as *Capparis lasiantha*, *Cissus opaca*, *Jasminum didymum* subsp. *racemosum* and various asclepiads are widespread though most individuals rarely climb into the canopy layer.

Grasses are characteristic of the ground layer, but as the canopy of the mid-storey and upper layers closes, forbs become relatively more prominent. *Paspalidium caespitosum* is usually predominant with *P. gracile*, *Chloris* spp. and *Enteropogon acicularis* also abundant, while *Sporobolus* spp., mainly *S. caroli* and *S. scabridus*, *Leptochloa* spp. and *Eriochloa* spp. are characteristic species. On texture-contrast soils, *Aristida ramosa*, *A. personata*, *Cymbopogon refractus* and *Calyptochloa gracillima* are prominent. The sedges *Cyperus gracilis*, and on heavy clay soils, *C. bifax* are often present. The most abundant forbs are *Abutilon oxycarpum*, *Einadia nutans*,

Sclerolaena tetracuspis, Brunoniella australis, Enchylaena tomentosa and various *Sida* spp.

Gidgee (*Acacia cambagei*)

This *Acacia* replaces *A. harpophylla* as rainfall decreases to below 550 mm a year. The eastern extent of its range lies between Charters Towers and Clermont where it forms open-forest communities with a canopy up to 15 m high. In the interface, *A. harpophylla* extends along run-on sites in drier areas with *A. cambagei* occupying the drier elevated sites in wetter areas. With increasing aridity the canopy becomes lower and more open, and west of the Great Dividing Range most communities are woodlands (Fig. 9.2) or shrublands.

Tongues of *A. cambagei* communities extend westward around the margin of the arid centre. In the northern tongue, *A. cambagei* occupies light clay and loamy soils, but as annual rainfall approaches 250 mm it is most often found on red earths and calcareous desert loams in low-lying sites. Scattered stands occur in the southeastern part of the Barkly Tableland extending into the Northern Territory.

Fig. 9.2. Woodland of *Acacia cambagei* in southeastern Queensland. (Photo: R.W. Johnson.)

In the south, it extends into northwestern New South Wales and across far northern South Australia.

The open-forest communities are similar, structurally and floristically, to the adjacent *A. harpophylla* communities. The suite of *Eucalyptus* species, which occur as emergents or co-dominants in the canopy layer, is similar to that occurring in the brigalow communities. *Eucalyptus populnea*, *E. cambageana* and *E. coolibah* occupy their characteristic habitats. Because *A. cambagei* is found in more arid habitats than *A. harpophylla*, other eucalypts become prominent. *Eucalyptus terminalis*, in tropical areas, and *E. ochrophloia* in alluvial plains in southwestern Queensland occur in ecotonal communities with *A. cambagei*.

In tropical areas, *Terminalia oblongata* and *Eremophila mitchellii* are the characteristic mid-storey species with *Geijera parviflora* replacing *T. oblongata* in southern areas of the mesic fringe. Because *A. cambagei* occupies a drier environment than *A. harpophylla* the more xeric *E. mitchellii* tends to be the most characteristic mid-storey species. Other mid-storey shrubs characteristic of *A. harpophylla* open-forests such as *Eremophila deserti*, *Carissa ovata*, *Rhagodia spinescens* and *Alectryon diversifolius* extend into *A. cambagei* open-forests. *Senna* spp. are frequent components of the mid-storey shrub layer with *S. nemophila* common in mesic areas and varying mixtures of other species in more arid zones. In southern areas which receive predominantly winter rainfall *Atriplex nummularia* is the most abundant understorey shrub.

The ground layer in the more mesic areas is predominantly grassy with *Paspalidium* spp., *Enteropogon acicularis* and *Sporobolus* spp. the most frequent species and *Enchylaena tomentosa*, *Abutilon* spp. and *Einadia nutans* widespread forbs. With increasing aridity *Paspalidium* become infrequent and species of *Enneapogon* and *Eragrostis* increase in importance. In the shrublands, *Dactyloctenium radulans* becomes more prominent whilst *Sclerolaena* spp., *Atriplex* spp. and other chenopodiaceous forbs become the major components of the ground stratum. Hence the gradient from open-forest to shrubland is paralleled by gradual replacement of grass by forbs.

In central-western Queensland, *A. cambagei* woodlands are closely associated with the Mitchell grass (*Astrebla* spp.) downs on heavy cracking clays. They often fringe or form mosaics with the tussock grassland and species of the grasslands become associated in the ground layer.

Blackwood (*Acacia argyrodendron*)

Acacia argyrodendron forms an almost unispecific canopy of open-forest communities which lie roughly in the overlap of the more mesic *A. harpophylla* and more xeric *A. cambagei* communities to the south of Charters Towers in the basins of the Cape, Suttor and Belyando Rivers. Outliers occur southwards in an arc along the western slopes of the Great Dividing Range and into the Isaacs River basin as far south as about 22° 30'S. Average rainfall varies from 475 to 655 mm. These communities occupy lowlands covered in Tertiary-weathered sediments.

Associated species are similar to those in adjoining *A. harpophylla* open-forests with *Eremophila mitchellii*, *Terminalia oblongata* and *Carissa ovata* occurring in the woody understorey and *Eucalyptus cambageana*, *E. coolibah* and *E. brownii* in particular habitats occurring as emergents. *Paspalidium caespitosum* is the most abundant herb; *Sporobolus caroli*, *Brunoniella australis* and *Dipteracanthus australasicus* subsp. *corynothecus* are moderately frequent.

Boree (*Acacia tephrina*)

Boree forms woodland and open-woodland communities in central-western Queensland from Hughenden to Tambo in the 425 to 550 mm annual rainfall belt. In the more arid areas, communities are reduced to open-shrublands. At their northern limit scattered stands are found eastward to the coast at 20° S.

Acacia tephrina occurs frequently on cracking clay soils and is often found in association with *A. cambagei*. Communities of *A. tephrina* and *A. tephrina-A. cambagei* often occur on the fringes of and form mosaics with *Astrebla* spp. grasslands. *Lysiphyllum carronii*, *Alectryon oleifolius*, *Atalaya hemiglauca* and *Flindersia maculosa* may be found scattered through the canopy layer particularly in more mesic areas. *Eremophila mitchellii*, *E. maculata* and in more western areas *E. dalyana* form an often very sparse low shrub layer. *Astrebla*, *Sporobolus* and many of the grass species abundant in the adjacent *A. cambagei* communities contribute to an open ground-layer. With increasing aridity the typical trend in forbs towards dominance of chenopodiaceous species, such as *Atriplex*, commonly *A. lindleyi* and *A. spongiosa*, and *Sclerolaena* spp. occurs.

Georgina gidgee (*Acacia georginae*)

Skirting the shrublands of the central arid area in the tropical 200 to 250 mm annual rainfall belt, *A. georginae* forms woodlands and open-woodlands up to 8 m tall from the Georgina River basin in western- central Queensland through the Barkly Tableland region to northeast of Alice Springs. It is best developed on medium to heavy grey and brown calcareous clays, although it also occurs on lighter soils including somewhat acid red earths.

On heavy clays it forms a very open canopy over *Astrebla* spp. and other grasses and forbs characteristic of the Mitchell grass downs, whilst on better drained, lighter soils, a sparse mid-storey shrub layer with *Senna* spp. and *Eremophila* spp. becomes evident. *Enneapogon* spp. are prominent in the short-grass ground layer in these situations.

Prickly acacia (*Acacia nilotica*)
This species native to India and Africa was introduced into Australia in 1890 as a source of gum arabic and as a shade and fodder tree. It was deliberately planted along bore drains in the Mitchell grass grasslands of northeastern Australia and now exists over 6 m. ha (Burrows *et al.*, 1990). These grasslands are found on alkaline cracking clays. Whilst initially a valuable source of shade and protein during the dry season, in many areas extensive woodlands and thickets have formed causing a substantial loss in pasture production and management problems. Since the 1950s, invasion has followed an exponential pattern with stepwise increases in density driven by above-average summer rainfall and an increase in the cattle/sheep ratio (Burrows *et al.*, 1990). It is now a declared plant under the Rural Lands Protection Act in Queensland.

Utilisation of Acacia *open-forests and woodlands of northeastern Australia*

The relatively fertile clay soils supporting *Acacia* open-forests and woodlands provide a sound base for a productive agricultural and pastoral industry. Because of the density of standing trees and shrubs, productivity of the undisturbed communities is low and it has generally been necessary to modify severely or replace the existing vegetation. Clearing has occurred over large areas, particularly in higher rainfall zones, by ringbarking and in more recent decades by pulling trees down with a heavy chain dragged between high-powered bulldozers.

Because of the high potential agricultural and pastoral productivity of lands supporting *A. harpophylla* open-forest, the forests have been replaced by crops and pastures of introduced species such as buffel grass (*Cenchrus ciliaris*), Rhodes grass (*Chloris gayana*) and green panic (*Panicum maximum* var. *trichoglume*) (Johnson, 1964). Few large areas of undisturbed vegetation remain. *Acacia argyrodendron* open-forests have also been converted to sown pastures. Carrying capacity may be increased initially ten to twentyfold. However, continued production is partly dependent on management to control regrowth of *A. harpophylla* and associated woody species. Both *A. harpophylla* and *A. argyrodendron* sucker from horizontal roots and other species such as *Eremophila mitchellii*, *Terminalia oblongata* and various *Eucalyptus* spp. regenerate from butts and lignotubers. Ploughing, chemical control, fire and severe grazing with sheep have been used to bring regrowth of *A. harpophylla* under control.

Acacia cambagei open-forests have also been replaced by sown pastures (Purcell, 1964). Clearing has mainly occurred along the wetter eastern fringe where more prevalent understorey trees and shrubs limit the growth of native grasses and where conditions are more favourable for establishment of sown pastures. Regrowth from *Eremophila mitchellii* is often a problem after clearing (Beeston & Webb, 1977).

In contrast, *Acacia* open-forests on the shallow acid soils have been largely left undisturbed. *Acacia shirleyi* is used locally to provide rails for fencing but soil fertility is too low to warrant expenditure on clearing. Limited clearing of *A. catenulata* open-forests has occurred on deeper red and yellow earths where pulling without burning and sowing buffel grass (*Cenchrus ciliaris*) is recommended (Tiller, 1971).

In more recent years a high premium has been placed on reservation of undisturbed areas of *A. harpophylla* and allied open-forests for conservation.

Acacia open-forests and woodlands of southern Australia

In areas to the south and southeast of the central Australian *Acacia* shrublands, most open-forests and woodlands are dominated by *Eucalyptus* spp. North of the Great Australian Bight as far west as Spencer Gulf, however, *Eucalyptus* open-forests and woodlands are infrequent and there are large areas of woodlands and open-woodlands characterised by trees of western myall (*Acacia papyrocarpa*), *Casuarina cristata* and *Myoporum platycarpum*.

Apart from *A. papyrocarpa*, myall (*A. pendula*) is the only other species of *Acacia* which forms woodland communities over considerable areas in southern Australia. Both species of *Acacia* are members of the Microneurae group of the Plurinerves and occur on similar soils and topography as related species in northeastern Australia.

Western Myall (*Acacia papyrocarpa* (syn. *A. sowdenii*))

The most extensive *Acacia* woodland formation excluding the *A. aneura* woodlands and those of northeastern Australia is that characterised by *A. papyrocarpa*. It forms woodlands to 10 m in height in the 200 to 250 mm rainfall belt to the northwest of Port Augusta and Whyalla in South Australia (Specht, 1972) and extends into Western Australia as a fringe skirting the Nullarbor Plain (Beard, 1975). However, in western South Australia and Western Australia the canopy is lower and very open and western myall often occurs merely as very scattered trees in a semi-succulent shrubland.

A. *papyrocarpa* is found mainly on shallow uniform loam to clay-loam soils or red and brown earths, both of which are calcareous and similar to those occupied by *A. cambagei* and *A. georginae* in the drier parts of their range. In woodland communities, *Myoporum platycarpum* is a common canopy associate. *Casuarina cristata* may also occur. A well-defined semi-succulent mid-storey shrub layer is characteristic with *Maireana sedifolia* predominant and *Atriplex vesicaria* and *A. stipitata* commonly present. Tall shrubs associated with the *Acacia* woodlands of northeastern Australia such as *Alectryon oleifolius* and *Senna* spp., such as *S. sturtii* and *S. phyllodinea*, extend into these southern communities. Low chenopodiaceous shrubs and forbs and, in favourable seasons a wide variety of annuals, occupy the ground layer with *Enneapogon* and temperate grasses such as *Danthonia* and *Stipa*.

A related species *A. loderi* forms similar communities in similar situations on solonised brown soils in southwestern New South Wales and extending into Victoria and South Australia. A low shrub layer with *Maireana sedifolia* and other chenopods together with sclerophyllous species such as *Grevillea huegellii* and *Templetonia egina* is usually present. The grassy ground layer becomes more prominent under heavy grazing.

Myall (*A. pendula*)

Acacia pendula forms woodland and open-woodland communities, rarely widespread but locally important, from Clermont in central Queensland to the Riverina (34°30'S; 145°00'E) area of southern New South Wales. They are or formerly were best developed in New South Wales, southwest of Hay, and on the northwestern plains in the Macquarie District (Beadle, 1948; Moore, 1953) as well as on the western Darling Downs in Queensland. Annual rainfall varies from 375 to 550 mm with communities occupying the wetter end of the range in Queensland and the drier end in southern areas. They occupy similar habitats to the *Acacia* communities of the lowland clays of northeastern Australia and represent the southern attenuation of this complex.

In northern areas *A. pendula* woodlands are mainly restricted to clay soils. Occasionally *Eucalyptus coolibah* and *E. populnea* occur as emergents and other trees include *Alectryon diversifolius* and *Acacia stenophylla*. No well-defined understorey occurs but scattered *Eremophila maculata* is a common associate. The ground flora resembles that of the *Astrebla* spp. and *Dichanthium sericeum* grasslands. In southern areas, *Atriplex nummularia* may form a conspicuous low shrub layer with other associated shrubs such as *Rhagodia spinescens* and *Enchylaena tomentosa*. Temperate grasses, such as *Danthonia* and *Stipa*, and *Chloris truncata* are prominent in the herbaceous layer. Communities also occur on better drained red earth soils in southern areas.

In southwestern Western Australia there is a great diversity of *Acacia* spp. They rarely form distinctive open-forests or woodlands, however. *Acacia rostellifera* does form low open-forests and thickets in coastal areas from Esperance to Shark Bay on recent or Pleistocene sand dunes, often in association with or replaced by *A. cyclops*, particularly in southern areas. In places it represents a fire subclimax following the destruction of *Callitris preissii* low forest (Beard, 1990). *A. cochlearis* is a common associate whilst *Melaleuca cardiophylla* is often common in mixed *Acacia* thickets. Inland from the coast, usually on limestone soils, the dominants become shrubby and the community merges with *A. acuminata* shrublands and *Eucalyptus loxophleba* woodlands.

Utilisation of Acacia *open-Forests and woodlands of southern Australia*

Acacia papyrocarpa woodlands are used for sheep and wool production from native pastures and the timber is cut locally for fence posts and also for firewood (Lange & Purdie, 1976). Present usage coupled with the infrequency of germination events threatens the survival of existing communities. Similarly, *A. pendula* woodlands have been largely replaced by grasslands in southern areas as a result of felling associated with grazing by sheep (Moore, 1953).

Acacia low woodlands & shrublands of central and western Australia

Acacia shrublands are the dominant woody vegetation of Australia's semi-arid and arid interior. The most characteristic species is mulga (*Acacia aneura*). Mulga communities together with mixed mulga-hummock grass communities and mulga mid-height grass communities occupy about 1 500 000 km² or about 20 per cent of the total area of the continent (Everist, 1971). *Acacia aneura* is by far the most ubiquitous *Acacia* within this range but it may be replaced by other localised dominants such as *A. cibaria* (syn. *A. brachystachya*), *A. stowardii*, *A. kempeana* and *A. ramulosa* (syn. *A. linophylla*), largely in response to topographic and edaphic factors. These communities show relationships with the open-forests of the shallow coarse-textured soils of northeastern Australia with the major dominants also being members of the Section Juliflorae.

At the time of completion of this account, Randell (1992) published a taxonomic account of the mulga complex which substantially modifies the taxonomic concept of *A. aneura* and its allies. It has not been possible in the time available to reinterpret the published literature and the traditional broad concept has been followed in this account.

Two types of *Acacia* low woodlands and shrublands can be recognised: one with tussock grasses and forbs characterising the ground layer; the other with hummock grasses predominant.

Acacia *with tussock grasses*

Mulga (*Acacia aneura*)

Acacia aneura associations extend in a discontinuous belt from near the Western Australian coast, across the southern edge of the central deserts, to western New South Wales and southwest Queensland, with another substantial occurrence in the north of the central arid area in the Northern Territory. *Acacia aneura* is adapted to environments where the soil water regime is almost always limiting for growth, but where there is some possibility of recharge at all seasons. *Acacia aneura* is mostly found in areas receiving from 200 to 500 mm mean annual rainfall, but it is conspicuously absent from the semi-arid regions with a regular summer or winter drought (Nix & Austin, 1973).

In Central Australia, Perry & Lazarides (1962) consider that climate has been the primary factor in selection of species and life forms and is the controlling factor influencing the structure and density of the communities. Soil characteristics are of major importance, however, in determining floristics and distribution of communities within the area.

The most extensive occurrence of *A. aneura* is on plains and sand plains which often receive some run-on water from adjacent hills and low ranges. Nevertheless a great diversity of habitat can be identified from dissected residual soils on ridges to desert sandhills and solonised brown soils, red earths and texture-contrast soils of the flats and plains.

Acacia aneura is commonly found on red earth soils throughout most of its Australian distribution. These soils are light textured with a hard coherent subsoil strongly impregnated with ferruginous compounds. It rarely occurs in calcareous habitats, especially in Queensland and the Northern Territory, although it can be found in areas with neutral to alkaline subsoils, mainly in the south and west of the continent. In the latter habitats, *A. aneura* is usually sparser and more stunted. *A. aneura* rarely occurs on clay soils of heavy texture. Although there is some variation evident in the tolerance to soil acidity levels, a general feature of all soils supporting *A. aneura* is that they have a very low available phosphorus content (< 20 ppm measured by acid extraction).

Extreme morphological variation within *A. aneura* is well documented (Pedley, 1973; Fox, 1986; Cody, 1989, 1991) and this has contributed to taxonomic problems. The variation appears genetically controlled (Beard, 1974; Fox, 1986; Cody, 1989) and could be related to polyploidy within the species (Fox, 1986); diploid and tetraploid races of *A. aneura* have been identified (I. DeLacy, personal communication). Variation is thought to be maintained through niche differentiation, which is more prominent in drier areas (Cody, 1989, 1991). A number of morphological variants have been described (Pedley, 1973; Maslin, 1981; Fox, 1986). Pedley (1973) suggested that the complex pattern of variation in growth form and phyllode dimensions of *A. aneura* is possibly because of the retreat of the species to refugia during arid periods at the end of the Tertiary, followed by a recent expansion of its range. Specht (1972) considered that calcifuge and calcicole forms of the complex would well repay investigations.

Acacia aneura communities (Fig. 9.3) vary widely in density and composition. Highest *A. aneura* densities are achieved in the eastern mesic extremity where up to 8000 stems/ha have been recorded. Over most of its range densities of the order of 100 to 300 stems/ha would be more common. In more xeric habitats, *A. aneura* is quite sparse.

Fig. 9.3. Tall shrubland of *Acacia aneura* in southern Queensland. (Photo: W.H. Burrows.)

It occurs as a tree (10–15 m tall) in the mesic areas, but only exists as a stunted low shrub (2–3 m tall) in very xeric habitats or where it occurs on very shallow or calcareous soils. *Eucalyptus* spp. occur comparatively infrequently in *A. aneura* communities.

Whatever the habit of the dominant *A. aneura*, multi-layering characterises almost all these communities. The physiognomic complexity of *A. aneura* associations decreases along a gradient from more favourable to harsh environments (Boyland, 1973). A low, rarely continuous, shrub stratum is often present. Associated shrubs may be either sclerophyllous or with hairy semi-succulent leaves (Specht, 1972). The most ubiquitous genus is *Eremophila* with over 100 species represented in the *A. aneura* areas of Western Australia (Speck, 1963) and 19 species recorded in *A. aneura* areas of Queensland (Boyland, 1974). Since many of the *Eremophila* species have distinctive habitat preferences they are often used in community identification. Widespread among the shrub layer genera are *Senna*, *Dodonaea* and *Maireana*, the latter being more common on saline soils and rare in Queensland. In general, *A. aneura* associations of Western Australia are more shrubby than those in the east.

A herbaceous layer of perennial or seasonal forbs and grasses is usually well developed, although it may be sparse. A feature of mulga lands is the floristic independence of ground-storey and upperstorey communities. Although mulga communities are relatively poor, floristically, there is considerable variation in ground layer grass and forb composition, more so on a north–south rather than an east–west axis. Perry & Lazarides (1962) recognised 18 distinct ground-storey communities associated with *A. aneura* in the Northern Territory. Speck (1963) identified some 75 communities, most of which contained *A. aneura*, in the Wiluna–Meekatharra area of Western Australia. Additional distinctive community-types are found in the remaining States where *A. aneura* occurs, although those of Queensland and the Northern Territory have much in common.

Acacia aneura is frequently associated with short-grass communities in which *Eragrostis eriopoda* and *Monochather paradoxa* are characteristic perennial tussock grass species. Genera of forbs are predominantly from the families Asteraceae, Chenopodiaceae, Amaranthaceae and Malvaceae. There is a notable paucity of leguminous forbs in mulga shrubland.

In southwestern Queensland, communities dominated by *Acacia aneura* are characteristic of the transported detritus and dissected residuals of the laterite and silcrete land surface. Mulga is often associated in less arid areas with *Eucalyptus populnea* with which it may be dominant, co-dominant or form an understorey tree stratum. On its eastern margin, *A. aneura* grades into *E. populnea*-dominated communities (see previous chapter and Gillison & Walker, 1981). These communities are usually found on plains of deep sandy red earths and loamy texture-contrast soils and are equally as extensive in New South Wales as in Queensland. Other associated upperstorey species can include *E. intertexta* and *E. melanophloia*. *Callitris glaucophylla* frequently occurs as a tree, mainly in New South Wales. A shrubby understorey is usually present though it is more prominent in southern areas. In drier environments, *E. populnea* is replaced by *E. terminalis* and less frequently by *E. papuana*.

Senna spp., such as *S. artemisioides* and *S. nemophila*, and *Eremophila* spp., such as *E. gilesii*, *E. bowmanii* and *E. mitchellii*, particularly where *Eucalyptus populnea* occurs, are characteristic understorey shrubs. *Themeda triandra* and *Dichanthium sericeum* are important grasses on the mesic fringe but *Eragrostis eriopoda* is the most widespread and abundant species. Other widespread species include *Aristida contorta* and *A. jerichoensis*, *Monachather paradoxa*, *Thyridolepis mitchelliana* and *Enneapogon*. Temperate species such as *Stipa variabilis* are more abundant in New South Wales. *Sclerolaena*, *Sida*, *Ptilotus* and *Euphorbia* are among the most frequent forbs with the ferns *Cheilanthes sieberi* and *C. tenuifolia* abundant locally.

On dune fields, *Acacia aneura* often forms open shrublands with *Atalaya hemiglauca*. Scattered trees of *Hakea leucoptera* and *Grevillea juncifolia* may be present and other *Acacia* such as *A. tetragonophylla*, *A. ligulata*, *A. murrayana* and *A. calcicola* are frequently associated where the habitat is suitable. *A. tetragonophylla* is found on shallow soils whilst *A. ligulata* and *A. murrayana* are frequent on the extended flanks of dunes. *A. calcicola* occurs with mulga on low eroded dunes. *Senna desolata*, *Dodonaea angustissima*, *Eremophila duttonii* and *E. sturtii* occur as low shrubs and *Aristida contorta*, *Enneapogon*, *Eragrostis* and *Eriachne* are the most frequent grasses. These dunal communities also extend into northwestern New South Wales.

Mulga communities extend westward into northern South Australia and into the Alice Springs region of the Northern Territory. In northern arid South Australia, *Acacia cibaria* is frequently associated with *A. aneura* forming tall shrublands on sands, earths and duplex soils with a sandy and loamy surface. A low shrub layer is usually present though it is usually not dense. *Senna* spp., with *S. nemophila* are frequent on the deeper sands, and *Eremophila* spp., such as *E. latrobei* and *E. glabra*, are found in the understorey shrub layer. On shallow soils, *S. sturtii* and *Acacia kempeana* become more frequent. *Atriplex vesicaria* and *Maireana sedifolia*, except on the deepest sands, are widespread whilst *Aristida contorta* and *Enneapogon* are frequent herbs. On deeper sands, *Monochather paradoxa* and *Eragrostis eriopoda* occur.

Mulga associations in the Northern Territory are best developed on plains adjacent to mountains and hills where they are extensive on coarse- to medium-textured red earth soils (Perry & Lazarides, 1962). Other *Acacia*, such as *A. cibaria*, *A. kempeana* and *A. tetragonophylla*, are frequent associates. *Hakea* and *Acacia estrophiolata* may occur as scattered trees whilst *Senna* and *Eremophila*, particularly *E. gilesii*, form a sparse to medium-dense low shrub layer. *Eragrostis eriopoda* is the most abundant species in the ground layer of most communities and is usually associated with *Thyridolepis mitchelliana*. Short grasses, such as *Aristida contorta* and *Enneapogon*, and forbs, such as *Helipterum*, particularly *H. floribundum*, *Ptilotus helipteroides* and *P. sessilifolius*, grow between the perennial grass tussocks or in some communities form the ground layer. *A. aneura* is also found on rolling quartzite and sandstone ranges around Alice Springs and in the Simpson Desert on shallow soil. *A. kempeana* and

A. tetragonophylla are frequent associates with sparse shrubs of *Senna* and *Eremophila*.

In Western Australia, *Acacia aneura* low woodlands and tall shrublands dominate the area south of the Tropic of Capricorn, extending as far south as about 30° S. They are best developed on plains covered with red earth and loamy soils over siliceous hardpans, between coastal areas on the west and southwest and the deserts on the east. Other *Acacia* are frequently present, the specific composition of the communities depending on habitat. *A. pruinocarpa* is widespread though it does not extend as far south as *A. aneura*. *A. ramulosa* and *A. ligulata* are more frequent in sandy habitats, *A. grasbyi* on granite or in creek beds, *A. sclerosperma* on calcrete and *A. victoriae* on salty flats. Towards the northwest, *A. aneura* and *A. xiphophylla* communities merge.

Where the surface is sandy, mallees such as *Eucalyptus kingsmillii* and *E. oleosa*, may be found. Other associated trees and large shrubs include *Brachychiton gregorii*, *Alectryon oleifolius*, *Hakea suberea* and *Canthium latifolium*. *Senna* spp., such as *S. helmsii*, *S. sturtii*, *S. artemisioides*, *S. nemophila*, *S. chatelainiana* and *S. desolata* and *Eremophila* are prominent in the small shrub layer. Species of *Eremophila* are useful indicators of habitat type with *E. foliosissima* and *E. margarethae* frequent on deep loams, *E. fraseri* more prominent on stony ground and *E. granitica* on hills. Other frequent species are *E. forrestii*, *E. cuneifolia* and *E. platycalyx*.

About the Tropic where rainfall has a larger summer component, annual grasses are important in the ground layer; forbs are favoured by winter rainfall. Frequent ephemerals include *Ptilotus*, *Helipterum*, *Goodenia*, *Schoenia cassiniana* and *Waitzia acuminata* whilst the most characteristic grasses are *Monochather paradoxa* and *Eragrostis eriopoda* with *Aristida*, *Eriachne*, *Enneapogon* and *Eragrostis*.

Throughout the Western Australian mulga zone, *Acacia aneura* forms shrublands on stony hills of granite and gneiss and on lateritic scarps and breakaways. *A. quadrimarginea* and *A. grasbyi* are frequent associates in western areas, whilst *A. ramulosa* and *A. cibaria* may also be present. *Senna*, *Eremophila* spp. (particularly *E. latrobei*) and *Ptilotus obovatus* are among the more widespread lower shrubs. *A. aneura* shrublands are also found in saline areas where they often occupy sandy patches. In these situations *A. ramulosa* is abundant and *A. sclerosperma* and *Hakea preissii* frequent associates. *Eremophila oldfieldii* becomes more prominent and halophytes such as *Maireana pyramidata*, *Atriplex* and *Sclerolaena* more abundant among the forbs and grasses. Similar communities are associated with greenstone hills.

On the lateritic plains of the Great Victoria Desert and extending into the south of the Great Sandy Desert, Beard (1968) describes a formation known as mulga parkland which is a mosaic of mulga scrub and *Hakea-Acacia* scrub steppe. The former occurs on loamy soil in draws and depressions while the latter occupies the rises of hard ironstone.

In the area separating the Great Victoria Desert from the succulent shrublands of southern Australia, *A. aneura* forms a low tree or shrubby layer over semi-succulent and succulent halophytic shrubs and forbs. The soils are saline red sands and loams with a siliceous hardpan. Characteristic associates are *Casuarina cristata* and *Myoporum platycarpum*. It merges with *A. papyrocarpa* open woodland which fringes the Nullarbor Plain, although the mulga communities are usually found on drier and sandier soils. *Eucalyptus* spp., such as *E. oleosa*, are locally present. The understorey may be predominantly *Maireana sedifolia*, particularly on calcareous soils, or *Atriplex* spp., such as *A. vesicaria*, particularly in saline areas. In South Australia, *A. cibaria* becomes frequent. Similar communities with an understorey of semi-succulent shrubs are associated with the *Atriplex* and *Maireana* shrublands of northeastern South Australia and northwestern New South Wales.

Within its distribution range *Acacia aneura* is sometimes replaced by other shrubby *Acacia* species such as *A. stowardii*, *A. cibaria*, *A. kempeana*, and *A. ramulosa*. This replacement is usually related to topographic and soil features.

Bastard mulga (*Acacia stowardii*)

This occurs as a sparse low open-shrubland 1 to 3 m high. It is confined mainly to shallow red earths and red lithosols on dissected plains and low hills within the mulga lands of eastern Australia, although communities extend into Western Australia. *A. stowardii* occasionally forms pure stands or it may be co-dominant with *A. aneura* or *Eucalyptus exserta*. The lower shrub layer is usually well developed. In addition to species of *Dodonaea*, *Eremophila*, *Acacia* and *Canthium*, shrubs from genera not normally associated with the surrounding *A. aneura* communities, such as *Phebalium glandulosum* and *Westringia rigida* also occur. The ground cover in these communities is usually quite sparse. *Eriachne pulchella* is prominent amongst the grasses that are present.

Turpentine mulga (*Acacia cibaria* (syn. *A. brachystachya*))

This species occurs as a low to tall open-shrubland 1 to 3 m in height. It is usually associated with *A. aneura* and occurs on shallow red earths in Queensland and on sand dunes in South Australia. There may be a well defined lower shrub layer of *A. stowardii*, *Canthium latifolium*, *Senna* spp. and *Dodonaea* spp. The ground cover is usually very sparse with grasses in the genera *Aristida*, *Eragrostis* and *Eriachne* and some forbs such as *Maireana*, *Ptilotus* and *Sida* being most frequent.

Witchetty Bush (*Acacia kempeana*)

Acacia kempeana is a very frequent shrub of arid areas, particularly in the Northern Territory and northwestern South Australia. It occurs infrequently in Queensland and Western Australia. It is a shrub 2 to 3 m high with many stems from the base. Perry & Lazarides (1962) claimed that *A. kempeana* has a lower water requirement than *A. aneura* which leads to *A. kempeana* occupying the more droughty habitats. However, *A. kempeana* also appears almost exclusively on the more calcareous soils, whereas *A. aneura* occurs mainly on acid to neutral soils. *Senna* spp. and *Eremophila* spp., are usually present in the shrub layer. Associated ground-layer species with *A. kempeana* are mainly short grasses and forbs. The commonest include species of *Aristida*, *Enneapogon*, *Sclerolaena*, *Ptilotus* and *Helipterum*.

Sandhill mulga (*Acacia ramulosa* (syn. *A. linophylla*))

This community mainly occurs in South Australia, occupying the area between the mallee in the south and the *A. aneura* communities of the drier north. *Acacia ramulosa* forms a tall shrubland formation 3 to 5 m high which is mainly found on deep, red, non-calcareous sand dunes. In more southerly regions *Casuarina cristata* is a common associate in the upper layer while in the north *Callitris glaucophylla* may partly or wholly replace the sand mulgas. *Senna nemophila* is a common undershrub. Grasses in the genera *Aristida* (particularly *A. holathera*) and *Enneapogon* are well represented and annual forbs in the family Asteraceae are prevalent.

In Western Australia south of the Tropic and west of the Great Victoria Desert, *A. ramulosa* is also a prominent species in tall shrublands on sandy rises in claypans and on sand dunes in sand plains. Other *Acacia* spp. including *A. murrayana*, *A. acuminata*, *A. acuminata* subsp. *burkittii* and *A. sclerosperma* are frequent. Associated shrubs include *Eremophila* spp., such as *E. forrestii* and *E. clarkei*, *Grevillea* spp., such as *G. eriostachya*, *G. stenobotrya* and *G. stenos-*

tachya, Melaleuca uncinata and *Thryptomene johnsonii*. In deeper sand, mallees such as *Eucalyptus leptopoda* and *E. oldfieldii* may be conspicuous. Under favourable seasonal conditions, ephemerals are abundant especially composites such as *Myriocephalus guerinae, Brachyscome ciliocarpa, Helichrysum davenportii* and *Podolepis auriculata,* and *Ptilotus*. Grasses such as *Monochather paradoxa, Eragrostis* and *Eriachne helmsii* also occur. In saline situations, halophytes are more abundant.

Snakewood (*Acacia xiphophylla*)

In the Ashburton region of Western Australia which receives rainfall of summer and bimodal incidence, *Acacia xiphophylla* forms extensive shrublands, often merging with *A. aneura* shrublands. They occur mainly on alluvial plains with clay soils carrying surface stone. *A. victoriae* and *A. tetragonophylla* are frequent associates; in southern areas *A. sclerosperma* becomes more prominent. Scattered small shrubs of *Eremophila cuneifolia* and *Senna* may occur and the ground vegetation is usually sparse. Ephemerals such as *Ptilotus* and tussock grasses are present but often halophytes such as *Maireana, Sclerolaena* and *Atriplex* may be more abundant. Occasionally *Triodia basedowii* is dominant on deeper sands.

Other *Acacia* shrublands

In a brief survey of *Acacia* shrublands it is not possible to cover all the local variants that occur. For instance *A. victoriae* and *A. tetragonophylla* are common understorey shrubs in the drier regions of areas dominated by *A. aneura*. Occasionally both of the former species may be dominant in their own right, particularly in situations with deeper soils and adjacent to water courses. *A. victoriae* is usually found on heavier, often saline, soils but it is also frequent on shallow stony soils. *A. estrophiolata* forms low open-woodlands with *Atalaya hemiglauca* on sandy undulating plains in the ranges around Alice Springs.

In western areas of Western Australia south of the Pilbara, *A. sclerosperma* with the same two species forms shrublands which are usually associated with limestone and calcrete. *Hakea preissii* may be prominent. In more mesic areas in southwestern Western Australia on the fringe of the *Acacia* shrublands, tall shrublands dominated by *Acacia* spp., such as *A. resinomarginea*, occur. They are characterised by a low heath-like layer of sclerophyllous shrubs. Jam (*Acacia acuminata* subsp. *acuminata*) forms shrublands in undulating to rugged country to the northeast and east of Perth. *Eucalyptus loxophleba* often occurs as an emergent and becomes dominant as the annual rainfall increases. Other *Acacia* spp. such as *A. ramulosa, A. quadrimarginea* and *A. grasbyi* are frequent associates whilst in the understorey *Melaleuca uncinata* or sclerophyllous low shrubs of *Calothamnus, Grevillea* and *Dryandra* may be common. In this area, patterning similar to that described for *Acacia* communities in northeastern Australia occurs (B. Maslin, personal communication). Members of the Section Juliflorae occur mainly on the sandy or sandy-lateritic rises; Section Phyllodineae on the intervening heavier soil flats. Members of the Microneurae group of the Section Plurinerves favour heavier soils also, but often not in association with the Phyllodineae. *Acacia – Casuarina – Melaleuca* thickets are prominent in the southwest on deep sand plains. *Acacia acuminata, A. longisperma* and *A. stereophylla* are the most common *Acacia* spp. with *Melaleuca uncinata* and *Casuarina* spp. Low myrtaceous shrubs form a ground layer.

Acacia *spp. over hummock grasses*

Throughout Central Australia, hummock grasses (*Triodia, Plectrachne*) are important constituents of the vegetation of sand plains, dune fields and many rocky hills (Groves & Williams, 1981). True hummock grasslands are however rare and there is usually a sparse cover of shrubs and low trees. Individual species of hummock grasses vary in importance according to geographical location and habitat. *T. pungens* is dominant north of the Tropic whilst *T. basedowii* replaces *T. pungens* in southern areas. *Plectrachne schinzii* predominates on deeper and coarser sands. In subtropical areas *A. aneura* is the most common *Acacia* found in the overstorey but in northern areas other *Acacia* become prominent. These communities are more abundant in western parts of the continent and in subtropical rather than in temperate areas.

Kanji (*Acacia pyrifolia*)

To the north of the Tropic in summer rainfall areas, *Acacia pyrifolia* shrublands with hummock grasses occupy large areas of hard alkaline red soils on granite plains and basaltic rises in the Pilbara Region. Frequently associated shrubs are *Hakea suberea* and *Grevillea pyramidalis* whilst *A. ancistrocarpa* and *A. tetragonophylla* may also be present. *Triodia pungens* is the most abundant hummock grass although in more stony ground on foot-slopes of hills *T. wiseana* becomes prominent, whilst in more southerly areas *T. basedowii* replaces *T. pungens*. Small understorey shrubs of *Senna* and *Acacia* spp. such as *A. bivenosa* and *A. translucens*, may be present. The latter, together with *A. inaequilatera*, may form low open shrubland in wetter coastal areas with *Eragrostis* and *Eriachne* replacing the hummock grasses. On heavier soils *A. xiphophylla* becomes dominant.

Acacia ancistrocarpa

To the west of the Great Sandy Desert, *Acacia ancistrocarpa* shrublands with hummock grasses occur on shallow earthy sands over laterite. *A. monticola* is a characteristic species with *Grevillea wickhamii*, *G. eriostachya*, *G. refracta* and *Hakea suberea* frequent associates. *Triodia pungens* is more abundant in northern areas and *T. basedowii* in southern areas whilst in deeper sands *Plectrachne schinzii* replaces *Triodia*. On sandplains and between sandhills the above association forms a mosaic with *A. coriacea – Hakea suberea* shrubland with hummock grasses. The latter usually occupies more low-lying sandy areas.

To the south of the Tropic, *A. aneura* becomes prominent but in the sandy interdunal areas north of the Great Victoria Desert, *A. aneura* is rare and similar *A. ancistrocarpa* communities occur.

Pindan wattles (*Acacia eriopoda*, *A. colei* and *A. tumida*)

These wattles occupy red earthy sand plains in coastal areas of the Dampier Region in the northwest of Western Australia. *Acacia* spp. form dense thickets with scattered emergent *Eucalyptus* spp. such as *E. dampieri*, *E. miniata*, *E. zygophylla*, *E. polycarpa*, *E. papuana* and *E. tectifera* and other sclerophyllous species of *Hakea*, *Grevillea* and *Petalostigma*. The prominent grasses are *Plectrachne* spp., *Triodia pungens*, *Eragrostis eriopoda*, *Sorghum* spp. and *Chrysopogon* spp.. Fire and disturbance cause fluctuations in dominance between grass and *Acacia* thicket. In southern parts of the pindan, *A. pachycarpa* becomes the dominant shrub (Beard, 1990).

Mixed Mallee – *Acacia*

In the deserts south of the Great Sandy Desert in the mulga region, *Acacia aneura* is not common in sandy interdunal areas. Various *Acacia* spp. such as *A. helmsiana*, *A. pachyacra*, *A. grasbyi* and *A. ramulosa* form mixed shrublands with mallees such as *Eucalyptus gamophylla* and *E. kingsmillii* and *Triodia basedowii* is the dominant hummock grass. *Plectrachne schinzii* is often present on deeper sands. *Hakea*, *Eremophila forrestii* and *Grevillea* spp. including *G. juncifolia* and *G. eriostachya* are frequent associates with *Newcastelia cephalantha, Dicrastylis exsuccosa* and *Thryptomene maisonneuvii*.

Similar mixed communities occur on sand plains and dunefields in regions of the Northern Territory on red earthy sand soils. Trees and shrubs are sparse. Numerous *Acacia* spp. including *A. maitlandii, A. murrayana, A. kempeana, A. dictyophleba, A. coriacea* and *A. tenuissima* occur in the shrublayer with mallees *E. gamophylla* and *E. pachyphylla*. Lower shrubs of *Senna, Grevillea* spp. (including *G. juncifolia*), *Eremophila* spp. (including *E. longifolia* and *E. latrobei*), *Hakea* and *Keraudrinia integrifolia* are widespread. In addition to *Triodia pungens* in northern areas and *T. basedowii* in the south, there is a sparse cover of grasses and forbs.

A. coriacea, A. dictyophleba and *Hakea macrocarpa* are prominent in *Triodia pungens* hummock grasslands on gently undulating plains covered with red earthy sands. *Plectrachne schinzii* frequently occurs in the ground layer. This community is widespread in the Tanami Desert area but extends eastward to the Queensland border. *Grevillea* spp. are common associates particularly in interdunal areas. In northern parts of the desert, *A. stipuligera* is the common overstorey shrub and emergent *Eucalyptus* spp., such as *E. pruinosa* and *E. opaca*, become more prominent.

To the south, northwest of Alice Springs *A. pruinosa* becomes dominant in the tall shrub layer and *Triodia basedowii* replaces *T. pungens*. Scattered mallee *Eucalyptus* spp. occur and *Senna* spp. are common in the sparse lower shrub layer. In the Simpson Desert in the southeastern part of the Northern Territory, *Acacia ramulosa, A. ligulata* and *A. aneura* occur as scattered shrubs in *Triodia basedowii* hummock grassland in interdunal areas.

The *Acacia* hummock grassland communities extend into Queensland where mallees are absent and *Eucalyptus terminalis* and *E. papuana* becomes more prominent. In southern Queensland and the arid areas of northern New South Wales and South Australia, *A. aneura* becomes more prominent in these communities. *Hakea, Grevillea, Acacia, Senna* and *Eremophila* are still characteristic though the specific composition varies with location. *Triodia basedowii* is the most abundant hummock grass. Grasses such as *Aristida, Enneapogon, Eragrostis* and *Eriachne* together with forbs such as *Helipterum* and *Ptilotus* occur between the hummocks.

Mulga—mallee—hummock grassland communities are common in the west, mainly south of the Tropic of Capricorn in the Wiluna-Warburton area. They are found on sandplains with or without dunes and in the latter case are often restricted to the interdunal areas.

Triodia basedowii is the common hummock grass. *A. aneura* is often associated with *A. pruinocarpa, Hakea suberea* and *Grevillea juncifolia* and the most widespread mallees are *Eucalyptus kingsmillii* and *E. gamophylla*. Occasionally the tree *E. gongylocarpa* occurs as an emergent. Smaller shrubs such as *Eremophila forrestii* and *Alyogyne pinoniana* and forbs such as *Helipterum stipitatum* and *Ptilotus polystachyus* are frequent associates.

Utilization of Acacia *shrublands*

Large areas of *Acacia* shrublands are unoccupied or not in commercial use, largely because of aridity, lack of surface water or the unsatisfactory nature of the understorey species (e.g. *Triodia* hummock grassland) as pasture plants. In the areas used commercially, beef cattle raising is almost exclusively practised in the northwest of Western Australia, the north of South Australia, the far west of Queensland and in the Northern Territory. In the southern portion of *Acacia* shrublands, and in eastern Australia, sheep raising, along with some cattle, is predominant.

Acacia aneura shrublands are by far the most important associations utilised commercially. *Acacia aneura* is Australia's premier fodder shrub, not because it is the most nutritious, but because it is palatable, abundant and widespread (Everist, 1971). During recurrent and cyclical droughts *A. aneura* phyllodes confer stability on an otherwise fragile grazing system by virtue of stock browsing the phyllodes within reach, or particularly in eastern Australia, trees and tall shrubs being felled to provide stock access. In *A. aneura*-tussock grassland communities, sheep have been observed to select up to 70 per cent of *A. aneura* phyllodes in their diet when grass is dry and dormant (Beale, 1975). Methods of utilising *A. aneura* phyllodes as stock feed were discussed by Everist (1949).

Since *Acacia aneura* is such a useful 'drought' reserve there has been a tendency for it to become depleted, especially in the marginal pastoral zones of southern and western Australia. Legislative controls over the cutting of *A. aneura* shrub have been instigated in Western Australia. Hall, Specht & Eardley (1964) and Preece (1971) have reported on the infrequency of *A. aneura* regeneration in the south of the continent. However, at the more mesic end of its range, regeneration of *A. aneura* is adequate (Burrows, 1973).

In areas of Queensland receiving greater than 450 mm mean annual rainfall *A. aneura* densities are so high that means of thinning stands to promote growth of understorey pasture species have been studied

(Beale, 1973; Pressland, 1975). Both authors reported an inverse relationship between shrub density and pasture yield. Artificial thinning of *A. aneura* shrublands becomes questionable in the more arid areas as disturbance may quickly lead to a disclimax of unpalatable shrubs such as species of *Eremophila* and *Dodonaea*, or at worst a denuded landscape from which shrubs and grasses are both absent.

The potential for replacement of *Acacia* shrublands with more productive pasture species is limited. It is likely, however, that some 40 000 km^2 of *A. aneura* shrublands will be converted to pasture based on the introduced species, *Cenchrus ciliaris*, within the next 30 to 50 years.

A broader perspective on the distribution, utilisation and potential of *Acacia aneura* shrublands as a vegetation resource in Australia's semi-arid and arid regions can be found in Sattler (1986) and the proceedings of a symposium published in *Tropical Grasslands* (1973), Volume 7 (1). *Acacia* shrublands other than *A. aneura* can be expected to be of little commercial importance and should retain their floristic integrity for the foreseeable future.

References

Aplin, T.E.H. (1980). The vegetation of Western Australia. In *Western Australian Year Book*, ed. W.M. Bartlett, pp. 63–77. Perth: Government Printer.

Beadle, N.C.W. (1948). *The Vegetation and Pastures of Western New South Wales with Special Reference to Soil Erosion*. Sydney: Government Printer.

Beadle, N.C.W. (1981). *The Vegetation of Australia*. Cambridge University Press.

Beale, I.F. (1973). Tree density effects on yields of herbage and tree components in south west Queensland mulga (*Acacia aneura* F. Muell) scrub. *Tropical Grasslands*, 7, 135–42.

Beale, I.F. (1975). Forage intake and digestion by sheep in the mulga zone of Queensland, Australia. PhD thesis, Colorado State University.

Beard, J.S. (1968). Drought effects in the Gibson Desert. *Journal & Proceedings of the Royal Society of Western Australia*, 51, 39–50.

Beard, J.S. (1974). *Vegetation Survey of Western Australia: Great Sandy Desert*. Perth: University of Western Australia Press.

Beard, J.S. (1975). *Vegetation Survey of Western Australia: Nullarbor*. Perth: University of Western Australia Press.

Beard, J.S. (1990). *Plant Life of Western Australia*. Kenthurst: Kangaroo Press.

Beeston, G.R. & Webb, A.A. (1977). *The ecology and control of* Eremophila mitchellii. Queensland Department of Primary Industries, Botany Branch Technical Bulletin No. 2.

Boyland, D.E. (1973). Vegetation of the mulga lands with special reference to south-western Queensland. *Tropical Grasslands*, 7, 35–42.

Boyland, D.E. (1974). Vegetation. In *Western Arid Region Land Use Study*, Part I, pp. 47–74. Queensland Department of Primary Industries, Division of Land Utilization Technical Bulletin No. 12.

Burrows, W.H. (1973). Regeneration and spatial patterns of *Acacia aneura* in southwest Queensland. *Tropical Grasslands*, 7, 57–68.

Burrows, W.H., Carter, J.O., Scanlan, J.C. & Anderson, E.R. (1990). Management of savannas for livestock production in north-eastern Australia: contrasts across the tree-grass continuum. *Journal of Biogeography*, 17, 503–12.

Cody, M.L. (1989). Morphological variation in mulga. I. Variation and covariation within and among *Acacia aneura* populations. *Israel Journal of Botany*, 38, 241–57.

Cody, M.L. (1991). Morphological variation in mulga. II. Covariation among morphology, phenology and spatial patterns in *Acacia*-dominated vegetation. *Israel Journal of Botany*, 40, 41–59.

Connor, D.J. & Tunstall, B.R. (1968). Tissue water relations for brigalow and mulga. *Australian Journal of Botany*, 16, 487–90.

Doing, H. (1981). Phytogeography of the Australian Floristic Kingdom. In *Australian Vegetation*, ed. R.H. Groves, pp. 3–25. Cambridge University Press.

Everist, S.L. (1949). Mulga (*Acacia aneura* F. Muell.) in Queensland. *Queensland Journal of Agricultural Science*, 6, 87–139.

Everist, S.L. (1971). Continental aspects of shrub distribution, utilization and potentials: Australia. In *Wildland Shrubs – Their Biology and Utilization*, pp. 16–25. US Forest Service General Technical Report INT-1.

Farmer, J.N., Everist, S.L. & Moule, G.R. (1947). Studies in the environment of Queensland. I. The climatology of semi-arid pastoral areas. *Queensland Journal of Agricultural Science*, 4, 21–59.

Fox, J.E.D. (1986). Vegetation: Diversity of the mulga species. In *The Mulga Lands*, ed. P.S. Sattler, pp. 27–32. Brisbane: Royal Society of Queensland.

Francis, W.D. (1925). Observations on the plants of Charleville: characteristics of the western flora. *Queensland Agricultural Journal*, 24, 598–602.

Gillison, A.N. & Walker, J. (1981). Woodlands. In *Australian Vegetation*, ed. R.H. Groves, pp. 177–97. Cambridge University Press.

Groves, R.H. & Williams, O.B. (1981). Natural grasslands. In *Australian Vegetation*, ed. R.H. Groves, pp. 293–316. Cambridge University Press.

Hall, E.A.A., Specht, R.L. & Eardley, C.M. (1964). Regeneration of the vegetation on Koonamore vegetation reserve, 1926–1962. *Australian Journal of Botany*, 12, 205–64.

Isbell, R.F. (1962). *Soils and vegetation of the brigalow lands, eastern Australia*. CSIRO, Australia, Division of Soils, Soil & Land Use Series No. 43.

Johnson, R.W. (1964). *Ecology and Control of Brigalow in Queensland*. Brisbane: Queensland Department of Primary Industries.

Lange, R. & Purdie, R. (1976). Western myall (*Acacia sowdenii*), its survival prospects and management needs. *Australian Rangelands Journal*, 1, 64–9.

Maslin, B.R. (1981). Mimosaceae. 3. *Acacia*. In *Flora of Central Australia*, pp. 115–42. Adelaide: Australian Systematic Botany Society.

Moore, A.W., Russell, J.S. & Coaldrake, J.E. (1967). Dry matter and nutrient content of a subtropical semiarid forest of *Acacia harpophylla* F. Muell. (brigalow). *Australian Journal of Botany*, **15**, 11–24.

Moore, C.W.E. (1953). The vegetation of the south-eastern Riverina, New South Wales. I. The climax communities. *Australian Journal of Botany*, **1**, 485–547.

Moore, R.M. (1973). Australian arid shrublands. In *Arid Shrublands*, ed. D.N. Hyder, pp. 6–11. Denver: Society for Range Management.

Nix, H.A. & Austin, M.P. (1973). Mulga: a bioclimatic analysis. *Tropical Grasslands*, **7**, 9–22.

Pedley, L. (1973). Taxonomy of the *Acacia aneura* complex. *Tropical Grasslands*, **7**, 3–8.

Pedley, L. (1978). A revision of *Acacia* Mill. in Queensland. *Austrobaileya*, **1**, 75–337.

Pedley, L. (1986). Derivation and dispersal of *Acacia* (Leguminosae) with particular reference to Australia, and the recognition of *Senegalia* and *Racosperma*. *Biological Journal of the Linnean Society*, **92**, 219–54.

Perry, R.A. & Lazarides, M. (1962). *Vegetation of the Alice Springs area.* CSIRO, Australia, Land Research Series No. 6, pp. 208–36.

Preece, P.B. (1971). Contributions to the biology of mulga. II. Germination. *Australian Journal of Botany*, **19**, 39–49.

Pressland, A.J. (1973). Rainfall partitioning by an arid woodland (*Acacia aneura* F. Muell.) in south western Queensland. *Australian Journal of Botany*, **21**, 235–45.

Pressland, A.J. (1975). Productivity and management of mulga in south-western Queensland in relation to tree structure and density. *Australian Journal of Botany*, **23**, 965–76.

Purcell, D.L. (1964). Gidyea to grass in the central west. *Queensland Agricultural Journal*, **90**, 548–58.

Randell, B.R. (1992). Mulga. A revision of the major species. *Journal of the Adelaide Botanic Garden*, **14**, 105–32.

Russell, J.S., Moore, A.W. & Coaldrake, J.E. (1967). Relationships between subtropical and semiarid forest of *Acacia harpophylla* (brigalow), microrelief, and chemical properties of associated gilgai soil. *Australian Journal of Botany*, **15**, 481–98.

Sattler, P.S. (ed.) (1986). *The Mulga Lands*. Brisbane: Royal Society of Queensland.

Slatyer, R.O. (1961). *Principles and problems of plant production in arid regions*. CSIRO, Australia, Division of Land Research Regional Survey Technical Memorandum No.61/22.

Slatyer, R.O. (1965). Measurements of precipitation interception by an arid plant community (*Acacia aneura* F. Muell.). *Arid Zone Research*, **25**, 181–92.

Specht, R.L. (1972). *The Vegetation of South Australia*, 2nd edn. Adelaide: Government Printer.

Speck, N.H. (1963). *Vegetation of the Wiluna-Meekatharra area.* CSIRO, Australia, Land Research Series No. 7, pp. 143–61.

Stebbins, G.L. (1972). Evolution and diversity of arid-land shrubs. In *Wildland Shrubs – Their Biology and Utilization,* pp. 111–20. US Forest Service General Technical Report INT-1.

Tiller, A.B. (1971). Is bendee country worth improving? *Queensland Agricultural Journal,* 97, 258–61.

Wilson, B.A., Brocklehurst, P.S., Clark, M.J. & Dickinson, K.J.M. (1990). *Vegetation survey of the Northern Territory.* Conservation Commission of the Northern Territory, Technical Report No. 49.

10

Eucalyptus *scrubs and shrublands*

R.F. PARSONS

EUCALYPTUS scrubs and shrublands are defined here as vegetation where the tallest stratum is made up of eucalypt shrubs 2 to 10 m high. For scrubs, the dominant shrubs are denser than in shrublands and have a projective foliage cover greater than 30 per cent (*Atlas of Australian Resources*, 1990). By far the major portion of these vegetation types is dominated by eucalypts having many stems arising from a large, underground, woody swelling composed of stem tissue called a lignotuber (syn. 'burl'). Eucalypts with this growth habit are commonly called mallees (an Aboriginal word). This term is also widely used to describe the plant communities and regions where these plants predominate and will be so used here for conciseness.

Mallee communities extensive enough to be mapped at a scale of 1:5 000 000 (*Atlas of Australian Resources*, 1990) extend across Australia from Western Australia (longitude 117° E) to New South Wales (longitude 147° E) with a latitudinal range of from 22° to 37° S. The great majority are located, however, between 25° and 36° S (*Atlas of Australian Resources*, 1990).

In this main area of mallee occurrence, climate is broadly of Mediterranean-type with predominantly winter rainfall. Significant falls of summer rain can occur, however, and it is possible that there is a generally higher proportion of summer rainfall than in otherwise comparable climates in other countries (Rowan & Downes, 1963; Specht, 1981). Nevertheless, the summer rainfall is highly erratic and summer droughts are characteristic (Leeper, 1970).

Using Köppen's classification, the wetter mallee areas are 'Csb' climates and most of the drier ones are 'BSfk' with smaller areas of 'BWk' and 'BWh' (*Atlas of Australian Resources*, 1973). Although the widespread BSfk climates have uniform rainfall by Köppen's definition, the ratio of May–October to November–April median rainfall at Australian sites exceeds 1.3 (*Atlas of Australian Resources*, 1973) and these climates are regarded as 'modified Mediterranean'

(Leeper, 1970). Coastal mallee areas, such as at Ceduna, South Australia, can have frosts (screen minima 2 °C or less) on average for 18 nights per year, whilst for inland stations, e.g. Kalgoorlie in Western Australia, the figure rises to 27 (*Atlas of Australian Resources*, 1986). A comparison of the vegetation and rainfall maps in *Atlas of Australian Resources* (1986, 1990) shows that virtually all mallee is located between the 130 mm and the 800 mm isohyets for median annual rainfall. It is most common and often predominant between the 200 mm and 550 mm isohyets. At their upper rainfall limit, mallee communities are usually replaced by woodlands dominated by single-stemmed eucalypts (Gillison, Chapter 8, this volume), and at their lower rainfall limit by *Acacia* shrublands (Johnson & Burrows, Chapter 9, this volume), *Myoporum* woodlands and other arid communities lacking eucalypts. On going north from the major mallee areas, the proportion of rain falling in summer increases. North of latitude 22° S, at sites with a definite peak in summer rainfall, mallee eucalypts are much less common. Sites with soils and annual rainfall levels suitable for mallee instead usually carry various single-stemmed subtropical eucalypts, such as *Eucalyptus dichromophloia* (Beard, 1974). Overall, mallee can be regarded as the most arid of the eucalypt-dominated communities of temperate Australia. Two exceptions to this are the areas of *E. odontocarpa* and *E. normantonensis* mallee found between 18 and 22° S in Western Australia, Northern Territory and Queensland, including areas with a very marked peak in summer rainfall (Chippendale, 1963; Hall & Brooker, 1974*a,b*; Noble & Bradstock, 1989*b*).

Eucalypts with a mallee growth form can also be found in areas wetter than those already considered and where single-stemmed eucalypts predominate. This occurs in a variety of unfavourable habitats, either by normally single-stemmed eucalypts assuming a mallee habit (e.g. *E. baxteri*) or by the occurrence of distinct wet-country mallees of restricted distribution specific to such habitats. Examples of the latter include the subalpine *E. kybeanensis* on exposed sites in southern Australia at altitudes up to 1600 m (Hall & Brooker, 1973), *E. rupicola* from sandstone cliff faces in the Blue Mountains of New South Wales (Kleinig & Brooker, 1974) and *E. codonocarpa* from sites with very infertile soils on rhyolite and a mean annual rainfall of 1520 mm in Queensland (Jones, 1964).

From these examples and others, is it clear that the mallee growth-habit can occur in response to a variety of stress conditions. The wet-country mallee communities will not be considered further in this chapter, but a brief account of some of them is given in Williams & Costin (Chapter 16, this volume); for the main areas of mallee vegetation considered in this chapter, the major stress involved is shortage of water. Noble & Bradstock (1989*a*) and Noble, Joss & Jones (1990) are useful general references on the ecology of mallee communities.

The mallee growth habit

Mallees are usually 3 to 9 m tall, but can exceptionally reach heights up to 18 m (e.g. some stands of *E. diversifolia* on Kangaroo Island, South Australia). The lignotubers (which occur in most single-stemmed species of *Eucalyptus* as well) arise as swellings in the axils of the cotyledons and first few leaves. They become large, woody, convoluted swellings often 0.3–0.6 m in diameter and sometimes up to 1.5 m. The largest recorded is 10 m across, with 301 living stems, in *E. gummifera* (Mullette, 1978).

Lignotubers have the same anatomical characteristics as normal stems but with greatly contorted xylem elements (Chattaway, 1958; Bamber & Mullette, 1978). Lignotuber wood can have almost twice the proportion of storage tissue as stem wood and thus a larger potential for starch storage (Bamber & Mullette, 1978). Lignotubers also contain a very large number of concealed dormant buds (Carrodus & Blake, 1970).

The frequent fires which occur in mallee areas rarely damage the largely buried lignotubers. Usually all aerial stems and leaves are killed and new shoots are produced from the dormant buds in the lignotuber. Mallee lignotubers may carry up to 70 shoots six months after fire; this can diminish to about 20–30 seven years later and to less than ten by 100 years (Holland, 1969*c*).

The stems of mallees usually branch sparingly and bear leaves only at the end of the branches so that the canopy resulting from many plants is often very narrow, even and horizontal, giving typical mallee communities a very distinctive appearance (Fig. 10.1; Noble & Bradstock, 1989*a*).

Whilst radiocarbon ages show that extant mallee lignotuber tisssues are less than 200 years old, progressive tissue replacement could occur so that mallees may be like other eucalypts in having life-spans of several hundred years (Wellington & Noble, 1985*a*).

Many mallee species occur occasionally as single-stemmed trees. Conversely, many eucalypt species which are usually single-stemmed, occur in multi-stemmed form under adverse site conditions or after destruction of the main stem by fires, termites or felling. The multi-stemmed character of mallee eucalypts is under partial genetic control

Fig. 10.1. Mature stand of mallee, about 8m tall, dominated by *Eucalyptus viridis*, *E. dumosa* and *E. calycogona* at Kiata Lowan Sanctuary, northwestern Victoria. (Photo: T. Pescott.)

(Mullette, 1976); most plants in an undisturbed even-aged stand were multi-stemmed at an age of 16 years (Myers, Ashton & Osborne, 1986).

Variation in the extent to which the multi-stemmed character is expressed means that it can be difficult to classify any given species as a mallee or non-mallee. This problem is compounded by the existence in dry areas of Western Australia of shrubby species similar to mallees but with the lignotuber absent or poorly developed. These are called 'marlocks' (Burbidge, 1952; Chippendale, 1973). Despite these problems, rough estimates of the numbers of species of mallee eucalypts are provided below.

Evolutionary history

Very little is known about the evolutionary history of mallee vegetation and the few general points that can be made refer almost entirely to the mallee eucalypts themselves. In 1978, there were about 515 generally accepted species of eucalypts of which about 108 were mallee eucalypts, excluding 'wet country' species. Of the 108 species, 71 occur only in Western Australia, 16 occur only east of Western Australia and 21 are shared by both areas (Parsons, 1981). Since then, a surge in taxonomic research has caused big changes. For example, in the *Eucalyptus foecunda* group, 15 species are now recognised rather than the former 5 and in series *Levispermae* 28 species are now recognised rather than 5 (Hopper, 1990). For some mallee species complexes, 'a reliable inventory is still many years away' (Hopper, 1990). From current estimates it may be suggested that there are 200–210 species of mallee eucalypts (excluding subalpine and 'wet country' species) in a genus of 700–800 species (M.I.H. Brooker, personal communication).

Mallee species are found in all eucalypt subgenera; from this and their floral morphology, it is assumed that the mallee habit is a secondary development in *Eucalyptus* which has arisen independently many times (Burbidge, 1952; Hill, 1989). The highest mallee species richness is in southwest Western Australia, where many mallee eucalypts may have arisen from the older, more stable, 'relic' flora of the wetter parts of southwest Western Australia in response to increasing climatic dryness. The mallee areas of southwest Western Australia are seen as the 'primary centre' for speciation of mallee eucalypts from which a number of species migrated to eastern Australia (Burbidge, 1960; Hill, 1989). As many as three-quarters of the eucalypts of southern Western Australia are mallees (Hopper, 1990).

General ecology

Much of this section has to centre around the ecology of the dominant *Eucalyptus* species as little detailed community ecology has been done and little is known of the relationship of the understorey species to habitat factors in many areas.

The 'core areas' of mallee

In the rainfall zone from 250 to 400 mm, mallee is predominant on a range of calcareous soils; it is this combination which traditionally has been regarded in Australia as constituting the 'typical' mallee country to be dealt with in this section.

Eastern Australia

The major areas are on Quaternary aeolian deposits (see e.g. Blackburn & Wright, 1989). Soil texture is very variable and topsoil textures range from sand to clay. Although the soils are predominantly calcareous, some of the deep sands like the Berrook sands can have siliceous surface horizons (Rowan & Downes, 1963; Victoria: Land Conservation Council, 1987). Nevertheless, these are mapped as brown sands with calcareous horizons rather than as leached sands (Northcote, 1960). Whilst topsoil pH for these brown sands is around neutral, for the other main soil groups it is usually strongly alkaline with average values higher than pH 8 (Rowan & Downes, 1963).

The ecologically best-known area is northwestern Victoria (Rowan & Downes, 1963; Connor, 1966; Parsons & Rowan, 1968; Noy-Meir, 1971, 1974; Victoria: Land Conservation Council, 1987; Cheal & Parkes, 1989) and adjoining parts of South Australia and New South Wales (Noy-Meir, 1971, 1974); accordingly this area will be emphasised here (Table 10.1). The same general relationships seem to apply in the other main areas of mallee in eastern Australia (Sparrow, 1989). A somewhat simplified community classification is used below (Table 10.1) to facilitate discussion of habitat factors; for a more detailed one using quadrat data on all vascular plant species to define and map 14 mallee communities in Victoria, see Victoria: Land Conservation Council (1987) and Cheal & Parkes (1989).

The literature frequently refers to the marked differences in mallee eucalypt size between communities reflected in the use of terms like 'big mallee', 'small mallee', etc. (Rowan & Downes, 1963; Noy-Meir, 1971). These have never been standardised or quantified. Partly after Rowan & Downes (1963), the following will be used here: big mallee usually has three to four stems per plant, stem diameters more than 15 cm at maturity, and height over 6 m; mallee has many thinner stems per plant and height about 3.5–6 m; small mallee is like mallee but with height less than 3.5 m at maturity.

In the region being discussed, soil texture is usually the most important single factor affecting the distribution of native plants (Noy-Meir, 1974). This is related both to increasing levels of macronutrients with increasing clay content (Parsons & Rowan, 1968) and to the 'inverse texture effect' (Rowan & Downes, 1963; Noy-Meir, 1974), whereby soil water supply to plants decreases with increasing clay content. This effect is a result of, firstly, the small depth of penetration of rainfall into the soils with higher clay contents, with subsequent increases in evaporation losses, and secondly, the larger amount of rainfall needed to bring the clayier soils from air-dry condition up to

Table 10.1. *Main mallee communities in northwestern Victoria and adjoining areas; annual rainfall 230 to 380 mm*

Main eucalypt species	Eucalypt growth form	Understorey type	Characteristic species	Predominant topsoil texture	Main references
E. incrassata *E. leptophylla*	Small mallee	Sclerophyllous shrubs	*E. incrassata* *Callitris verrucosa* *Aotus ericoides*	Sand	Parsons & Rowan (1968); Noy-Meir (1971)
E. socialis *E. dumosa*	Small mallee	Hummock grasses (*Triodia*)	*E. socialis* *Triodia* spp. *Sclerolaena parviflora*	Sand[a]	Parsons & Rowan (1968); Noy-Meir (1971)
E. oleosa *E. gracilis* *E. dumosa*	Mallee	Mixed shrubs	*Acacia colletioides* *Eremophila glabra* *Senna artemisioides*	Probably loamy sand to sandy loam	Parsons & Rowan (1968); Noy-Meir (1971)
E. oleosa *E. gracilis* *E. dumosa*	Mallee, big mallee	Semi-succulent shrubs (chenopods, etc.)	*E. oleosa* *E. gracilis* *Sclerolaena diacantha*	Sandy loam to clay loam	Parsons & Rowan (1968); Noy-Meir (1971)
E. calycogona *E. dumosa*	Mallee, big mallee	Not known	Not known	Clay	Parsons & Rowan (1968); Litchfield (1956)
E. behriana[b]	Mallee, big mallee	Chenopods, low shrub *Acacia*	Not known	Clay	Connor (1966); Litchfield (1956)

[a]These sands are usually more fertile than those carrying *E. incrassata*.
[b]This species mostly occurs where annual rainfall exceeds 330 mm.

the 'available water' range of soil water potential, so that light showers can make water available on dry sandy soils but not on dry clayey ones (Rowan & Downes, 1963; Noy-Meir, 1974). These effects will be discussed in more detail later.

The small mallee, mallee and big mallee categories mentioned above show some obvious general relationships to soil texture, with size of eucalypt plants usually increasing with clay content (Table 10.1). In Victoria, big mallee can occur on clays in the wetter part of the area, but in the driest parts (annual rainfall less than 280 mm), grassland often occupies these soils (Rowan & Downes, 1963). Big mallee is particularly prominent in these dry areas, however, on fertile sandy loams to clay loams. It seems likely that, on these soils, height of mallee eucalypts increases with decreasing rainfall and that this change can occur even when soil fertility stays approximately constant. One possible explanation is that in the wetter, fertile areas, water supply is sufficient to allow ample regeneration, thereby causing high densities of mature eucalypts so that competition between trees restricts the maximum height attained. In contrast, in the drier, fertile areas, lower rainfall may restrict regeneration, thereby causing lower densities of mature eucalypts and so allowing individual plants to attain greater heights than in dense stands. Data on density and other parameters are badly needed to confirm or reject this hypothesis. Certainly, general observation suggests that densities of mature eucalypts are usually lower in big mallee than in mallee (Rowan, 1971). Low eucalypt density in big mallee on clay plains in wetter areas may be related to inherent subsoil salinity (Rowan, 1971).

Regarding overall floristics in the mallee communities of the area, the most marked discontinuity is between the *Eucalyptus incrassata* type (Table 10.1) with its species-rich, sclerophyllous understorey (*Hibbertia*, *Aotus*, *Leptospermum*, etc.) of 'Southern Temperate' affinities and all the other types, in which semi-succulent shrubs, especially chenopods, are prominent (*Sclerolaena*, *Maireana*, *Zygophyllum* etc.). This floristic series is said to have 'semi-arid, Eremaean' affinities (Noy-Meir, 1971). The same discontinuity is reflected in the 'Chenopod Mallee' and 'Mallee Heath' units of Cheal & Parkes (1989) and it is also highlighted by Tiver, Sparrow & Lange (1989).

From a correlation of floristic data with habitat factors using multiple regression techniques (including both mallee and non-mallee communities in the area), Noy-Meir (1974) suggested that the floristics are determined mainly by variables related to soil texture. The strongest correlations were with depths of soil wetting by various typical falls of rain (calculated as the ratio of amounts of rain to soil-water capacity) and with topsoil water capacity. Other variables contributing consistently but in smaller amounts to floristic variation were the relative importance of calcium in the soil exchange complex, topsoil salinity and subsoil phosphorus (Noy-Meir, 1974). In summary, Noy-Meir (1974) saw the major cause of vegetation variation in the area as an interaction between rainfall and texture, with increased sandiness operating in the same direction as increased rainfall (as expressed by the depth of wetting variables). This view has since been supported by Sparrow (1989), who uses the same interaction to explain the major north-south changes in composition of mallee communities in South Australia.

The fine degree of control operated by texture-related factors is illustrated by the very closely related species *Eucalyptus socialis* and *E. oleosa*. Sandy soils on crests and slopes of small dunes carry *E. socialis*, but this species gives way completely to *E. oleosa* with only slight increases in clay content on the intervening flats (Parsons & Rowan, 1968).

Some detailed autecological work is available on the mallees *E. incrassata* and *E. socialis* (Parsons, 1968, 1969). *Eucalyptus socialis* has drier upper and lower rainfall limits than *E. incrassata* (200–460 mm compared to 250–560 mm). Where the rainfall ranges overlap, *E. socialis* is absent from the sandiest soils lowest in nitrogen, phosphorus and calcium, which carry *E. incrassata*. Relatively fertile, less sandy soils carry *E. socialis* whilst intermediate soils may carry both species. Thus the relative distribution of the two could be controlled by soil nutrients, soil water supply (inverse texture effect) or both.

Pot experiments with seedlings (Parsons, 1968, 1969) showed no consistent differences between species in nutrient response but that *E. socialis* has consistently higher root : shoot ratios. *E. incrassata* outcompeted *E. socialis* on all soils used at optimal water levels, but this advantage could be nullified on the fertile *E. socialis* soil by droughting. The higher percentage of plants wilting and the greater drought damage of *E. incrassata* than *E. socialis* on the *E. socialis* soil may be caused, at least partly, by inferior drought avoidance of *E. incrassata* because of its faster growth in competition (faster water depletion rate) and its lower root : shoot ratio (Parsons, 1968, 1969). Thus, it may be that *E. incrassata* can outcompete *E. socialis* on infertile soils with a high water-supplying capacity, whilst drier, more fertile soils carry *E. socialis* in part because of its superior drought avoidance (Parsons, 1969).

The finding that increased soil fertility has an important effect on drought susceptibility of some species by promoting faster growth can be compared with Noy-Meir's (1974) results. That is, not only does increased clay content reduce the amount of soil water available in the area (the inverse texture effect), but it is also correlated with higher fertility which may increase growth rate and water consumption and thus increase drought susceptibility in some species, thereby enhancing the inverse texture effect (Parsons, 1969; Noy-Meir, 1974).

Whilst the detailed information given in this section applies to mallee in southwestern New South Wales (Noy-Meir, 1971, 1974), much less is known about the large, disjunct area of mallee around Mt Hope (*Atlas of Australian Resources*, 1990) on the eastern edge of Noy-Meir's area. Here it appears that mallee dominated by *E. socialis*, *E. oleosa* and *E. dumosa* is widespread over acidic, non-calcareous loamy soils with a topsoil pH of about 5.6 (Stannard, 1958; Holland, 1968, 1969b). This atypical relationship deserves further attention.

Wood (1929) has provided life-form spectra (Raunkiaer, 1934) for the type of mallee dealt with in this section, stressing the high percentage of ephemeral species which occur only after rain. The percentage of ephemerals is higher still in the arid communities beyond the lower rainfall limit of mallee (Wood, 1929). Whittaker, Niering & Crisp (1979) provide comparisons of both life-form spectra and species richness with North American vegetation.

From detailed seasonal sampling of vascular plants, it is known that the percentage of species which are annuals is about 26 per cent in a mature *E. socialis–E. dumosa* mallee community and can reach 44 per cent in a mature *E. incrassata* community. Whilst some of these annuals can germinate in autumn (April), most germinate in winter (July–August). Maximum biomass for annuals occurs in spring or summer (October–early January). By March, towards the end of the dry season, biomass of annuals is insignificant. The normal life-span of the annual species involved is from two to ten months (Holland, 1968, 1969b).

Perennial species in the field layer of these communities usually grow rapidly in the spring–early summer period (September–January), but produce no new shoots in the late autumn and winter period. By contrast, shoot production by mallee eucalypts and other tall shrubs is from mid-December to early May (Holland, 1968).

There have been three recent community classification studies using full floristic data. The Victorian one showed that the composition of eucalypt species was no use for defining some communities; full floristic data were needed for adequate classification (Cheal & Parkes, 1989). Similarly, the South Australian studies showed appreciable independence between the distribution of the eucalypts and the understorey species (Tiver *et al.*, 1989; Sparrow, 1990). The latter work also demonstrates an east–west biogeographic gradient because of attenuating Western Australian elements and presence of Eyre Peninsula endemics (Sparrow, 1989). Whilst Cheal & Parkes (1989) recognise 14 Victorian mallee communities, using a different approach in South Australia, Sparrow (1990) recognises five mallee complexes comprising 73 types. Finally, O'Brien (1989) describes the effects of severe frost on mallee eucalypts in northwestern Victoria.

Western Australia

The ecology of much Western Australian mallee is poorly understood, so that only a few general points can be made here. Mallee dominated by eucalypts including *Eucalyptus redunca* and *E. eremophila* is mapped as widespread (Beard, 1975) on areas of brown calcareous earths north of Esperance (Northcote *et al.*, 1967). Proceeding north on these soils into drier areas, where annual rainfall drops below about 300 mm, this vegetation is replaced by various communities dominated by appreciably taller eucalypts (more than 10 m tall) which are thus mapped as woodlands (Beard, 1975). One common type is *E. oleosa*–*E. flocktoniae* woodland up to 18 m tall. There is continuous variation from this to typical mallee shrubland, *E. oleosa* varying greatly in size and occurring in both structural types (Beard, 1975).

Similar woodlands occur in the same climatic region on other loamy soils, for example, the *E. transcontinentalis*–*E. flocktoniae* community commonly 12–18 m tall and the *E. salmonophloia* community commonly 18–27 m tall, in both cases the dominant eucalypts being single-stemmed. The lowest of these woodlands tend to have the highest eucalypt density and the tallest, the lowest density (Beard, 1969). These woodlands and similar ones make up the very large Goldfields area of Western Australia mapped as woodland at 1:5 000 000 by *Atlas of Australian Resources* (1990), which had been mapped as mallee on previous maps of Australian vegetation.

There is a similar transition from mallee shrubland to woodland east of the one just described. On the Nullarbor limestone southwest of Caiguna, there is a coastal strip of *E. socialis* mallee about 15 km wide. With decreasing rainfall inland on the same substrate, this changes to *E. oleosa*–*E. flocktoniae* woodland up to 18 m high (Beard, 1975).

These increases in eucalypt height with declining rainfall on comparatively fertile soils are strongly reminiscent of the similar change from mallee to big mallee on such soils already mentioned in eastern Australia. It is again possible that one cause is less regeneration, lower eucalypt density and therefore greater height of individual eucalypts in the drier areas. Such an hypothesis has added credence in Western Australia, because in the examples described, soil uniformity gives greater assurance that fertility is similar throughout the sequence, and also, it is clear that eucalypt height and density are inversely correlated (Beard, 1969). In neither area is it known how community biomass changes with decreasing rainfall; clearly, more soil and vegetation data are desirable to clarify the position.

Some of the shorter vegetation mapped by Beard (1975) and subsequently by *Atlas of Australian Resources* (1990) as woodland in the 250 to 300 mm rainfall belt of Western Australia is likely to have a structure identical to 'big mallee' in eastern Australia; for example, the community named as woodland partly dominated by apparently multi-stemmed *Eucalyptus oleosa* in Plate 24, Beard (1975). The widespread dominance of eucalypt woodlands 15–27 m tall on non-floodplain sites in this rainfall belt, however, is completely without parallel in eastern Australia, where similar soils in the same climate carry shrublands of the 'big mallee' type with a maximum height of about 9 m.

Whilst some of the smaller Western Australian woodland species, like *Eucalyptus flocktoniae* and *E. oleosa*, spread to eastern Australia, where their stature appears to be somewhat reduced, the tallest Western Australian species, like *E. dundasii*, *E. salmonophloia* and *E. transcontinentalis*, are endemic there (Beard, 1969; Chippendale, 1973). The reasons for the evolution of these species of strikingly larger stature are unknown.

In the Western Australian core area considered in this section there is a range of mallee communities not discussed here (Beard, 1975 and references therein). Their detailed environmental relationships are not known.

At the drier rainfall limits of the core areas of mallee, species such as the widespread mallee *E. socialis* become confined to deep sandy soils along watercourses and occasionally skeletal soils on rocky slopes (Carrodus, Specht & Jackman, 1965), presumably because of the superior water-supplying characteristics of such soils in this climate (the inverse texture effect of Noy-Meir, 1974).

Other areas of mallee

Other types of mallee occur in wetter regions, often on acidic, non-calcareous soils and, especially in eastern Australia, often in regions where significant amounts of non-mallee vegetation occur, especially on the most productive sites.

Where average annual rainfall ranges from 380 mm to 430 mm, mallee usually predominates. In even wetter areas up to 660 mm, mallee becomes scarcer and is often restricted to a range of relatively infertile soils (Table 10.2).

The commonest of these types, and the one which usually adjoins the core areas of mallee on their wetter side, is mallee on either deep siliceous sands or on siliceous sands over sandy clays (Table 10.2). All the other communities listed in Table 10.2 can be found in comparatively wet areas as well, except for the final, arid category. Some Western Australian types not listed there include *Eucalyptus preissiana–E. lehmannii* on the Barren Ranges quartzites and *E. gardneri–E. nutans* mallee on the basic, igneous 'greenstones' of the Ravensthorpe area (Beard, 1972a). For an account of other mallee communities, see Beadle (1981), Hill (1989), Sparrow (1989), Tiver *et al.* (1989), Beard (1990 and references therein) and Pressey, Bedward & Nicholls (1990).

Large, continuous belts of mallee are usually absent once annual rainfall drops below about 200–230 mm and mallee most often occurs as discontinuous patches in a matrix of arid zone communities usually lacking eucalypts. These mallee communities are mapped as far north as 22°S (*Atlas of Australian Resources*, 1990), mainly in areas with annual rainfalls from 170 to 200 mm. They contain a distinctive suite of species mostly absent from the wetter mallee communities to the south (Table 10.2).

Production ecology, nutrient cycling and hydrology

Biomass–time curves from *E. incrassata* mallee-broombush for 12 years after fire reflect rapid regeneration from lignotubers, and the eucalypts are still increasing in biomass after 12 years, with an average aerial biomass increment of 8–9 per cent per year. The broombush species (*Baeckea behrii* and *Melaleuca uncinata*) contribute 20–30 per cent of standing community biomass, reaching peak biomass eight years after fire and then declining (Specht, 1966). In other types of mallee, average aerial biomass increment for eucalypts is between 6 and 8 per cent per year for stands up to about 35 years after burning (Holland, 1969a).

Productivity comparisons of 15-year-old mallee regeneration after clearing with mature mallee show above-ground net primary productivity of 5406 kg ha^{-1} yr^{-1} and 2379 kg ha^{-1} yr^{-1} respectively. The corresponding leaf area indices are 0.57 and 0.73. Understorey

Table 10.2. *Some mallee communities other than those listed in Table 10.1*

Category	Soil characteristics	Eucalypt species[a]	Understorey type	Main references
Mallee on calcareous coastal soils				
(a) Eastern Australia	Calcareous beach sands; soils on aeolian calcarenite	*E. diversifolia* *E. rugosa*	Coastal shrubs	Parsons & Specht (1967); Beadle (1981); Sparrow (1989).
(b) Western Australia	Calcareous beach sands; soils on aeolian calcarenite	*E. calcicola* *E. oraria*	Coastal shrubs	Brooker (1974); Beard (1976).
Mallee on siliceous dunes and sandplains				
(a) Eastern Australia	Deep siliceous sand; siliceous sand over sandy clay	*E. incrassata* *E. leptophylla*	Sclerophyllous shrubs including 'broombush' type	Coaldrake (1951); French (1958); Specht (1966); Cheal & Parkes (1989); Sparrow (1989).
(b) Western Australia	Deep siliceous sand; siliceous sand over sandy clay	*E. incrassata* *E. tetragona* *E. redunca* *E. eremophila*	Sclerophyllous shrubs, including 'broombush' type	Beard (1972a, b, 1973a, b); Beadle (1981).

Mallee on ironstone soils

Habitat	Soils	Species	Vegetation	References
(a) Eastern Australia	Various ironstone gravel soils; a gravelly sand over clay type is common	E. cosmophylla, E. diversifolia, E. remota, E. incrassata, E. viridis	Sclerophyllous shrubs, including 'broombush' type	Cleland (1928); Baldwin & Crocker (1941); Northcote & Tucker (1948); Litchfield (1956); French (1958).
(b) Western Australia	Various ironstone gravel soils; a gravelly sand over clay type is common	E. tetragona, E. tetraptera, E. incrassata, E. eremophila	Sclerophyllous shrubs, including 'broombush' type	Beard (1969, 1972a, 1973a).
Mallee on rocky, sandstone soils Eastern Australia	Skeletal or shallow loam-over-clay soils on Ordovician sandstone	E. viridis, E. polybractea, E. behriana, E. froggattii	Sclerophyllous shrubs, including 'broombush' type	Biddiscombe (1963); Rowan (1963); Beadle (1981); Myers et al. (1986); Cheal & Parkes (1989).
Arid mallee communities[b]	Sandplains; interdune sands; skeletal, rocky soils	E. gamophylla, E. kingsmillii, E. oleosa complex, E. oxymitra, E. pachyphylla, E. youngiana, E. gillii	Shrubs and Triodia hummock grasses	Chippendale (1963); Mabbutt et al. (1973); Beard (1975); Beadle (1981).

[a] The species listed for each habitat do not necessarily cohabit.
[b] Mallee communities in areas where mean annual rainfall is less than 200–230mm.

shrubs make up only 2.3 per cent of the total standing biomass by about 55 years after fire. Peak productivity following destruction of all aerial plant parts by clearing or burning may be reached after about 15 years. Standing biomass may not reach a plateau for at least 30 years (Burrows, 1976). In the fire management context, it has been suggested that this time period to reach a plateau gives a rough guide to the length of time needed between fires to ensure maximal levels of species survival for species regenerating from underground lignotubers and rootstocks, as this time interval should be sufficient to prevent depletion of lignotuber and rootstock food reserves (Groves, 1977).

Regarding nutrient cycling, nutrient pool size for 15-year-old mallee regeneration after clearing has been compared with that for mature mallee. The results (Table 10.3) show that where the surface soil remains intact, as in this case, there is little difference in the total pool sizes, suggesting high resilience of the nutrient pool to massive disturbance. The ability to regenerate from lignotubers may exert a stabilising influence by allowing rapid initial uptake, after which community requirements (at least for mobile elements) may be largely met by recycling (Burrows, 1976).

Litter production is strongly seasonal with a pronounced mid-summer maximum, and in this regard is similar to other eucalypt communities in southern Australia. Regarding nutrient withdrawal from leaves prior to abscission, the percentage weight loss per unit area is between 52 and 57 per cent for phosphorus and between 37 and 48 per cent for nitrogen. Whilst such figures indicate some conservation in the use of these nutrients, it is not yet clear whether they indicate any special adaptation to infertile soils or whether similar values also occur in vegetation on soils of higher fertility (Burrows, 1976).

Mallee eucalypt root systems reach depths of 28 m and are extremely efficient at extracting soil-water. Values for mallee stem-flow, throughfall and interception are now available (Mallee Vegetation Management Working Group, 1991 and references therein). Subterranean termites may consume a large fraction of the dead organic matter in mallee communities and may have important hydrological effects (Whitford, Ludwig & Noble, 1992).

Regeneration

Eucalypt seed availability

In the mallee eucalypts examined, seed supply is unlikely to limit recruitment. Heavy flowering can result in abundant seed every 2–5

Table 10.3. *Organic matter, nitrogen and phosphorus distribution in two stands of mallee dominated by Eucalyptus socialis, after Burrows (1976). 1, stand regenerated from clearing 15 years earlier; 2, stand undisturbed for about 55 years*

	Organic matter (kg/ha)		Total nitrogen (kg/ha)		Total phosphorus (kg/ha)	
	1	2	1	2	1	2
Above-ground						
Total standing	19 553	40 164	75.4	88.1	5.8	5.0
Total litter	7 547	11 372	41.2	54.2	2.1	3.1
Total above-ground	27 100	51 536	116.6	142.3	7.9	8.1
Below-ground (to 1 m depth)						
Eucalypt lignotubers	15 288	13 860	25.4	23.1	2.5	2.2
Total lignotubers and roots	20 501	28 533	47.3	103.9	3.8	6.7
Soil	79 800	73 336	5074	5804	3423	4035
Total	127 401	153 405	5238	6050	3455	4050

years (Noble, 1985; Wellington, 1989). After fire killed the adult stems, regrowth from the lignotubers of all five species studied took four years to initiate floral buds. *Eucalyptus gracilis* flowered four months later whilst the last species to flower, *E. leptophylla*, flowered 21 months later. Seed would probably be fully ripe within one year of flowering (Noble, 1985).

The seed is stored in the canopy in woody capsules (seed load $7-9 \times 10^6$ seeds/ha). Without fire, some seeds are released as a slow trickle ($1-3\times10^6$ seeds/ha/y). Most seeds survive the recurrent fires typical of mallee areas and are shed post-fire following capsule death (Wellington, 1989). It is estimated that large fires recur at about 20 year intervals in some mallee core areas (Cheal, Day & Meredith, 1979), most adult mallees surviving them by producing regrowth from the lignotuber (Wellington, 1989).

Eucalypt seed predators and soil storage of seeds

On the soil surface, most seeds (60–100 per cent) are soon removed by harvester ants, such seeds having a half-life of about five days. The harvested seed is used as food and does not provide established seedlings. Mass release of canopy-stored seed post-fire may satiate the ant seed predators and allow appreciable seed storage in the soil. Most of this seed germinates soon after the advent of suitable climate in late autumn and winter; it does not form a persistent seed bank (Wellington, 1989).

Eucalypt germination

Field germination of mallee eucalypts has been recorded from April to August; it may possibly occur whenever soil moisture is adequate and has been recorded five years out of six in an area with 367 mm mean annual rainfall. Some mallees may show a degree of seed dormancy, the ecological significance of which is unknown (Wellington & Noble, 1985b; Wellington, 1989).

Eucalypt seedling establishment

Young mallee eucalypt seedlings can be found sparingly (1 seedling per 300 m²) in unburnt, undisturbed mallee stands in winter. They do not persist, probably most being killed by summer drought exacerbated by water use by established plants. By contrast, in the spring after a summer fire, 3–10 000 eucalypt seedlings per hectare can be found, such recruitment probably being related to fire-induced mass seed release, seed-eating ant satiation, increased mortality of adult mallees, reduced community transpiration and soil heating effects on nutrient availability. In one study, 75 per cent of such seedlings died within two years, especially in the first summer (Wellington, 1989). In another, 10 per cent of post-fire seedlings were still alive after six years and appeared well established (Noble, 1985). Whilst detailed studies found no seedling mortality because of rabbits and other browsing vertebrates, this is likely to be very important in some contexts (Wellington, 1989).

Successful seedling regeneration is found following less than 10 per cent of fires, suggesting that both fires and above-average spring-summer rainfall are needed (Wellington, 1989). Particularly in the drier mallee areas, it seems plausible that spring-summer rainfalls may only rarely be sufficient for this, leading to erratic pulses of seedling recruitment. In dry areas adjacent to mallee, rainfall has been sufficient to allow seedling regeneration of *Acacia sowdenii* only three times this century (Lange & Purdie, 1976). In mallee core areas, Bishop (1990) estimates that climate would allow prolific mallee eucalypt regeneration less than once every 16 to 20 years but at least some regeneration once every two to four years.

Prescribed fires of high intensity in spring can produce 3350 eucalypt seedlings per ha while cooler autumn fires can produce only 140 seedlings per ha (Noble, 1985). The reasons for this difference do not seem to be understood.

Eucalypt mortality and age structure

Adult eucalypt mortality rates can change from 0.3 per cent per year in long-unburnt mallee to 2.5 per cent per year in the two years after fire. Successful seedling regeneration must occur at least every few hundred years for populations to be perpetuated. Such events could vary from 'one to one' replacement to establishment of large, even-aged cohorts so that mallee populations may have complex age structures varying greatly in space and time (Wellington, 1989). Such age structures have been documented in some *Eucalyptus behriana* communities, where long-unburnt mallee contains groups of 20–30 stunted probably suppressed eucalypts, suggesting gap regeneration at long intervals (Myers *et al.*, 1986).

Eucalypt lignotuber regrowth

In mallee eucalypts, the importance of lignotuber regrowth, rather than regeneration from seed after burning, stands in marked contrast to the behaviour of *Callitris* spp., the native conifers which dominate large areas of woodlands within and adjacent to the main mallee areas. These species are killed by fire and regenerate solely from seed.

An equally interesting contrast can be seen in many mallee areas of Western Australia, where, interspersed among mallee communities, there are frequent areas dominated by single-stemmed low eucalypts which are killed by fire and which regenerate only from seed. The best known examples are *Eucalyptus annulata*, *E. platypus* and *E. spathulata* 5 to 7 m tall on low-lying clay soils (Beard, 1967, 1972*a*). These are referred to as 'thicket-formers' or 'marlocks' (Beard, 1967, 1972*a*), the latter name implying that lignotubers are absent or poorly developed (Burbidge, 1952).

The same total dependence on seedling regeneration rather than lignotuber regeneration is seen in the tall, dry woodland species *E. salmonophloia* (Beard, 1972*b*), which occurs at low density over a sparse understorey, so that these woodlands 'burn only rarely and with difficulty if at all' (Beard, 1969). Beard (1969) indicated as a generalisation that the other eucalypt tree dominants of these arid woodlands in the Goldfields region of Western Australia also show this regeneration strategy, which is puzzling given that they include *E. oleosa*, which occurs widely as a mallee capable of lignotuber regrowth. Similarly, normally single-stemmed tree species, such as *E. falcata* and *E. gardneri*, are regarded as fire-tender seed-regenerators (Brockway & Hillis, 1955; Beard, 1972*b*), but are also said to occur in mallee forms capable of lignotuber regrowth after fire (Brockway & Hillis, 1955). Also, *E. diptera* and *E. eremophila* are said by Beard (1975) to be single-stemmed fire-tender, seed-regenerators. However, a burnt woodland stand of them has regenerated partly from seed and partly from lignotuber regrowth to produce a stand of mallee structure (Hopkins & Robinson, 1981). More data on lignotuber incidence and behaviour are badly needed for these and similar species.

In summary, it is striking that Western Australia has a number of apparently strictly seed-regenerating non-mallee eucalypts in habitats which in eastern Australia would be entirely mallee-dominated. With the exception of flood-plain tree species like *Eucalyptus camaldulensis*, all the eucalypts in the mallee regions of eastern Australia have well-developed lignotubers which readily produce new shoots after burning. The apparent unimportance or absence of lignotubers in *E. salmonophloia* and similar woodlands at and around the main dry limits for eucalypts in Western Australia is an important exception to the generalisation that lignotubers are best developed in *Eucalyptus* where the genus is close to its physiological limits (Gill, 1975).

Community studies

Of the studies of fire effects on community floristics (see Gill, 1990), one is from small mallee with a *Triodia* understorey in a 260 mm annual rainfall area in Victoria. Here, unburnt areas carry 18 vascular plant species, but a year after fire a burnt area carried 63 species, including 26 annuals (Zimmer, 1940). This appears to be the largest fire induced increase in species richness yet recorded in Australia (see Gill, 1977 and references therein). Noble (1989) describes the post-fire pioneers and other post-fire floristic changes of similar mallee, whilst Bradstock (1989) gives valuable life cycle data for a range of understorey species and predicts understorey response to a range of fire histories. Cheal *et al.* (1979) give further details on floristic changes with time since fire and on desirability of various fire frequencies.

Regarding re-colonisation of mallee areas previously cleared for farming, in one study unstabilised deep sands were sown with *Secale cereale* (cereal rye). In one area, in the second year, dominance was shared by self-sown *S. cereale* and the introduced annual *Brassica tournefortii*. In the third year, 95 per cent of the plants were *B. tournefortii*, whilst in the fourth year this species declined markedly and 80 per cent of the plants were the native annual composite *Myriocephalus stuartii* (Sims, 1949), also known as a post-fire pioneer (Zimmer, 1940). These very marked and rapid changes in dominance are strongly reminiscent of those recorded on abandoned farms in Idaho in a similarly semi-arid climate (Piemeisel, 1951) and alien crucifers and *Salsola* are important in both areas. From four to six years, the vegetation can be a mixture of Acacias (e.g. *A. brachybotrya* and *A. ligulata*) and native grasses and, from 11 to 33 years, dense shrublands of *A. brachybotrya* and *Dodonaea viscosa*. Depending on soil type, other Acacias, and *Senna artemisioides* can also be important. As all these shrubs are probably hard-seeded, soil storage of hard seed is likely to be important in the genesis of the secondary shrub communities. It seems that recolonisation of these shrublands by mallee eucalypts would require as a minimum parent trees within 10 m and fire or clearing to reduce shrub competition. In areas where farming predominates, mallee communities are now so fragmented and fires so scarce that the chances of most secondary shrublands reverting to mallee are negligible (Onans & Parsons, 1980).

Conservation

Pressey *et al.* (1990) give the best account of the reservation status of mallee communities. In the Western Australian mallee, such rapid geographic replacement of taxa occurs that reserves need to be established at least every 15 km for the same broad vegetation formation and soil type if they are to conserve a representative sample of the flora (Hopper, 1990).

A major threat to the future regeneration of mallee vegetation is likely to be the destruction of seedlings by grazing rabbits. This was not important in Wellington's (1989) study, presumably done when rabbit numbers were low. Inhibition of regeneration of a whole range of species from mallee and adjacent communities by rabbits and other grazing mammals is likely, however (Cooke, 1990; Parsons, 1990 and references therein), especially if rabbit numbers continue to increase as at present. Further data are urgently needed.

Concluding discussion

The overall relationship of understorey type to climate and soil can be summarised as follows (after Specht, 1972).

(a) Infertile soils in the wettest areas have a dense, species-rich understorey of sclerophyllous shrubs usually less than 2 m high ('mallee-heath' type).
(b) Infertile soils of areas of intermediate rainfall have a dense understorey of sclerophyllous shrubs with repeatedly-branching, erect stems terminating at about the same height, usually more than 2 m high and often species of *Melaleuca* ('mallee-broombush' type).
(c) Infertile soils of the driest areas have a stratum of hummock grasses (*Triodia* spp.) with or without a range of shrubs.
(d) Fertile soils have a range of understorey types including sparse grassy ones and various shrub mixtures. With increasing dryness, semi-succulent low shrubs (chenopods, *Cratystylis*, *Zygophyllum*) become progressively more important and finally predominant.

Regarding the general climatic and edaphic ranges of the eucalypts present within a given area, species are often very sensitive to environmental changes and show precise correspondence with particular microhabitats, as was shown for *Eucalyptus incrassata* and *E. socialis* in Victoria. When total range is considered, however, a species can be found in a very wide range of habitats. For example, *E. incrassata* can occupy soils ranging from excessively drained to seasonally waterlogged and from extremely infertile siliceous sands and ironstone soils to relatively fertile calcareous loams. This, with the experimental data cited earlier, suggests that a number of species have wide physiological tolerances and that competition plays an important part in determining the precise correspondence with particular microhabitats observed within given areas.

Considering total species range, wide edaphic ranges are not uncommon, with species such as *Eucalyptus diversifolia*, *E. incrassata* and *E. eremophila* occuring over wide ranges of soil texture, pH and fertility. In these and other cases, climate (probably rainfall in particular) seems to be the most important factor determining the overall distribution limits. No mallee eucalypt has yet been shown to be physiologically specific to a narrow range of edaphic conditions. Narrow endemic species exist, however, but have not been investigated experimentally.

Within the climatically determined overall range, in the only area studied in detail, soil texture is the most important determinant of species distribution. In most cases, this probably operates via water availability, fertility or an interaction between these two factors.

To give a rough idea of species richness in mallee eucalypts, north-western Victoria has 12 species of mallee eucalypts in an area of $22\,362$ km^2 which is mostly mallee-dominated, and any given community contains between one and four such species. Such species richness figures would be much higher in comparable areas of Western Australia (Hopper, 1990).

Many aspects of the ecology of mallee communities are in need of further work. For example, the very small amount of ecophysiological work available is virtually restricted to mineral nutrition, whilst so little is known of grazing effects on regeneration that grave concern should be felt concerning the management and conservation of these communities. Finally, it seems likely that work comparing mallee communities with the adjacent tall eucalypt woodlands of single-stemmed obligate seed-regenerators in very similar habitats in the Goldfields area of Western Australia would be capable of yielding a variety of valuable ecological insights.

References

Atlas of Australian Resources, Second Series. (1973). Climate. Canberra: Australian Department of Minerals & Energy.

Atlas of Australian Resources, Third Series. (1986). Vol. 4. Climate. Canberra: Australia, Division of National Mapping.

Atlas of Australian Resources, Third Series. (1990). Vol. 6. Vegetation. Canberra: Australian Surveying & Land Information Group.

Baldwin, J.G. & Crocker, R.L. (1941). The soils and vegetation of portion of Kangaroo Island, South Australia. *Transactions of the Royal Society of South Australia*, **65**, 263–75.

Bamber, R.K. & Mullette, K.J. (1978). Studies of the lignotubers of *Eucalyptus gummifera* (Gaertn. & Hochr.) II. Anatomy. *Australian Journal of Botany*, **26**, 15–22.

Beadle, N.C.W. (1981). *The Vegetation of Australia*. Cambridge University Press.

Beard, J.S. (1967). A study of patterns in some West Australian heath and mallee communities. *Australian Journal of Botany*, **15**, 131–9.

Beard, J.S. (1969). The vegetation of the Boorabbin and Lake Johnston areas, Western Australia. *Proceedings of the Linnean Society of New South Wales*, **93**, 239–68.

Beard, J.S. (1972a). *The Vegetation of the Newdegate and Bremer Bay Areas, Western Australia*. Sydney: Vegmap Publications.

Beard, J.S. (1972b). *The Vegetation of the Southern Cross Area, Western Australia*. Sydney: Vegmap Publications.

Beard, J.S. (1973a). *The Vegetation of the Ravensthorpe Area, Western Australia*. Perth: Vegmap Publications.

Beard, J.S. (1973b). *The Vegetation of the Esperance and Malcolm Areas, Western Australia*. Perth: Vegmap Publications.

Beard, J.S. (1974). *Vegetation Survey of Western Australia – Great Sandy Desert*. Perth: University of Western Australia Press.

Beard, J.S. (1975). *Vegetation Survey of Western Australia-Nullarbor*. Perth: University of Western Australia Press.

Beard, J.S. (1976) *The Vegetation of the Shark Bay and Edel Areas, Western Australia*. Perth: Vegmap Publications.

Beard, J.S. (1990). The mallee lands of Western Australia. In *The Mallee Lands: a Conservation Perspective*, eds J.C. Noble. P.J. Joss & G.K. Jones, pp. 29–33. Melbourne: CSIRO.

Biddiscombe, E.F. (1963). *A vegetation survey in the Macquarie region, New South Wales*, CSIRO, Australia, Division of Plant Industry, Technical Paper Number 18.

Bishop, G.C. (1990). Regeneration of mallee eucalypts. In *The Mallee Lands: a Conservation Perspective*, eds J.C. Noble, P.J. Joss & G.K. Jones, pp. 186–88. Melbourne: CSIRO.

Blackburn, G. & Wright, M.J. (1989). Soils. In *Mediterranean Landscapes in Australia*, eds J.C. Noble & R.A. Bradstock, pp. 35–53. Melbourne: CSIRO.

Bradstock, R.A. (1989). Dynamics of a perennial understorey. In *Mediterranean Landscapes in Australia*, eds J.C. Noble & R.A. Bradstock, pp. 141–54. Melbourne: CSIRO.

Brockway, G.E. & Hillis, W.E. (1955). Tan bark eucalypts of the semi-arid regions of south western Australia. *Empire Forestry Review*, 34, 31–41.

Brooker, M.I.H. (1974). Six new species of *Eucalyptus* from Western Australia. *Nuytsia*, 1, 297–314.

Burbidge, N.T. (1952). The significance of the mallee habit in *Eucalyptus*. *Proceedings of the Royal Society of Queensland*, 62, 73–8.

Burbidge, N.T. (1960). The phytogeography of the Australian region. *Australian Journal of Botany*, 8, 75–212.

Burrows, W.H. (1976). Aspects of nutrient cycling in semi-arid Mallee and Mulga communities. PhD thesis, Australian National University.

Carrodus, B.B. & Blake, T.J. (1970). Studies on the lignotubers of *Eucalyptus obliqua* L'Hérit. I. The nature of the lignotuber. *New Phytologist*, 69, 1069–72.

Carrodus, B.B., Specht, R.L. & Jackman, M.L. (1965). The vegetation of Koonamore Station, South Australia. *Transactions of the Royal Society of South Australia*, 89, 41–57.

Chattaway, M.M. (1958). Bud development and lignotuber formation in eucalypts. *Australian Journal of Botany*, 6, 103–15.

Cheal, P.D., Day, J.C. & Meredith, C.W. (1979). *Fire in the National Parks of North-west Victoria*. Melbourne: National Parks Service.

Cheal, D.C. & Parkes, D.M. (1989). Mallee vegetation in Victoria. In *Mediterranean Landscapes in Australia*, eds J.C. Noble & R.A. Bradstock, pp. 125–40. Melbourne: CSIRO.

Chippendale, G.M. (1963). Ecological notes on the 'Western Desert' area of the Northern Territory. *Proceedings of the Linnean Society of New South Wales*, 88, 54–66.

Chippendale, G.M. (1973). *Eucalypts of the Western Australian Goldfields*. Canberra: Australian Government Publishing Service.

Cleland, J.B. (1928). The plants of the Encounter Bay district, notes on the ecology. *South Australian Naturalist*, 9, 57–60.

Coaldrake, J.E. (1951). *The Climate, Geology, Soils and Plant Ecology of a Portion of the County of Buckingham (Ninety-Mile Plain), South Australia*. CSIRO, Australia, Bulletin Number 266.

Connor, D.J. (1966). Vegetation studies in north-west Victoria. I. The Beulah-Hopetoun area. *Proceedings of the Royal Society of Victoria*, 79, 579–95.

Cooke, B.D. (1990). Damage caused by introduced mammals in mallee ecosystems: priorities and methods for control. In *The Mallee Lands: a Conservation Perspective*, eds J.C. Noble, P.J. Joss & G.K. Jones, pp. 202–5. Melbourne: CSIRO.

French, R.J. (1958). *Soils of Eyre Peninsula*. South Australia Department of Agriculture Bulletin Number 457.

Gill, A.M. (1975). Fire and the Australian flora: a review. *Australian Forestry*, 38, 4–25.

Gill, A.M. (1977). Management of fire-prone vegetation for plant species conservation in Australia. *Search*, 8, 20–6.

Gill, A.M. (1990). Fire management of Mallee lands for species conservation. In *The Mallee Lands: a Conservation Perspective*, eds J.C. Noble, P.J. Joss & G.K. Jones, pp. 202–5. Melbourne: CSIRO.

Gillison, A.N. (1994). Woodlands. This volume.

Groves, R.H. (1977). Fire and nutrients in the management of Australian vegetation. In *Proceedings of the Symposium on Environmental Consequences of Fire and Fuel Management in Mediterranean Ecosystems*, eds H.A. Mooney & C.E. Conrad, pp. 220–9. Washington: US Department of Agriculture Forest Service General Technical Report WO-3.

Hall, N. & Brooker, I. (1973). *Kybean mallee ash*, Eucalyptus kybeanensis *Maiden et Cambage*. Australia, Forestry & Timber Bureau Forest Tree Series Number 95.

Hall, N. & Brooker, I. (1974a). *Normanton box*, Eucalyptus normantonensis *Maiden et Cambage*. Australia, Forestry & Timber Bureau Forest Tree Series Number 160.

Hall, N. & Brooker, I. (1974b). *Sturt Creek mallee*, Eucalyptus odontocarpa *F. Muell*. Australia, Forestry & Timber Bureau Forest Tree Series Number 161.

Hill, K.D. (1989). Mallee eucalypt communities: their classification and biogeography. In *Mediterranean Landscapes in Australia*, eds J.C. Noble & R.A. Bradstock, pp. 93–108. Melbourne: CSIRO.

Holland, P.G. (1968). Seasonal growth of field layer plants in two stands of mallee vegetation. *Australian Journal of Botany*, 16, 615–22.

Holland, P.G. (1969a). Weight dynamics of *Eucalyptus* in the mallee vegetation of southeast Australia. *Ecology*, 50, 212–19.

Holland, P.G. (1969b). The plant patterns of different seasons in two stands of mallee vegetation. *Journal of Ecology*, 57, 323–33.

Holland, P.G. (1969c). The maintenance of structure and shape in three mallee eucalypts. *New Phytologist*, 68, 411–21.

Hopkins, A.J.M. & Robinson, C.J. (1981). Fire induced structural change in a Western Australian woodland. *Australian Journal of Ecology*, 6, 177–88.

Hopper, S.D. (1990). Conservation status of mallee eucalypts in southern Western Australia. In *The Mallee Lands: a Conservation Perspective*, eds J.C. Noble, P.J. Joss & G.K. Jones, pp. 21–4. Melbourne: CSIRO.

Johnson, R.W. & Burrows, W.H. (1994). *Acacia* open-forests, woodlands and shrublands. This volume.

Jones, R. (1964). The mountain mallee heath of the McPherson Ranges. *University of Queensland, Department of Botany Papers*, 4, 159–220.

Kleinig, D. & Brooker, I. (1974). *Cliff mallee ash*, Eucalyptus rupicola *L. Johnson & D. Blaxell*. Australia, Forestry & Timber Bureau Forest Tree Series Number 169.

Lange, R. & Purdie, R. (1976). Western myall (*Acacia sowdenii*), its survival prospects and management needs. *Australian Rangelands Journal*, 1, 64–9.

Leeper, G.W. (1970). Climates. In *The Australian Environment*, ed. G.W. Leeper, 4th edn (rev.), pp. 12–20. Melbourne: CSIRO & Melbourne University Press.

Litchfield, W.H. (1956). Species distribution over part of the Coonalpyn Downs, South Australia. *Australian Journal of Botany*, 4, 68–115.

Mabbutt, J.A., Burrell, J.P., Corbett, J.R. & Sullivan, M.E. (1973). In *Lands of Fowler's Gap Station, New South Wales*, ed. J.A. Mabbutt, pp. 25–43. University of New South Wales Research Series Number 3.

Mallee Vegetation Management Working Group (1991). *Mallee Vegetation Management*. Canberra: Murray-Darling Basin Commission.

Mullette, K.J. (1976). Mallee and tree forms within *Eucalyptus* species. Ph.D. thesis, University of New South Wales.

Mullette, K.J. (1978). Studies of the lignotuber of *Eucalyptus gummifera* (Gaertn. & Hochr.) I. The nature of the lignotuber. *Australian Journal of Botany*, 26, 9–13.

Myers, B.A., Ashton, D.H. & Osborne, J.A. (1986). The ecology of the mallee outlier of *Eucalyptus behriana* F. Muell. near Melton, Victoria. *Australian Journal of Botany*, 34, 15–39.

Noble, J.C. (1985). Fires and emus: the population ecology of some woody plants in arid Australia. In *Studies on Plant Demography: a Festschrift for John L. Harper*, ed. J. White, pp. 33–49. London: Academic Press.

Noble, J.C. (1989). Fire studies in mallee (*Eucalyptus* spp.) communities of western New South Wales: the effects of fires applied in different season on herbage productivity and their implications for management. *Australian Journal of Ecology*, 14, 169–87.

Noble, J.C. (1985). Fires and emus: the population ecology of some woody plants in arid Australia. In *Studies on Plant Demography: a Festschrift for John L. Harper*, ed. J. White, pp. 33–49. London: Academic Press.

Noble, J.C. (1989). Fire studies in mallee (*Eucalyptus* spp.) communities of western New South Wales: the effects of fires applied in different seasons on herbage productivity and their implications for management. *Australian Journal of Ecology*, 14, 169–87.

Noble, J.C. & Bradstock, R.A. (eds) (1989*a*). *Mediterranean Landscapes in Australia*. Melbourne: CSIRO.

Noble, J.C. & Bradstock, R.A. (1989*b*). A historical overview of ecological studies. In *Mediterranean Landscapes in Australia*, eds J.C. Noble & R.A. Bradstock, pp. 3–10. Melbourne: CSIRO.

Noble, J.C., Joss, P.J. & Jones, G.K. (eds) (1990). *The Mallee Lands: a Conservation Perspective*. Melbourne: CSIRO.

Northcote, K.H. (1960). *Atlas of Australian Soils. Explanatory Data for Sheet 1 Port Augusta–Adelaide–Hamilton Area*. Melbourne: CSIRO.

Noy-Meir, I. (1971). Multivariate analysis of the semi-arid vegetation in south-eastern Australia: nodal ordination by component analysis. *Proceedings of the Ecological Society of Australia*, 6, 159–93.

Noy-Meir, I. (1974). Multivariate analysis of the semi-arid vegetation in south-eastern Australia. *Australian Journal of Botany*, 22, 115–40.

O'Brien, T.P. (1989). The impact of severe frost. In *Mediterranean Landscapes in Australia*, eds J.C. Noble & R.A. Bradstock, pp. 141–54. Melbourne: CSIRO.

Onans, J. & Parsons, R.F. (1980). Regeneration of native plants on abandoned mallee farmland in south-eastern Australia. *Australian Journal of Botany,* **28**, 479–93.

Parsons, R.F. (1968). Ecological aspects of the growth and mineral nutrition of three mallee species of *Eucalyptus. Oecologia Plantarum,* **3**, 121–36.

Parsons, R.F. (1969). Physiological and ecological tolerances of *Eucalyptus incrassata* and *E. socialis* to edaphic factors. *Ecology,* **50**, 386–90.

Parsons, R.F. (1981). *Eucalyptus* scrubs and shrublands. In *Australian Vegetation,* 1st edn, ed. R.H. Groves, pp. 227–52. Cambridge University Press.

Parsons, R.F. (1990). Plant conservation problems at species level in Mallee lands. In *The Mallee Lands: a Conservation Perspective,* eds J.C. Noble, P.J. Joss & G.K. Jones, pp. 25–8. Melbourne: CSIRO.

Parsons, R.F. & Rowan, J.N. (1968). Edaphic range and cohabitation of some mallee eucalypts in south-eastern Australia. *Australian Journal of Botany,* **16**, 109–16.

Parsons, R.F. & Specht, R.L. (1967). Lime chlorosis and other factors affecting the distribution of *Eucalyptus* on coastal sands in southern Australia. *Australian Journal of Botany,* **15**, 95–105.

Piemeisel, R.L. (1951). Causes affecting change and rate of change in a vegetation of annuals in Idaho. *Ecology,* **32**, 53–72.

Pressey, R.L., Bedward M. & Nicholls, A.O. (1990). Reserve selection in Mallee lands. In *The Mallee Lands: a Conservation Perspective,* eds J.C. Noble, P.J. Joss & G.K. Jones, pp. 167–78. Melbourne: CSIRO.

Raunkiaer, C. (1934). *The Life Forms of Plants and Statistical Plant Geography.* Oxford University Press.

Rowan, J.N. (1963). *A Study of the Mallee Lands around Wedderburn and Inglewood.* Melbourne: Soil Conservation Authority of Victoria (mimeographed report).

Rowan, J.N. (1971). *Salting on dryland farms in north-western Victoria.* Soil Conservation Authority of Victoria Technical Communication Number 7.

Rowan, J.N. & Downes, R.G. (1963). *A Study of the Land in North-western Victoria.* Soil Conservation Authority of Victoria Technical Communication Number 2.

Sims, H.J. (1949). Plant regeneration on stabilised sandhills in the Mallee. *Victorian Naturalist,* **66**, 37–9.

Sparrow, A. (1989). Mallee vegetation in South Australia. In *Mediterranean Landscapes in Australia,* ed. J.C. Noble & R.A. Bradstock, pp. 109–24. Melbourne: CSIRO.

Sparrow, A.D. (1990). Floristic patterns in South Australian mallee vegetation and some implications for conservation. In *The Mallee Lands: a Conservation Perspective,* eds J.C. Noble, P.J. Joss & G.K. Jones, pp. 12–15. Melbourne: CSIRO.

Specht, R.L. (1966). The growth and distribution of mallee-broombush (*Eucalyptus incrassata* – *Melaleuca uncinata* association) and heath vegetation near Dark Island Soak, Ninety-Mile Plain, South Australia. *Australian Journal of Botany,* **14**, 361–71.

Specht, R.L. (1972). *The Vegetation of South Australia*, 2nd edn. Adelaide: Government Printer.

Specht, R.L. (1981). Mallee ecosystems in southern Australia. In *Mediterranean-Type Shrublands*, eds F. di Castri, D.W. Goodall & R.L. Specht, pp. 203–31. Amsterdam: Elsevier Scientific Publishing Company.

Stannard, M.E. (1958). Erosion survey of the south-west Cobar peneplain. II. Soils and vegetation. *Journal of the Soil Conservation Service of New South Wales*, 14, 30–45.

Tiver, F., Sparrow, A.D. & Lange, R.T. (1989). The composition and distribution of the vegetation of north-west Eyre Peninsula. *Transactions of the Royal Society of South Australia*, 113, 47–61.

Victoria: Land Conservation Council (1987). *Report on the Mallee Area Review*. Melbourne: Land Conservation Council.

Wellington, A.B. (1989). Seedling regeneration and the population dynamics of eucalypts. In *Mediterranean Landscapes in Australia*, eds J.C. Noble & R.A. Bradstock, pp. 155–67. Melbourne: CSIRO.

Wellington, A.B. & Noble, I.R. (1985a). Post-fire recruitment and mortality in a population of the mallee *Eucalyptus incrassata* in semi-arid, south-eastern Australia. *Journal of Ecology*, 73, 645–56.

Wellington, A.B. & Noble, I.R. (1985b). Seed dynamics and factors limiting recruitment of the mallee *Eucalyptus incrassata* in semi-arid, south-eastern Australia. *Journal of Ecology*, 73, 657–66.

Whitford, W.G., Ludwig, J.A. & Noble, J.C. (1992). The importance of subterranean termites in semi-arid ecosystems in south-eastern Australia. *Journal of Arid Environments*, 22, 87–91.

Whittaker, R.H., Niering, W.A. & Crisp, M.D. (1979). Structure, pattern and diversity of a mallee community in New South Wales. *Vegetatio*, 39, 65–76.

Williams, R.J. & Costin, A.B. (1993). Alpine and sub-alpine vegetation. This volume.

Wood, J.G. (1929). Floristics and ecology of the mallee. *Transactions of the Royal Society of South Australia*, 53, 359–78.

Zimmer, W.J. (1940). Plant invasions in the Mallee. *Victorian Naturalist*, 56, 143–7.

11

Heathlands

R.L. SPECHT

IN EUROPEAN ecological literature, use of the term heathland has been restricted to describe the low sclerophyllous communities of *Calluna*, *Erica*, *Rhododendron* and *Vaccinium* found on nutrient-deficient soils from lowland to alpine habitats (Gimingham, 1972).

The term 'heath' (or more appropriately 'heathland' when applied to the plant community) is a vernacular word describing wasteland in northwestern Europe (Rübel, 1914). By coincidence, much of the nutrient-poor wasteland in this region was covered with a degraded community dominated by *Calluna vulgaris* (heath or heather); but on apparently similar soils, this sclerophyllous, ericaceous species was absent and a stunted grassland dominated by *Agrostis* and *Festuca* (sometimes termed 'grass-heath') occurred.

Heathlands are well developed on nutrient-deficient sandstone soils in the southwestern Cape Province of South Africa, where they were originally compared to the Mediterranean 'macchia', but are now termed more appropriately 'fynbos' (Kruger, 1979). Here the community contains a wealth of ericaceous species (including a great number of species of *Erica*) as an understorey to low sclerophyllous, often broader-leafed, shrubs belonging mainly to the families Proteaceae, Rhamnaceae and Rosaceae. Node-sedges (Restionaceae) and many members of the Liliales are common as a ground stratum.

The heathlands of Australia are closely allied to the fynbos of Cape Province, more than to the macchia of Mediterranean Europe or the chaparral of California (Specht, 1969). The family Ericaceae is virtually absent, except in the subalpine heathlands where the genera *Agapetes*, *Gaultheria*, *Pernettya* and *Rhododendron* are present. The closely allied heath-family Epacridaceae, with 25 genera, has taken the place of the family Ericaceae in lowland heath communities in southern Australia. As in South Africa, many broader-leafed, sclerophyllous shrubs (belonging to a wide range of families) overtop the epacridaceous understorey. Many genera of Cyperaceae, Liliales,

Orchidaceae, Restionaceae, etc. are also common in the ground stratum (see Fig. 11.3).

Community structure

Three distinct strata (with occasional emergent plants) can be distinguished in heathlands in southern Australia (Fig. 11.1, after Specht *et al.*, 1991).

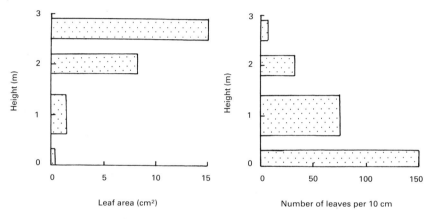

Fig. 11.1. Foliage profile (leaf area and number of leaves per 10 cm length of stem) of dicotyledonous species in the emergent, upper, mid and ground strata of Dark Island heathland, South Australia.

1. The upper stratum is composed of deep-rooted nanophanerophytes (up to 2 m tall) with evergreen, broad sclerophyllous leaves and internodes 2–5 mm long, longer in emergent species. Plants in this stratum produce new foliage shoots in late spring through summer.
2. The mid-stratum is composed of relatively shallow-rooted nanophanerophytes (0.5–1.5 m tall) with evergreen, narrow semisclerophyllous leaves and internodes less than 1 mm long. Plants in this stratum produce new foliage shoots during spring.
3. The ground stratum contains a mixture of shallow-rooted chamaephytes, with densely packed, small leaves, and evergreen, sclerophyllous hemicryptophytes, with varying numbers of culms/leaves in each clump (depending on the species). As the air temperature increases (when the sun is shining) within the densely packed

leaves on the foliage shoots of the chamaephytes, new growth results during late winter and early spring.

Annual growth in sclerophyllous, monocotyledonous tussocks of the ground stratum will range from late winter (very dense clumps) to summer (open clumps), depending on the density of the culms/leaves in each clump.

Australian ecologists have developed a two-way classification to describe the structure of Australian plant communities (Specht, 1970, 1981*d*, in press). This classification is based on two readily observable attributes of the plant community, life form and foliage projective cover of the uppermost stratum. The classification may be used to cover the same attributes of lower strata in the community.

By definition (Specht, 1979*b*, 1981*d*), Australian heathlands may be classified structurally as follows.

Tallest stratum: sclerophyllous shrubs 25 cm to 2 m tall.
    The height category may be subdivided if required into *tall* (1 to 2 m) and *low* (25 cm to 1 m).
    Foliage projective cover, 100 to 70 per cent *closed-heathland*.
    Foliage projective cover, 70 to 50 per cent *heathland*.
    Foliage projective cover, 50 to 30 per cent *open-heathland*.
Tallest stratum: low sclerophyllous shrubs, less than 25 cm tall.
    Foliage projective cover, 30 to 10 per cent *dwarf heathland* (fellfield).
    Foliage projective cover, <10 per cent *dwarf open-heathland* (open-fellfield).
Pertinent aspects of the lower strata can be included as follows.
    Sclerophyllous shrubs and graminoids co-dominant: *graminoid-heathland*.
    Sclerophyllous shrubs and *Sphagnum* moss co-dominant: *bog-heathland*.

The above, purely structural terms, used to describe the range of heathland communities observed in Australia, avoid the inclusion of habitat terms such as 'wet' and 'dry' which previously have been applied to heathlands developed, respectively, on seasonally water-logged and seasonally droughted soils deficient in most plant nutrients (Groves & Specht, 1965; Specht, 1972, 1979*c*). The altitudinal terms, such as coastal, lowland, montane or subalpine, are not used.

True heathlands, as thus defined, are found in most parts of humid to subhumid Australia, but usually only in small patches. These islands of heathland vegetation grade into structural formations in which the heathy strata are overtopped by a stratum of tall shrubs (a tall shrubland), low scattered trees (a woodland) or denser trees (an open-forest) usually of *Banksia*, *Allocasuarina* or *Eucalyptus*. Various terms, such as 'shrub-heath', 'mallee-heath', 'tree-heath', 'sclerophyll-woodland', 'sclerophyll-forest', 'kwongan' in Western Australia, 'wallum' in Queensland, have been used to describe these more complex communities, the terms 'heath' or 'sclerophyll' being used to designate the heathy understorey.

Observations made over several decades in southern Australia indicate that, apart from secondary successional changes following fire, the heathland and adjacent sclerophyll communities have not changed structurally unless the environment has been modified by humans (Groves & Specht, 1981). In effect, most Australian heathland communities must be regarded as climax, not seral, vegetation.

Biodiversity (species richness per hectare)

The number of species which are to be found in Australian heathlands varies from as few as 33 to as many as 131 species per hectare (Specht, 1979c; George, Hopkins & Marchant, 1979). The greatest number of species is to be found in heathlands where emergent shrubs and trees do not overshadow the heathland strata. As the foliage projective cover of emergent shrubs and trees increases, the species richness of the understorey heathland strata decreases (Fig. 11.2 *a*, after Specht & Morgan, 1981; Specht & Specht, 1989).

Even in the absence of emergent shrubs and trees, the upper stratum (Fig. 11.1) of the heathland gradually suppresses the lower strata by shading as it regenerates after fire (Specht, Rayson & Jackman, 1958; Specht, 1981*b*; Specht & Morgan, 1981; Specht & Specht, 1989). Of 38 species recorded in a heathland on deep sand in southeastern Australia after a fire, only 10 species survived 50 years later (Fig. 11.2*b*). The foliage cover of the upper stratum of heathlands growing on extremely infertile, lateritic residuals of southwestern Australia develops far slower than that of the sand-heathlands, enabling a higher species richness to be maintained in the lower strata throughout the post-fire succession (Specht & Specht, 1989).

In wet-heathlands on seasonally waterlogged soils, the rapid regeneration, after repeated fires, of a dense upper stratum has excluded many of the species of the ground stratum (Specht & Specht, 1989), reducing the species richness (11 to 25 species per 8 m^2) of the wet-heathland to almost half of that (22 to 36 species per 8 m^2) recorded in nearby dry-heathlands (Specht, 1979 *c*).

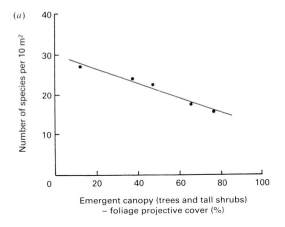

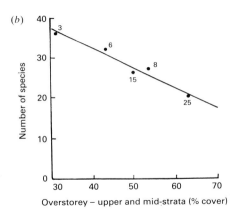

Fig. 11.2. The influence of foliage projective cover on the species richness (number of vascular species per unit area) of the understorey of: (*a*) the overstorey in related plant communities (heathland to open-forest) on Mt Hardgrave, North Stradbroke Island, Queensland. (After Clifford & Specht, 1979; Specht & Morgan, 1981; Specht, Clifford & Rogers, 1984.) (*b*) the upper stratum in a post-fire succession in Dark Island heathland, near Keith, South Australia. (After Specht, Rayson & Jackman, 1958; Specht & Specht, 1989.)

Attributes

Life forms

Heathlands are dominated by low shrubs (up to 2 m tall) which possess small (<225 mm²), evergreen, sclerophyllous leaves and extensive root systems, often arising from lignotubers (syn. 'burls'). Several shrubby species may be co-dominant, as shown in Figure 11.3, where, in Australia, *Banksia, Allocasuarina, Leptospermum* and *Xanthorrhoea* contribute the greatest biomass. In the open-heathland depicted in Figure 11.3, 33 of the 76 species recorded may be classed as shrubs (1 to 2 m) or subshrubs (25 cm to 1 m) at maturity. The genera *Allocasuarina* and *Phyllota*, which covered 28 per cent of the ground six years after a fire, are relatively short-lived (15 to 25 years), whilst the co-dominants *Banksia* and *Xanthorrhoea* survive for well over 50 years (Specht, Rayson & Jackman, 1958).

Many low evergreen plants (<25 cm tall) form a ground stratum together with evergreen hemicryptophytes (evergreen graminoid

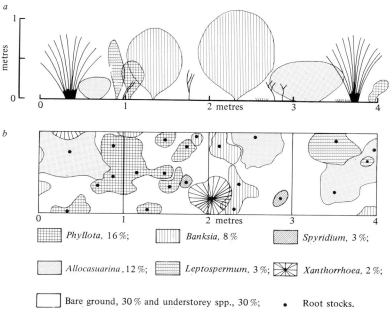

Fig. 11.3. Profile sketch (*a*) and coverage chart (*b*) of a typical dry-heathland near Dark Island, 16 km northeast of Keith, South Australia. (See Specht & Rayson, 1957.)

plants, a life form not recognised by Raunkiaer (1934), but introduced by Specht, 1979*b*). Seasonal grasses and geophytes are present, whilst annual herbs are rare, except where the community has been grazed and fired regularly.

Parasitic twining plants of the genus *Cassytha* may be common on the subshrubs.

Leaf characteristics

The leaves of most heathland species are sclerophyllous, with thick cuticles and sunken stomata. The cells of the leaf are usually thick-walled, often lignified (sometimes silicified) and contain tannins, resins and essential oils thus increasing their flammability during periods of water stress.

Many genera characteristic of the upper stratum (1 to 2 m tall) possess broader leaves which may be classed as microphyll (over 225 mm^2 in area) on the Raunkiaer scale; low shrubs of *Acacia*, *Banksia*, *Eucalyptus*, *Xanthorrhoea*, etc. fall into this leaf-size class. Most subshrubs (25 cm to 1 m tall) possess small leaves in the nano-phyll (25 to 225 mm^2) and leptophyll (<25 mm^2) size classes of Raunkiaer (1934); some of the smaller leaves are typically ericoid with the upper surface of the leaf hard and waxy and the lower surface deeply grooved.

The ground stratum (<25 cm tall) contains many evergreen hemi-cryptophytes with sclerophyllous, graminoid leaves, as well as dwarf shrubs usually with leptophyllous leaves. The leaves of the few seasonal plants are usually mesophyllous, rather than sclerophyllous, in texture.

Root systems

Underground organs such as lignotubers, rhizomes, bulbs and tubers are a distinctive attribute of heathlands. Although the above-ground part of the vegetation may be razed by fire, these underground organs survive and rapidly regenerate new aerial shoots. Heathlands in most humid parts of southern Australia possess so many root-stock regenerators that, within ten years from the fire, the above-ground vegetation contains about 95 per cent (by dry weight) of these species, with only 5 per cent of the biomass contributed by species regenerating by seed. As the environment becomes less humid, more gaps appear between the plants which form the upper stratum; this enables species which regenerate by seed to persist in the stand, as shown in the following equation (Specht, 1981*b*):

$$RSR = 0.645 \, FPC + 41.4$$

where RSR are root-stock regenerators (percentage total above-ground biomass), and FPC the foliage projective cover (per cent) of the upper stratum.

In the nutrient-deficient conditions in which heathlands are found, peculiar rootlets develop in the interface between decomposing litter and the soil surface. Proteoid rootlets are characteristic of the family Proteaceae; restioid rootlets of the family Restionaceae; cyperoid or dauciform rootlets of the family Cyperaceae; mycorrhizal rootlets in the family Myrtaceae and many other families (Jeffrey, 1967; Lamont, 1973, 1981, 1982; Malajcuk & Glenn, 1981; Malajcuk & Lamont, 1981; Pate & Beard, 1984).

A distinctive feature of all these different rootlets is their ability to synthesise polyphosphate from orthophosphate released in the decomposition of litter during spring when aerial growth of the heathland is minimal, and to release the polyphosphate as ortho-phosphate when shoot growth commences (Specht & Groves, 1966; Jeffrey, 1968; Coleman & Specht, 1981; Specht, 1981*e*).

Nitrogen-fixing, nodulated rootlets are found in the families Casuarinaceae, Fabaceae, Mimosaceae, and Zamiaceae.

Haustorial connections have been observed between the roots of the semi-parasitic plants (families Olacaceae and Santalaceae, and the genus *Euphrasia* in the Scrophulariaceae) and the roots of other heathland species.

Bradysporous fruits

A number of characteristic species within heath vegetation possess hard woody fruit which retain their seed for many years, usually until released by heat during a bush fire (Gill, 1976; Gill & Groves, 1981; Specht, 1981*b*; Gill & McMahon, 1986). Because of this brady-sporous habit, regeneration of these species by seedlings is largely dependent on fire. The frequency of fire is important. A fire before the species sets viable seed may lead to extinction of the species; a fire too long delayed may lead to viable seed being destroyed in the woody fruits by boring insects and fungi.

Seeds

Bradyspory (discussed above) is a characteristic of heathland species of the families Casuarinaceae, Myrtaceae and Proteaceae. Seeds eventually released from the woody fruits of these species are rela-

tively large and often winged in the families Casuarinaceae and Proteaceae, but are very small and wingless in the family Myrtaceae. These
small seeds form an important food-source for harvester ants which
abound in the heathlands.

The other heathland species, belonging to a wide range of families
(Table 11.1), release seed from their fruits as soon as it is mature.
From 21 to 32 per cent of the heath flora of southern Australia
produce reasonably large seed with a thick hard testa, with a smooth
and darkly coloured surface to which is attached a white or light-
coloured appendage termed an 'elaiosome' (Berg, 1975, 1981). This
relatively dry and hard appendage is attractive to ants which gather
the seed into their nests, eat the elaiosome, but usually cannot pierce
the hard testa to eat the embryo. The seed is then discarded either
outside the nest or, if too large to move, stored in galleries. Myrmecochorous dispersal of seed is a unique characteristic of Australian
heathland species. Approximately 1500 species of Australian vascular
plants, representing 87 genera in 24 families are probably myrmecochorous, compared with less than 300 myrmecochorous species
known for the rest of the world (Berg, 1981).

Distribution

Heathland vegetation, with or without an overstorey of sclerophyllous trees or tall shrubs, is widespread across Australia in all humid
and subhumid areas (Fig. 11.4). Remnants of the flora are found even
in the tropical north and the arid centre of Australia (where the
hummock grass, *Triodia*, becomes prominent on low nutrient soils).

By far the greatest concentration of heathland species is found on
the sandplain and lateritic soils of southwest Western Australia where
50 per cent of 3700 typical Australian heathland species are located
(Fig. 11.5). The Hawkesbury sandstone soils of the Sydney area of
New South Wales support 20 per cent of the Australian heath flora.
Both these areas have a warm temperate climate. The proportion of
heathland species falls to 3 to 6 per cent in the tropics, and 9 to 14
per cent in the cool temperate climate of southeastern Australia.

The heathland flora forms a distinct suite of plant communities
(Groups B and C, Fig. 11.6), clearly separated from closed-forest (rainforest) vegetation (Group A) and open-communities (Groups D–G)
containing a wealth of grasses, herbs, ground-ferns or halophytes.
Two heathland subgroups may be distinguished: Group B includes
communities with a preponderance of sclerophyllous plants (true

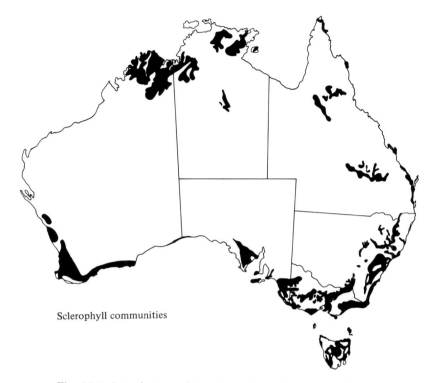

Sclerophyll communities

Fig. 11.4. Distribution of heathland flora (± trees or tall shrubs) in Australia. (After Carnahan, 1976.) The northwest vegetation contains a large percentage of hummock grasses.

heathlands, mallee–heath, sclerophyll–woodland and dry sclerophyll–forest); Group C includes communities in temperate Australia where more herbaceous elements (grass, herbs, ferns) are intermingled with the heathland genera (alpine heathlands, wet sclerophyll-forest on nutrient-poor soils, savannah/sclerophyll–woodland).

If the 591 genera recorded in Australian heathlands *sensu stricto* are analysed by the same classificatory program (DIVINF) which was used to show the relationships of the heathland flora to other Australian plant formations (Fig. 11.6), the relationships shown in Figure 11.7 result. The Australian heathlands fall into six natural groups: Group A in southwest Western Australia; Group B in South Australia extending into the drier side of the Great Dividing Range in southeastern Australia; Group C in warm temperate to subtropical eastern Australia; Group D in temperate areas of southeastern

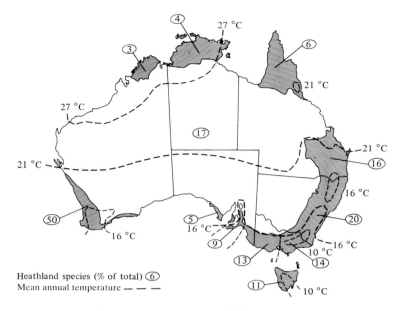

Fig. 11.5. Regional distribution of 3700 typical Australian heathland species. The circled numbers give the percentages of heathland species in each area, and the dashed lines indicate the mean annual temperature.

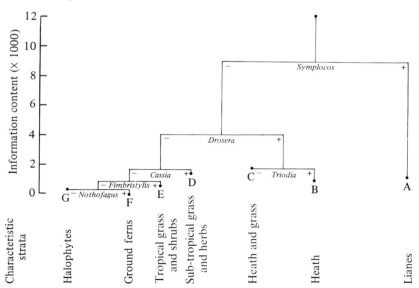

Fig. 11.6. Relationships of 32 major Australian plant formations classified on the basis of the 1398 component genera by the classificatory program DIVINF (Lance & Williams, 1968) as run by Dr M.B. Dale.

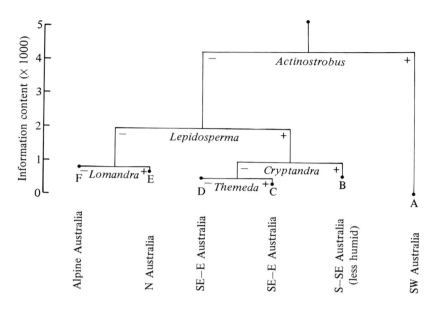

Fig. 11.7. Relationship of 32 Australian heathland communities classified on the basis of the 591 component genera by the classificatory program DIVINF (Lance & Williams, 1968) as run by Dr M.B. Dale. Group A, Southwest (Western Australia) dry-heathlands; Group B, Dark Island (South Australia) dry-heathland, Dark Island (South Australia) mallee-broombush, Billiatt (South Australia) mallee-broombush, Mt Compass (South Australia) mallee-heathland, Hincks-Murlong (South Australia) heathland/mallee-broombush; Group C, Shoalwater Bay (Qld) dry-heathland, Toolara (Qld) wet-heathland, Beerwah (Qld) wet-heathland, Sydney (NSW) dry- and wet-heathlands, Mt Hardgrave, North Stradbroke Island (Qld) dry-heathland; Group D, Dave's Creek, Lamington N.P. (Qld) wet-heathland, Bonny Hills (NSW) dry heathland, Tidal River, Wilson's Promontory (Victoria) dry-heathland, Darby River, Wilson's Promontory (Victoria) dry-heathland, Barry's Creek, Wilson's Promontory (Victoria) wet-heathland, Frankston (Victoria) dry-heathland, Lower South East (South Australia) wet-heathland; Group E, Oenpelli, Arnhem Land (NT) dry-heathland, Jardine River (north Qld) wet-heathland, Cape Flattery-Lizard Island (north Qld) dry-heathland, Burra Ranges (north Qld) dry-heathland, Forty-Mile Scrub (north Qld) dry-heathland; Group F, Mt Bellenden Ker (north Qld) alpine heathland, Mt Kosciusko (NSW) alpine heathland, Mt Kosciusko (NSW) alpine bog-heathland, Lake Mountain (Victoria) alpine heathland, Lake Mountain (Victoria) alpine bog-heathland, Cradle Mountain (Tasmania) alpine heathland, Mt Wellington (Tasmania) alpine heathland (for further site details see Specht, 1979c).

Australia extending into the highlands of eastern Australia; Group E in northern Australia; and Group F in alpine/subalpine areas of southeastern Australia with a small outlier on the mountain summits in northern Queensland (see Fig. 11.8).

Many of the heathland genera have a considerable distributional range throughout these six major suites; some show disjunct distribution patterns occurring in isolated pockets found either in eastern and southwestern heathlands or in northern and southwestern, sometimes also eastern heathlands (Specht, 1981*c*); other genera may be confined to one heathland suite; some are invaders from adjacent communities. The number of genera recorded in each of the heathland suites is listed under families in Table 11.1. As the heathland genera are not listed in Table 11.1, an idea of the genera in five major heathland families is given below.

Epacridaceae (28 genera): *Acrotriche, Andersonia, Archeria, Astroloma, Brachyloma, Choristemon, Coleanthera, Conostephium, Cosmelia, Cyathodes, Dracophyllum, Epacris, Leucopogon, Lissanthe, Lysinema, Melichrus, Monotoca, Needhamiella, Oligarrhena, Pentachondra, Prionotes, Richea, Rupicola, Sphenotoma, Sprengelia, Styphelia, Trochocarpa, Woollsia.*

Fabaceae (24 genera): *Aotus, Bossiaea, Brachysema, Burtonia, Chorizema, Daviesia, Dillwynia, Eutaxia, Gastrolobium, Gompholobium, Goodia, Hardenbergia, Hovea, Isotropis, Jacksonia, Kennedia, Mirbelia, Oxylobium, Phyllota, Platylobium, Pultenaea, Sphaerolobium, Templetonia, Viminaria.*

Myrtaceae (38 genera): *Actinodium, Agonis, Angophora, Astartea, Austromyrtus, Baeckea, Balaustion, Beaufortia, Callistemon, Calothamnus, Calytrix, Calythropsis, Chamaelaucium, Conothamnus, Darwinia, Eremaea, Eucalyptus, Eugenia sens. lat., Fenzlia, Homalocalyx, Homoranthus, Hypocalymma, Kunzea, Lamarchea, Leptospermum, Lhotskya, Melaleuca, Micromyrtus, Phymatocarpus, Pileanthus, Regelia, Scholtzia, Sinoga (syn. Asteromyrtus), Thryptomene, Tristania sens. lat., Verticordia, Wehlia, Xanthostemon.*

Proteaceae (22 genera): *Adenanthos, Banksia, Bellendena, Cenarrhenes, Conospermum, Dryandra, Franklandia, Grevillea, Hakea, Isopogon, Lambertia, Lomatia, Musgravea, Orites, Persoonia, Petrophile, Stirlingia, Strangea, Symphionema, Synaphea, Telopea, Xylomelum.*

Restionaceae (17 genera): *Alexgeorgia, Anarthria, Chaetanthus, Coleocarya, Dielsia, Empodisma, Harperia, Hopkinsia, Hypolaena, Lepidobolus, Leptocarpus, Lepyrodia, Loxocarya, Lyginia, Meeboldina, Onychosepalum, Restio.*

Table 11.1. *Number of genera, in major families and classes, recorded in the six Australian heathland suites (see Fig. 11.7). The subalpine suite has been subdivided further to show the differences between southeastern and northern Australia*

Class/family	Total number of genera	Subhumid temperate (SW Aust.)	Subhumid temperate (S and SE Aust.)	Humid temperate (SE Aust.)	Humid subtropical (SE Aust.)	Monsoonal (N Aust.)	Subalpine (SE Aust.)	Subalpine (N Aust.)
Bryophyta *(Sphagnum)*	1	–	–	–	–	–	1	–
Pteridophyta	19	1	–	9	3	1	9	6
Gymnospermae	8	3	2	1	1	1	5	–
Monocotyledoneae	164	92	41	45	38	37	39	10
Anthericaceae	15	15	5	4	5	1	2	–
Cyperaceae	23	6	4	7	6	8	8	2
Orchidaceae	29	18	13	13	3	2	6	3
Poaceae	27	6	5	7	8	11	7	–
Restionaceae	17	15	5	4	6	2	2	–
Dicotyledoneae	399	195	118	81	95	151	111	31
Asteraceae	28	7	10	2	–	5	16	–
Epacridaceae	28	14	9	8	9	1	10	2
Euphorbiaceae	17	9	5	3	7	9	1	1
Fabaceae	24	24	16	13	14	12	4	–
Myrtaceae	38	30	9	11	12	16	6	3
Proteaceae	22	15	8	8	11	4	8	3
Rubiaceae	12	1	1	1	1	6	4	1
Rutaceae	12	9	8	4	6	3	3	1

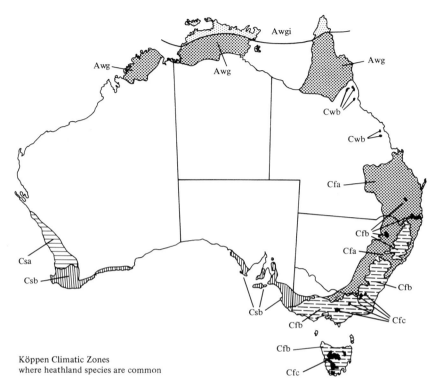

Fig. 11.8. Distribution of the major Köppen climatic types (after Dick, 1975) in which the heathland vegetation groups (defined in the classificatory programme, Fig. 11.7) are found in Australia.

Ecological Relationships

Climate

Six major suites of heathland communities were distinguished by the classificatory program depicted in Figure 11.7. The distribution of these groups coincides with major Köppen climatic types defined for humid and subhumid areas of Australia by Dick (1975).

The Köppen climatic types, in which each heathland group is developed, are shown on a map of Australia (Fig. 11.8) and may be described as follows.

Group A: Csa/Csb (warm climate with a long dry summer, ranging from hot (over 22 °C) to mild (below 22 °C)).

Group B: Csb (warm climate with a long mild (under 22 °C) dry summer).

Group C: Cfa (warm climate, with uniform rainfall, long hot summer (over 22 °C)).

Group D: Cfb (warm climate, with uniform rainfall, long mild summer (under 22 °C)).

Group E: Aw (hot climate with a long dry winter).

Group F: Cfc in southeastern Australia (warm climate, with uniform rainfall, short mild summer (over 10 °C, under 22 °C)). Cwb in northern Queensland (warm climate, with a dry winter and a long mild summer (over 10 °C, under 22 °C)).

In the Mediterranean-type climate of southern Australia, heathlands experience seasonal drought during summer, and yet this season is the period of active foliage growth of the upper stratum (Specht & Rayson, 1957; Specht, Rogers & Hopkins, 1981). Sclerophylly (with all its water-conservation attributes strongly developed in response to low nutrient levels (Specht & Rundel, 1990)) has been considered as the main physiological adaptation to drought (Connor & Doley, 1981), but this operates only during the dry season. Control of water loss operates throughout the year, and ensures conservation of water during the wet season and use throughout the dry season. The foliage projective cover of the plant community has 'adjusted', in relation to the evaporative power of the atmosphere, to ensure this annual water balance (Specht, 1981g).

The establishment and growth of tall shrubs and trees is inhibited in both seasonally droughted heathlands and in seasonally waterlogged heathlands. The anaerobic condition of waterlogged soils inhibits root growth of many plants in wet-heathlands, with resultant reduction in water-transfer from the waterlogged soils, through the plant community, to the atmosphere (Armstrong, 1981; Specht, 1981f; Bolton, 1986).

Soils

Heathland vegetation is found from temperate to tropical areas of Australia, from lowland to alpine sites. Wherever the vegetation occurs, the major ecological factor controlling its distribution appears to be the nutrient status of the soil.

As shown in Figure 11.9, heathland soils are invariably low in essential plant nutrients such as phosphorus and nitrogen. Other plant nutrients (potassium, sulphur, copper, zinc, manganese, molybdenum) may be of limited availability in some heathland soils. These soils may be developed either on parent material (sandstone, quartz-ite, acid granites, wind-distributed sand) low in plant nutrients, or on highly leached soils (podzols, lateritic podzols and solonised sandy soils). Poorly drained, peaty soils, also low in available nutrients, may support a bog-heathland.

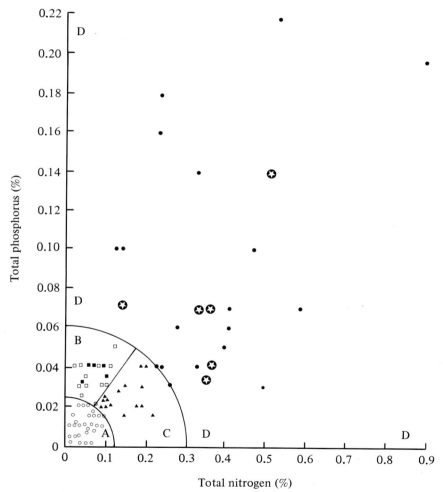

Fig. 11.9. Edaphic relationships (based on the levels of total nitrogen and total phosphorus in the surface soil, Stace *et al.*, 1968) of heathy and *Triodia* communities (open circles; zone A) compared with grassy communities (triangles; zone C), chenopod (open squares) and *Stipa* (filled squares) communities (zone B) and grassy (filled circles) and ferny (asterisks) communities (zone D).

The soil analyses presented in Figure 11.9 illustrate the tenuous interface between heath and savannah floras. Intense leaching of plant nutrients will lead to the development of heath vegetation.

The continual growth of heathlands (Specht, Rayson & Jackman, 1958; Groves & Specht, 1965; Clifford & Specht, 1979; Specht, 1979*b*; Groves, Hocking & McMahon, 1986; Low & Lamont, 1990) on these nutrient-poor soils appears to depend on subtle nutrient conservation (Specht & Groves, 1966; Jeffrey, 1968; Coleman & Specht, 1981). Such nutrient-conservation strategies appear to be upset by increasing soil acidity and aluminium toxicity (Woolhouse, 1981), often associated with water-logging (Armstrong, 1981); in coastal areas, salt spray can also play a detrimental role (Parsons, 1981).

Slightly more fertile soils, or nutrient-poor soils contaminated with nutrients, will lead to the extinction of the heath flora and the development of a savanna flora, as shown by manipulation experiments reported by Specht (1963), Heddle & Specht (1975), and Specht, Connor & Clifford (1977). At first, on the application of phosphatic fertiliser, all heathland plants show increased growth (species characteristic of the early phase of the post-fire succession showing most growth; species persisting into the mature phase of the post-fire succession responding the least (Specht, 1972)). The increased growth leads to early death of the plants (rootstock as well as seedling regenerators). In soils high in phosphates, low in nitrogen, post-fire regeneration of heathland seedlings often leads to 'phosphorus toxicity' (Specht, 1963, 1981*b*; Ozanne & Specht, 1981), which may be the major factor limiting the survival of heath species on more fertile soils.

Fire

Fire is an integral factor in Australian heathlands. The dense structure of the mature community and the presence of flammable volatile oils and resins in the sclerophyllous leaves of the component species make the heathlands a potential fire hazard during periods of drought.

The vegetation is well adapted to the ravages of fire (Gill & Groves, 1981). The aerial stems of many species may be destroyed by fire, but many plants sprout again from lignotubers, rhizomes, tubers and bulbs protected in the soil (Specht, 1981*b*). Other species release large numbers of seed held above-ground in hard woody fruits which spring open during a fire. A wealth of regenerating plants and seedlings soon appears (Fig. 11.10), thereby repairing the catastrophic effect of the fire.

Fig. 11.10. Dry-heathland, 12 years since the last fire, near Keith, South Australia. The taller (2 m), large-leafed shrubs are *Banksia ornata*; the specimens with the grass-like leaves are *Xanthorrhoea caespitosa*; the low, small-leafed shrubs are *Allocasuarina pusilla*, *Phyllota pleurandroides* and *Leptospermum myrsinoides*.

Animals

Native animals (Edmonds & Specht, 1981; Catling, 1988) such as the kangaroo (*Macropus* spp.), wallaby (*Macropus* spp.) and emu (*Dromaius novaehollandiae*), and a few introduced rabbits (*Oryctolagus cuniculus*), all essentially visitors from adjacent communities, exert a minor grazing pressure on heathlands in southern Australia. Black cockatoos (*Calyptorhynchus funereus*) may visit the ecosystem to extract grubs from the inflorescences of *Banksia* spp. or *Xanthorrhoea* spp., the latter species tending to flower only after a bushfire. The introduced house mouse (*Mus musculus*) invades the heathlands after a fire, but tends to disappear as the community ages.

In mature stands of heath, honeyeaters (Meliphagidae) and wrens (Maluridae) are resident birds, feeding on nectar and insects and probably aiding cross-fertilisation of many of the heathland species.

Insects inhabiting the heathland fall into classes such as foliage-feeders, nectar-feeders, seed-gatherers, wood-suckers and wood-borers; other insects and spiders are predators and parasites on these first-order consumers. The populations of each of these classes of insects fluctuate seasonally in response to the seasonal growth and flowering rhythms of the heathland species (Edmonds & Specht, 1981; Majer & Greenslade, 1988). Ants are particularly active in seed removal and dispersal, especially attracted by the large number of seeds which possess fleshy elaiosomes (Berg, 1975).

Collembola, mites, worms, nematodes and protozoa act with decomposing fungi and bacteria in the decomposition of litter, returning mineral nutrients to the soil where they tend to be trapped by a network of fine rootlets at the interface of litter and soil.

References

Armstrong, W. (1981). The water relations of heathlands: General physiological effects of waterlogging. In *Ecosystems of the World*. vol. 9B, ed. R.L. Specht, pp. 111–21. Amsterdam: Elsevier Scientific Publishing Company.

Berg, R.Y. (1975). Myrmecochorous plants in Australia and their dispersal by ants. *Australian Journal of Botany*, **23**, 475–508.

Berg, R.Y. (1981). The role of ants in seed dispersal in Australian lowland heathland. In *Ecosystems of the World*, vol. 9B, ed. R.L. Specht, pp. 51–60. Amsterdam: Elsevier Scientific Publishing Company.

Bolton, M.P. (1986). Community dynamics and productivity in a subtropical wet heathland. PhD thesis, University of Queensland.

Carnahan, J.A. (1976). Natural vegetation. In *Atlas of Australian Resources*, Second Series. Canberra: Department of National Resources.

Catling, P.C. (1988). Vertebrates. In *Mediterranean-type Ecosystems. A Data Source Book*, ed. R.L. Specht, pp. 171–94. Dordrecht: Kluwer.

Clifford, H.T. & Specht, R.L. (1979). *Vegetation of North Stradbroke Island*. St Lucia: Queensland University Press.

Coleman, R.G. & Specht, R.L. (1981). Mineral nutrition of heathlands: The possible role of polyphosphate in the phosphorus economy of heathland species. In *Ecosystems of the World*. vol. 9B, ed. R.L. Specht, pp. 197–207. Amsterdam: Elsevier Scientific Publishing Company.

Connor, D.J. & Doley, D. (1981). The water relations of heathlands: Physiological adaptation to drought. In *Ecosystems of the World*. vol. 9B, ed. R.L. Specht, pp. 131–41. Amsterdam: Elsevier Scientific Publishing Company.

Dick, R.S. (1975). A map of the climates of Australia: According to Köppen's principles of definition. *Queensland Geographical Journal*, **3**, 33–69.

Edmonds, S.J. & Specht, M.M. (1981). Dark Island heath, South Australia: Faunal rhythms. In *Ecosystems of the World*, vol. 9B, ed. R.L. Specht, pp. 15–28. Amsterdam: Elsevier Scientific Publishing Company.

George, A.S., Hopkins, A.J.M. & Marchant, N.G. (1979). The heathlands of Western Australia. In *Ecosystems of the World*, vol. 9A, ed. R.L. Specht, pp. 211–30. Amsterdam: Elsevier Scientific Publishing Company.

Gill, A.M. (1976). Fire and the opening of *Banksia ornata* F. Muell. follicles. *Australian Journal of Botany*, **24**, 329–35.

Gill, A.M. & Groves, R.H. (1981). Fire regimes in heathlands and their plant ecological effects. In *Ecosystems of the World*, vol. 9B, ed. R.L. Specht, pp. 61–84. Amsterdam: Elsevier Scientific Publishing Company.

Gill, A.M. & McMahon, A. (1986). A post-fire chronosequence of cone, follicle and seed production in *Banksia ornata*. *Australian Journal of Botany*, **34**, 425–33.

Gimingham, C.H. (1972). *Ecology of Heathlands*. London: Chapman & Hall.

Groves, R.H. (1981). Nutrient cycling in heathlands. In *Ecosystems of the World*. vol. 9B, ed. R.L. Specht, pp. 151–63. Amsterdam: Elsevier Scientific Publishing Company.

Groves, R.H., Hocking, P.J. & McMahon, A. (1986). Distribution of biomass, nitrogen, phosphorus and other nutrients in *Banksia marginata* and *B. ornata* shoots of different ages after fire. *Australian Journal of Botany*, **34**, 709–25.

Groves, R.H. & Specht, R.L. (1965). Growth of heath vegetation. I. Annual growth curves of two heath ecosystems in Australia. *Australian Journal of Botany*, **13**, 261–80

Groves, R.H. & Specht, R.L. (1981). Seral considerations in heathland. In *Vegetation Classification in the Australian Region*, eds A.N. Gillison & D.J. Anderson, pp. 78–85. Canberra: CSIRO & Australian National University Press.

Heddle, E.M. & Specht, R.L. (1975). Dark Island heath (Ninety-Mile Plain, South Australia). VIII. The effect of fertilizers on composition and growth, 1950–1972. *Australian Journal of Botany*, **23**, 151–64.

Jeffrey, D.W. (1967). Phosphate nutrition of Australian heath plants. I. The importance of proteoid roots in *Banksia* (Proteaceae). *Australian Journal of Botany*, **15**, 403–11.

Jeffrey, D.W. (1968). Phosphate nutrition of Australian heath plants. II. The formation of polyphosphate by five heath species. *Australian Journal of Botany*, **16**, 603–13.

Kruger, F.J. (1979). South African heathlands. In *Ecosystems of the World*, vol. 9A, ed. R.L. Specht, pp. 19–80. Amsterdam: Elsevier Scientific Publishing Company.

Lamont, B.B. (1973). Factors affecting the distribution of proteoid roots within the root systems of two *Hakea* species. *Australian Journal of Botany*, **21**, 165–87.

Lamont, B.B. (1981). Specialized roots of non-symbiotic origin in heathlands. In *Ecosystems of the World*. vol. 9B, ed. R.L. Specht, pp. 183–95. Amsterdam: Elsevier Scientific Publishing Company.

Lamont, B.B. (1982). Mechanisms for enhancing nutrient uptake in plants, with particular reference to mediterranean South Africa and Western Australia. *Botanical Review*, **48**, 597–689.

Lance, G.N. & Williams, W.T. (1968). A note on a new information statistic classificatory program. *Computing Journal*, **11**, 195.

Low, A.B. & Lamont, B.B. (1990). Aerial and below-ground phytomass of *Banksia* scrub-heath at Eneabba, south-western Australia. *Australian Journal of Botany*, 38, 351–9.

Majer, J.D. & Greenslade, P. (1988). Soil and litter invertebrates. In *Mediterranean-type Ecosystems. A Data Source Book*, ed. R.L. Specht, pp. 197–226. Dordrecht: Kluwer.

Malajcuk, N. & Glenn, A.R. (1981). *Phytophthora cinnamoni* – A threat to the heathlands. In *Ecosystems of the World*, vol. 9B, ed. R.L. Specht, pp. 241–47. Amsterdam: Elsevier Scientific Publishing Company.

Malajcuk, N. & Lamont, B.B. (1981). Specialized roots of symbiotic origin in heathlands. In *Ecosystems of the World*, vol. 9B, ed. R.L. Specht, pp. 165–82. Amsterdam: Elsevier Scientific Publishing Company.

Ozanne, P.G. & Specht, R.L. (1981). Mineral nutrition of heathlands: Phosphorus toxicity. In *Ecosystems of the World*, vol. 9B, ed. R.L. Specht, pp. 209–13. Amsterdam: Elsevier Scientific Publishing Company.

Parsons, R.F. (1981). Salt-spray effects in heathlands. In *Ecosystems of the World*, vol. 9B, ed. R.L. Specht, pp. 225–30. Amsterdam: Elsevier Scientific Publishing Company.

Pate, J.S. & Beard, J.S. (eds) (1984). *Kwongan. Plant Life of the Sandplain*. Nedlands: University of Western Australia Press.

Raunkiaer, C. (1934). *The Life Form of Plants and Statistical Plant Geography*, Oxford University Press.

Rübel, E.A. (1914). Heath and steppe, macchia and garigue. *Journal of Ecology*, 2, 232–7.

Specht, R.L. (1963). Dark Island heath (Ninety-Mile Plain, South Australia). VII. The effect of fertilizers on composition and growth, 1950–1960. *Australian Journal of Botany*, 11, 67–94.

Specht, R.L. (1969). A comparison of the sclerophyllous vegetation characteristic of mediterranean type climates in France, California and southern Australia. I. Structure, morphology and succession. II. Dry matter, energy and nutrient accumulation. *Australian Journal of Botany*, 17, 277–92, 293–308.

Specht, R.L. (1970). Vegetation. In *The Australian Environment*, 4th edn (rev.), ed. G.W. Leeper, pp. 44–67. Melbourne: CSIRO & Melbourne University Press.

Specht, R.L. (1972). *The Vegetation of South Australia*. 2nd edn. Adelaide: Government Printer.

Specht, R.L. (ed.) (1979a). *Ecosystems of the World*, vol. 9A. *Heathlands and Related Shrublands, Descriptive Studies*. Amsterdam: Elsevier Scientific Publishing Company.

Specht, R.L. (1979b). Heathlands and related shrublands of the world. In *Ecosystems of the World*, vol. 9A, ed. R.L. Specht, pp. 1–18. Amsterdam: Elsevier Scientific Publishing Company.

Specht, R.L. (1979c). The sclerophyllous (heath) vegetation of Australia: The eastern and central States. In *Ecosystems of the World*, vol. 9A, ed. R.L. Specht, pp. 125–210. Amsterdam: Elsevier Scientific Publishing Company.

Specht, R.L. (ed.) (1981a). *Ecosystems of the World*, vol. 9B. *Heathlands and*

Related Shrublands. Analytical Studies. Amsterdam: Elsevier Scientific Publishing Company.

Specht, R.L. (1981*b*). Responses of selected ecosystems: heathlands and related shrublands. In *Fire and the Australian Biota*, eds A.M. Gill, R.H. Groves & I.R. Noble, pp. 395–415. Canberra: Australian Academy of Science.

Specht, R.L. (1981*c*). Major vegetation formations in Australia. In *Ecological Biogeography in Australia*, ed. A. Keast, pp. 163–298. The Hague: W. Junk.

Specht, R.L. (1981*d*). Structural attributes – foliage projective cover and standing biomass. In *Vegetation Classification in the Australian Region*, eds A.N. Gillison & D.J. Anderson, pp. 10–21. Canberra: CSIRO & Australian National University Press.

Specht, R.L. (1981*e*). Nutrient release from decomposing leaf litter of *Banksia ornata*, Dark Island heathland, South Australia. *Australian Journal of Botany*, **6**, 59–63.

Specht, R.L. (1981*f*). The water relations of heathlands: Seasonal waterlogging. In *Ecosystems of the World*, vol. 9B, ed. R.L. Specht, pp. 99–106. Amsterdam: Elsevier Scientific Publishing Company.

Specht, R.L. (1981*g*). The water relations of heathlands: Morphological adaptations to drought. In *Ecosystems of the World*, vol. 9B, ed. R.L. Specht, pp. 123–9. Amsterdam: Elsevier Scientific Publishing Company.

Specht, R.L. (1981*h*). Conservation: Australian heathlands. In *Ecosystems of the World*, vol. 9B, ed. R.L. Specht, pp. 235–40. Amsterdam: Elsevier Scientific Publishing Company.

Specht, R.L. (1988). Climatic control of ecomorphological characters and species richness in mediterranean ecosystems in Australia. In *Mediterranean-type Ecosystems. A Data Source Book*, ed. R.L. Specht, pp. 149–55. Dordrecht: Kluwer.

Specht, R.L. (ed.) (in press). Major plant communities in Australia: An objective assessment, *Australian Journal of Botany*, **41**.

Specht, R.L., Clifford, H.T. & Rogers, R.W. (1984). Species richness in a eucalypt open-woodland on North Stradbroke Island, Queensland. The effect of overstorey and fertilizer, 1965–1984. In *Focus on Stradbroke*, eds R.J. Coleman, J. Covacevitch & P. Davie, pp. 267–77. Brisbane: Boolarong Press.

Specht, R.L., Connor, D.J. & Clifford, H.T. (1977). The heath-savannah problem: the effect of fertilizer on sand-heath vegetation of North Stradbroke Island, Queensland. *Australian Journal of Ecology*, **2**, 179–86.

Specht, R.L. & Groves, R.H. (1966). A comparison of the phosphorus nutrition of Australian heath plants and introduced economic plants. *Australian Journal of Botany*, **14**, 201–21.

Specht, R.L. & Moll, E.J. (1983). Mediterranean-type heathlands and sclerophyllous shrublands of the world: an overview. In *Mediterranean-type Ecosystems. The Role of Nutrients*, eds F.J. Kruger, D.T. Mitchell & J.U.M. Jarvis, pp. 120–32. Berlin: Springer-Verlag.

Specht, R.L. & Morgan, D.G. (1981). The balance between the foliage projective covers of overstorey and understorey strata in Australian vegetation. *Australian Journal of Ecology*, **6**, 193–202.

Specht, R.L. & Rayson, P. (1957). Dark Island heath (Ninety-Mile Plain, South Australia). I. Definition of the ecosystem. *Australian Journal of Botany*, 5, 52–85.

Specht, R.L., Rayson, P. & Jackman, M.E. (1958). Dark Island heath (Ninety-Mile Plain, South Australia). VI. Pyric succession: changes in composition, coverage, dry weight, and mineral nutrient status. *Australian Journal of Botany*, 6, 59–88.

Specht, R.L., Rogers, R.W. & Hopkins, A.J.M. (1981). Seasonal growth and flowering rhythms: Australian heathlands. In *Ecosystems of the World*, vol. 9B, ed. R.L. Specht, pp. 5–13. Amsterdam: Elsevier Scientific Publishing Company.

Specht, R.L. & Rundel, P.W. (1990). Sclerophylly and foliar nutrient status of mediterranean-climate plant communities in southern Australia. *Australian Journal of Botany*, 38, 459–74.

Specht, R.L. & Specht, A. (1989). Species richness of sclerophyll (heathy) plant communities in Australia – the influence of overstorey cover. *Australian Journal of Botany*, 37, 337–50.

Specht, R.L., Yates, D.J., Sommerville, J.E.M. & Moll, E.J. (1991). Foliage structure and shoot growth in heathlands in the mediterranean-type climate of southern Australia and South Africa. *Ecologia Mediterranea*, 16, 195–207.

Stace, H.C.T., Hubble, G.D., Brewer, R., Northcote, K.H., Sleeman, J.R., Mulcahy, M.J. & Hallsworth, E.G. (1968). *A Handbook of Australian Soils*. Glenside. S. Aust.: Rellim Technical Publications.

Woolhouse, H.W. (1981). Soil acidity, aluminium toxicity and related problems in the nutrient environment of heathlands. In *Ecosystems of the World*, vol. 9B, ed. R.L. Specht, pp. 215–24. Amsterdam: Elsevier Scientific Publishing Company.

12

Chenopod shrublands

J.H. LEIGH

Distribution

Approximately 7 per cent (5 000 000 km²) of the total land area of mainland Australia is occupied by Chenopod shrublands dominated by various annual and perennial species of the family Chenopodiaceae. The terminology of these communities has varied over the years, having been variously referred to as 'shrub steppe' (Williams, 1955; Wood & Williams, 1960), 'saltbush–xerophytic midgrass grazing lands' (Moore, 1969), 'low shrubland' (Specht, 1970, 1981) and 'low shrublands', 'low open shrublands' and 'chenopod shrublands' (AUSLIG, 1990). In this chapter they are referred to as Chenopod shrublands.

These communities lie mostly south of the Tropic of Capricorn in an environment receiving rainfall ranging from 125 to 266 mm per annum, 30–50 per cent of which falls in winter. The most extensive areas occur in South Australia (247 500 km²), western New South Wales (77 000 km²) and southeast Western Australia (75 800 km²). Smaller areas (approximately 16 000 km²) are found in Queensland and Northern Territory. Victoria contains less than 1500 km² of this vegetation type. The shrubland is composed of xeromorphic halophytes which are drought- and salt-tolerant. These areas were mapped by Dr John Carnahan (AUSLIG, 1990) and their distribution is shown, together with isohyets, in Figure 12.1.

Floristics

The shrubland is composed of xeromorphic halophytes which are drought-and salt-tolerant. There are 293 taxa (species plus subspecies) in the family Chenopodiaceae occurring in Australia classified

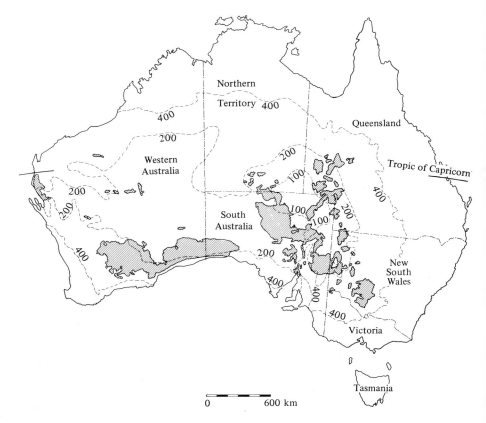

Fig. 12.1. Distribution of Chenopod shrublands (hatched areas) in Australia in relation to the isohyets representing mean annual rainfalls of 100, 200 and 400 mm (dashed lines). (Figure modified from Atlas of Australian Resources, 1980.)

within 33 genera, 28 of which are endemic (Hnatiuk, 1990). Within these genera there are 289 endemic species plus 81 subspecies. The naturalised genera contain 20 species plus 2 subspecies. The most widespread and economically important genera, with the number of endemic species in parentheses are *Sclerolaena* (66), *Atriplex* (59), *Maireana* (57), *Chenopodium* (15), *Rhagodia* (11) and *Dysphania* (10).

Of less importance and with more restricted occurrence to areas such as salt marshes are *Halosarcia* (23), *Sclerostegia* (5) and *Sarcocornia* (3). Most naturalised species are forbs and belong in the genus *Chenopodium* (8); they mostly occur as weeds of habitation, waste places and crop lands.

Occurrence and present condition

The great majority of the endemic species enumerated above occur as shrubs or shrublands. These shrubs may occur as extensive monospecific stands presenting a simple and uniform appearance, or they may occur (cf. *Atriplex vesicaria*) as vegetated bands aligned along topographical contours, or as mixed stands of two or more species. The most common species (e.g. *Atriplex vesicaria*) are less than 1.5 m high and have a projected foliage cover of between 10 and 30 per cent (Specht, 1970). Some however, such as *Atriplex nummularia*, may attain a height of 3 m and the bushes may be congregated in groups or occur as individuals separated by areas of bare soil (Fig. 12.2). The bluebushes (e.g. *Maireana astrotricha*, *M. pyramidata* and *M. sedifolia*) usually occur sparsely with a projected foliage cover less than 10 per cent. Most species may also be present as an understorey to low woodlands with an overstorey of species of *Eucalyptus*, *Casuarina*, *Acacia*, *Heterodendrum* and *Myoporum* (see Chapters 8, 9 & 10; Gillison & Walker, 1981; Johnson & Burrows, 1981; Parsons, 1981).

Newman & Condon (1969) assessed the extent to which the original rangeland types may have degenerated over the last 100 to 150 years, since settlement by Europeans and the introduction of domestic livestock, particularly sheep, as well as the introduction of the rabbit, now feral and widespread over most of southern Australia. There is much evidence to suggest that many Australian rangelands have altered and degenerated. In particular, the shrublands have fared badly. Newman & Condon assessed the extent of deterioration by relating present pastoral value to expected pristine condition. They estimated that 25 per cent of Chenopod shrublands were severely degenerated (< 40 per cent of original pristine condition), 40 per cent were moderately deteriorated (40 to 60 per cent of original condition), 25 per cent showed minor degeneration (60 to 80 per cent of original condition) and only 10 per cent showed little or no deterioration (80 to 100 per cent of original condition).

In considering the conservation of Chenopod species individually, data collated by Briggs & Leigh (1988) indicate that some 9 per cent (26) of all endemic species (289) are considered to be rare or threatened. Of those, two species (*Atriplex kochiana* and *Scleroleana napiformis*) are endangered (in serious risk of disappearing from the wild state within one or two decades if present land use or other causal factors continue to operate). A further nine species (*Halosarcia bulbosa* and *H. flabelliformis*, *Maireana cheelii* and *M. melanocarpa*, *Malococera gracilis*, *Rhagodia acicularis*, *Roycea pycnophylloides*, *Scleroleana blakei* and *S. walkeri*) are vulnerable (at risk of disappearing from the wild over a longer period of 20–50 years through continued depletion, or which largely occur on sites likely to experience changes in land use that would threaten the survival of the species in the wild). Many (11 species) of the rare or threatened Chenopods have a very limited distribution, with four species being known from the site of the original collection only and seven species from an area with a maximum geographic range of less than 100 km. Because of their narrow range, this group could be at great risk from factors such as cropping, fire, grazing, or mining. Most of the rare or threatened species occur in Western Australia (eight species) with lesser numbers (four species each) occurring in New South Wales, the Northern Territory and Queensland.

Description and distribution of economically important species

The original distribution of individual species was determined largely by factors such as physical and chemical properties of the soil, and soil moisture status, particularly as influenced by topography. A secondary, but nevertheless significant factor affecting the present-day occurrence of the shrubs, has been the introduction by humans of domestic and feral animals. The saltbush country was most easily settled and was one of the first regions to be occupied in the early days of pastoral expansion. At the present time, Chenopod shrublands support approximately 2.75×10^6 sheep at an average stocking rate of 6 sheep/km² and 84 000 cattle at an average stocking rate of 2 head/km² (R.D. Graetz, personal communication).

From the viewpoint of resource stability and the pastoral industry, the six most valued shrubs are *Atriplex nummularia*, *A. vesicaria*, *Maireana aphylla*, *M. astrotricha*, *M. pyramidata* and *M. sedifolia*. A brief description of these six species follows.

Atriplex nummularia (Oldman saltbush)

A blue-grey shrub *A. nummularia* grows to a height and diameter of about 3 m, and has large, oval or almost circular leaves 1 to 2 cm long. The leaves are covered with a scaly layer; their margins may be uninterrupted or slightly toothed. The plants are usually of one sex, the female flowers occurring in dense clusters in the leaf axles at the end of the branchlets. The fruiting body consists of two flat paper-like bracteoles, hemispherical in outline, toothed or smooth-edged, pressed closely together but united only at the base, where the seed is enclosed, the whole being 5 to 8 mm across.

Atriplex nummularia is the largest endemic species of *Atriplex*, and is more common in southeastern Australia, becoming restricted to local drainage areas in more inland regions. This shrub occurs either associated with *Acacia pendula* (boree or myall) (an association considered to be once far more widespread than it is today), or as an understorey to *Eucalyptus largiflorens* (black box) or in association with *A. vesicaria*. Although it shows a preference for clay and clay-loam soils, *A. nummularia* can be found on most soil types.

Atriplex vesicaria (Bladder saltbush)

This perennial shrub, with woody brittle stems, reaches a height and diameter of 60 to 70 cm. Its leaves are roughly oval in outline, up to 25 mm in length, and have a silvery-grey appearance because of vesicular hairs on the surface. The seed is enclosed within two bracteoles, on each of which generally there occurs a large membranous bladder-like appendage, the whole being up to 12 mm in diameter.

The most extensive tracts of *A. vesicaria* occur in parts of western New South Wales and the northern and eastern regions of South Australia. Other areas are found on the Nullarbor Plain, extending into Western Australia (Fig. 12.2), in northwestern Victoria, south-western Queensland and southern-central Australia.

This shrub grows on a wide range of soil types and in a variety of habitats. On the riverine plain of central southwestern New South Wales it is found on flat terrain of brown and grey soils of heavy texture. Here it may occur in pure stands or in association with *Maireana aphylla* or *A. nummularia*. In South Australia and south-western New South Wales, *A. vesicaria* occurs on red-brown soils, some with a clay subsoil, others with a subsoil containing limestone nodules. In the latter case, the associated shrubs are usually *Maireana sedifolia* or *M. astrotricha*. In northwestern New South Wales and northern South Australia, *A. vesicaria* is found on the loams of the gibber plains and on the stony desert soils of the undulating and hilly country surrounding the Barrier and Grey Ranges (29°03′S; 142°57′E). The communities in most of these areas are in a severely degraded state. Possibly the most vigorous stands of *A. vesicaria* occur in the vicinity of Hay, in central New South Wales.

Fig. 12.2. *Atriplex vesicaria* shrubland near Meekathara, Western
Australia, with small trees of *Acacia aneura* in the background.
(Photo: C.J. Totterdell.)

Maireana aphylla (Cotton bush)

Maireana aphylla is a much-branched perennial shrub growing to a
height and diameter of about 1 m, with slender striate branches which
become rigid and spiny with age. The leaves, short (1–2 mm long),
thick, and crowded along the branchlets, fall off at an early stage.
A membranous wing, pink when young, later maturing to brown,
surmounts the short tube of the fruiting body, the whole being up to
12 mm across. Clusters of white cotton-like galls, caused by small
grubs, often occur along the branches, which give the shrub its
common name of 'cotton bush'.

 Maireana aphylla is distributed widely throughout much of arid
and semi-arid Australia. It prefers clay and clay-loam soils, and
reaches its maximum density on the southern riverine plain of New
South Wales where it may be the dominant species or it may be
associated with other Chenopod shrubs, particularly *A. vesicaria*.
Elsewhere, associated plants may be *M. pyramidata* or *Eremophila
maculata*.

Maireana astrotricha (Southern bluebush)

Similar in habit and appearance to *M. sedifolia*, although greyish rather than blue, *M. astrotricha* can be distinguished by its flatter leaves, which taper to a conspicuous petiole. The lower leaves may be up to 15 mm long and 3 to 4 mm broad. The winged fruits are 6 to 15 mm across, with the prominently lobed wing surmounting a large hard obconical tube 3 to 5 mm long.

The major areas of *M. astrotricha* occur on sandy soils in the far west of New South Wales and in South Australia (where the similar *M. planifolia* also occurs). Minor areas of *M. astrotricha* are also found in central Australia, Queensland and Western Australia. This shrub may be associated with *A. vesicaria* and various other *Maireana* species.

The relative abundance of each of the *Maireana* shrubs so far discussed is largely determined by the depth at which the limestone layer occurs beneath the soil surface. With the layer close to the surface *M. sedifolia* is the predominant species, and *M. pyramidata* and then *M. astrotricha* take precedence as the depth to the layer increases.

Maireana pyramidata (Black bluebush)

Maireana pyramidata is the largest of the Australian *Maireana* species with a height of almost 2 m. Its greenish-grey, fleshy leaves are more or less obovoid, 2 to 4 mm long, and shorter than the other species. The characteristic fruiting body has a pyramidal spongy extension above the narrow horizontal wing which encircles the short basal tube. Before maturity the fruits are bright green; when mature and dry they turn black.

The distribution of *M. pyramidata* is similar to that of *M. sedifolia*, being confined to the calcerous red and red-brown soils. *Maireana pyramidata* is perhaps the more prevalent species of the two in the eastern areas and is a more common constituent of the shrub component of communities dominated by *Eucalyptus* (mallee), *Casuarina cristata*-(belah)-*Heterodendrum oleifolium* (inland rosewood), and *Acacia aneura* (mulga).

Maireana sedifolia (Pearl bluebush)

A much-branched perennial with thick woody stems, *M. sedifolia* grows to a height and diameter of about 1 m. The young branches and foliage have a dense covering of white hairs which impart a pale-blue, almost white appearance to the bush. The leaves are soft, thick, almost cylindrical in outline and 3 to 10 mm in length. The short tube of the fruiting body is encircled by a reddish or light-brown horizontal wing, 8–10 mm in diameter, finely veined, unbroken or with one radial slit.

Maireana sedifolia extends over much of western New South Wales and South Australia and is one of the major shrubs on the Nullarbor Plain. It occurs either in pure stands or in association with *A. vesicaria*, *M. pyramidata* or *M. astrotricha*, or as an understorey to *Myoporum platycarpum* (sugarwood) or *Casuarina cristata* (belah [New South Wales], black oak [South Australia]); it occupies possibly the largest area of any of the Chenopod browse shrubs in Australia.

Maireana sedifolia prefers calcareous soils in which the limestone lies close to the surface. These soils are particularly susceptible to erosion if the surface cover is removed, either by wind in the case of the level alluvial desert loams or by both wind and water in the case of the undulating stony desert-soils.

Additional Chenopods which may be widespread but of little economic value, or may be of pastoral value but of minor occurrence in relation to the associated vegetation, or may be only locally common but pastorally important, include the following species; *Atriplex semibaccata*, *Chenopodium auricomum*, *C. nitrariaceum*, *Enchylaena tomentosa*, *Maireana brevifolia*, *M. georgei*, *M. tomentosa*, *M. triptera*, *M. villosa*, and *Rhagodia spinescens*. Many species of *Sclerolaena* (copperburrs) have also been shown to be of pastoral importance.

Community relationships

As mentioned previously it is generally accepted, mostly in the absence of sound scientific evidence, that the distribution of Chenopod shrubs is determined by factors such as the physical and chemical properties of the soil and the soil moisture status.

The most thorough studies on factors affecting interspecies relationships, namely those which relate to the relative distribution of *Atriplex vesicaria* and *Maireana sedifolia*, were investigated by Carrodus & Specht (1965). They showed that the distribution of the two species was correlated with the depth to which the soil was wetted by normal rainfall, *M. sedifolia*, the deeper-rooted species, occurring on soils which could be wetted to a depth of 50 cm or more, and *A. vesicaria*, the shallow-rooted species, on soils in which a heavy clay subsoil or hardpan impeded root penetration beyond 30 cm. Both species could grow successfully in pots of the surface soil associated with the other species. Plants of *A. vesicaria* were also shown to be capable of reducing the percentage moisture in the soil to a significantly lower level than could *M. sedifolia* when subjected to drought. The practical consequence of these attributes therefore is that *M. sedifolia* occupies deep soils and, being long-lived, once established is capable of exploiting all the soil moisture available, whereas *A. vesicaria* with its shallower root system can exploit a smaller soil volume. With the onset of drought *A. vesicaria* could be expected to succumb more rapidly and so *M. sedifolia* is capable of maintaining the community in a closed state. In shallower soil, however, *A. vesicaria* has the advantage of more rapid re-establishment when drought breaks. If a community of *M. sedifolia* is destroyed, it takes a very long time to re-establish itself (Hall, Specht & Eardley, 1964) and in the meantime, the community may be occupied by *A. vesicaria*.

On superficial examination of most Chenopod-dominated communities, particularly in the summer, the overriding impression is of Chenopod shrubs. With the possible exception of some of the salt marsh communities, all Chenopod shrubland communities characteristically support a ground cover, the main grass genera of which include *Danthonia*, *Stipa*, *Eragrostis* and *Aristida*. In the autumn, winter and spring months a number of annual grasses, mostly naturalised, including *Lolium*, *Hordeum*, *Vulpia* and *Bromus*, are common, as are a number of forbs, both native (e.g. *Brachycome*, *Calotis*, *Helichrysum*, *Helipterum*) and naturalised (e.g. *Medicago* and *Trifolium*). These naturalised grasses and forbs have changed significantly the productivity of such communities, and the legumes probably the soil fertility. The importance to the pastoral industry of these associated groups of cover grasses and forbs will be discussed more fully later in this chapter.

A high intensity of sheep grazing has led to a reduction, in some instances, in the number of desirable Chenopod shrubs and an increase in the numbers of unpalatable spiny shrubs, chiefly species of *Nitraria* and *Sclerolaena*. Under extreme conditions, complete baring or 'scalding' of the surface may occur, so that only bare soil remains.

A characteristic pattern of vegetated bands develops on areas of gently sloping pediments aligned along topographical contours and separated by bare interbands (Valentine & Nagorcka, 1979). This creates a number of interesting microtopographical features which enable some rain to run off the bare areas and infiltrate the vegetated areas. Should the bare areas be enlarged, and the bands be breached by overgrazing, then the fetch of the run-off water may be sufficient to encourage water erosion (Valentine & Nagorcka, 1979). Charley & Cowling (1968) described a process for the lateral distribution of salt in a similar shrub community, where chlorides were leached out of the bands and accumulated in the interbands, thereby allowing the growth of salt-intolerant species and stabilisation along the bands.

The results of early studies of Osborn, Wood & Paltridge (1932) and more recently, of Lange (1969), Barker & Lange (1969, 1970), Graetz & Ludwig (1978), Barker (1979) and Lay (1979) have shown clearly the influence that location of watering points, and associated pattern of grazing by domestic stock, have on the biomass, density and species composition of Chenopod shrubs and associated species located in the Chenopod shrubland communities. For example, Osborn *et al.* (1932) showed that *A. vesicaria* populations were denuded in the first 400 m from the watering point but beyond this, growth was stimulated. Similarly, Barker (1979) showed that, within a wide range of Chenopod shrub-steppe communities, only one species (*A. vesicaria*) disappeared close to watering points as a result of sheep grazing, whereas two species (*M. pyramidata* and *Sclerolaena patenticuspis*) were stimulated and spread into the area of the watering point from other parts of the paddock, one species (*Maireana excavata*) was stimulated to grow at an intermediate distance from the watering point and a further seven species (*Atriplex eardleyae, A. spongiosa, Carthamus lanatus, Inula graveolens, Maireana breviflora, Marrubium vulgare* and *Sclerolaena paradoxa*) invaded from outside the paddock into the region of the watering point.

Trumble & Woodroffe (1954) described the effects of eleven years of sheep grazing in a study of the *Maireana sedifolia* association in South Australia. Food intake was measured in a number of grazing treatments by assessing the amounts of green edible forage present in spring. Moderate to heavy grazing, particularly in favourable seasons, was more beneficial to growth of *M. sedifolia* and animal production from it was greatest under this regime.

Palatability, pastoral value and response to grazing

Palatability

No single index of acceptability to livestock in general can be given for a particular species because acceptability is dependent on a number of interacting factors. The more important of these are class of livestock, and level of availability and growth stage, both of the species in question and that of other associated species.

In general, it has been observed that most Chenopod shrubs are of moderate to low palatability and that stock prefer associated grasses and forbs when these are available. On the riverine plain of south-western New South Wales it was found that *Atriplex vesicaria* and *Maireana aphylla* shrubs were ungrazed or only lightly grazed (10 per cent of the diet or less) by sheep when grasses and forbs were present, even though these grasses and forbs were often only minor constituents of the pastures (Leigh, Wilson & Mulham, 1968; Wilson, Leigh & Mulham, 1969). In times of feed scarcity and an absence of associated species, *A. vesicaria* comprised 70–90 per cent of the diet eaten, whereas *M. aphylla* seldom exceeded 10 per cent of the diet eaten.

More recently, a comparison has been made into the diets selected by sheep and goats grazing identical low woodland communities (*Casuarina cristata–Heterodendrum oleifolium*) near Ivanhoe, New South Wales, at light stocking rates (Wilson, Leigh & Mulham, 1975). This community characteristically has a tree overstorey, many species of large shrub (chiefly species of *Acacia*, *Apophyllum*, *Cassia* and *Eremophila*) and a ground stratum of grasses (chiefly *Stipa variabilis*), forbs and Chenopod shrublands (chiefly *Sclerolaena diacantha*). Results of diet analyses over four seasons between May 1971 and February 1972 showed that trees and large shrubs (excluding Chenopods) made up between 58 and 80 per cent of the goat diets, compared to 3–24 per cent of the sheep diets, grasses made up less than 3 per cent of goat diets and up to 49 per cent of sheep diets, and *Sclerolaena* species (chiefly *S. diacantha*) less than 8 per cent in goat diets and up to 48 per cent in sheep diets.

Whilst palatability comparisons as presented above may be generally valid, it is also true that in certain circumstances considerable differences in palatability within individual species may exist, even when the plants are growing on the same soil type. In *A. vesicaria* shrublands growing on the riverine plain of New South Wales, Williams, Anderson & Slater (1978) demonstrated changes in the sex ratio between shrub populations subject to different grazing intensities. Sheep were observed to graze selectively on the green bladder fruits of the female plants, and thereby reduce their reproductive vigour with consequential losses to seed production, and they tended to avoid the granular catkins of the male plants.

Reaction to grazing

The effects of grazing on various plants differ from species to species but it is generally agreed that some degree of grazing is beneficial. The point at which defoliation by grazing has a detrimental effect on the plant varies with species. The degree to which *A. vesicaria* is grazed has a direct bearing on its chances of survival. Moderate grazing leads to a more compact, leafier bush (Osborn *et al.*, 1932); complete defoliation will almost certainly result in its death (Leigh & Mulham, 1971). On the other hand, *A. nummularia* and *Maireana* species appear able to withstand periods of total defoliation. None of the plants will survive extended periods of continual defoliation, however.

The comparatively long-term (1980–87) population dynamic study of the three Chenopod shrubs *Atriplex vesicaria*, *Maireana astrotricha* and *M. pyramidata* growing in western New South Wales under conditions of moderate grazing, showed that grazing did not affect the population dynamics (Eldridge *et al.*, 1990). Density of *A. vesicaria* declined on both grazed and ungrazed transects, but that of *Maireana* species remained almost constant. Too often, conclusions about the alleged deleterious effects of grazing are drawn without there being adequate controls in the form of ungrazed plants. Shrub size was the only population parameter significantly affected by grazing. Heights and diameters for both grazed and ungrazed populations declined during dry periods, but this decline was greatest for grazed shrubs. After rain, however, grazed shrubs quickly returned to sizes similar to ungrazed. This study concluded that the grazing strategies currently maintained in that area were having no adverse effects on the population of the three chenopod shrubs examined.

The stems of the two major *Atriplex* species, *A. vesicaria* and *A. nummularia*, are particularly brittle, and they may be severely damaged by livestock trampling. Large stands of *A. nummularia* have been destroyed by the combined effects of trampling, especially by cattle, and grazing.

Selectivity differences between classes of livestock and subsequent regrowth pattern may be of some significance for pasture management and weed control. Wilson (1976) observed, in a comparison between the effects of cattle and sheep grazing, that immature plants of *Salsola kali* (soft roly-poly) were grazed closely by sheep and thereby prevented from forming their normal upright habit and prickly stems characteristic of mature bushes, whereas cattle ignored the young plants so that many mature plants of *S. kali* occurred.

Mineral content and forage value

The forage value of Chenopod shrubs has been assessed in laboratory and field trials. Results from these have been reviewed by Leigh (1986). The laboratory trials have taken the form of chemical analyses and pen feeding trials, and the field trials have assessed the value of the Chenopod shrubs both as natural stands of shrubs and on areas to which the shrubs have been introduced.

In comparison with non-Chenopods, these shrubs are high in nitrogen, sodium, potassium and chloride salts (Wilson, 1966a). Typical results of analyses showing the range in these constituents, determined over a number of localities, are presented in Table 12.1. Species of *Atriplex* accumulate ions to approximately the same concentration despite wide variations in the mineral composition of the soils on which they grow. High sodium levels are usually accompanied by high nitrogen levels. From an animal husbandry viewpoint these high mineral contents may have a number of deleterious effects. High sodium levels reduce the water use of Chenopods per unit of dry weight and increase the water needs of grazing animals. Salt intakes approaching 200 g NaCl/sheep/day have been measured for sheep grazing *A. vesicaria* in summer (Wilson, 1967). Penned sheep fed on a diet of *A. vesicaria* leaf were shown to consume 11.3 l water/day (Wilson, 1966b), sheep grazing on a natural *A. vesicaria* community drank 7.5 l/day (Wilson, 1967). The increased water intake is necessary for urinary excretion of the high intakes of salt. Substantial reductions in feed intake also resulted when sheep consuming Chenopod leaf had only saline water to drink. This situation, where sheep grazing saltbush pastures in summer have no feed other than saltbush leaf and have access only to highly saline bore water, is by no means uncommon in Australia. It has been suggested for such situations, because of the need to make frequent visits to water, that the grazing range of sheep is likely to be severely decreased and that this may result in over-exploitation of the vegetation growing nearest to the water source (Squires & Wilson, 1971).

Chenopod shrubs generally have been regarded over the years as fodder of high quality on the basis of analyses of the crude protein content and dry matter digestibility of the type given in Table 12.1. It is true that the crude protein values are high relative to those of pasture grasses and legumes and are maintained throughout the year. The ability of these species to remain green under moisture stress, together with their high crude protein contents, has led numerous

Table 12.1. *Published values for sodium, potassium, chloride and crude protein content and dry matter digestibility of several chenopod shrubs, expressed as a percentage of dry weight. After Leigh, 1986*

	Sodium	Potassium	Chloride	Crude protein	Digestibility
Atriplex angulata	3.3	1.8	—	17.5	58
A. nummularia	3.8–8.2	1.14–3.9	7.8–13.6	17.0–21.9	68–74
A. semibaccata	3.5	4.2	—	10.0–19.9	—
A. vesicaria	3.2–6.5	1.2–7.2	3.9–14.3	11.1–18.4	52–54
Dissocarpus paradoxa	—	—	—	10.9	57
Maireana astrotricha	3.9	2.4	—	12.7	68
M. brevifolia	—	—	—	24.8	—
M. pyramidata	4.6–6.6	1.5–2.45	3.4	15.0–22.0	58–69
M. sedifolia	3.8	2.15	—	15.1–17.2	69
Rhagodia spinescens	—	—	—	9.7–12.6	52–64
Sclerolaena diacantha	—	—	—	12.5	41–60

workers to suggest that Chenopod shrubs could serve as valuable protein supplements to sheep and cattle when the associated grasses were dry and possibly deficient in protein. Weight losses of pen-fed sheep have been reduced when *A. nummularia* and *A. vesicaria* were given as a supplement to a protein-deficient diet (Wilson, 1966*a*), but similar supplements have not been shown to improve weight gain or wool growth of animals grazing natural grasslands (Wilson 1966*c*). It may be concluded that protein levels of the species available within the pasture have been adequate or that digestion processes have been altered.

Some evidence from impaired digestion has come from three pen-feeding experiments. In the first of the pen-feeding trials conducted by Weston, Hogan & Hensley (1970), it was demonstrated that diets containing appreciable quantities of *A. nummularia* leaf failed to provide sufficient nutrients for the maintenance of energy and nitrogen equilibrium. This failure was shown to be related to altered digestion processes. The stomach plays a less important role in the digestion of fibre with *A. nummularia* than with grasses and legumes. In addition, ruminal absorption of fatty acids is impaired, and protein is extensively degraded to ammonia. Thus the protein value of an *A. nummalaria* diet (and possibly other Chenopod shrubs) may be much lower than is indicated by either its crude protein content or its digestible crude protein content. Two further pen-feeding trials were conducted in Egypt using *A. nummularia*. In the first trial (Hassan *et al.*, 1979), *A. nummularia*, *Chloris gayana*, and *Pennisetum clandestinum* were fed separately and compared. *A. nummularia* was higher in crude protein and minerals, but lower in soluble carbohydrates than *C. gayana* and *P. clandestinum*. It was concluded that the lack of readily available carbohydrates in *A. nummularia* is its main defect; better nutritive value might be achieved either by mixing *A. nummularia* with grasses (which are high in soluble carbohydrates), or by supplementing it with cereal grains. In addition, low levels of readily available carbohydrates and the rapid fermentation of crude protein in the rumen might be responsible for the poor utilisation of proteins as suggested by the large losses of nitrogen in urine (Weston *et al.*, 1970). In the third trial, which lasted seven weeks, *A. nummularia* and various amounts of barley (a supplementary energy source) were fed to rams (Hassan & Abdel-Aziz, 1979). Each ram was fed *A. nummularia ad libitum* and one of four levels of barley (0, 50, 100 or 150 g/d). During the first three weeks, rams fed *A. nummularia* and 0, 50, 100 and 150 g barley lost 4.1, 3.7, 3.5 and 2.8 kg live weight respectively, possibly due to changes occurring in the gut fill and losses in body substances. During the following four weeks, rams given *A. nummularia* as the sole diet continued to lose weight at an average of 80.4 g/d. Adding 100 g barley to the feed, retarded rates of weight loss ($P < 0.05$) but liveweights still decreased. The only positive weight gain (62.5 g/d) was for animals given the highest level of barley. These results contrast with those of Wilson (1966a) who found that sheep could attain slightly positive weight gain when pen-fed only *A. nummularia* or *A. vesicaria*. The barley supplementation had significant effects ($P < 0.01$) on nitrogen balance. Animals fed *A. nummularia* supplemented with either 0 or 50 g barley were in negative nitrogen balance, but those supplemented with 100 and 150 g barley, attained positive nitrogen balance. Despite similar digestible nitrogen intake by sheep fed 150 g and 100 g barley the efficiency of nitrogen utilisation (nitrogen retained as a percentage of that digested) was 30.4 per cent when 150 g barley was fed, but only 3.6 per cent for sheep given 100 g barley. These results demonstrate that the nutritive value of *A. nummularia* can be improved by barley supplementation; 150 g/d seems to be sufficient.

Field experiments designed to assess the pastoral value of *Atriplex nummularia*, *A. vesicaria* and *Maireana aphylla* have been conducted near Deniliquin in southwestern New South Wales (Leigh, Wilson & Williams, 1970). Comparisons were made of growth rates of wool and changes in body weight of flocks of sheep grazing established native perennial grass communities in which Chenopod shrubs were either dominant constituents or absent. It was found that the presence of the three Chenopod shrub species was of little or no value in reducing seasonal fluctuations in animal production and it was concluded that in terms of wool production the introduction of these shrubs into a perennial grassland was of doubtful economic value. The reason for the lack of a positive response would, in all cases, be because of a relatively low intake of the Chenopod shrubs themselves in relation to the total nutritional needs of the animals. In areas like the riverine plain of New South Wales, where elimination of Chenopods leads to the establishment of a stable subclimax grassland with a high component of perennial grasses (Love, 1981), the disclimax grasslands have been shown to be capable of sustaining higher levels of animal production than adjacent climax communities dominated by Chenopod shrubs (Wilson & Leigh, 1970).

In other areas of Australia where the bushes are not replaced by stable disclimax communities, the Chenopod shrubs are regarded as valuable to ensure an amount of feed for use during dry periods, as an aid to increased rainfall infiltration, for nutrient cycling and also for the possible reduction in wind erosion.

Reaction to fire

Wildfire is a rare event in Chenopod-dominated communities because the bushes themselves have low flammability because of their succulence and high salt content and also because there is usually little combustible grass growing in between the bushes. When fires do occur the response between- and within-species varies considerably.

One of the earliest records concerning the effects of fire on Chenopod shrublands is that by Murray (1931) when she reported on the 1922 wildfire in the Lake Torrens district of South Australia. She found large areas of what was originally good quality shrubland to be covered with dead bushes of myall, bluebush, mulga and saltbush, although some scattered plants of black bluebush (*Maireana pyramidata*) had survived the fire.

More recently, Lay (1976) found most species of bluebush including *Maireana pyramidata*, *M. sedifolia* and *M. astrotricha*, to be susceptible to wildfires in 1975 and 1976 although, in one case, he found 50 per cent had survived a low-intensity fire. He also recorded *A. vesicaria* as being extremely fire-sensitive, particularly when there was complete scorching of the canopy, even though no part of the plant was actually burnt. Graetz & Wilson (1984) and Mitchell (1986) have also observed total kills to *A. vesicaria* after fire. Mitchell (1976, 1986) and Graetz & Wilson (1984) report similarly that stands of *Maireana sedifolia* and *M. pyramidata* may be thinned out. Mitchell (1986) observed that the proportion of *M. sedifolia* killed depended on fire intensity and that severely burnt shrubs recovered at a slower rate than did those shrubs subjected to a less intensive fire. Lightly burnt shrubs with little bark damage developed new shoots from aerial buds. If, as in a high intensity fire, the aerial branches were killed, calluses formed on the hypocotyl region and these produced aerial shoots. This hypocotyl region is usually covered by 5 cm soil, thereby insulating it from the fire.

As most regeneration takes place from seed, and these regeneration events are often infrequent and can be many years apart (as in species of *Maireana*), it is considered prudent to either withdraw stock or graze only very lightly to allow young plants to establish (Hodgkinson & Griffin, 1982; Mitchell, 1986).

'Dieback'

From time to time, the distribution and density of *A. vesicaria* stands on the riverine plain and associated areas of southern Australia are severely reduced and depleted by a complex set of as yet not understood factors to which the descriptive term 'dieback' has been applied (Clift, Semple & Prior, 1987; Clift, Dalton & Prior, 1989). Symptoms are initially ill-thrift of bushes followed by leaf death and a rapid defoliation. Often only part of the plant is affected. Even before defoliation is complete the shrubs turn black. Possible causes include 'overgrazing', stem-boring insects, root and/or stem pathogens, drought (in the latter stages), lack of effective summer rain, abundance of winter rains, senescence of even-aged stands, competition from annual plants and defoliation by caterpillars (Clift *et al.*, 1987).

It has been shown (Clift *et al.*, 1989) that the regeneration, initially as vegetative regrowth and later from seedling establishment, was closely related to variations in soil types and the effects of pre- and post-dieback grazing.

Conclusions

Chenopod shrublands have significantly captured the curiosity, interest and imagination of a great number of research scientists over the past 60 years. In comparison to other Australian plant communities, they are relatively well understood and have received a disproportionate amount of study in relation to their land area or value. Proceedings of various symposia held in Australia (Jones, 1970; Graetz & Howes, 1979; Harrington, Wilson & Young, 1984), have summarised much of our knowledge. From those symposia, and from a digest of the many other published papers, it could be concluded that relatively little remains to be learnt of their physiology, nutrition, general ecology, pastoral value or reaction to fire and grazing.

As a broad generalisation, it can be assumed that pastoralists with significant areas of Chenopod shrublands know how to manage and utilise them, that adequate extension guidance is available from State personnel, and that the future of Australia's Chenopod shrublands very largely depends on economic pressures, rather than any lack of vital biological knowledge. It is imperative that properties remain large enough to be economically viable. Owners or lessees must be able to manage their Chenopod shrublands in a conservative way, free from economic pressures which might otherwise force them to overstock for financial reasons. Perhaps the most important lesson which has been learnt is that such lands once they have been depleted of vegetation and once erosion has occurred are often both costly and difficult to rehabilitate. No one wishes to see a return to the eroded state of these shrublands in the 1930s so eloquently described by Ratcliffe (1936).

References

Atlas of Australian Resources (1980). 3rd Series. Pastures. Canberra: Department of National Development.

AUSLIG (1990). Chenopod shrubland. In *Vegetation: Atlas of Australian Resources*, Third Series, Volume 6, p. 42. Canberra: Australian Surveying & Land Information Group, Department of Administrative Services.

Barker, S. (1979). Shrub population dynamics under grazing – within paddock studies. In *Chenopod Shrublands, Studies in the Australian Arid Zone IV*, eds R.D. Graetz & M.W. Howes, pp. 83–106. Perth: CSIRO, Australia, Division of Land Resources Management.

Barker, S. & Lange, R.T. (1969). Effects of moderate sheep grazing on plant populations of a black oak-bluebush association. *Australian Journal of Botany*, **17**, 527–37.

Barker, S. & Lange, R.T. (1970). Population ecology of *Atriplex* under sheep grazing. In *The Biology of Atriplex. Studies in the Australian Arid Zone I*, ed. R. Jones, pp. 105–20. Canberra: CSIRO, Australia, Division of Plant Industry.

Briggs, J.D. & Leigh, J.H. (1988). *Rare or Threatened Plants: 1988 Revised Edition*. Australian National Parks & Wildlife Service Special Publication No. 14.

Carrodus, B.B. & Specht, R.L. (1965). Factors affecting the relative distribution of *Atriplex vesicaria* and *Kochia sedifolia* (Chenopodiaceae) in the arid zone of South Australia. *Australian Journal of Botany*, **14**, 419–33.

Charley, J.L. & Cowling, S.W. (1968). Changes in soil nutrient status resulting from overgrazing and their consequence in plant communities of semi-arid areas. *Proceedings of the Ecological Society of Australia*, **3**, 28–38.

Clift, D.K., Dalton, K.L. & Prior, J.C. (1989). Bladder saltbush (*Atriplex vesicaria* Howard ex Benth) regeneration on the Riverine Plain of south-eastern Australia since 1983. *Australian Rangelands Journal*, **11**, 31–9.

Clift, D.K., Semple, W.S. & Prior, J.C. (1987). A survey of bladder saltbush (*Atriplex vesicaria* Howard ex Benth) dieback on the Riverine Plain of southeastern Australia from the late 1970s to 1983. *Australian Rangelands Journal*, **9**, 39–48.

Eldridge, D.J., Westoby, M. and Stanley, R.J. (1990). Population dynamics of the perennial rangeland shrubs *Atriplex vesicaria, Maireana astrotricha* and *M. pyramidata* under grazing, 1980–87. *Journal of Applied Ecology*, **27**, 502–12.

Gillison, A.N. & Walker, J. (1981). Woodlands. In *Australian Vegetation*, ed. R.H. Groves, pp. 177–97. Cambridge University Press.

Graetz, R.D. & Howes, K.M.W. (eds) (1979). *Chenopod Shrublands. Studies in the Australian Arid Zone IV*. Perth: CSIRO, Australia, Division of Land Resources Management.

Graetz, R.D. & Ludwig, J.A. (1978). A method for the analysis of piosphere data applicable to range assessment. *Australian Rangelands Journal*, **1**, 126–36.

Graetz, R.D. and Wilson, A.D. (1984). Saltbush and bluebush. In *Management of Australia's Rangelands*, eds G.M. Harrington, A.D. Wilson & M.D. Young, pp. 209–22. Melbourne: CSIRO.

Hall, E.A.A., Specht, R.L. & Eardley, C.M. (1964). Regeneration of the vegetation on Koonamore Vegetation Reserve, 1926–1962. *Australian Journal of Botany*, **12**, 205–64.

Harrington, G.N., Wilson, A.D. & Young, M.D. (1984). *Management of Australia's Rangelands*. Melbourne: CSIRO.

Hassan, N.I. & Abdel-Aziz, H.M. (1979). Effect of barley supplementation on the nutritive value of saltbush (*Atriplex nummularia*). *World Review of Animal Production*, **15** (4), 47–55.

Hassan, N.I., Abd-Elaziz, H.M. & El-Tabbakh, A.E. (1979). Evaluation of some forages introduced to newly reclaimed areas in Egypt. *World Review of Animal Production*, **15**(2), 31–5.

Hnatiuk, R.J. (1990). *Census of Australian Vascular Plants*. Australian Flora and Fauna Series Number 11. Canberra: Australian Government Publishing Service.

Hodgkinson, K.C. & Griffin, G.P. (1982). Adaptation of shrub species to fires in the arid zone. In *Evolution of the Flora and Fauna of Arid Australia*, eds W.R. Barker & P.J.M. Greenslade, pp. 145–52. Adelaide: Peacock.

Johnson, R.W. & Burrows, W.H. (1981). *Acacia* open-forests, woodlands and shrublands. In *Australian Vegetation*, ed. R.H. Groves, pp. 198–226. Cambridge University Press.

Jones R. (ed.). (1970). *The Biology of Atriplex. Studies in the Australian Arid Zone I*. Canberra: CSIRO, Australia, Division of Plant Industry.

Lange, R.T. (1969). The piosphere: sheep track and dung patterns. *Journal of Range Management*, **22**, 396–400.

Lay, B.G. (1976). Fire in the pastoral country. *Journal of Agriculture of South Australia*, **79**, 9–14.

Lay, B.G. (1979). Shrub population dynamics under grazing – longterm study. In *Chenopod Shrublands. Studies of the Australian Arid Zone IV*, eds R.D. Graetz & K.M.W. Howes, pp. 107–24. Perth: CSIRO, Australia, Division of Land Resources Management.

Leigh, J.H. (1972). Saltbush and other browse shrubs. In *The Use of Trees and Shrubs in the Dry Country of Australia*, ed. N. Hall, pp. 284–98. Canberra: Australian Government Publishing Service.

Leigh, J.H. (1986). Forage value and utilization of Chenopod dominated shrubland. *Reclamation and Revegetation Research*, **5**, 387–402.

Leigh, J.H. & Mulham, W.E. (1971). The effect of defoliation on the persistence of *Atriplex vesicaria*. *Australian Journal of Agricultural Research*, **22**, 239–44.

Leigh, J.H., Wilson, A.D. & Mulham, W.E. (1968). A study of merino sheep grazing a cotton-bush (*Kochia aphylla*) – grassland (*Stipa variabilis Danthonia caespitosa*) community of the Riverine Plain. *Australian Journal of Agricultural Research*, **19**, 947–61.

Leigh, J.H., Wilson, A.D. & Mulham, W.E. (1979). A study of sheep grazing a Belah (*Casuarina cristata*) – Rosewood (*Heterodendrum oleifolium*) shrub woodland in western New South Wales. *Australian Journal of Agricultural Research*, **30**, 1223–36.

Leigh, J.H., Wilson, A.D. & Williams, O.B. (1970). An assessment of the value of three perennial chenopodiaceous shrubs for wool production of sheep grazing semi-arid pastures. *Proceedings of the XIth International Grassland Congress*, pp. 55–9.

Love, L.D. (1981). Mangrove swamps and salt marshes. In *Australian Vegetation*, ed. R.H. Groves, pp. 319–34. Cambridge University Press.

Mitchell, A.A. (1976). *Regeneration of shrubs after fire in the Goldfields district of Western Australia*. Australian Arid Zone Research Conference, Kalgoorlie Working Papers 3 c, pp. 21–3.

Mitchell, A.A. (1986). The effects of fire on a bluebush (*Maireana sedifolia*) shrubland in the Western Australian Goldfields. In *Rangelands: A Resource Under Siege*, Proceedings of the Second International Rangelands Congress, Adelaide, 1984, pp. 606–7. Canberra: Australian Academy of Science.

Moore, R.M. (1969). Grazing lands of Australia. In *Australian Grasslands*, ed. R.M. Moore, p. 86. Canberra: Australian National University Press.

Muir, S.J. (1990). The nutritive value of selected forage species in western New South Wales. Final report to Australian Wool Corporation. NSW Agriculture and Fisheries, Cobar, Agdex 130/81, pp. 1–118.

Murray, B.J. (1931). A study of the vegetation of the Lake Torrens plateau, South Australia. *Transactions of the Royal Society of South Australia*, **55**, 91–112.

Newman, J.C. & Condon, R.W. (1969). Land use and present condition. In *Arid Lands of Australia*, eds R.O. Slatyer & R.A. Perry, pp. 105–32. Canberra: Australian National University Press.

Osborn, T.G.B., Wood, J.G. & Paltridge, T.B. (1932). On the growth and reaction to grazing of the perennial saltbush, *Atriplex vesicaria*: an ecological study of the biotic factor. *Proceedings of the Linnean Society of New South Wales*, **57**, 377–402.

Parsons, R.F. (1981). Eucalyptus scrubs and shrublands. In *Australian Vegetation*, ed. R.H. Groves, pp. 227–52. Cambridge University Press.

Ratcliffe, F.N. (1936). *Soil Drift in and Pastoral Areas of South Australia*. Council for Scientific and Industrial Research, Australia, Pamphlet No. 64.

Specht, R.L. (1970). Vegetation. In *The Australian Environment*, 4th edn (rev.), ed. G.W. Leeper, pp. 44–67. Melbourne: CSIRO & Melbourne University Press.

Specht, R.L. (1981). Structural attributes – foliage projective cover and standing biomass. In *Vegetation Classification in the Australian Region*, ed. A.N. Gillson & D.J. Anderson, pp. 10–21. Canberra: CSIRO & Australian National Univeristy Press.

Squires, V.R. & Wilson, A.D. (1971). Distance between food and water supply and its effect on drinking frequency, and food and water intake of Merino and Border Leicester sheep. *Australian Journal of Agricultural Research*, **22**, 283–90.

Trumble, H.C. & Woodroffe, R. (1954). The influence of climatic factors in the reaction of desert shrubs to grazing by sheep. In *Biology of Deserts*, ed. J.L. Cloudsley-Thompson, pp. 129–47. London: Institute of Biology.

Valentine I. & Nagorcka, B.N. (1979). Contour patterning in *Atriplex vesicaria* communities. In *Studies in the Australian Arid Zone IV*, eds R.D. Graetz & R.M.W. Howes, pp. 61–74. Perth: CSIRO, Australia, Division of Land Resources Management.

Weston, R.H., Hogan, J.P. & Hensley, J.A. (1970). Some aspects of the digestion of *Atriplex nummularia* (Saltbush) by sheep. *Proceedings of the Australian Society of Animal Production*, **8**, 517–21.

Williams, D.G., Anderson, D.J. & Slater, K.R. (1978). The influence of sheep on pattern and process in *Atriplex vesicaria* populations from the Riverine Plain of New South Wales. *Australian Journal of Botany*, **26**, 381–92.

Williams, R.J. (1955). Vegetation regions of Australia. In *Atlas of Australian Resources*, 2nd series. Canberra: Department of National Development.

Wilson, A.D. (1966a). The value of *Atriplex* (saltbush) and *Kochia* (bluebush) species as food for sheep. *Australian Journal of Agricultural Research*, **17**, 147–53.

Wilson, A.D. (1966b). The intake and excretion of sodium by sheep fed on species of *Atriplex* (saltbush) and *Kochia* (bluebush). *Australian Journal of Agricultural Research*, **17**, 155–63.

Wilson, A.D. (1966c). Saltbush and irrigated pasture as supplements to a native pasture. *CSIRO, Australia, Division of Plant Industry, Field Station Record*, **5**, 71–6.

Wilson, A.D. (1967). Observations on the adaption of sheep to saline drinking water. *Australian Journal of Experimental Agriculture and Animal Husbandry*, **7**, 321–4.

Wilson, A.D. (1976). Comparison of sheep and cattle grazing on a semi-arid grassland. *Australian Journal of Agricultural Research*, **27**, 155–62.

Wilson, A.D., & Leigh, J.H. (1970). Comparisons of the productivity of sheep grazing native pastures of the Riverine Plain. *Australian Journal of Experimental Agriculture and Animal Husbandry*, **10**, 549–54.

Wilson, A.D., Leigh, J.H. & Mulham, W.E. (1969). A study of Merino sheep grazing a saltbush (*Atriplex vesicaria*) – cotton-bush (*Kochia aphylla*) community on the Riverine Plain. *Australian Journal of Agricultural Research*, **20**, 1123–36.

Wilson, A.D., Leigh, J.H. & Mulham, W.E. (1975). Comparison of the diets of goats and sheep on a *Casuarina cristata* – *Heterodendrum oleifolium* woodland community in western New South Wales. *Australian Journal of Experimental Agriculture and Animal Husbandry*, **15**, 45–53.

Wood, J.G. & Williams, R.J. (1960). Vegetation. In *The Australian Environment*, 3rd edn (rev.), pp. 67–84. Melbourne: CSIRO & Melbourne University Press.

13

Natural and derived grasslands

J.J. MOTT & R.H. GROVES

MOORE (1970) defined a grassland as any plant community, whether natural or modified by humans, in which grasses provide a substantial proportion of the feed for domestic stock. Based on this definition, such lands cover a large proportion of Australia. For the purposes of this chapter, and in keeping with Groves & Williams (1981), we shall restrict its coverage to those treeless communities dominated by indigenous perennial and annual grasses, but which nearly always include some introduced and alien species. Other definitions used in this chapter will be the same as those used in Groves & Williams (1981).

Four basic types of grassland are described below.

Arid tussock grassland
Areas of Mitchell grasses (*Astrebla* spp.) extensively distributed through inland Queensland, the Northern Territory and northern Western Australia in a zone receiving between 200 and 500 mm average annual rainfall and characterised by predominantly summer rain.

Arid hummock grasslands (Fig. 13.1)
Dominated by the genera *Triodia* and *Plectrachne* in areas with less than 200 mm average annual summer and/or winter rainfall. They have an extensive distribution throughout arid Australia and cover about one-third of the Australian land mass.

Coastal grasslands
Dominated by species of *Sporobolus* and *Xerochloa* and confined to the tropical summer-rainfall region.

Subhumid grasslands
These are subdivided further.

TROPICAL   Grasslands dominated by *Dichanthium* and *Eulalia*, and sometimes by *Bothriochloa* and *Heteropogon*, in eastern and northern Queensland with predominantly summer rain.

369

Fig. 13.1. Hummock grassland of *Triodia* in central Australia. (Photo: C.J. Totterdell.)

TEMPERATE   Grasslands with an irregular distribution from north of Adelaide around the zone of 500 to 1000 mm average annual winter and/or summer rainfall of southeastern Australia to northern New South Wales. Dominant genera are *Themeda*, *Poa* and *Stipa*.

SUBALPINE   Confined to the cold and wet mountain regions of the Monaro region of southern New South Wales, northeastern Victoria and the Central Plateau of Tasmania. Dominant genera are *Poa* and *Danthonia*.

In addition to these areas of natural grassland, there are vast areas of derived grasslands in the grassy understorey to open *Eucalyptus* woodlands (see Chapter 8) or savannas in both southern and northern Australia (Moore, 1970; Mott *et al.*, 1985). Much of the initial tree cover has been removed from the southern regions and, in many cases, the grasslands are not always natural, although they may now consist of native perennial grasses, especially species of *Danthonia*, *Poa* and *Themeda*. In lower rainfall areas of southern Australia, however, these understorey species mainly consist of annual grasses and herbs introduced from Mediterranean Europe. The dominant grasses in the understorey of the northern communities are *Themeda* and

Sorghum near Katherine, Northern Territory, and *Heteropogon* in Queensland (Mott *et al.*, 1985).

Groves & Williams (1981) pointed to the difficulty of applying classical successional concepts and philosophies to the dynamics of Australian grassland communities. They advanced the concept of grasslands existing in several 'aspects' which could consist of nearly pure swards of one of several short-lived perennials on a seemingly random basis. Westoby (1980) also reported variation in composition of grass swards over time and emphasised the importance of the interaction between different growth forms and the environment that may cause changes in the composition of vegetation at any time.

The above concepts have been expressed by a number of other authors and represent the basis of the 'non-equilibrium' hypothesis for ecosystem functioning (e.g. Noy-Meir, 1975; DeAngelis & Waterhouse, 1987; Westoby, Walker & Noy-Meir, 1989). This hypothesis maintains that many natural ecosystems are not in equilibrium and controlled by a series of biological feedbacks (e.g. grazing, nutrient limitations), but rather are the resultant of various historical and chance factors; thus the systems may exist in multiple states at any time.

The importance of abiotic factors in controlling the dynamics of arid, semi-arid and savanna grasslands in Australia has been emphasised by other authors (Noble, 1986; Walker, 1988). In addition, the greater availability of models for grassland plant production has emphasised the critical role that episodic drought can play as a major factor affecting the overgrazing of grasslands (Pressland & McKeon, 1989).

The importance of the non-equilibrium model for grasslands has been argued cogently in Australia by Westoby *et al.* (1989) and is also being employed for essentially similar grasslands in southern Africa (Ellis & Swift, 1988; Mentis *et al.*, 1989). In an attempt to develop a useful model for managers, Westoby *et al.* proposed the state-and-transition model for grassland ecosystems (Fig. 13.2). In this model, defined aspects or 'states' are separated by discrete 'transitions' between those states. Many of the hypothesised transitions can only occur given an appropriate climatic sequence together with a specific management option (e.g. grazing, fire). Such a model can accommodate the impacts of climate and management on life history noted in Groves & Williams (1981), together with the catastrophic climatic events (droughts, floods, etc.) which characterise much of Australia's extensive grasslands.

In parallel with a growing awareness of the importance of episodic

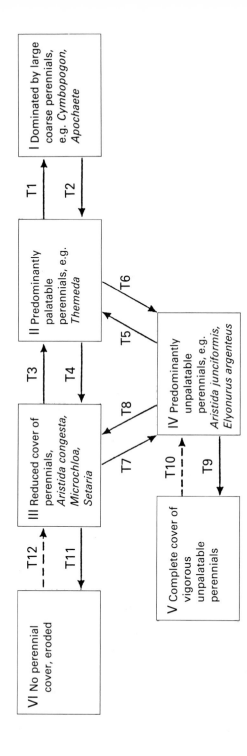

Fig. 13.2. The state-and-transition model for grassland ecosystems as developed by Westoby et al. (1989).

events in determining both the composition and survival of grassland ecosystems, the extent of degradation within Australia's grassland systems has been emphasised in a number of reviews (see, e.g. Woods, 1983; Burch, Graetz & Noble, 1987; Williams, 1991; Mott & Tothill, 1994). This degradation affects both the soil and vegetation base of the ecosystems. In some cases, up to 90 per cent of certain regions have suffered degradation from overgrazing (De Corte *et al.*, 1991).

The increasing demand for sustainable utilisation of Australia's natural resources (Australian Government Publishing Service, 1989) has also meant that both the spatial and temporal relationships of grass plants as individuals, and of grasslands as landscape elements, must be taken into account (Morrisey & O'Connor, 1990; Scanlan *et al.*, 1990; Williams, 1991).

There have been several major thrusts in grassland science in Australia over the last ten or so years. Study of the biology of individual grassland species has continued, not just as a basis for exploitative grazing systems but also in relation to conservation biology of the community, and especially of rare or endangered forbs in grasslands. Another thrust has been to start to quantify the impact and importance of spatial heterogeneity and climatic variability on the productivity and long-term sustainability of grassland systems. Additional to this latter thrust has been the acknowledgment of an increasing need to develop information structures to support sustainable management practices in relation to the grazing of domesticated animals. In this chapter, we shall attempt to review and draw together recent research results that have been forthcoming to support these three major thrusts in grassland science.

Conservation biology of grasslands

Natural grasslands in Australia consist generally of various proportions of four main floristic elements:

1. a group of C4 grasses of which the major species are *Themeda triandra* (syn. *T. australis*) and *Bothriochloa ambigua*, but grasses such as *Chloris gayana* and *Sorghum leiocladum* may also be present.
2. a group of C3 grasses of which the major genera are *Danthonia*, *Poa* and *Stipa*.
3. Many perennial indigenous forbs from a wide range of plant

families but especially species of the genera *Helichrysum*, *Helipterum*, *Gnaphalium* and *Leptorhynchos* (Asteraceae), *Wahlenbergia* (Campanulaceae), *Convolvulus* (Convolvulaceae), *Bulbine* (Liliaceae) and *Stylidium* (Stylidiaceae).

4. a diverse group of introduced grasses and forbs, many of which are annual species of European origin. The perennial C4 grass *Eragrostis curvula* from southern Africa is especially invasive in *Themeda* grasslands in southern Australia currently.

There is a considerable amount of information on the biology of several of the dominant C4 and C3 grasses in natural and derived grasslands in southern Australia, e.g. for *Themeda triandra* (Groves, 1975; Hagon, 1976; Stuwe & Parsons, 1977; Groves, Hagon & Ramakrishnan, 1982; McDougall, 1989), *Danthonia caespitosa* (Williams, 1961, 1968; Hodgkinson & Quinn, 1976) and *Poa labillardieri* (Groves, Keraitis & Moore, 1973). In summary, this information shows that growth and development of the different grasses depends mainly on temperature and photoperiod – the C3 element grows mainly in spring and autumn and the C4 element mainly in late spring–early summer, as soil moisture levels permit.

Less is known, however, of the population biology of the native forbs which grow in between the grass tussocks and add to grassland diversity. A start has been made recently on studying the germination and seedling establishment of certain selected species (see, e.g Hitchmough, Berkeley & Cross, 1989; McIntyre, 1990; Willis & Groves, 1991). Knowledge of the biology of certain rare and endangered forbs is accumulating for species such as *Rutidosus leptorrhynchoides* and *Swainsona recta* (Scarlett & Parsons, 1982). Most of the species in this group flower in spring, but some (e.g. *Wahlenbergia*) may also flower in autumn. The regeneration biology of such species is still poorly known at present, however, especially as it relates to field performance. No generalisations are yet tenable.

Plant production

In a geographic evaluation of the factors controlling grassland distribution, Fitzpatrick & Nix (1970) formulated a grass growth index by multiplying the separate effects of temperature, solar radiation and soil moisture. Simple water balance models allow simulation of seasonal plant growth patterns from relatively simple climatic inputs. Together with an evaluation of species composition, Mott *et al.*

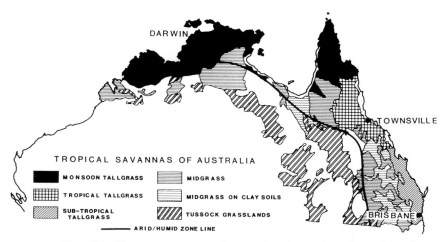

Fig. 13.3. The savanna types of vegetation in northern Australia (after Mott *et al.*, 1985).

(1985) used these models as the basis for a general functional separation of the semi-arid and subhumid northern grasslands into a number of separate savanna types (Fig. 13.3).

The generalised plant responses incorporated into these models are somewhat simplistic since they do not accommodate the fact that different tropical grass species may have a wide range of growth responses to temperature, with species in the same habitat having different responses (McKeon *et al.*, 1990). Broad-scale analyses necessarily use long-term average temperature and evaporation rates, thereby causing errors in estimates of grass production. In addition to climatic variation affecting the prediction of plant production, soil type can have a major effect on the distribution and productivity of the resultant system. A change in soil texture from fine to coarse over a short distance with rainfall and temperature constant is associated with a significant change in vegetation structure and production from low scrubland to woodland, for instance.

To address some of the above problems, and to include factors such as sward structure, senescence and variation in species composition and phenological development, plant densities and other variables, a compromise between information derived empirically and physiologically has been incorporated into a grass production model called GRASP (McKeon *et al.*, 1982). Driven by water use efficiency, the model has been validated against sites in most of the major vegetation communities in northern Australia. It provides a powerful tool for the long-term monitoring of plant production over different grassland communities and soil types (McKeon *et al.*, 1990). In addition to production, grassland composition can also be related to broad climatic and soil conditions (see, e.g. Table 13.1).

Whilst these and other models are proving useful to determine impacts of such factors as climate change on grasslands (Graetz *et al.*, 1988; McKeon *et al.*, 1988), we still do not have a general relationship for the effect of the supply of available nutrients on grassland production. Australian grassland soils are of very low fertility, particularly of nitrogen and phosphorus, by comparison with those of grasslands elsewhere. Small variation in the availability of nutrients between soil types will be manifested as large changes in vegetation composition (Beard, 1976). This sensitivity was emphasised in an experiment where both fertiliser and nutrients were applied to existing vegetation (Table 13.2). Fertiliser and irrigation were applied both separately and together to a subtropical tall grassland for two years. Swards were burnt annually in the spring; yield was measured throughout the growing season (Mott *et al.*, unpublished observations). Even with severe water limitation (transpiration ratio > 0.2), available nutrients limited growth on this soil of low fertility with a sandy surface. When production was expressed in terms of radiation use efficiency, nutrient limitation resulted in 80 per cent of potential grass production at low levels of water availability and 60 per cent of potential if water were not limiting. Empirical measurements are continuing and it is hoped to incorporate a module for nutrient availability within the overall GRASP framework.

Results from a number of independent and empirical measurements of plant variability have shown that major degradation of pastures occurs when more than 30 per cent of the annual production is removed by grazing animals (Orr, 1988). In drought years, the much lower plant production can be modelled using rainfall figures and validated models such as GRASP. General relationships exist that relate utilisation to animal numbers (Minson & McDonald, 1987). Based on long-term rainfall figures, Pressland & McKeon (1989) were able to show for a number of areas, using the simple criterion of percentage utilisation of annual production, that the slow increase in cattle numbers carried in drought years were above those judged to be 'safe' (Fig. 13.4). Under these circumstances, major degradation occurred in the upper Burdekin catchment during the drought period of the late 1980s.

Table 13.1. *Distribution of grass genera and Tribes within savanna systems of northern tropical Australia along gradients of soil water, annual rainfall and soil texture (after Johnson & Tothill, 1985)*

SOIL–WATER RELATIONS: DRY → WET | SOIL TEXTURE: SAND → CLAY

ANNUAL RAINFALL (mm)	SAND →			→ CLAY
1500	ANDROPOGONEAE (Sorghum (annual), Heteropogon); ERAGROSTIDEAE (Plectrachne)	ANDROPOGONEAE (Sorghum (perennial), Themeda)	ANDROPOGONEAE (Themeda)	ORYZEAE (Oryza, Leersia); CYPERACEAE (Eleocharis); PANICEAE (Panicum, Hymenachne)
1000	ANDROPOGONEAE (Sorghum (annual), Heteropogon); ERAGROSTIDEAE (Plectrachne)	ANDROPOGONEAE (Sorghum (perennial), Themeda, Sehima)	DANTHONIEAE (Eriachne) — ANDROPOGONEAE (Sorghum (perennial), Themeda, Sehima, Bothriochloa)	ANDROPOGONEAE (Dichanthium, Eulalia)
750	ARISTIDEAE (Aristida); ANDROPOGONEAE (Cymbopogon); ERAGROSTIDEAE (Plectrachne)	ANDROPOGONEAE (Themeda, Heteropogon, Sehima); ARISTIDEAE (Aristida)	ANDROPOGONEAE (Themeda, Heteropogon, Sehima) — ANDROPOGONEAE (Themeda, Heteropogon, Sehima, Bothriochloa); ARISTIDEAE (Aristida)	ANDROPOGONEAE (Dichanthium, Eulalia)
625	ERAGROSTIDEAE (Triodia, Plectrachne); ARISTIDEAE (Aristida)	ARISTIDEAE (Aristida); ANDROPOGONEAE (Sehima, Cymbopogon)	ARISTIDEAE (Aristida) — ARISTIDEAE (Aristida); PAPPOPHOREAE (Enneapogon)	ANDROPOGONEAE (Dichanthium, Eulalia)
500	ERAGROSTIDEAE (Triodia)	ARISTIDEAE (Aristida); ERAGROSTIDEAE (Eragrostis, Triodia)		CHLORIDEAE (Astrebla)
375				

Table 13.2. *Growth and radiation use efficiency (RUE) of subtropical tallgrass savanna community to fertiliser and irrigation over 2 years (J.J. Mott, unpublished data)*

	No fertiliser		Added fertiliser	
	Growth	RUE	Growth	RUE
No irrigation				
Year 1	1376	1.66	2453	2.00
Year 2	597	1.62	1082	1.76
Irrigation				
Year 1	2916	2.62	9049	4.44
Year 2	2640	2.40	8760	3.91

Growth is maximum dry matter production (kg/ha) by April (1986 and 1987). RUE is radiation use efficiency (kg/ha/MJ m^2) where radiation is total solar radiation and percentage interception = $100 -$ per cent transmission. Fertiliser is N (600 kg/ha/season). Irrigation is >75 per cent pan evaporation.

Climatic variability

The simple growth models used by Fitzpatrick & Nix (1970) and Mott *et al.* (1985) showed that, for similar levels of rainfall, patterns of growth were different for northern and subtropical savanna areas. The high temperatures that exist following the first rains of the wet season in the northern savanna lands, led to a rapid growth in the early wet season compared to the slower growth, limited by temperature, that occurs for grasses in the subtropical zone. Mott (1986) postulated that this variation in growth pattern could differentiate the grazing sensitivity evident between northern and subtropical savanna areas. From the results of a detailed experiment that tested this postulate, Mott *et al.* (1992) simulated the difference between these two environments and found that defoliated plants of *Themeda triandra* died rapidly under the conditions representing the seasonal patterns that exist in the northern grasslands. This interrelationship between sward structure and defoliation was similar to that observed in grazed pastures (Norman, 1965; Mott, 1987); it explains the historical disappearance of *Themeda triandra* from the subtropical savannas following winter drought and rapid early growth in the wet season

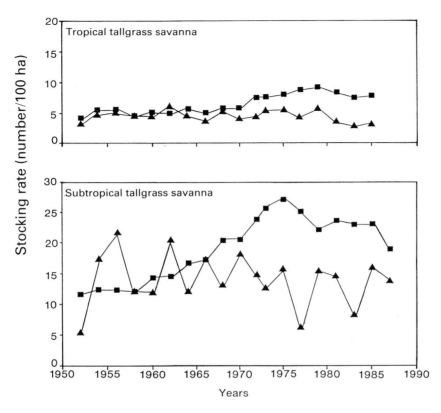

Fig. 13.4. Actual stocking rates (■) of beef cattle compared with the 'safe' stocking rates calculated from the GRASP model for pasture production (▲) (after Pressland & McKeon, 1989).

which plant production models predict for subtropical grasslands (McKeon *et al.*, 1990).

In the more arid tussock grasslands, episodic winter rainfall produces growth from the group of C3 forbs (Orr, 1986) that inhabit the intertussock spaces (see earlier). If substantial winter rains occur, then after the following summer rains, subsequent growth of forbs and C4 grasses can support high liveweight gains in grazing animals (McKeon, Rickert & Scattini, 1986; Orr, 1988).

Whilst growth patterns within seasons can affect grass growth, the major impact of climatic variability on plant productivity is caused by the large difference in the amount of rainfall between seasons. This high variability in rainfall between seasons in the northern savanna lands has long been realised (Dick, 1958), together with the large variation occurring in patterns over longer time scales (decades and generations). Analyses of climatic records for this region have shown that for many northern systems the annual variability of rainfall is greater than for other similar continental regions (Mott *et al.*, 1985). McKeon *et al.* (1990) and others have shown the close relationship that exists between changes in the southern oscillation index (SOI) and rainfall patterns; they comment that, within northern Australia, such variability is likely to be a result of the stronger control of meteorological mechanisms associated with the phases of the SOI.

Inspection of rainfall records for many of Australia's grassland areas shows that drought periods are common. Where drought incidence is defined as less than 60 per cent of mean summer rainfall (Daly & Dudgeon, 1987), then drought occurs in more than 25 per cent of years over much of the northern grasslands. The impact of drought depends strongly, however, on grazing pressure on the grasslands. Results from a number of independent and empirical measurements of plant viability have shown that most degradation in pastures occurs when more than 30 per cent of the production is removed by grazing animals (Orr, 1988). Pressland & McKeon (1989) used this simple criterion to show for a number of areas that the slow increase of stocking rate for beef cattle during good years did not decrease under a situation where 'safe' stocking rates had dropped substantially (Fig. 13.4). Under these circumstances, major degradation occurred during the drought period of the late 1980s and a survey of the region conducted at the same time showed that in some areas over 90 per cent of the landscapes were seriously degraded (De Corte *et al.*, 1991).

Other biological parameters, in addition to grass production, are affected by episodic climatic events. Taylor & Tulloch (1985) examined a range of biological events and found that management recommendations for grazing were impossible to formulate from short-term experience, thereby necessitating methodologies which allowed for extrapolation of both research results and field experience. Walker (1988) also discussed the importance of relating grazing results obtained in the short term to longer-term rainfall patterns.

Spatial heterogeneity

Much of the early research on northern grasslands was based on the concept developed by the CSIRO Division of Land Research that landscapes are organised hierarchically, such that the small and relatively uniform land components (land units) are arranged in complexes or land systems (Christian & Stewart, 1953). These can sometimes, but not always, be related to a watershed- or catenary-based evaluation of landscapes. Other agencies adopted this approach (e.g. Aldrick & Robinson, 1970) and over large areas there were descriptive evaluations of spatial interrelationships within a landscape. This work formed the basis for considerable ecological work on the northern grasslands (Norman, 1966).

After this approach there was a long period when the focus was on the individual plant or small plot as the experimental unit in the wetter subhumid savannas (Norman & Begg, 1973; Eyles, Cameron & Hacker, 1985). Only in the more arid areas, which in Australia were defined as 'rangelands', did a continuation of research effort persist on the ecological implications of spatial relationships of the landscape. This divergence in philosophy occurred at the same time as much of the research on Australia's northern grasslands was also being split into that of a 'replacement philosophy' that aimed at maximising short-term animal production by the use of introduced species (e.g. Eyles *et al.*, 1985) and more long-term rangeland management (Harrington, Wilson & Young, 1984). A number of early reviews pointed to the potential problems for sustainability with a philosophy oriented towards short-term production in the wetter savanna areas (Mott, Bridge & Arndt, 1979; Mott & Tothill, 1984; Mott, 1986). The two opposing philosophies dominated research on grasslands in the two regions until a series of drought years in the 1980s revealed the extent of overgrazing which had been engendered by the focus on short-term animal production (Pressland & McKeon, 1989; De Corte *et al.*, 1991; Williams & Chartres, 1991).

The need to consider the spatial relationships of landscapes in sustainable land use management has been emphasised recently for savanna lands (Williams, 1991; Williams & Chartres, 1991; Mott & Tothill, 1993). These evaluations parallel the concepts expressed by those scientists studying arid zone grasslands; for instance, Stafford Smith & Pickup (1990) have claimed that 'the development of techniques which come to grips with spatial heterogeneity and the interdependence of points in the landscape is one of the several great advances which Australian ecology can offer the world'. To evaluate the importance of these aspects to grassland ecology we shall next consider both land systems and land units within the landscape.

Within land systems

At this level of landscape there can be major differences between grass growth on different systems. Productivity can be affected by surface landscape processes such as run-off and erosion. But it is also important to consider subsurface processes such as deep drainage and movements in the water table. In the case of savanna systems, tree removal can be accompanied by catastrophic degradation at the landscape level, with a rising water table thereby increasing the chances for salinity at the surface (Gillard, Williams & Monypenny, 1989).

An increase in the effectiveness of rainfall by redistribution to run-on areas has been noted for many arid grassland areas (Noy-Meir, 1982). In the arid Murchison region of Western Australia, water re-distribution after rainfall can produce dramatic variation in the productivity of annual plants between the broad run-off plains and the run-on drainage areas (Mott & McComb, 1974; Mott, 1978b). For the grassy understorey to mulga lands (see Chapter 9), Mott & McComb found that, with little long-term seed store in the region, recruitment of new plants would occur on run-on areas from wind-blown seed (Mott, 1978b). In eastern Australia, an overstorey of tree and/or shrub cover can also be related to distinct bands of perennial grasses within the landscape.

Superimposed on this basic grassland pattern in the landscape are the impacts caused by grazing animals. Detailed studies in the arid zone of central Australia have shown variable grazing patterns and, in some cases, more than half the grazing by domestic stock occurred on less than a quarter of the area of the paddock (Low, Dudzinski & Muller, 1981). Specific grazing patterns are affected by distance from water and by prevailing wind and fenceline patterns, as well as by vegetation type (Stafford Smith, 1988; Stafford Smith & Pickup, 1990). Such patterns interact with preferred vegetation types, such as riparian areas, to result in the final degradation pattern occurring within the paddock. In the more humid areas, the siting of watering points may not have the over-riding effect that occurs in arid areas. Soil type can play an important part in grazing preference and in the Katherine area the avoidance of the moister vertisol areas during the wet season means that cattle congregate on the red-earth soils with better drainage. Because of the sensitivity of these grasslands to wet-season grazing (see Mott *et al.*, 1985), severe over-grazing of the red-earth section of a paddock can take place even when the mean stocking rate for the whole paddock is well within safe limits (McCosker *et al.*, 1988; Mott, unpublished data).

Within land units

The low fertility status of soils of many arid and savanna grasslands (see, e.g. Mott *et al.*, 1985) means that small differences in fertility can cause major differences in floristic composition. The concept of 'fertile islands' was put forward by Garcia-Moya & McKell (1970) and has been reported subsequently for many arid lands (Noy-Meir, 1975). In a detailed study of small patches of the understorey to mulga (*Acacia aneura*, see Chapter 9), Mott & McComb (1974) found that these patches usually formed around shrubs and the soil could have much higher fertility and moisture storage than any of the surrounding areas. Friedel, Foran & Stafford Smith (1990) commented on the importance of fertile patches as being a major site for grassland production in these arid systems. Wilson, Tongway & Tupper (1988) recorded, although not in a grassland but in a semi-arid *Eucalyptus populnea* community (see Chapter 8), as much as a ten-fold difference in yield of the grassy understorey at the same site within a land unit and related these changes to variation in overstorey cover.

As well as the influence of other vegetation types on the productivity and composition of the grassy understorey within a land unit, invertebrates can also be important in determining small-scale spatial patterning. Termites can play a major role in redistributing nutrients and thereby increasing herb productivity. Results of recent work in the arid tropics suggest that microbial activity and levels of nutrients and organic matter are all greater within termite mounds than outside them (Beckman, 1988). If the termite colony survives 20 to 40 years and then weathers away over a similar time period, an island occupied by fast-growing, nutrient-demanding species is created on a nutrient-enriched patch (Spain & McIvor, 1988).

Grazing animals can also prefer small areas of selected species within a land unit (Stafford-Smith & Pickup, 1990). In many cases, selectivity can be related to a more nutrient-rich species, but in the rank tall grasslands of the monsoon region, Mott (1987) found that the low quality of the rapidly growing ungrazed grasses meant that grazed patches form at random and once formed, cattle continued to graze preferentially these patches of regrowth that were both younger and of better quality compared to the more rank forage available elsewhere in the paddock. Because these grass species, such as *Sorghum plumosum* or *Themeda triandra*, are sensitive to continuous grazing, such a grazing pattern could lead to death of individual plants in patches after two seasons of summer growth.

Grassland monitoring

If an objective evaluation of the impact of pastoralism on grasslands is to be obtained, then some form of monitoring is needed to separate landscape changes arising from natural processes (e.g. species change because of episodic events) from those solely arising from the effects of grazing. Whilst the close coupling of water availability with soil degradation makes it tempting to develop simple relationships between grass growth and water availability (e.g. Pressland & McKeon, 1989), the low fertility of much of the landscape means that the availability of water must also be linked to the timely availability of nutrients (Gutierrez & Whitford, 1987). In addition to the temporal constraints existing between water and nutrients, the spatial variability that exists in grassland landscapes means that a particular monitoring site chosen at random could give grossly misleading results.

Environmental variation is not hard to recognise, although it can be a major task to describe it in a repeatable way. Analysis of the gross spatial patterns at a landscape level is essential prior to the initiation of any monitoring program to measure grassland change. The concept of land systems advanced by CSIRO is based on the grouping of areas throughout which there is a recurring pattern of topography, soils and vegetation (Perry, 1962). Usually, the initial selection of sites for monitoring is made at this scale (Bastin, 1984; Holm, Burnside & Mitchell, 1987). However, as argued above, there can be a complex matrix of smaller-scale variation within land units, caused by such impacts as erosion cell structures (Pickup, 1985) and smaller-scale moisture and nutrient sinks. All these are overlain by a grazing pattern which, in itself, may be related to independent factors such as wind direction and distance from water.

In addition to spatial variability, it has become apparent that for effective monitoring, ecological factors driving the dynamics of the grasslands must be incorporated into the measurement base. Foran, Bastin & Shaw (1986) developed procedures to assess and monitor arid rangelands, including grasslands, which enabled a broad range of influences to be included in the interpretation of vegetation change. Using multivariate methods, range types could be classified, condition states identified and trends in grassland change noted. Friedel, Chewings & Bastin (1988) took this approach a step further by using the vital attributes of a number of important species to form a small number of functional groups or 'guilds'. The relative proportions of the functional groups at monitoring sites spread throughout a particular grassland type are used to classify the sites in terms of their 'condition state'. Thus, sites in a poor condition state for grazing might have predominantly 'unpalatable forbs' and few 'palatable grasses', for instance. These two approaches provide a repeatable method which also minimises subjectivity.

Whilst it is widely accepted that the distribution of grassland components is strongly influenced by landscape elements, there is often considerable subjectivity in the definition of these elements. Spatial heterogeneity may also mean that a random siting of linear quadrats may lead to high variability of results. Gillison & Brewer (1985) advanced the concept of 'gradsects' or belt transects which are purposely located along pre-defined environmental gradients derived from an hierarchy of environmental parameters. Analyses of this information enables the frequency and distribution of spatial heterogeneity to be quantified and appears to be a better measure biologically of elements within a landscape.

Ludwig & Tongway (1990) proposed that any effective method for monitoring should meet six requirements:

1. be applicable at the landscape scale for managers;
2. use indicators of ecosystem processes and driving forces;
3. make clear which plant and soil indicators are linked;
4. discriminate between short-term system changes arising from seasonality and long-term changes because of land management;
5. have statistical rigour;
6. be reproducible and transferable between observers.

Taking these requirements as a whole they have combined many of the parameters discussed above; Ludwig & Tongway suggest use of a four-step monitoring method applicable to many extensive grassland systems, viz.:

1. link monitoring points through a landscape gradient and the establishment of a permanent 'gradsect';
2. select the most informative indicators of ecosystem processes;
3. identify important soil and vegetation zones through boundary analysis along the gradsect;
4. establish permanent ecological monitoring points within important vegetation/soil zones.

The combination of such a monitoring system with use of plant production models and long-term climatic records would enable a powerful and analytical treatment of the long-term risks involved in any method of land management for grasslands. Because of the complexity of the issues involved there is a growing demand to incorporate the relevant variables into Decision Support Systems which enable rapid analysis and quantification of the impact of land management decisions on grassland ecosystems (Scanlan *et al.*, 1990).

Conclusion

The wide acceptance recently of the scale of degradation which has occurred in many of the natural and derived grassland systems in both southern and northern Australia has led to a number of changes in legislation that addresses the management of these lands. In recent legislation concerning the pastoral lands of South Australia, for instance, there is the requirement written into the legislation that leaseholders maintain the 'condition' of their land. Recent administrative reviews of the land acts of both the Northern Territory and Queensland (Wolfe, Wright & Murphy, 1990) have also emphasised the need to move towards sustainable systems for use of these lands. As well, there is an increasing concern to conserve a representative range of grassland types and some of the rare and endangered species within those grasslands in Australia, especially in southeastern Australia (see above and Chapter 18). The present challenge is to develop both management and legislative structures which take into account both the biological and socio-economic constraints to the wise use of extensive areas of grasslands in both high and low rainfall regions of southern and northern Australia (see, e.g. Holmes & Mott, 1992). These natural and derived grasslands form the cover, after all, for a very large part of the Australian continent.

References

Aldrick, J.M. & Robinson, C.S. (1970). Report on the land units of the Katherine-Douglas area, N.T. Land Conservation Series No. 1. Canberra: Australian Government Publishing Service.

Australian Government Publishing Service (1989). *Our Country, Our Future. Statement on the Environment. The Hon. R.J.L. Hawke, A.C.* Canberra: Australian Government Publishing Service.

Bastin, G. (1984). Centralian range assessment program. Northern Terrritory Department of Primary Production Mimeographed Report.

Beard, J.S. (1976). The evolution of Australian desert plants. In *Evolution of Desert Biota*, ed. D.W. Goodall, pp. 51–64. Austin: University of Texas Press.

Beckman, R. (1988). Termites and nutrient cycles in the north. *Ecos*, **54**, 15–18.

Burch, G., Graetz, D. & Noble, I. (1987). Biological and physical phenomena in land degradation. In *Land Degradation – Problems and Policies*, eds A. Chisolm & R. Dumsday, pp. 27–48. Cambridge University Press.

Christian, C.S. & Stewart, G.A. (1953). Survey of the Katherine-Darwin region, 1946. CSIRO Australia Land Research Series No. 1.

Daly, J.J. & Dudgeon, G.S. (1987). Drought management reduces degradation. *Queensland Agricultural Journal*, **113**, 45–9.

De Angelis, D.L. & Waterhouse, J.C. (1987). Equilibrium and nonequilibrium concepts in ecological models. *Ecological Monographs*, **57**, 1–21.

De Corte, M., Cannon, M., Barry, E., Bright, M. & Scanlon, J. (1991). *Land Degradation in the Dalrymple Shire: A Preliminary Assessment*. Melbourne: CSIRO.

Dick, R.S. (1958). Variability of rainfall in Queensland. *Journal of Geography*, **11**, 32–42.

Dudgeon, G.S. & Fry, W.B. (1990). Rangeland vegetation monitoring using NOAA AVHRR Data 3. An NDVI based GIS for pastoral management. In *Proceedings of the 5th Australasian Remote Sensing Conference, Perth, 1990*, Volume 2, pp. 868–71. Perth: University of Western Australia.

Ellis, J.E. & Swift, D.M. (1988). Stability of African pastoral ecosystems: alternate paradigms and implications for development. *Journal of Range Management*, **41**, 450–9.

Eyles, A.C. & Cameron, D.G. (1985). Pasture research in northern Australia – its history, achievements and future emphasis, ed. J.B. Hacker. Research Report No 4. CSIRO Division of Tropical Crops and Pastures: Brisbane.

Eyles, A.G., Cameron, D.G. & Hacker, J.B. (1985). Pasture ecology. In *Pasture Research in Northern Australia – Its History, Achievements and Future Emphasis*, ed. J.B. Hacker, pp. 95–105. CSIRO Division of Tropical Crops & Pastures Research Report No. 4.

Fitzpatrick, E.N. & Nix, H.A. (1970). The climatic factor in Australian grassland ecology. In *Australian Grasslands*, ed. R.M. Moore, pp. 1–26. Canberra: Australian National University Press.

Foran, B.D., Bastin, G. & Shaw, K.A. (1986). Range assessment and monitoring in arid lands: the use of classification and ordination in range survey. *Journal of Environmental Management*, **22**, 67–84.

Friedel, M.H., Chewings, V.H. & Bastin, G.N. (1988). The use of comparative yield and dry-weight-rank techniques for monitoring arid rangeland. *Journal of Range Management*, **41**, 430–5.

Friedel, M.H., Foran, B.D. & Stafford Smith, D.M. (1990). Where the creek runs dry or ten feet high: pastoral management in arid Australia. *Proceedings of the Ecological Society of Australia*, **16**, 185–94.

Garcia-Moya, E. & McKell, C.M. (1970). Contribution of shrubs to the nitrogen economy of a desert wash community. *Ecology*, **51**, 81–8.

Gillard, P., Williams, J. & Monypenny, R. (1989). Clearing trees from Australia's semi-arid tropics. *Agricultural Science*, **2**, 34–9.

Gillison, A.N. & Brewer, K.R.W. (1985). The use of gradient directed transects or gradsects in natural resource surveys. *Journal of Environmental Management*, **20**, 103–27.

Graetz, R.D., Walker, B.H. & Walker, P.A. (1988). The consequences of climate change for seventy percent of Australia. *Nature & Environment*, **4**, 399–420.

Groves, R.H. (1975). Growth and development of five populations of *Themeda australis* in response to temperature. *Australian Journal of Botany*, **23**, 951–63.

Groves, R.H., Hagon, M.W. & Ramakrishnan, P.S. (1982). Dormancy and germination of seed of eight populations of *Themeda australis*. *Australian Journal of Botany*, **30**, 373–86.

Groves, R.H., Keraitis, K. & Moore, C.W.E. (1973). Relative growth of *Themeda australis* and *Poa labillardieri* in pots in response to phosphorus and nitrogen. *Australian Journal of Botany*, **21**, 1–11.

Groves, R.H. & Williams, O.B. (1981). Natural grasslands. In *Australian Vegetation*, ed. R.H. Groves, pp. 293–316. Cambridge University Press.

Gutierrez, J.R. & Whitford, W.G. (1987). Responses of Chihuahuan Desert herbaceous annuals to rainfall augmentation. *Journal of Arid Environments*, **12**, 127–39.

Hagon, M.W. (1976). Germination and dormancy of *Themeda australis*, *Danthonia* spp., *Stipa bigeniculata* and *Bothriochloa macra*. *Australian Journal of Botany*, **24**, 319–27.

Harrington, G.N., Wilson, A.D. & Young, M.D. (1984). Management of rangeland ecosystems. In *Management of Australia's Rangelands*, ed. G.N. Harrington, A.D. Wilson & M.D. Young, pp. 3–14. Melbourne: CSIRO.

Hitchmough, J., Berkeley, S. & Cross, R. (1989). Flowering grasslands in the Australian landscape. *Landscape Australia*, **4/89**, 394–403.

Hodgkinson, K.C. & Quinn, J.A. (1976). Adaptive variability in the growth of *Danthonia caespitosa* Gaud. populations at different temperatures. *Australian Journal of Botany*, **24**, 391–6.

Holm, A. McR., Burnside, D.G. & Mitchell, A.A. (1987). The development of a system for monitoring trend in range conditions in the arid shrublands of Western Australia. *Australian Rangeland Journal*, **9**, 14–20.

Holmes, J. & Mott, J.J. (1992). Australia's tropical savannas: development or custodial use. In *Economic Driving Forces and Constraints on Savanna Land Use*, eds M. Young & O. Solbrig, pp. 283–320. Lancashire: Parthenon.

Johnson, R.W. & Tothill, J.C. (1985). Definition and broad geographic outline of savanna lands. In *Ecology and Management of the World's Savannas*, eds J.C. Tothill & J.J. Mott, pp. 1–13. Canberra: Australian Academy of Science.

Low, W.A., Dudzinski, M.L. & Muller, W.J. (1981). The influence of forage and climatic conditions on range community preference of Shorthorn cattle in central Australia. *Journal of Applied Ecology*, **18**, 11–26.

Ludwig, J. & Tongway, D.J. (1990). Monitoring the condition of Australian arid lands: linked plant-soil indicators. In *E.P.A. Symposium on Ecological Indicators* (in press).

McCosker, T.H., O'Rourke, P.K., Eggington, A.R. & Doyle, F.W. (1988). Soil and plant relationships with cattle production on a property scale in the monsoonal tallgrass tropics. *Australian Rangeland Journal*, **10**, 18–29.

McDougall, K.L. (1989). The re-establishment of *Themeda triandra* (kangaroo grass): implications for the restoration of grassland. Arthur Rylah Institute for Environmental Research Technical Report Series No. 89.

McIntyre, S. (1990). Germination in eight native species of herbaceous dicots and implications for their use in revegetation. *Victorian Naturalist*, **107**, 154–8.

McKeon, G.M., Day, K.A., Howden, S.M., Mott, J.J., Orr, D.M., Scattinit, W.J. & Weston, E.J. (1990). Northern Australian savannas: management for pastoral production. *Journal of Boiogeography*, **17**, 355–72.

McKeon, G.M., Howden, S.M., Silburn, D.M., Carter, J.O., Hammer, G.L.; Johnson, P.W., Lloyd, P.L., Mott, J.J., Walker, B., Weston, E.J. & Willcocks, J.R. (1988). The effect of climate change on crop and pastoral production in Queensland. In *Greenhouse: Planning for climate change*, ed. G.I. Pearman pp. 546–63. Melbourne: CSIRO

McKeon, G.M., Rickert, K.G., Ash, A.J., Cooksley, D.G. & Scattini, W.J. (1982). Pasture production model. *Proceedings of the Australian Society of Animal Production*, **14**, 202–4.

McKeon, G.M., Rickert, K.B. & Scattini, W.J. (1986). Tropical pastures in the farming system: cast studies of modelling integration through simulation. *Proceedings of the 3rd Australian Conference on Tropical Pastures*, pp. 92–100. Brisbane: Tropical Grassland Society.

Mentis, M.T., Grossman, D., Hardy, M.B., O'Connor, T.G. & O'Reagain, P.J. (1989). Paradigm shifts in South African range science, management and administration. *South African Journal of Science*, **85**, 684–7.

Minson, D.J. & McDonald, C.K. (1987). Estimating forage intake from the growth of beef cattle. *Tropical Grasslands*, **21**, 116–22.

Moore, R.M. (ed.) (1970). *Australian Grasslands*. Canberra: Australian National University Press.

Morrisey, J.G. & O'Connor, R.E.Y. (1990). Twenty-eight years of station management: fair use and a fair go. In *Working Papers of the 5th Biennial Conference, Australian Rangeland Society, Longreach, Queensland*, pp. 179–80. (Mimeo)

Mott, J.J. (1974). Mechanisms controlling dormancy in the arid zone grass *Aristida contorta* F. Muell. I. Physiology and mechanisms of dormancy. *Australian Journal of Botany*, **22**, 635–45.

Mott, J.J. (1978a). Dormancy and germination in five native grass species from savanna woodland communities of the Northern Territory. *Australian Journal of Botany*, **26**, 621–31.

Mott, J.J. (1978b). The influence of moisture stress on the germination and phenology of some annual species from an arid zone. *Studies in the Australian Arid Zone*, **3**, 122–9.

Mott, J.J. (1979). High temperature contact treatment of hard seed in *Stuylosanthes*. *Australian Journal of Agricultural Research*, **30**, 847–54.

Mott, J.J. (1986). Planned invasions of Australian tropical savannas. In *Ecology of Biological Invasions: An Australian Perspective*, eds J. Bourbon & R.G. Groves, pp. 89–96, Canberra: Australian Academy of Science.

Mott, J.J. (1987). Patch grazing and degradation in native pastures of the tropical savannas of northern Australia. In *Grazing Lands Research of the Plant–Animal Interface*, eds F.P. Horne, J. Hodgson, J.J. Mott & R.W. Brougham, pp. 153–62. Arkansas: Winrock International.

Mott, J.J., Bridge, B.J. & Arndt, W. (1979). Soil seals in tropical tallgrass pastures of northern Australia. *Australian Journal of Soil Research*, **30**, 483–94.

Mott, J.J., Ludlow, M.M., Richards, J.H. & Parsons, A.D. (1992). Causes of variation in seasonal response to defoliation in three tropical savanna grasses. *Australian Journal of Agricultural Research*, **44**, 241–60.

Mott, J.J. & McComb, A.J. (1974). Patterns in annual vegetation and microrelief in an arid region of Western Australia. *Journal of Ecology*, **62**, 115–26.

Mott, J.J. & Tothill, J.C. (1984). Tropical and subtropical woodlands. In *Management of Australia's Rangelands*, eds G.N. Harrington, A.D. Wilson & M.D. Young, pp. 255–69. Melbourne: CSIRO.

Mott, J.J. & Tothill, J.C. (1993). Degradation of savanna woodlands in Australia. In *Conservation Biology in Australia and Oceania*, ed. C. Moritz, (in press). Chipping Norton, NSW: Surrey Beatty & Sons.

Mott, J.J., Williams, J., Andrew, M.H. & Gillison, A.N. (1985). Australian savanna ecosystems. In *Ecology and Management of the World's Savannas*, eds J.C. Tothill & J.J. Mott, pp. 56–82. Canberra: Australian Academy of Science.

Noble, I.R. (1986). The dynamics of range ecosystems. In *Rangelands: A Resource Under Siege*, eds P.W. Joss, P.W. Lynch & O.B. Williams, pp. 3–5. Canberra: Australian Academy of Science.

Norman, M.J.T. (1965). Post-establishment grazing management of Townsville lucerne on uncleared land at Katherine, N.T. *Journal of the Australian Institute of Agricultural Science*, **31**, 311–13.

Norman, M.J.T. (1966). Katherine research station 1956–64: a review of published work. CSIRO Division of Land Research Technical Paper No. 28.

Norman, M.J.T. & Begg, J.E. (1973). Katherine Research Station: A review of published work, 1965–72. CSIRO Division of Land Research Technical Paper No. 33.

Noy-Meir, I. (1975). Stability in grazing systems : an application of predator-prey graphs. *Journal of Ecology*, **63**, 459–81.

Noy-Meir, I. (1982). Stability of plant–herbivore models and possible applications to savanna. In *Ecology of Tropical Savannas*, eds B.J. Huntley & B.H. Walker, pp. 591–609. New York: Springer-Verlag.

Orr, D.M. (1980). Effects of sheep grazing *Astrebla* grassland in central western Queensland. I. Effects of grazing pressure and livestock distribution. *Australian Journal of Agricultural Research*, **31**, 797–806.

Orr, D.M. (1986). Factors affecting the vegetation dynamics of *Astrebla* grasslands. PhD thesis, University of Queensland.

Orr, D.M. (1988). Interaction of rainfall and grazing on the demography of *Astrebla* spp. in north western Queensland. *Proceedings 3rd International Rangelands Congress, New Delhi*, pp. 192–4.

Perry, R.A. (1962). General report on lands of the Alice Springs area, Northern territory, 1956–57. CSIRO Land Research Series No. 6.

Pickup, G. (1985). The erosion cell – a geomorphic approach to landscape classification in range assessment. *Australian Rangeland Journal*, 7, 114–21.

Pressland, A.J. & McKeon, G.M. (1989). Monitoring animal numbers and pasture condition for drought administration – an approach. *Proceedings 5th Australian Soil Conservation Conference, Perth*, pp. 17–27.

Russell, J.S. (1981). Geographic variation in seasonal rainfall in Australia. *Journal of the Australian Institute of Agricultural Science*, 47, 59–66.

Scanlan, J.C., Mott, J.J., McKeon, G.M., Day, K.D. & Lawes, D. (1990). Linking process information to grazing management needs: a systems analysis approach. In *Proceedings of a Native Grass Workshop, Dubbo*, pp. 89–107.

Scarlett, N.H. & Parsons, R.F. (1982). Rare plants of the Victorian Plains. In *Species at Risk: Research in Australia*, eds R.H. Groves & W.D.L. Ride, pp. 89–105. Canberra: Australian Academy of Science.

Spain, A.V. & McIvor, J.G. (1988). The nature of herbaceous vegetation associated with termitaria in north-eastern Australia. *Journal of Ecology*, 76, 181–91.

Stafford Smith, D.M. (1988). Modelling three approaches to predicting how herbivore impact is distributed in rangelands. New Mexico Agriculture Station Regular Research Report No. 628.

Stafford Smith, D.M. & Pickup, G. (1990). Pattern and production in arid lands. *Proceedings of the Ecological Society of Australia*, 16, 195–200.

Stuwe, J. & Parsons, R.F. (1977). *Themeda australis* grasslands on the Basalt Plains, Victoria: floristics and management effects. *Australian Journal of Ecology*, 2, 467–76.

Taylor, J.A. & Tulloch, D. (1985). Rainfall in the wet-dry tropics: extreme events at Darwin and similarities between years during the period 1870–1983 inclusive. *Australian Journal of Ecology*, 10, 281–295.

Walker, B.H. (1988). Autecology, synecology, climate and livestock as agents of rangeland dynamics. *Australian Rangeland Journal*, 10, 69–75.

Westoby, M. (1980). Elements of a theory of vegetation dynamics in arid rangelands. *Israel Journal of Botany*, 28, 169–94.

Westoby, M., Walker, B. & Noy-Meir, I. (1989). Opportunistic management for rangelands not at equilibrium. *Journal of Range Management*, 42, 266–74.

Williams, J. (1991). Search for sustainability: agriculture and its place in the natural ecosystem. *Agricultural Science*, 4, 32–9.

Williams, J. & Chartres, C.J. (1991). Sustaining productive pastures in the tropics. I. Managing the soil resource. *Tropical Grasslands*, 25, 73–84.

Williams, O.B. (1961). Studies in the ecology of the Riverine Plain. III. Phenology of a *Danthonia caespitosa* Gaudich. grassland. *Australian Journal of Agricultural Research*, 12, 247–59.

Williams, O.B. (1968). Studies in the ecology of the Riverine Plain. IV. Basal area and density changes of *Danthonia caespitosa* Gaudich. in a natural pasture grazed by sheep. *Australian Journal of Botany*, 16, 565–78.

Willis, A.J. & Groves, R.H. (1991). Temperature and light effects on the germination of seven native forbs. *Australian Journal of Botany*, **39**, 219–28.

Wilson, A.D., Tongway, D.J. & Tupper, G.J. (1988). Factors contributing to differences in forage yield in the semi-arid woodlands. *Australian Rangeland Journal*, **10**, 13–17.

Wolfe, P.M., Wright, R.G. & Murphy, D.G. (1990). Report of a review of land policy and administration in Queensland. Brisbane: Government Printer.

Woods, L.E. (1983). *Land Degradation in Australia*. Canberra: Australian Government Publishing Service.

PART 3

VEGETATION OF EXTREME HABITATS

14

Saltmarsh and mangrove

P. ADAM

INTERTIDAL communities dominated by vascular plants occur extensively on the shores of estuaries and embayments and on some open, low wave energy coasts. These communities fall into two categories: saltmarsh and mangrove. Mangroves are communities of trees and shrubs, whereas saltmarshes are dominated by herbs and low shrubs. Around most of the Australian coastline there is a very sharp distinction, both floristically and structurally, between mangrove and saltmarsh but, at the southern limit of its geographic distribution in Victoria the mangrove *Avicennia marina* forms only a low shrub, which may be of lesser stature than the chenopod *Sclerostegia arbuscula* on adjacent saltmarshes; communities dominated by *A. marina* are, by convention, always mangroves regardless of the height of the vegetation. The word mangrove has two different meanings, referring to tree species occurring in the intertidal and to communities formed by these species. Thus *Avicennia marina* is a mangrove, and communities of *A. marina* are mangrove communities, or often, simply mangroves. To avoid potential confusion, the term mangal has been proposed for mangrove communities (Macnae, 1968), but has not come into general usage. In this account, mangrove will be used in both senses.

Distribution and abundance

The total area of mangrove in Australia is about 11 500 km² (Galloway, 1982), the majority of which occurs north of the Tropic of Capricorn. South of the Tropic mangroves are found in most estuaries and inlets on the southern Queensland and New South Wales coast but are absent from the Gippsland coast, although they are found at Corner Inlet (Wilsons Promontory), Westernport Bay and Port Phillip Bay. These Victorian mangroves are at the highest latitudes of any mangroves in the world. In South Australia, mangroves are found at

395

sheltered sites on the coasts of the Spencer and St Vincents Gulfs and on the west of the Eyre Peninsula, around Ceduna. On the west coast, temperate mangroves are very restricted in occurrence being found near Bunbury, on the Abrolhos Islands, at Shark Bay, Lake Macleod and in the estuary of the Gascoyne River at Carnarvon. There are no mangroves on Tasmania.

The total area of saltmarsh on the Australian coast is unknown. The national survey of the coast by Galloway *et al.* (1984) did not record saltmarsh as a distinct category of vegetation and whilst saltmarsh would be included in several vegetation categories recognised in their report the areal estimates also include a large proportion of supratidal habitat. It is commonly held that saltmarshes are the temperate equivalent of mangroves and that, therefore, saltmarsh is rare or absent from tropical shores. Tropical saltmarshes in Australia are almost certainly much more extensive than those on temperate shores. Tidal amplitude is greater on most tropical shores than in southern Australia and this, coupled with flat topography, is conducive to the development of a broader vegetated intertidal zone. In New South Wales, although saltmarsh and mangrove occur in virtually all estuaries, stands are rarely extensive and the total area of intertidal wetland is only approximately 16 500 ha, of which saltmarsh makes up less than 6000 ha (West *et al.*, 1985).

Adjacent and related vegetation types

The seaward limit of mangrove (or in the absence of mangroves, saltmarsh) is normally well defined. Estuarine and coastal waters in Australia support a great abundance and diversity of seagrasses (Larkum, McComb & Shepherd, 1989) but only rarely are these exposed to the air. On the landward side, mangroves may be bordered by mudflats (Thom, Wright & Coleman, 1975), saltmarsh (see Fig. 14.1) or they may merge into dryland forest or some form of fresh to brackish forested wetland. These latter communities also fringe saltmarshes and are widespread around estuaries and lagoons. Characteristic canopy dominants include *Melaleuca* spp. and, in southeast Australia, *Casuarina glauca*, *Eucalyptus botryoides* and *E. robusta*. The study of these coastal swamp forests has been much neglected, and their composition and distribution poorly known.

Stunted mangroves, particularly *Avicennia marina*, are widely scattered in crevices in the upper part of intertidal rock platforms but are rare in non-tidal habitats, although following reclamation of inter-

Fig. 14.1. Boundary between mangrove and saltmarsh, Hexham, Hunter River, New South Wales. *Avicennia marina* in the upper canopy has been defoliated by caterpillars (see West & Thorogood, 1985).

tidal wetlands individual mangroves may persist for many years, if not subject to other disturbance, in freshwater conditions. The most inland stand of mangroves known is in northwest Western Australia over 20 km from the sea where *Avicennia marina* fringes a permanent saline creek (Beard, 1967; Burbidge, McKenzie & Kenneally, 1991). On the Australian territory of Christmas Island, in the Indian Ocean, a small stand of tall *Bruguiera gymnorhiza* occurs around a freshwater spring (Hosnies Spring), some over 30 m elevation above the current high water mark (Woodroffe, 1988a). (This unusual mangrove community is one of 40 sites in Australia currently listed as wetlands of international significance under the Ramsar Convention.)

Although there are extensive inland saline wetlands, these do not generally support forest or woodland vegetation although a number of trees (particularly *Eucalyptus* spp.) are moderately salt-tolerant (Blake, 1981; Sands, 1981; Pepper & Craig, 1986).

The upper limit of saltmarsh vegetation may be very difficult to determine. On dry coastlines with no marked topographic change at high water mark, the intertidal saltmarsh vegetation may merge imperceptibly with inland grassland or chenopod shrublands. On higher rainfall coasts, saltmarshes more frequently abut tall shrubland or forest vegetation, although the boundary rarely coincides exactly with the current highwater mark. In brackish lagoons and in the upper reaches of estuaries, saltmarsh may be replaced by tall reedswamp communities (dominated by *Phragmites australis, Schoenoplectus* spp., *Bulboschoenus* spp. or *Typha* spp.).

On very exposed headlands, above the intertidal zone but frequently drenched by breaking waves, grasslands indistinguishable in species composition from those on intertidal saltmarsh occur (Adam, Wilson & Huntley, 1988; Adam *et al.*, 1989). Seepage zones on headlands and seacliffs in southeastern Australia are characterised by an assemblage of species also found in brackish saltmarshes (*Apium prostratum, Cotula coronopifolia, Lobelia alata, Samolus repens, Isolepis cernua, Paspalum vaginatum*) (Adam *et al.*, 1988, 1989); a very similar community is also found on seacliffs in New Zealand (Wilson & Cullen, 1986).

On sand spits at the entrances to estuaries and lagoons there may be a transition between saltmarsh and sand dune vegetation where boundaries between the two habitats are indistinct.

Extensive areas of inland Australia have either naturally saline soils or are undergoing salinisation as a result of human activity (Bettenay, 1986). The vegetation of these areas shares species in common with intertidal coastal saltmarsh and many of the plant communities are very similar. The vegetation of these inland saline habitats is not discussed further in this chapter, however.

Intertidal mud and sandflats may be found to seaward of saltmarsh and mangrove, and particularly in northern Australia, above the mangrove zone. Such flats are devoid of vascular plants but the sediment's surface (down to a depth of several millimetres) may support abundant growth of microalgae. The contribution of these flats to estuarine productivity has not been studied in detail but may be considerable; around the Gulf of Carpentaria high level hypersaline flats release considerable quantities of salt and nutrients during tidal flooding (Ridd, Sandstrom & Wolanski, 1988).

The difficulty of defining the landward boundary of intertidal communities relates in part to the differential responses of species along the gradient of environmental conditions between fully marine and fully terrestrial habitats and partly because the theoretical boundary (the tidal limit) is in itself inherently variable in position. Where intertidal wetlands are in regions of low relief a small vertical change in the tidal limit may have an extensive horizontal expression, resulting in a broad zone of uncertainty in the position of vegetation boundaries.

Tide heights vary from day to day and from spring-neap cycle to spring-neap cycle on an annual basis (the highest predicated spring tides – king tides – occurring around the equinoxes). In addition there is a longer term cycle (of about 18 years) in the height of the highest tides. Tidal height is also affected by meterological conditions so that, if a high tide coincides with a storm, the actual level reached by seawater may be well above predictions. Along the saltmarsh or mangrove fringe there may be a zone subject to tidal inundation extremely infrequently which might support non-salt-tolerant species for periods of years between tidal incursions but which can still properly be regarded as part of the intertidal zone. Clark (1990) has shown how fluctuations in the upper boundaries of an American saltmarsh resulting from the 18-year component of the tidal cycle are reflected in the fossil record.

Superimposed on tidal fluctuations are movements in boundaries associated with changes in the relative levels of land and sea.

Composition and biogeography

There are a number of accounts which provide national or regional overviews of the composition of mangrove and saltmarsh vegetation including, *inter alia*, for mangroves: Macnae (1966, 1968), Saenger *et al.* (1977), Lear & Turner (1977), Semeniuk, Kenneally & Wilson (1978), Clough (1982), Craig (1983), Barson & Heatwole (1985), West (1985), Tomlinson (1986), Davie (1987), Hutchings & Saenger (1987), Wightman (1989), Adam & King (1990), King, Adam & Kuo (1990), and for saltmarsh: Saenger *et al.* (1977), Adam (1981*a,b*), Bridgewater, Rosser & de Corona (1981), Kirkpatrick & Glasby (1981), Love (1981), Bridgewater (1982), Craig (1983), Pen (1983), Adam *et al.* (1988), Adam & King (1990). The existence of such a large body of relatively recent information may create the impression that the flora of intertidal wetlands is well known. This would be misleading; although the composition of the flora is well established our knowledge of its distribution, whilst better than that for many terrestrial habitat types, is still inadequate. There are many additional data in the 'grey literature' of Environmental Impact Statements and planning studies but there are many estuaries for which species lists of even the more common plants are not available.

Table 14.1. *Genera of mangroves in Australasia (nomenclature after Tomlinson, 1986)*

Family	Genus
Acanthaceae	*Acanthus*
Avicenniaceae	*Avicennia*
Bombacaceae	*Camptostemon*
Combretaceae	*Lumnitzera*
Ebenaceae	*Diospyros*
Euphorbiaceae	*Excoecaria*
Leguminosae	*Cynometra*
Lythraceae	*Pemphis*
Meliaceae	*Xylocarpus*
Myrsinaceae	*Aegiceras*
Myrtaceae	*Osbornia*
Palmae	*Nypa*
Plumbaginaceae	*Aegialitis*
Rhizophoraceae	*Bruguiera*
	Ceriops
	Rhizophora
Rubiaceae	*Scyphiphora*
Sonneratiaceae	*Sonneratia*
Sterculiaceae	*Heritiera*

Mangroves

The mangrove flora is defined as those plants associated with, and more or less exclusive to, mangrove habitats (Saenger, Hegerl & Davie, 1983). This definition is inherently circular, but there is a wide consensus as to which species are conventionally regarded as part of the mangrove flora (see Tomlinson, 1986). The consensus is not, however, unanimous and Mepham & Mepham (1985) have argued that the category of mangrove includes a very large number of species, although many are only casual and marginal denizens of the habitat. There are a number of species characteristic of the uppermost limit of mangroves and authors differ as to whether they should be regarded as true mangroves. The most widespread species in this category in Australia, *Hibiscus tiliaceus*, is by convention, not listed as a mangrove.

The flora can be further subdivided by growth form. The predominant element is the trees (i.e. the mangroves). The term mangrove has no taxonomic significance, the 50–60 species which have been recognised as true mangroves (Tomlinson, 1986) coming from 20 families. Biogeographically, the species fall into two major provinces: a rich flora (of 40–50 species) in the Indian Ocean and west Pacific region (to which the Australian mangroves belong) and a depauperate flora (of about 10 species) in the Atlantic and eastern Pacific (Davies, 1980; Tomlinson, 1986).

In Australia there are about 40 species of mangroves in some 17 families (see Table 14.1). Only one of these species is regarded as endemic, *Avicennia integra* (Duke, 1988, 1991) (a taxon previously referred to *A. officinalis* by Wells, 1982). (*Lumnitzera rosea*, interpreted by Tomlinson *et al.* (1978) as the hybrid *L. racemosa* x *L. littorea*, is known with certainty only from northeastern Queensland – Tomlinson, 1986).

At a global scale mangrove species' diversity is strongly correlated with temperature. Many species appear to be very intolerant of low temperatures and the latitudinal ranges of species can be explained by different tolerances to low temperatures (see Markley, McMillan & Thompson, 1982; Saenger & Moverley, 1985). On the east coast of Australia the decline in mangrove species richness with increasing latitude can be explained by a sorting of species on the basis of their temperature optimum for growth (Saenger & Moverley, 1985). From more than 30 species in northeastern Queensland richness declines to only five species in northern New South Wales whilst south of the Manning River estuary there are only two species, *Avicennia marina* and *Aegiceras corniculatum*, of which the latter reaches its southern limit at Merimbula, close to the New South Wales/Victoria border (West, 1985).

Whilst it can be argued that temperature is the determinant of distribution amongst current mangrove species, it does not explain the low diversity of mangroves at higher latitudes. Mangroves are a taxonomically diverse assemblage, and the necessary physiological and morphological adaptations to the mangrove habitat appear to have evolved independently a number of times. In non-tidal habitats there are a number of trees in temperate climates which exhibit a degree of both salt and flooding tolerance. Why have these species not given rise to temperate mangroves? Nothing currently known about the physiology of mangroves would appear to preclude the development of temperate mangroves.

In northern Australia there is considerable variation in mangrove floras between estuaries, with estuaries east of Cape York generally containing more species than those to the west. Smith & Duke (1987) have examined the correlation between species richness of an estuary and a number of environmental variables. Both mean minimum temperature of the coldest month and mean maximum temperature of the warmest month were correlated with species richness across northern Australia, but minimum temperature was a more important determinant of species number than maximum temperature. Tidal amplitude was inversely correlated with species richness but for a given increase in amplitude there was a greater reduction in species richness in the west than in the east. The greater the tidal amplitude the greater the velocity of tidal currents and hence the larger the mechanical forces on plants – these forces may be particularly important during the establishment of seedlings. Why the effect of tides should differ between the northwestern and northeastern coasts is unknown.

A number of factors was significantly correlated with species richness in the east but not in the west. In the east, estuaries which are long and have large catchments tend to have more species than short estuaries with small catchments. Big estuaries tend to have greater habitat diversity than small ones and so may provide niches for a greater range of species. Interannual rainfall variability and cyclone frequency were negatively correlated with species richness in the east but showed no significant relationship in the west. However, the volume of freshwater runoff from the catchment was inversely correlated with species richness in the west but not in the east. Contrary to the suggestion of Saenger & Moverley (1985), there was no correlation between mean annual rainfall and species richness.

The analysis by Smith & Duke (1987) indicates that whilst there are similarities in patterns of species richness east and west of Cape York there are also major differences, for which causal explanations are lacking, in the response of species to different environmental factors in the two regions.

Whilst cyclone frequency did not explain variation in species richness between estuaries in the western region, this does not imply that cyclones are unimportant in the west. Cyclones are major natural disturbance factors. During cyclones mangroves experience both defoliation and wind throw and the impacts of wave action and storm surge. The response of individual species to cyclones may vary. Following cyclone Kathy in 1984, Bardsley (1985) reported that mortality was greater in genera in the Rhizophoraceae (*Bruguiera*, *Ceriops* and *Rhizophora*) which do not resprout from epicormic buds than in genera such as *Avicennia* or *Excoecaria* in which rapid regrowth from epicormics was possible. Frequent recurrence of cyclones may result in a change in species composition of the mangrove community with a permanent decline in representation of Rhizophoraceae.

Mangrove communities in Australia, as elsewhere in the world, are generally devoid of an understorey of vascular plants (Janzen, 1985). Locally, where the canopy is open, saltmarsh plants may form a patchy ground cover. In northeastern Australia, particularly where the uppermost mangroves have permanently brackish conditions the ferns *Acrostichum aureum* and *A. speciosum* may be locally common. In southeastern Australia, the boundary between saltmarsh and mangroves may be diffuse with small groves of trees interspersed with saltmarsh.

In addition to the canopy trees, mangrove stands provide a habitat for a range of other life forms, particularly in the tropics. Lianes, vascular epiphytes (including ferns and orchids) and semi-parasitic mistletoes may be locally abundant. In northern Queensland the epiphytic anthouse plants *Myrmecodia antoinii* and *Hydnophytum formicarium*, although not restricted to mangroves, are characteristic of the habitat (Huxley, 1982).

Saltmarsh

The flora of saltmarshes is not as well documented as that of mangroves, nor is there a consensus as to those species which can be regarded as being saltmarsh species. The lower and mid-elevation saltmarsh zones support a relatively small flora, easily recognised as saltmarsh. However, the upper saltmarsh zones support a variable, but sometimes species-rich, range of communities. Saenger *et al.* (1977) presented separate floristic tables for saltmarsh and the fringe zone. However, a number of species listed as being from the fringe zone may extend well down into the intertidal zone (notably *Juncus kraussii* – recorded in Saenger *et al.* (1977) as *J. maritimus* var. *australiensis*). Even if all the fringe species were included as saltmarsh species the list presented by Saenger *et al.* (1977) would still underestimate the size of the flora. Recent surveys in southern Australia (Kirkpatrick & Glasby, 1981; Bridgewater, 1982; Adam *et al.*, 1988) have added considerably to the list of widespread saltmarsh species.

At a global scale saltmarsh vegetation can be divided into a small number of categories which show strong similarities, both structurally and floristically, over a wide geographical range (Adam, 1990).

At this very broad scale of analysis Australian saltmarshes fall into three major groups, although there is considerable intergradation between groups. On tropical coasts saltmarsh vegetation at sites with frequent and extended periods of hypersalinity is extremely sparse, with succulents (chenopods, *Sesuvium* and *Batis*) and grasses being the major species. In more favourable conditions extensive, species-poor grasslands (chiefly *Sporobolus virginicus*) occur. These tropical saltmarshes are very similar to those described from the coasts of other continents at similar latitudes.

On coasts experiencing a strongly seasonal mediterranean climate, saltmarsh vegetation is more species-rich than in the tropics and is characterised by shrubby chenopods (*Halosarcia*, *Sclerostegia*); the subshrub *Frankenia* is also characteristic of these marshes. Although the vegetation cover is high it is not continuous and there are usually small bare patches between shrubs. Very similar vegetation is found on other shores with similar climates such as the Mediterranean Basin and southern California (Adam, 1990).

Saltmarshes on temperate coasts with predictable rain during most of the year have closed vegetation. The lower marsh zones are characterised by dominance of *Sarcocornia quinqueflora* and the upper marsh by tall rush or sedge lands. Marshes of very similar structure and floristics (similar in some cases down to species level) are found in southern Africa, New Zealand and South America and have analogues in temperate marshes in Europe, northeast Asia and western North America (Adam, 1990). Along the east coast of Australia there is a very gradual transition from more species-rich and diverse marshes in the south to the much simpler grass-dominated (*Sporobolus virginicus*) communities of the tropical coasts.

The increase in species richness in the saltmarsh flora with increasing latitude (Saenger *et al.*, 1977) is not unique to Australia. An explanation of this pattern is still to be provided. However, in tropical regions saltmarsh occupies habitats where mangroves are unable to grow. These are normally sites experiencing prolonged hypersalinity and the number of species which can tolerate these conditions may be very limited. In temperate regions, in the absence of mangroves, a greater diversity of microhabitats is available for colonisation by saltmarsh species and this environmental diversity may provide a greater range of niches and thus permit a greater species richness.

Despite the structural, physiognomical and floristical similarities with marshes elsewhere there are some floristic elements, the presence of which permits the recognition of an Australasian province amongst saltmarsh floras. Amongst the Salicornieae there are several genera endemic to Australia, whilst *Selliera radicans*, *Baumea juncea* and *Stipa stipoides* are among a number of species common to Australia and New Zealand. The endemic genus *Wilsonia* (Convolvulaceae) with three species in southern Australia is without close parallels elsewhere.

More detailed studies of saltmarsh in southern Australia permit recognition of regions of particular biogeographic interest. Bridgewater (1982) identified the Coorong in South Australia as an area with a particularly high diversity of both communities and species; the climate of this stretch of coastline is transitional between the strongly seasonal mediterranean climate to the west and the more equable climate of the southeast, and a number of species whose distributions are centred in the different climatic regimes reach distributional limits in the region. On the New South Wales coast, Jervis Bay marks the northern limit of a number of species and communities characteristic of southeastern Australia (Adam & Hutchings, 1987; Adam *et al.*, 1988).

Non-vascular flora of intertidal wetlands

Visually, intertidal wetlands are overwhelmingly dominated by angiosperms. Gymnosperms are totally absent and pteridophytes only locally present in mangroves. This triumph of the flowering plants is surprising in view of the abundant fossil flora, which is usually interpreted as being derived from pre-flowering plant analogues of mangrove communities, in estuarine deltaic sediments of Mesozoic age.

Intertidal wetlands, however, support a diversity of non-vascular plants. The most important of these, in terms of their probable contribution to ecosystem productivity, are the algae and cyanobacteria. These occur in a number of microhabitats; on mud surfaces, in shallow pools and creeks (both permanent and ephemeral), as epiphytes on aerial roots and trunks and as epiphytes on upper branches and leaves (Hutchings & Saenger, 1987).

The standing crop of microalgae on, and within the surface layers of, mud can be high, both in mangroves and saltmarsh, but may vary considerably on a seasonal basis (Hutchings & Saenger, 1987). Many of the unicellular algae have mucilaginous sheaths and they may play an important role in promoting sedimentation and stabilising the sediment surface (Coles, 1979) although this has not been studied in Australia. At high tide surface films of microalgae may be grazed by mullet. Mats of filamentous cyanobacteria which may form a rubbery skin to the sediment, occupy extensive areas on high level saltflats and open areas in saltmarshes. A number of cyanobacteria have the capacity for nitrogen fixation but their contribution to the nutrient economy of Australia coastal wetlands has been studied at few sites.

Boto & Robertson (1990) did not detect nitrogen fixation on saltflats and recorded relatively low levels of fixation in mangroves.

Multicellular algae may also be common on sediment surfaces, particularly greens such as *Enteromorpha* and *Rhizoclonium*. At a few localities in southern Australia free living populations of the brown alga *Hormosira banksii*, which is usually attached on rocky shores, have been recorded (King, 1981).

Pools and creeks within wetlands have a diverse microflora but may also have dense free-floating macroalgae, particularly *Enteromorpha* and *Ulva*. These are often very prominent at sites near settlements and their abundance may be a response to eutrophication.

The aerial roots and lower trunks of mangroves frequently support dense growths of an assemblage of red algae referred to by Post (1936) as the *Bostrychia–Caloglossa* association. The assemblage is characterised by four genera; *Bostrychia* (including *Stictosiphonia*), *Caloglossa*, *Catenella* and *Murrayella*, but the species composition varies both within sites and between localities (King *et al.*, 1990).

A number of algae have been recorded growing on the upper stems and leaves of mangroves (Cribb, 1979). These are well above direct tidal influence and, with the exception of some influence of salt spray, occupy an essentially terrestrial habitat.

Mangroves support a considerable diversity of epiphytic lichens (Stevens, 1979, 1981; Stevens & Rogers, 1979), growing on the trunk above the normal tidal limit. As lichens absorb water (and nutrients) over their surface, they are dependent on rainwater and thus species on mangroves occupy a microhabitat essentially similar to that on the trunks of terrestrial trees, and almost all the species also occur on nearby terrestrial vegetation. On saltmarshes, particularly on arid coasts, lichens occur as epiphytes on the stems of shrubby chenopods.

Bryophytes are found occasionally as epiphytes on mangroves but not as abundantly or as widespread as lichens. At high latitudes in the northern hemisphere, bryophytes may be a major component of saltmarsh vegetation (Adam, 1990) but, even in southeastern Australia, bryophytes are never more than a very minor component of the uppermost fringe of brackish saltmarshes.

Fungi, although inconspicuous, may be an important component of intertidal ecosystems although they have been little studied in Australia (the available information is reviewed by Hutchings & Saenger, 1987). There have been a number of reports of vesicular–arbuscular mycorrhizae in saltmarsh plants in the northern hemisphere (Adam, 1990), but it is not known whether any Australian saltmarsh or mangrove species are mycorrhizal.

In the anaerobic soils of intertidal wetlands bacteria play the major role in nutrient recycling and transformation. However, little is known in Australia about the composition or functioning of these bacteria in the ecosystem.

PAMPAS GRASS

Introduced species

Mangroves have the distinction of being one of the very few Australian plant communities lacking introduced species. Saltmarshes, however, have been invaded by a large number of alien species, although currently only four constitute a serious threat, *Spartina anglica*, *Cortaderia selloana*, *Juncus acutus* and *Baccharis halimifolia*.

Spartina anglica, cord grass, is a low marsh species capable of surviving prolonged submergence in seawater. The species is the presumed hybrid between the European *S. maritima* and the American *S. alterniflora* which was first observed in Southampton Water, England in the late nineteenth century. The vigorous growth of the species as a coloniser of mudflats led to it being widely planted to promote marsh establishment and to stabilise mobile mudflats.

The date of introduction of *Spartina* to Australia is unknown (Boston, 1981), but it is believed that the first unsuccessful plantings were at Corner Inlet, Victoria. In the late 1920s there were at least 14 recorded introductions and by the early 1930s plantings had been attempted in all states except the Northern Territory (Boston, 1981). Most of these plantings failed to establish but there are now vigorous populations in Tasmania (particularly in the estuary of the Tamar River) and in southeastern Victoria. Very small populations survive, but have not spread, in South Australia and southern New South Wales (Boston, 1981). In Victoria, *Spartina* has established to seaward of *Avicennia marina*, changing the zonation pattern. The long-term effects on the viability of mangroves are unknown.

Cortaderia selloana, pampas grass, is an aggressive weed of disturbed bushland in southeastern Australia. It has considerable salt tolerance and has invaded saltmarshes at a number of sites on the New South Wales coast.

The rush *Juncus acutus* is native to saltmarshes and dune slacks in the Mediterranean Basin, extending northwards on the Atlantic seaboard to Wales. In southeastern Australia it has invaded a range of coastal wetland communities and in saltmarshes has displaced the native *J. kraussii* at a number of sites.

The groundsel bush *Baccharis halimifolia* is native to upper saltmarsh communities in the eastern United States. In Australia it is a major weed species in coastal southern Queensland and northern New South Wales and appears to be spreading south. It can form dense stands in disturbed saltmarshes and adjacent communities such as *Casuarina glauca* woodland.

There are a very large number of small annual, or short-lived, species which can be found in the upper saltmarsh in southern Australia, particularly on sandy soils where they occur as part of an open community and do not appear to displace native species. Widespread aliens in this zone include *Plantago coronopus* and the grasses *Lagurus ovatus*, *Parapholis incurva* and *Polypogon monspeliensis* (Adam, 1981a).

Features of the saltmarsh and mangrove environment

Intertidal wetlands are found in geomorphologically suitable sites around the whole Australian coast and thus experience a great diversity of climatic regimes. As has been indicated above, climate appears to be the major determinant of the distribution of mangrove vegetation whilst the geographical distribution of individual species within both saltmarsh and mangrove may also be determined by climate. Nevertheless, we know very little about the climatic conditions within intertidal wetlands, and correlations based on regional climatic data may be misleading; proximity to the sea will result in substantial modifications to the climate in comparison to that only a few kilometres inland (Oliver, 1982). The temperature range will be significantly less: relative humidity, rainfall and wind patterns may also differ markedly.

The factor which distinguishes saltmarsh and mangroves from other vascular plant communities is tide. Whilst tidal fluctuations are, in themselves, highly predictable the interactions between tides, climate and vegetation (Fig. 14.2) produce spatially and temporally complex patterns of environmental variation.

A brief review of tidal regimes around the Australian coast is provided by Womersley & King (1990). In southwest Western Australia, and parts of the Gulf of Carpentaria, tides are diurnal, with one low and one high tide a day on a cycle of 24 hours 50 minutes. Elsewhere tides are semi-diurnal, with two highs and two lows of similar height a day, or mixed, when the two highs (and two lows) per day have a markedly different height.

The tidal range varies considerably around the coast. In southern Australia tidal ranges are mostly low (micro- to mesotidal) although

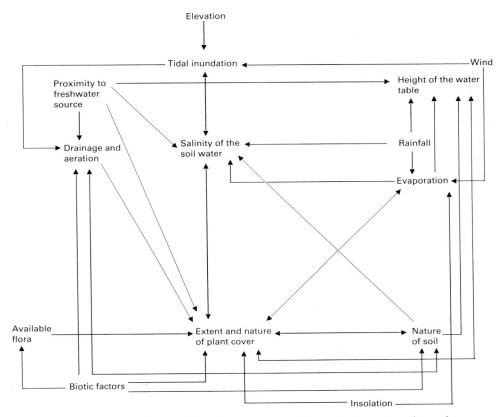

Fig. 14.2. The complex of environmental factors influencing saltmarsh vegetation (from Adam & King, 1990, modified from Clarke & Hannon, 1969).

within bays and inlets much higher ranges are experienced (for example 4 m at the head of Spencer Gulf and 3 m at the head of Gulf St Vincent). In northern Australia tidal ranges are generally high, with a maximum in excess of 8 m around Broome.

The tidal range will determine the vertical extent of intertidal communities but the relationship between lateral (horizontal) extent and tidal range is complex and will depend on local topography and geomorphology. For a given surface gradient the greater the tidal range the wider the saltmarsh or mangrove but it is possible to find extensive stands in regions of low tidal range and narrow fringes along coasts of high tidal range.

In southern Australia there are many coastal lagoons which are only intermittently open to the sea. When closed, water level (and salinity) fluctuations reflect the balance between precipitation (and/ or inflow) and evaporation; when open to the sea the tidal range is normally very much less than that on the open coast. Such lagoons are nevertheless frequently fringed with saltmarsh (less frequently mangroves) which at times of high water level may be continuously submerged for periods of weeks.

In consequence of tidal flooding the soils in saltmarsh and mangroves will be saline. The more frequent the tidal flooding the less the variation in soil salinity. At higher elevations, with less frequent tidal flooding salinities can vary considerably depending upon the balance between rainfall and evaporation. Around much of the Australian coast periods of hypersalinity are commonly experienced in saltmarshes.

Saline soils have low soil water potentials, but the consequences to plants are not simply osmotic. Seawater is predominantly a solution of sodium, magnesium, chloride and sulphate ions. The balance between ions in the soil is very different from that regarded as ideal for plant growth; plants growing in such soils have to contend with an excess of some ions (which are potentially toxic and which generate low water potentials) but also with a deficiency of some essential plant nutrients. Available nitrogen and phosphorus may be in short supply, whilst potassium and calcium although present in what are, in comparison with many inland soils, high amounts are relatively scarce, and there may be problems of uptake in the presence of a super-abundance of competing ions.

The soil in saltmarshes and mangroves varies considerably between sites, from coarse sand to soft clays. This results in different physical characteristics but would also, *a priori*, lead to variation in soil nutrient status. The extent to which any substratum effect interacts with the tidal input, and its consequences for plant growth are unknown.

Poorly drained soils will tend to become anaerobic and even relatively sandy saltmarsh soils show poor drainage characteristics (Clarke & Hannon, 1967). Sulphate is present in high concentrations in seawater and under anaerobic conditions undergoes bacterially mediated reduction to sulphide. In estuaries suffering industrial pollution this may result in heavy metals being 'locked up' as insoluble sulphides in saltmarsh or mangrove soils. There appear to be considerable interactions between salinity and water logging on the availability of nutrients for plant growth (Ball, 1988a) but these are poorly understood.

In addition to creating saline and anaerobic soils, tides have a number of other effects on vegetation. Tidal currents may dislodge seedlings thereby preventing establishment. The velocity of currents is a function of tidal amplitude. Estuarine water is normally turbid so that after submergence vegetation will be coated with a sediment film, which may reduce photosynthesis (mature mangrove canopies are normally above the tidal limit but seedlings might be affected). During the period of submergence photosynthesis will be impaired; the effect of flooding is to alter day length but whether this affects growth processes is unknown. For much of the year sudden immersion will constitute a temperature shock, but again the effects of this have not been studied.

Although most saltmarshes and mangroves can be characterised as having, at least periodically, high soil salinities (high is a vague term but in this context can be interpreted as being greater that the equivalent of 300 mM NaCl) there are some areas which are reliably brackish. These include areas subject to freshwater inputs (either as surface or as groundwater flows) or those towards the tidal limit in estuaries with consistent freshwater river flows. A number of species are characteristic of brackish marshes in southern Australia (including *Cotula coronopifolia* and *Mimulus repens*). Fresh groundwater is probably a feature of many saltmarshes and mangroves, although this aspect of their hydrology has been little studied. In Florida, Sternberg & Swart (1987) found evidence that, even though the surface soil was regularly inundated with seawater, some species in the mangroves obtained much of their water from deeper fresh groundwater, and were thus physiologically insulated from the fluctuating soil salinities of the surface layers. This phenomenon may be more widespread. If it is, it has implications for activities adjacent to intertidal wetlands which interfere with groundwater hydrology.

Coping with the environment: features of the flora and vegetation of saltmarsh and mangroves

Whilst each species has its own niche, and hence some unique biological features which enable occupation of that niche, there are sufficient features in common to enable some broad generalisations to be made.

The vegetation of both saltmarsh and mangroves is, compared with many terrestrial communities, species-poor and there is a tendency for most communities to be clearly dominated by a single species. The species which are community dominants nearly always have a wide geographical distribution, and within a single site show a wide ecological amplitude.

Jefferies & Rudmik (1991) have pointed out that the majority of saltmarsh species are perennials, many of which disperse by vegetative propagules, and have a considerable capacity for clonal growth. Although the seeds of many species can tolerate prolonged immersion in seawater few species will germinate under saline conditions (Adam, 1990) and even when germination does occur seedling survival is poor (Ungar, 1978). Jefferies & Rudmik (1991) suggested that vegetative propagation and clonal growth were less susceptible to high salinities than germination and early growth of seedlings. There have been no studies on the demography of saltmarsh species in Australia but field observation indicates that these generalisations hold.

Long-lived clonally growing species endure a great range of environmental variation and must therefore be phenotypically plastic (particularly in their physiology). A number of species also show modified breeding systems which will limit the development and expression of genetic variability. In some populations of *Halosarcia indica* and *H. pergranulata* ssp. *queenslandica*, male organs have been lost and reproduction is parthenogenic (Wilson, 1980; Jefferies & Rudmik, 1991).

Despite phenotypic plasticity and mechanisms to limit genetic variability many saltmarsh plants display considerable genotypic variation, both within and between populations (Adam, 1990). Few Australian species have been studied but Smith-White (1981, 1988) has shown that the widespread grass *Sporobolus virginicus* is extremely variable in morphology, physiology and cytology. This information is of considerable practical importance. With an increasing interest in rehabilitating sites which have been damaged, there have been a number of proposals to replant damaged areas. For such an exercise to succeed, it is important that genetically appropriate material be planted. The taxon for planting has to be a selected genotype and not the species at large.

Mangroves are, by definition, relatively long-lived and thus will require a degree of phenotypic plasticity. A number of species can spread vegetatively by suckers and can coppice following damage. *Avicennia marina*, with anomalous secondary thickening and multiple phloem (Tomlinson, 1986), is particularly resistant and recovers vigorously from damage.

A conspicuous feature of many, but not all, mangrove species is vivipary – the precocious development of seedlings still attached to the parent. A distinction is drawn by some authors between 'true' vivipary in which the embryo ruptures the pericarp and cryptovivipary, as in *Avicennia marina* and *Aegiceras corniculatum*, in which the embryo enlarges considerably but does not normally break the pericarp. The adaptive significance of vivipary has been much debated (Tomlinson, 1986; Hutchings & Saenger, 1987) and its value remains uncertain. However, like the vegetative propagules of saltmarsh plants it may permit regeneration while avoiding the difficulties of true seedling establishment in a saline environment.

Plants which grow in saline soils are referred to as halophytes (in contrast to non-salt tolerant species – glycophytes). There is a spectrum of salt tolerance amongst flowering plants from species intolerant of very low salinities to those which can survive and grow in salinity to complete their lifecycle – and facultative halophytes which the literature as to the threshold along this spectrum at which plants can be called halophytes, but 300 mM NaCl would be an appropriate (if approximate) value. Distinction is sometimes attempted between obligate halophytes – species with an absolute requirement for high salinity to complete their lifecycle and facultative halophytes which tolerate, but do not require, high salinity. Barbour (1970) argued persuasively that no angiosperm was an obligate halophyte, and more recent physiological research has supported this conclusion.

The term halophyte draws attention to the importance of salt in the life of saltmarsh and mangrove species. However, whilst salt tolerance is a *sine qua non* of being a halophyte, attempts to explain the distribution of species within intertidal wetlands in terms of differential salt tolerance have been, at best, only partially successful. Absence of salt tolerance explains the exclusion of glycophytes from intertidal habitats, but within such habitats segregation of species reflects selection for a range of traits within a generally salt-tolerant flora.

The mechanisms of salt tolerance have long attracted the attention of physiologists and we now have a reasonable, if not complete, understanding of them (see Flowers, Hajibagheri & Clipson, 1986; Ball, 1988*a*, *b*; Adam, 1990; King *et al.*, 1990).

For a plant to grow its cells must be turgid in order to maintain turgor; the water potential within the tissues must be lower than that in the soil surrounding the roots. The problem for halophytes is how to generate low water potentials without impairing metabolism.

Halophytes in general accumulate large quantities of inorganic ions in their tissues, sufficient to generate the low water potentials required. [For this reason they have a high ash content – *Salicornia* in England was known as glasswort because the ash was used in glassmaking and in Australia *Avicennia marina* was burnt to provide ash for soap making during the early years of European settlement (Bird, 1981)]. However, the enzymes of halophytes do not appear to be salt tolerant so that there must be compartmentation in the tissues to separate the high salt content from centres of metabolism. If compartmentation occurs then the ions in one compartment (presumably the vacuole) must be osmotically balanced by solutes (so-called compatible solutes) in the other compartments. Compatible solutes must permit enzyme function at high concentrations, and preferably be uncharged, neutral compounds of low molecular weight. A number of putative compatible solutes have been identified. These fall into a small number of classes of compound: the amino acid proline; quaternary ammonium compounds (particularly glycine betaine); and sugar alcohols. Saltmarsh plants are mainly either proline or glycine betaine accumulators (Poljakoff-Mayber *et al.*, 1987; Adam, 1990). Amongst mangroves, glycine betaine is the compatible solute in *Avicennia marina* (Poljakoff-Mayber *et al.*, 1987) but the most widely distributed compatible solutes are sugar alcohols (pinitol and mannitol) (Popp 1984*a*, *b*; Popp, Larher & Weigel, 1984, 1985). The prevalence of sugar alcohols in mangroves may reflect the trees' high biomass, since although the quantity of compatible solute per cell is small the total amount per plant will be high in a mangrove. If proline or quaternary ammonium compounds were utilised, this would impose a very high demand for nitrogen.

Even though halophytes accumulate ions there must be a balance between the supply of ions to the shoots and the capacity of the plant to accommodate the influx of salt.

There are a number of mechanisms employed by halophytes to regulate their ionic content.

In all species in which it has been studied, the ionic concentration of the xylem sap in plants growing at high salinities is much less than that in the root environment. The mechanisms by which ions are excluded from the xylem are imperfectly known but one of the essential features of halophytes is clearly a capacity for selective ion uptake. Even with a very high degree of exclusion, other mechanisms are required to regulate ionic content. Reduction in transpiration rate will lower the rate of supply of ions to leaves. The xeromorphic features of many saltmarsh plants may be adaptations to reduce transpiration and hence salt uptake. Mangroves, although all employing the C_3 photosynthetic pathway, have rates of water loss which are unusually low for trees, and have correspondingly high water use efficiencies (Ball, 1988*a*). Low transpiration rates in environments with high insolation incur the risk of leaf temperatures reaching damaging levels. However, outer canopy leaves are frequently oriented so as to minimise the heat load. Ball, Cowan & Farquhar (1988) showed that amongst members of the Rhizophoraceae in northern

Queensland the leaf angle was smaller in those species with the greatest salt tolerance.

A number of species, from both mangrove and saltmarsh, have glandular structures on their leaves (and in some cases also the stems) which secrete a concentrated salt solution. Evaporation of the exudate results in the leaves being studded with visible salt crystals (Ball, 1988*a*; Adam, 1990; King *et al.*, 1990). Exclusion of ions from the xylem sap is less efficient in species with salt glands (Ball, 1988*a*).

Maintenance of tissue salt concentration can be achieved through growth and/or the development of succulence. Some species do not appear to regulate tissue salt concentration; rather salt accumulates to fatal levels, and the dead, salt-laden organ is shed to be replaced by new growth. This is the mechanism employed, for example, by halophytic *Juncus* spp.

Although much is known about salt regulation in leaves there is still much to be learned about integration mechanisms at a whole plant level and about the physiology of roots under saline conditions (Adam, 1990).

There are considerable metabolic costs associated with salt tolerance: synthesis of compatible solutes represents a drain of carbon (and frequently also nitrogen) which could otherwise be used for growth; selective ion uptake and maintenance of tissue ionic balance involves energy-consuming mechanisms for ion transport. Reductions in transpiration rates inevitably involve a reduction in carbon dioxide uptake and hence growth. Halophytes tend to have large root/shoot ratios so that for every unit of carbon gained only a small portion is invested in new photosynthetically active tissue. These costs of salt tolerance appear to be the major reasons excluding facultative halophytes from a wider range of habitats. Although the growth rate of many halophytes is maximal under nonsaline conditions the rates are less than those achieved by glycophytes. Under nonsaline conditions halophytes succumb to competition from more vigorous glycophytes.

The potential problems for plants growing in waterlogged soils are lack of oxygen and the presence of phytotoxins. Survival of higher plants in waterlogged soil is dependent upon avoidance of the need for prolonged anaerobic respiration. This is made possible by an extensive system of air spaces (aerenchyma) connecting the roots to the above-ground parts of the plant. In *Avicennia marina* 70 per cent of the total root volume may be aerenchyma (Curran, 1985). Aerenchyma reduce the resistance to gas flow within plants and also result in a lower demand for oxygen per unit volume of tissue. Many mangrove species have obviously modified root systems to facilitate the entry of air, for example, pneumatophores (*Avicennia*), knee roots (*Bruguiera*), cone roots (*Sonneratia*), buttresses (*Xylocarpus*) or stilt roots (*Rhizophora*). In *A. marina* the root aerenchyma contain sufficient oxygen to maintain aerobic respiration during normal tidal flooding (Curran, Cole & Allaway, 1986). Changes to the hydrology of mangrove stands (such as through bunding or the installation of flood gates on creeks) which lead to the retention of water above the pneumatophores for prolonged periods may result in the death of trees.

Under anoxic conditions a number of potentially phytotoxic compounds accumulate in the soil. Iron and manganese will be present in the toxic ferrous and manganous forms and sulphide may be very abundant; various organic products of bacterial metabolism may also be phytotoxic. Higher plants avoid contact with toxins through oxidation which can be achieved by diffusion of oxygen from roots; enzymatic oxidation either at the root surface or internally and microbial oxidation within the rhizosphere (Armstrong, 1982). Differential tolerance/susceptibility of species to various toxins may also be a factor determining the local abundance of species.

Zonation

At many sites saltmarsh and mangrove species are strikingly zoned, with communities arranged in belts parallel to the shoreline. The zonation patterns commonly recognised are determined by the distribution of community dominants. The extent to which other elements in the biota have congruent distribution patterns is still unclear.

Despite the ubiquity of zonation, there is considerable variation between sites in the ordering of species within the zonation. Bunt & Williams (1980) analysed the distribution of species within the species-rich stands of mangroves in northeastern Queensland and were unable to detect a consistent sequence of species. Buckley (1982) re-examined the same data and concluded that the various species assemblages recognised by Bunt & Williams (1980) could be assigned to different zones, but within each zone a number of different assemblages can be found and most particular sequences were site specific.

Our understanding of the factors responsible for the distribution patterns of individual species, which in aggregate are perceived as zonation, is incomplete. There has been a tendency to assume that zonation is a reflection of the partitioning of a relatively simple (or at least monotonic) environmental gradient (often presumed to be salinity). However, the interaction of environmental factors (Fig. 14.2) will produce complex and temporally variable gradients. In addition environmental effects will be mediated by competitive interactions between species. Clarke & Hannon (1967, 1969, 1970, 1971) investigated experimentally the growth responses of a number of saltmarsh species in the Sydney region and proposed hypotheses to account for the observed distribution of species. These hypotheses are amenable to testing through experimental manipulation in the field (using techniques such as reciprocal transplants and habitat modification) but there is a surprising paucity (world wide) of experiments of this type. In general terms, however, it is probable that the spread of species upshore is limited by competitive interactions whilst seaward limits are physiological, relating to tolerance of some factor associated with repeated tidal flooding.

Zonation patterns are also interpreted dynamically as spatial expressions of succession. Conceptual models have been developed in which species colonise mudflats and promote the accretion and stabilisation of sediment. As the soil surface rises, environmental conditions change thereby permitting the entry of other species which displace the primary colonists. Continued expansion of the primary colonist seawards results in zonation. This model has been supported by much empirical evidence (but not from Australian sites) although the ability to extend seawards is limited by a requirement for shelter from wave and current erosion, and in some cases zonation patterns reflect particular local circumstances and are not constant through time (Adam, 1990). The interpretation of zonation as a reflection of succession in Australian conditions is less certain.

Pidgeon (1940) proposed that the zonation of intertidal communities on the New South Wales coast could be interpreted as a succession (Fig. 14.3) and this interpretation has become part of the received wisdom in basic ecology texts. If true, it would be a very unusual sequence in that the initial colonists are trees, subsequently replaced by herbs. The model also postulates that succession will continue above the tidal limit. In the absence of a change in land/sea level it is difficult to see how a succession driven by accretion of tide-borne sediment could achieve this. Mitchell & Adam (1989a) were unable to find evidence supporting Pidgeon's (1940) model from saltmarsh and mangroves in the Sydney region. There were no detectable mangrove remains under saltmarsh (on the other hand Herbert (1951) reports the presence of remains of *Avicennia* stumps in saltmarsh in the Brisbane region). There are very few sites in the Sydney region at which active colonisation is occurring at the present time. Mitchell

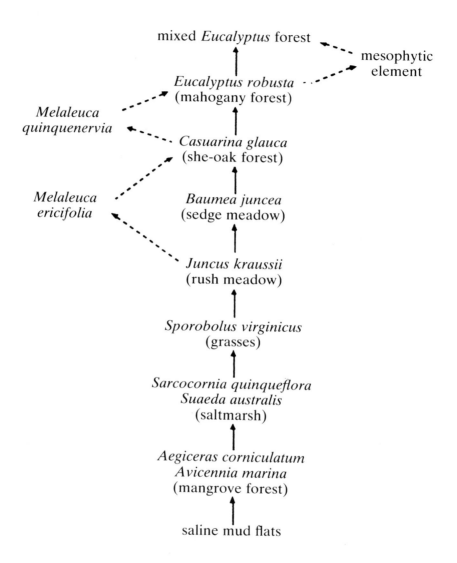

Fig. 14.3. The succession model (Pidgeon, 1940) for intertidal wetlands on the New South Wales coast (from Mitchell & Adam, 1989*a*).

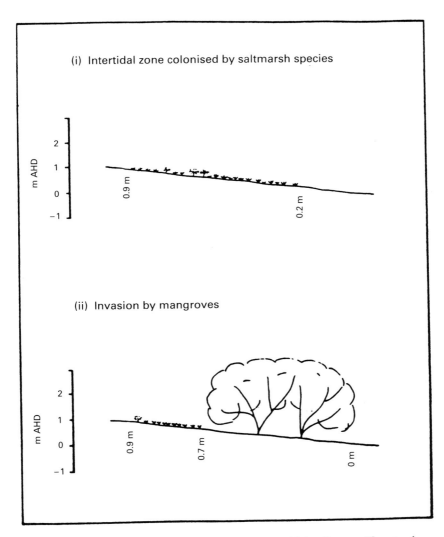

Fig. 14.4. Model for succession on intertidal saline mudflats in the Sydney region proposed by Mitchell & Adam (1989a).

& Adam (1989*a*) suggested from observation of two sites that initial colonisation was by saltmarsh species, followed by invasion of the lower intertidal by mangroves (Fig. 14.4). More long-term studies are needed, from both saltmarshes and mangroves, to clarify the relationships between zonation and succession.

The changing coastline

The present coastline is, geologically speaking, a very recent phenomenon. Sea level reached its current position about 6000 years ago as water levels rose following the melting of ice caps at the end of the last glacial period. During the preceding two million years of global glacial/interglacial cycles there was a parallel series of dramatic fluctuations in sea level. In Australia the coast line remained ice-free although the lowering in temperature would have caused major changes in the distribution of species. The position and shape of the coastline has changed frequently; at times of low sea level Bass and Torres Straits and much of the continental shelf were dry land, and the habitats for intertidal wetland were thus far removed from their present position.

Even during the past 6000 years there have been major changes to the shape of the coast. Soft coastal landforms undergo continual changes because of the effects of wind, waves and current, and estuaries have been subject to infilling by sediment from both land and sea sources. In the last 200 years the Australian coast has been further modified, deliberately or inadvertently, by human activity. The rates of change along the coast are not constant, either in space or time, so the present coastline is in varying stages of geomorphological evolution (Roy, 1984).

Coastal plant communities are therefore necessarily transient in their occupation of a particular locality. Selection is likely to have favoured dispersal traits (although we are woefully ignorant as to the natural dispersal potential of most species) so it is not surprising that many coastal plants have wide geographic distributions.

An understanding of the history of shoreline development is essential to any explanation of species' distribution patterns; geomorphological changes to coastal wetlands also have important implications for the function of these habitats as ecosystems.

Studies on the stratigraphy and geomorphology of the Alligator Rivers region demonstrate that around 6000 years ago the region was occupied by a vast mangrove stand (Woodroffe, Thom & Chappell, 1985; Woodroffe *et al.*, 1986; Woodroffe, 1988*b*). A similar 'Big Swamp' stage has also been identified in the Ord and Fitzroy Rivers and was probably a feature of most northern estuaries at this time (Woodroffe, 1988*b*). In the Alligator Rivers today mangroves now form a relatively narrow fringe along river channels and the extensive supratidal plain is occupied by grass and sedgelands; in the Ord and Fitzroy estuaries narrow bands of mangroves fringe the low tide channels and the mid and upper intertidal is dominated by broad hypersaline mudflats. If there is a relationship between mangroves and secondary productivity in estuaries, was the 'Big Swamp' phase a period of very high secondary productivity? What was the response of the estuarine biota (particularly those elements which utilise mangrove habitats directly) to the decline in area of mangroves? Answers to these questions could help us predict the consequences of future changes to intertidal wetlands.

Superimposed on the broad sweep of geomorphological evolution various episodic events impact upon intertidal wetlands. Meandering of channels can result in sites of erosion and accretion changing over time and consequently considerable heterogeneity in structure and composition of the fringing plant communities at different positions along the estuary (as has been demonstrated for the mangroves along the South Alligator River by Davie, 1985). Disturbances caused by storms (particularly cyclones in northern Australia) may also be important factors determining community composition.

Saltmarsh and mangroves as ecosystems

Saltmarsh and mangroves occupy the interface between land and sea. Although there are some exchanges across the upper tidal boundary of intertidal wetlands, those associated with surface and ground-water runoff being most important, the more important exchanges are those with adjacent waterways. The linkages between intertidal wetlands and estuaries and offshore waters have been the subject of much research. Evaluation of the role of wetlands may be assisted by the use of modelling techniques (Knox, 1986) but, as yet, there are few reliable data to which models could be applied. Obtaining appropriate data is complicated by the spatial and temporal heterogeneity within intertidal wetlands. If data are to be representative and permit the drawing of valid generalisations then considerable attention needs to be given to experimental design. Given the global threats to the survival of intertidal wetlands it is important that there is a greater understanding of their function. Australia is well placed to carry out research on the topic and work on mangrove ecosystems is a major part of the activities of the Australian Institute of Marine Science (see, for example, Clough, 1982; Field & Dartnall, 1987; Boto & Robertson, 1990; Robertson & Duke, 1990; Wolanski *et al.*, 1990).

There have been a number of studies of litter fall in Australian mangroves (see Allaway, 1987; Clough, 1987). Litter fall does not always provide a good estimate of primary productivity (Clough, 1987), but measurements of litter fall are useful for estimating the potential transfer of organic material out of the ecosystem. Allaway (1987) has estimated that the *c.* 107 km^2 of mangrove in New South Wales produce about 46 000 tonnes of litter a year. There have been few studies of productivity of saltmarshes in Australia, but the values reported for above-ground productivity in *Juncus kraussii* in the Blackwood estuary in Western Australia (0.3 –1.3 kg dry wt m^2 y^{-1}) by Congdon & McComb (1980) are within the range, although below the maximum values, from northern hemisphere marshes.

The importance of intertidal wetland productivity for adjacent waters has been the subject of much speculation (Nixon, 1980). The concept, developed in the eastern United States in the 1960s, of coastal wetlands as sources of detritus for offshore waters (the out-welling hypothesis) has been used to promote the case for wetland conservation. However, few of the early studies attempted to measure directly export from marshes. The current understanding (Jansson *et al.*, 1988) is that there is substantial internal recycling of energy in both saltmarshes and mangroves together with some fluxes with adjacent systems (which may vary considerably in both magnitude and direction between sites). Even where export of organic carbon does occur it is unlikely to be of great significance outside the immediate estuary.

There are large fluxes of various nutrients associated with tidal flows in intertidal wetlands but the net movement may not be large. Boto & Robertson (1990) demonstrated that for mangroves in north-eastern Queensland the overall nitrogen budget was virtually balanced with loss of nitrogen by tidal export of particulate matter being matched by the tidal import of dissolved nitrogen plus nitrogen fixation within the mangrove system. Nixon (1980) summarised studies on nutrient fluxes in northern hemisphere saltmarshes and concluded for both nitrogen and phosphorus that in most cases the nutrient budgets balanced. These findings should not be interpreted as meaning that coastal wetlands are insignificant in the broader scheme of things. The transformation of chemical form of nutrient elements within wetlands may be a very important mechanism for increasing the bioavailability of nutrients.

Whilst mangroves and saltmarshes may not be significant contributors to offshore ecosystems, wetland detritus is clearly an important input to creeks and estuaries, and in addition wetlands provide habitat for numerous organisms.

Both commercial and recreational fisheries are of considerable economic value. A high proportion of the catch is of estuarine dependent species (species which require estuarine habitat for at least some stage in their life cycle). Currently, for example, about 60 per cent by weight and 70 per cent by value of the commercial catch landed in New South Wales is estuarine-dependent and the proportion in the recreational catch is probably greater than 60 per cent (Leadbitter & Doohan, 1991).

Many of the estuarine species utilise intertidal wetlands for part of their lives. Mangroves are not major fish-breeding grounds but creeks within mangroves are utilised by many species as nursery areas (Bell *et al.*, 1984; Blaber, Young & Dunning, 1985; Blaber, 1986; Blaber, Brewer & Salini, 1989). Juveniles of some species recruit first to seagrass beds before moving into mangroves. Many saltmarshes in Australia have few permanent creeks and pans but studies at a number of sites (Gibbs, 1986; Morton, Pollock & Beumer, 1987; Morton, Beumer & Pollock, 1988) indicate that saltmarshes may also be important fish habitats.

Coastal wetlands are popularly identified as being important habitats for birds. Many land-based species use wetlands and a smaller number of species are resident. The resident avifauna of mangrove, although small in absolute terms, is nevertheless large by world standards. The number of birds breeding in saltmarshes is small but upper marsh rush and sedge communities provide nest sites for some species. A large part of the population of one of the rarest birds in Australia, the Orange-bellied Parrot (*Neophema chrysogaster*), overwinters on saltmarshes in Victoria, feeding on seeds of *Halosarcia*. Migratory waders feed largely on invertebrates in intertidal sand and mudflats but saltmarshes may provide secure high tide roosts. Conservation of waders is a matter of international concern and Australia is signatory to three agreements (the Ramsar convention, the Japan-Australia Migratory Birds Agreement and the China-Australia Migratory Birds Agreement) which seek to protect habitats utilised by migratory waders.

The interactions between fauna and vegetation in saltmarsh and mangroves are still largely unknown. In mangroves crabs may be a conspicuous element, and through burrowing may alter microtopography (Warren & Underwood, 1986) and soil aeration. Crabs are major consumers of fallen litter (Robertson 1986, 1987), although this has been ignored in many litter fall studies, and through consuming fruit may influence reproduction of mangroves (Smith 1987*a,b*). Herbivory on mangrove canopies by insects is normally relatively low compared with that in terrestrial Australian forests (Robertson & Duke, 1987) but mass defoliation of *Avicennia marina* by caterpillars in the Hunter River estuary occurred in 1985 (West & Thorogood, 1985), although subsequently there was full recovery.

Management issues

For most of the period since European colonisation Australian intertidal wetlands have not been regarded highly. Sites close to habitation have been substantially modified, and in many cases, destroyed. More recently they have been a focus for conservation and measures have been implemented to limit future losses. Popular support for such action recognises the fish habitat importance of intertidal wetlands.

Direct exploitation of saltmarsh and mangroves in Australia has been limited. Allaway (1987) discusses uses of mangroves but these have been local and on a small scale. In Indonesia and elsewhere in southeast Asia mangroves have been regarded as important forest resources, harvested for timber, firewood and tannin and more recently for woodchips. The remoteness from population centres of Australia's extensive northern mangroves has probably prevented similar exploitation here.

In the northern hemisphere grazing by domestic stock is a factor determining the structure and composition of saltmarsh communities (Adam, 1990). In northern Australia many saltmarshes and mangroves are accessible to livestock but the extent of utilisation and its effects on the vegetation are unknown. In southern Australia, a number of saltmarshes are heavily grazed and some mangrove stands are browsed by cattle. Although this causes obvious disturbance in the form of trampling, the long-term effects on species composition are unknown.

In settled areas, coastal wetlands have been reclaimed for port, industrial and housing development, and for parks and sports fields. In rural areas, some intertidal wetlands have been reclaimed for agriculture but the greatest losses to agriculture have been of coastal freshwater wetlands and estuarine floodplains. In more recent decades, coastal wetlands have been threatened by developments for recreation and tourism (marinas, resorts and canal estates). Construction of solar salt production ponds has resulted in loss of mangroves and saltmarsh (Semeniuk, 1987).

In southeast Asia, considerable loss of mangroves has occurred as a result of conversion to aquaculture ponds. With the current interest in aquaculture in Australia, it will be important to ensure that construction of facilities does not involve destruction of intertidal habitats.

Although there have been considerable losses of intertidal wetlands in settled areas there have also been some localised gains. Increased sedimentation as a result of erosion from cleared catchments has provided habitat for the seaward expansion of mangroves (McLoughlin, 1985; Dunstan, 1990). In the Towra Point Nature Reserve in Botany Bay, Mitchell & Adam (1989*b*) have documented the spread of mangroves into saltmarsh, a process which has been continuing for at least the past hundred years. Similar loss of saltmarsh through invasion by mangroves is occurring at other localities on the New South Wales coast, although the factors responsible for this spread are unknown.

Even where intertidal wetlands are not directly threatened with destruction, they are vulnerable to degradation from a number of causes.

In urban areas, coastal wetlands are still subject to illegal rubbish dumping, and to disturbance through the construction and maintenance of easements for pipelines and powerlines. Vehicular use (4-WD, trail bikes and BMX bikes) alters the microtopography of saltmarshes leading to changes in vegetation: even if the disturbance factor is removed, recovery is extremely slow. Stormwater drains frequently discharge into saltmarshes and mangroves. Apart from introducing gross pollutants and nutrients, freshwater discharge can cause local erosion and through altering the salinity regime can promote the spread of *Phragmites australis* and *Typha* spp. at the expense of more salt-tolerant species (Pen, 1983; Zedler, Paling & McComb, 1990).

Intertidal wetlands are depositional sinks, and pollutants, from both terrestrial and marine sources, may accumulate in them. Very little is known about the pollution load, and its effects, in intertidal wetlands. The most publicised pollution threat is that from oil. Fortunately, the number of major oil spills to date in Australian waters has been small. The intertidal wetlands of Towra Point have been oiled on several occasions. Allaway (1987) has shown that, in areas subject to oiling, there has subsequently been mangrove mortality, although the cause of death is not yet established.

Sewage discharge into estuaries, and runoff from agricultural catchments increase the nutrient levels in estuarine waters, which may promote algal productivity. In a number of systems this has resulted in excessive growth of algae (Hodgkin *et al.*, 1985; Norman, 1988). In the Peel-Harvey Inlet, which is the most seriously affected estuary, there has been damage to saltmarshes from smothering by algae, as well as further damage caused by the use of vehicles to remove the decaying algal mass (Hodgkin *et al.*, 1985). The problems of the Peel-Harvey inlet are caused by inputs of phosphorus from the catchment, and are exacerbated by the poor flushing characteristics of the estuary. A large proportion of the phosphorus entering the estuary is from the store of phosphorus in the soil, built up from fertiliser input over the past 30 years, and not from the current application (Hodgkin *et al.*, 1985). Given the long period of time over which the problem in the Peel-Harvey has developed, it is possible that, even if changes to catchment management were implemented now, other estuaries may show similar symptoms at some future date.

With the population in the coastal zone increasing, more areas close to mangrove and saltmarshes are being developed for housing. This may create management problems for the future. Mosquitos and sandflies constitute a nuisance and, in some circumstances, a definite threat to human health (through spread of diseases for which the insects are vectors). Control of insects, through spraying and hydrological modification, is already practised, particularly in southern Queensland and northern New South Wales. Little is known about the wider ecological consequences of these control programs.

In the relatively near future there may be major changes to intertidal wetlands as a result of the greenhouse effect. Considerable uncertainty surrounds the possible effects of increased atmospheric carbon dioxide and other greenhouse gases but the potential for global warming is well established. If the climate were warmer, it would be predicted that mangrove diversity at higher latitudes would increase and that the absolute limit of mangroves would extend further south.

As a consequence of global warming there may also be an increase in sea level although the magnitude of any possible rise is much debated. Intertidal wetlands have adjusted to previous fluctuations in sea level but the effects of a rise at the present time may differ from those in the past, at least in southern Australia.

As sea level rises a regression of the seaward boundary of intertidal wetlands would occur, accompanied where topography and other circumstances permitted, by an extension landward (Bird, 1988; Vanderzee, 1988). For much of the Australian coastline adjustment would be possible, and although there would be local losses of wetland these would probably be balanced by gains elsewhere. In southern Australia, however, much of the coastline has been modified, and the response to a rising sea level may well be further modification to protect valuable real estate. In these circumstances, there could be net loss of intertidal wetland, and in particular, loss of saltmarsh as the saltmarsh/mangrove boundary adjusted upwards.

References

Adam, P. (1981*a*). Australian saltmarshes. *Wetlands (Australia)*, **1**, 8–10.

Adam, P. (1981*b*). Saltmarsh plants of New South Wales. *Wetlands (Australia)*, **1**, 11–19.

Adam, P. (1990). *Saltmarsh Ecology*. Cambridge University Press.

Adam, P. & Hutchings, P. (1987). The saltmarshes and mangroves of Jervis Bay. *Wetlands (Australia)*, **6**, 58–64.

Adam, P. & King, R.J. (1990). Ecology of unconsolidated shores. In *Biology of Marine Plants*, eds M.N. Clayton & R.J. King, pp. 296–309. Melbourne: Longman Cheshire.

Adam, P., Stricker, P., Wiecek, B.M. & Anderson, D.J. (1989). The vegetation of seacliffs and headlands in New South Wales, Australia. *Australian Journal of Ecology*, **14**, 515–47.

Adam, P., Wilson, N.C. & Huntley, B. (1988). The phytosociology of coastal saltmarshes in New South Wales, Australia. *Wetlands (Australia)*, **7**, 35–84.

Allaway, W.G. (1987). Exploitation and destruction of mangroves in Australia. In *Mangrove Ecosystems of Asia and the Pacific. Status, Exploitation and Management*, eds C.D. Field & A.J. Dartnall, pp. 183–92. Townsville: Australian Institute of Marine Science.

Armstrong, W. (1982). Waterlogged soils. In *Environment and Plant Ecology*, 2nd ed., ed. J.R. Etherington, pp. 290–330. Chichester: John Wiley.

Ball, M.C. (1988*a*). Ecophysiology of mangroves. *Trees*, **2**, 129–42.

Ball, M.C. (1988*b*). Organisation of mangrove forests along natural salinity gradients in the Northern Territory: an ecophysiological perspective. In *Northern Australia: Progress and Prospects, Vol. 2. Floodplain Research*, eds D. Wade-Marshall and P. Loveday, pp. 84–100. Darwin: Northern Australia Research Unit.

Ball, M.C., Cowan, I.R. & Farquhar, G.D. (1988). Maintenance of leaf temperature and the optimisation of carbon gain in relation to water loss in a tropical mangrove forest. *Australian Journal of Plant Physiology*, **15**, 263–76.

Barbour, M.G. (1970). Is any angiosperm an obligate halophyte? *American Midland Naturalist*, **89**, 105–20.

Bardsley, K.N. (1985). The effects of Cyclone Kathy on mangrove vegetation. In *Coasts and Tidal Wetlands of the Australian Monsoon Region*, eds K.N. Bardsley, J.D.S. Davie & C.D. Woodroffe, pp. 167–85. Darwin: Northern Australia Research Unit.

Barson, M. & Heatwole, H. (1985). Seashore vegetation. In *Seashore Ecology*, ed. T.M. Carefoot, modified by R.D. Simpson, pp. 194–216. St. Lucia: University of Queensland Press.

Beard, J.S. (1967). An inland occurrence of mangrove. *West Australian Naturalist*, 10, 112–15.

Bell, J.D., Pollard, D.A., Burchmore, J.J., Pease, B.C. & Middleton, M.J. (1984). Structure of a fish community in a temperate tidal mangrove creek in Botany Bay, New South Wales. *Australian Journal of Marine & Freshwater Research*, 35, 33–46.

Bettenay, E. (1986). Salt affected soils in Australia. *Reclamation & Revegetation Research*, 5, 167–79.

Bird, E.C.F. (1988). Physiographic indications of a sea-level rise. In *Greenhouse. Planning for Climate Change*, ed. G.I. Pearman, pp. 60–73. Melbourne: CSIRO.

Bird, J.F. (1981). Barilla production in Australia. In *Plants and Man in Australia*, eds D.J. Carr & S.G.M. Carr, pp. 274–80. Sydney: Academic Press.

Blaber, S.J.M. (1986). Feeding selectivity of a guild of piscivorous fishes in mangrove areas of north west Australia. *Australian Journal of Marine & Freshwater Research*, 37, 329–36.

Blaber, S.J.M., Brewer, D.T. & Salini, P. (1989). Species composition and biomass in different habitats of a tropical northern Australian estuary: their occurrence in the adjoining sea and estuarine dependence. *Estuarine, Coastal & Shelf Science*, 29, 509–31.

Blaber, S.J.M., Young, J.W. & Dunning, M.C. (1985). Community structure and zoogeographic affinities of the coastal fishes of the Dampier region of north-western Australia. *Australian Journal of Marine & Freshwater Research*, 36, 246–66.

Blake, T.J. (1981). Salt tolerance of eucalypt species grown in saline solution culture. *Australian Forest Research*, 11, 179–83.

Boston, K.G. (1981). The introduction of *Spartina townsendii* (s.l.) to Australia. *Melbourne State College Dept of Geography Occasional Papers*, 6, 1–57.

Boto, K.G. & Robertson, A.I. (1990). The relationship between nitrogen fixation and tidal exports of nitrogen in a tropical mangrove system. *Estuarine, Coastal & Shelf Science*, 31, 531–40.

Bridgewater, P.B. (1982). Phytosociology of coastal salt-marshes in the mediterranean climatic region of Australia. *Phytocoenologia*, 10, 257–96.

Bridgewater, P.B., Rosser, C. & de Corona, A. (1981). *The Saltmarsh Plants of Southern Australia*. Clayton: Dept. of Botany, Monash University.

Buckley, R.C. (1982). Short note – patterns in north Queensland mangrove vegetation. *Australian Journal of Ecology*, 7, 103–6.

Bunt, J.S. & Williams W.T. (1980). Studies on the analysis of data from Australian tidal forests ('Mangroves'). I. Vegetational sequences and their graphic representation. *Australian Journal of Ecology*, 5, 385–90.

Burbidge, A.A., McKenzie, N.L. & Kenneally, K.F. (1991). *Nature Conservation Reserves in the Kimberley, Western Australia*. Perth: Department of Conservation & Land Management.

Clark, J.S. (1990). Population and evolutionary implications of being a coastal plant: long-term evidence from the North Atlantic coasts. *Aquatic Sciences*, 2, 509–33.

Clarke, L.D. & Hannon, N.J. (1967). The mangrove swamp and salt marsh communities of the Sydney district. I. Vegetation, soils and climate. *Journal of Ecology*, 55, 753–71.

Clarke, L.D. & Hannon, N.J. (1969). The mangrove swamp and salt marsh communities of the Sydney District. II. The holocoenotic complex with particular reference to physiography. *Journal of Ecology*, 57, 213–34.

Clarke, L.D. & Hannon, N.J. (1970). The mangrove swamps and salt marsh communities of the Sydney District. III. Plant growth in relation to salinity and waterlogging. *Journal of Ecology*, 58, 351–69.

Clarke, L.D. & Hannon, N.J. (1971). The mangrove swamp and salt marsh communities of the Sydney District. IV. The significance of species interaction. *Journal of Ecology*, 59, 535–53.

Clough, B.F. (ed.) (1982). *Mangrove Ecosystems in Australia. Structure, Function and Management*. Canberra: Australian Institute of Marine Science/Australian National University Press.

Clough, B.F. (1987). Measurement of mangrove productivity. In *Mangrove Ecosystems of Asia and the Pacific. Status, Exploitation and Management*, eds C.D. Field & A.J. Dartnall, pp. 256–63. Townsville: Australian Institute of Marine Science.

Coles, S.M. (1979). Benthic microalgal populations on intertidal sediments and their role as precursors to salt marsh development. In *Ecological Processes in Coastal Environments*, eds R.L. Jefferies & A.J. Davy, pp. 25–42. Oxford: Blackwell Scientific Publications.

Congdon, R.A. & McComb, A.J. (1980). Productivity and nutrient content of *Juncus kraussii* in an estuarine marsh in south-western Australia. *Australian Journal of Ecology*, 5, 221–34.

Craig, G.F. (1983). *Pilbara Coastal Flora*. Perth: Western Australian Department of Agriculture.

Cribb, A.B. (1979). Algae associated with mangroves in Moreton Bay, Queensland. In *Northern Moreton Bay Symposium*, eds A. Bailey & N.C. Stevens, pp. 63–9. Brisbane: Royal Society of Queensland.

Curran, M. (1985). Gas movements in the roots of *Avicennia marina* (Forsk.) Vierh. *Australian Journal of Plant Physiology*, 12, 97–108.

Curran, M., Cole, M. & Allaway, W.G. (1986). Root aeration and respiration in young mangrove plants (*Avicennia marina* (Forsk.) Vierh.). *Journal of Experimental Botany*, 37, 1225–33.

Davie, J.D.S. (1985). The mangrove vegetation of the South Alligator river, Northern Australia. In *Coasts and Tidal Wetlands of the Australian Monsoon Region*, eds K.N. Bardsley, J.D.S. Davie & C.D. Woodroffe, pp. 133–51. Darwin: Northern Australia Research Unit.

Davie, J.D.S. (1987). Mangrove ecosystems in Australia. In *Mangrove*

Ecosystems of Asia and the Pacific, Status, Exploitation and Management,
eds C.D. Field & A.J. Dartnall, pp. 3–23. Townsville: Australian Institute
of Marine Science.

Davies, J.L. (1980). *Geographical Variation in Coastal Development,* 2nd ed.
London: Longman.

Duke, N.C. (1988). An endemic mangrove species, *Avicennia integra* sp. nov.
(Avicenniaceae), in northern Australia. *Australian Systematic Botany,* 1,
177–80.

Duke, N.C. (1991). A systematic revision of the mangrove genus *Avicennia*
(Avicenniaceae) in Australasia. *Australian Systematic Botany,* 4, 299–324.

Dunstan, D.J. (1990). Some early environmental problems and guidelines in
New South Wales estuaries. *Wetlands (Australia),* 9, 1–6.

Field, C.D. & Dartnall, A.J. (eds) (1987). *Mangrove Ecosystems of Asia and
the Pacific. Status, Exploitation and Management.* Townsville: Australian
Institute of Marine Science.

Flowers, T.J., Hajbagheri, M.A. & Clipson, N.J.W. (1986). Halophytes.
Quarterly Review of Biology, 61, 313–37.

Galloway, R.W. (1982). Distribution and physiographic patterns of Australian
mangroves. In *Mangrove Ecosystems in Australia. Structure, Function and
Management,* ed. B.F. Clough, pp. 31–54. Canberra: Australian Institute
of Marine Science/Australian National University Press.

Galloway, R.W., Story, R., Cooper, R. & Yapp, G.A. (1984). Coastal lands
of Australia. *CSIRO Division of Water & Land Resources Natural
Resources Series,* 1, 1–53.

Gibbs, P.J. (1986). The fauna and fishery of Wallis Lake. In *Wallis Lake –
present and future. Australian Marine Sciences Association (NSW),
Occasional Papers Series,* 86/2. Sydney: AMSA.

Herbert, D.A. (1951). The vegetation of south-eastern Queensland. In
Handbook of Queensland prepared for the 28th meeting of ANZAAS.
Brisbane: Government Printer.

Hodgkin, E.P., Birch, P.B., Black, R.E. & Hillman, K. (1985). *The Peel-Harvey
estuarine system. Proposals for management.* Western Australia, Department
of Conservation and Environment, Report No. 14.

Hutchings, P. & Saenger, P. (1987). *Ecology of Mangroves.* St Lucia: University
of Queensland Press.

Huxley, C.R. (1982). Ant-epiphytes of Australia. In *Ant–plant Interactions
in Australia,* ed. R.C. Buckley, pp. 63–73. The Hague: W. Junk.

Jansson, B.O., McIntyre, A.D., Nixon, S.W., Pamatmat, M.M., Zeitschell, B.
& Zijlstra, J.J. (1988). Coastal–offshore interactions – an evaluation of
presented evidence. In *Coastal–Offshore Ecosystem Interactions,* ed. B.-
O. Jansson, pp. 357–63. Berlin: Springer-Verlag.

Janzen, D.H. (1985). Mangroves. Where's the understorey? *Journal of Tropical
Ecology,* 1, 89–92.

Jefferies, R.L. & Rudmik, T. (1991). Growth, reproduction and resource
allocation in halophytes. *Aquatic Botany,* 39, 3–16.

King, R.J. (1981). The free-living *Hormosira banksii* (Turner) Descaine
associated with mangroves in eastern Australia. *Botanica marina,* 30, 341–50.

King, R.J., Adam, P. & Kuo, J. (1990). Seagrasses, mangroves and saltmarsh plants. In *Biology of Marine Plants*, eds M.N. Clayton & R.J. King, pp. 213–39. Melbourne: Longman Cheshire.

Kirkpatrick, J.B. & Glasby, J. (1981). Salt marshes in Tasmania – distribution, community composition and conservation. *Dept. of Geography, University of Tasmania, Occasional Paper*, 8, 1–62.

Knox, G.A. (1986). *Estuarine Ecosystems, A Systems Approach*. Boca Raton: CRC Press.

Larkum, A.W.D., McComb, A.J. & Shepherd, S.A. (ed.) (1989). *Biology of Seagrasses. A Treatise on the Biology of Seagrasses with special reference to the Australian Region*. Amsterdam: Elsevier.

Leadbitter, D. & Doohan, M. (1991). Wise use of wetlands – sustaining our fish harvest. In *Educating and Managing for Better Wetlands Conservation*, eds R. Donohue & W. Phillips, pp. 133–48. Canberra: Australian National Parks and Wildlife Service.

Lear, R. & Turner, T. (1977). *Mangroves of Australia*. St Lucia: University of Queensland Press.

Love, L.D. (1981). Mangrove swamps and salt marshes. In *Australian Vegetation*, ed. R.H. Groves, pp. 319–34. Cambridge University Press.

Macnae, W. (1966). Mangroves in eastern and southern Australia. *Australian Journal of Botany*, 15, 67–104.

Macnae, W. (1968). A general account of the fauna and flora of mangrove swamps and forests in the Indo-West Pacific region. *Advances in Marine Biology*, 6, 73–270.

Markley, J.L., McMillan, C. & Thompson, G.A. (1982). Latitudinal differentiation in response to chilling temperatures among populations of three mangroves, *Avicennia germinans*, *Laguncularia racemosa* and *Rhizophora mangle*, from the western tropical Atlantic and Pacific Panama. *Canadian Journal of Botany*, 60, 2704–15.

McLoughlin, L. (1985). *The Middle Lane Cove River. A History and a Future*. Sydney: Macquarie University, Centre for Environmental & Urban Studies.

Mepham, R.H. & Mepham, J.S. (1985). The flora of tidal forests – a rationalisation of the use of the term 'mangrove'. *South Africa Journal of Botany*, 51, 77–99.

Mitchell, M.L. & Adam, P. (1989a). The relationship between mangrove and saltmarsh communities in the Sydney region. *Wetlands (Australia)*, 8, 37–46.

Mitchell, M.L. & Adam, P. (1989b). The decline of saltmarsh in Botany Bay. *Wetlands (Australia)*, 8, 55–60.

Morton, R.M., Beumer, J.P. & Pollock, R.B. (1988). Fishes of a subtropical saltmarsh and their predation upon mosquitoes. *Environmental Biology of Fishes*, 21, 185–94.

Morton, R.M., Pollock, B.R. & Beumer, J.P. (1987). The occurrence and diet of fishes in a tidal inlet to a saltmarsh in southern Moreton Bay, Queensland. *Australian Journal of Ecology*, 12, 217–37.

Nixon, S.W. (1980). Between coastal marshes and coastal waters – a review of twenty years of speculation and research on the role of salt marshes in estuarine productivity and water chemistry. In *Estuarine and Wetland Processes*, eds P. Hamilton & K.B. MacDonald, pp. 437–526. New York: Plenum Press.

Norman, L. (1988). The 1987/88 Gippsland Lakes algal bloom. In *Gippsland Lakes Algal Bloom Seminar – Discussion Papers*, pp. 20–5. Melbourne: Dept. of Conservation, Forestry & Lands.

Oliver, J. (1982). The geographic and environmental aspects of mangrove communities: climate. In *Mangrove Ecosystems in Australia. Structure, Function and Management*, ed. B.F. Clough, pp 19–30. Canberra: Australian Institute of Marine Science/Australian National University Press.

Pen, L.J. (1983). Peripheral vegetation of the Swan and Canning estuaries 1981. *Western Australia: Dept. of Conservation & Environment Bulletin*, 113, 1–43.

Pepper, R.G. & Craig, G.F. (1986). Resistance of selected *Eucalyptus* species to soil salinity in Western Australia. *Journal of Applied Ecology*, 23, 977–87.

Pidgeon, I.M. (1940). The ecology of the Central Coast of New South Wales. III. Types of primary succession. *Proceedings of the Linnean Society of NSW*, 65, 221–49.

Poljakoff-Mayber, A., Symon, D.E., Jones, G.P., Naidu, B.P. & Paleg, L.G. (1987). Nitrogenous compatible solutes in native South Australian plants. *Australian Journal of Plant Physiology*, 14, 341–50.

Popp, M. (1984a). Chemical composition of Australian mangroves. I. Inorganic ions and organic acids. *Zeitscrifte für Pflanzenphysiologie*, 113, 395–409.

Popp, M. (1984b). Chemical composition of Australian mangroves. II. Low molecular weight carbohydrates. *Zeitshcrifte für Pflanzenphysiologie*, 113, 411–21.

Popp, M., Larher, F. & Weigel, P. (1984). Chemical composition of Australian mangroves. III. Free amino acids, total methylated onium compounds and total nitrogen. *Zeitshcrifte für Pflanzenphysiologie*, 114, 15–25.

Popp, M., Larher, F. & Weigel, P. (1985). Osmotic adaptation in Australian mangroves. *Vegetatio*, 61, 247–53.

Post, E. (1936). Systematische und pflanzengeographische Notizen zur *Bostrychia – Caloglossa* – Assoziation. *Revue algologie*, 9, 1–84.

Ridd, P., Sandstrom, M. & Wolanski, E. (1988). Outwelling from tropical tidal salt flats. *Estuarine, Coastal & Shelf Science*, 26, 243–53.

Robertson, A.I. (1986). Leaf-burying crabs: their influence on energy flow and export from mixed mangrove forests (*Rhizophora* spp.) in northeastern Australia. *Journal of Experimental Marine Biology & Ecology*, 102, 237–48.

Robertson, A.I. (1987). The determination of trophic relationships in mangrove – dominated systems: areas of darkness. In *Mangrove Ecosystems of Asia and the Pacific. Status, Exploitation and Management*, eds C.D. Field & A.J. Dartnall, pp. 292–304. Townsville: Australian Institute of Marine Science.

Robertson, A.I. & Duke, N.C. (1987). Insect herbivory on mangrove leaves in north Queensland. *Australian Journal of Ecology*, 12, 1–7.

Robertson, A.I. & Duke, N.C. (1990). Recruitment, growth and residence time of fishes in a tropical Australian mangrove system. *Estuarine, Coastal & Shelf Science*, 31, 723–43.

Roy, P.S. (1984). New South Wales estuaries: their origin and evolution. In *Coastal Geomorphology in Australia*, ed. B.G. Thom, pp. 99–121. Sydney: Academic Press.

Saenger, P., Hegerl, E.J. & Davie, J.D.S. (1983). Global status of mangrove ecosystems. *The Environmentalist*, 3, *Supplement 3*, 1–88.

Saenger, P. & Moverley, J. (1985). Vegetative phenology of mangroves along the Queensland coast. *Proceedings of the Ecological Society of Australia*, 13, 257–65.

Saenger, P., Specht, M.M., Specht, R.L. & Chapman, V.J. (1977). Mangal and coastal salt-marsh communities in Australasia. In *Wet Coastal Ecosystems*, ed. V.J. Chapman, pp. 293–345. Amsterdam: Elsevier.

Sands, R. (1981). Salt resistance in *Eucalyptus camaldulensis* Dehn. from three different seed sources. *Australian Forest Research*, 11, 93–100.

Semeniuk, V. (1987). Threats to, and exploitation and destruction of, mangrove systems in Western Australia. In *Mangrove Ecosystems of Asia and the Pacific. Status, Exploitation and Management*, eds C.D. Field & A.J. Dartnall, pp. 228–40. Townsville: Australian Institute of Marine Science.

Semeniuk, V., Kenneally, K.F. & Wilson, P.G. (1978). *Mangroves of Western Australia*. Perth: Western Australian Naturalists' Club.

Smith, T.J. (1987a). Seed predation in relation to tree dominance and distribution in mangrove forests. *Ecology*, 68, 266–73.

Smith, T.J. (1987b). Effects of seed predators and light levels on the distribution of *Avicennia marina* (Forsk) Vierh. in tropical tidal forests. *Estuarine, Coastal & Shelf Science*, 25, 43–51.

Smith, T.J. & Duke, N.C. (1987). Physical determinants of interestuary variation in mangrove species around the tropical coastline of Australia. *Journal of Biogeography*, 14, 9–19.

Smith-White, A.R. (1981). Physiological differentiation in a saltmarsh grass. *Wetlands (Australia)*, 1, 20.

Smith-White, A.R. (1988). *Sporobolus virginicus* (L.) Kunth in coastal Australia: the reproductive behaviour and the distribution of morphological types and chromosome races. *Australian Journal of Botany*, 36, 23–39.

Sternberg, L. da S.L. & Swart, P.K. (1987). Utilisation of freshwater and ocean water by coastal plants of southern Florida. *Ecology*, 68, 1898–905.

Stevens, G.N. & Rogers, R.W. (1979). The macrolichen flora from the mangroves of Moreton Bay. *Proceedings of the Royal Society of Queensland*, 90, 33–49.

Stevens, G.N. (1979). Distribution and related ecology of macrolichens on mangroves on the east Australian coast. *Lichenologist*, 11, 293–305.

Stevens, G.N. (1981). The macrolichen flora on mangroves of Hinchinbrook Island, Queensland. *Proceedings of the Royal Society of Queensland*, 92, 75–84.

Thom, B.G., Wright, L.D. & Coleman, J.M. (1975). Mangrove ecology and deltaic-estuarine geomorphology: Cambridge Gulf-Ord River, Western Australia. *Journal of Ecology*, **63**, 203–32.

Tomlinson, P.B. (1986). *The Botany of Mangroves*. Cambridge University Press.

Tomlinson, P.B., Bunt, J.S., Primack, R.B. & Duke, N.C. (1978). *Lumnitzera rosea* (Combretaceae) – its status and floral morphology. *Journal of the Arnold Arboretum*, **59**, 342–51.

Ungar, I.A. (1978). Halophyte seed germination. *Botanical Review*, **44**, 233–64.

Vanderzee, M.P. (1988). Changes in saltmarsh vegetation as an early indicator of sea-level rise. In *Greenhouse. Planning for Climate Change*, ed. G.I. Pearman, pp. 147–60. Melbourne: CSIRO.

Warren, J.H. & Underwood, A.J. (1986). Effects of burrowing crabs on the topography of mangrove swamps in New South Wales. *Journal of Experimental Marine Biology & Ecology*, **102**, 223–35.

Wells, A.G. (1982). Mangrove vegetation of northern Australia. In *Mangrove Ecosystems in Australia. Structure, Function and Management*, ed. B.F. Clough, pp. 57–78. Canberra: Australian Institute of Marine Science/Australian National University Press.

West, R.J. (1985). *Mangroves*. Agfact F2.0.1. Sydney: Dept. of Agriculture, New South Wales.

West, R.J. & Thorogood, C.A. (1985). Mangrove dieback in Hunter River caused by caterpillars. *Australian Fisheries*, **44** (9), 27–8.

West, R.J., Thorogood, C.A., Walford, T.R. & Williams, R.J. (1985). *An Estuarine Inventory for New South Wales, Australia*. Fisheries Bulletin, 2. Dept. of Agriculture, New South Wales, Sydney.

Wightman, G.A. (1989). *Mangroves of the Northern Territory*. Darwin: Conservation Commission of the Northern Territory.

Wilson, J.B. & Cullen, C. (1986). Coastal cliff vegetation of the Catlins region, Otago, South Island, New Zealand. *New Zealand Journal of Botany*, **24**, 567–74.

Wilson, P.G. (1980). A revision of the Australian species of Salicornieae (Chenopodiaceae). *Nuytsia*, **3**, 3–154.

Wolanski, E., Mazda, Y., King, B. & Gay, S. (1990). Dynamics, flushing and trapping in Hinchinbrook Channel, a giant mangrove swamp, Australia. *Estuarine, Coastal & Shelf Science*, **31**, 555–80.

Womersley, H.B.S. & King, R.J. (1990). Ecology of temperate rocky shores. In *Biology of Marine Plants*, eds M.N. Clayton & R.J. King, pp. 266–95. Melbourne: Longman Cheshire.

Woodroffe, C.D. (1988a). Relict mangrove stand on last interglacial terrace, Christmas Island, Indian Ocean. *Journal of Tropical Ecology*, **4**, 1–7.

Woodroffe, C.D. (1988b). Changing mangrove and wetland habitats over the last 8,000 years, Northern Australia and Southeast Asia. In *Northern Australia: Progress and Prospects. Vol. 2, Floodplains Research*, eds D. Wade-Marshall & P. Loveday, pp. 1–33. Darwin: Northern Australia Research Unit.

Woodroffe, C.D., Chappell, J.M.A., Thom, B.G. & Wallensky, E. (1986).

Geomorphological Dynamics and Evolution of the South Alligator Tidal River and Plains. Darwin: Northern Australia Research Unit.

Woodroffe, C.D., Thom, B.G. & Chappell, J. (1985). Development of widespread mangrove swamps in mid-Holocene times in northern Australia. *Nature*, 317, 711–13.

Zedler, J.B., Paling, E. & McComb, A. (1990). Differential responses to help explain the replacement of native *Juncus kraussii* by *Typha orientalis* in Western Australian salt marshes. *Australian Journal of Ecology*, 15, 57–72.

15

Aquatic vegetation of inland wetlands

M.A. BROCK

Plants in water

> for there are some plants which cannot live except in wet; and again these are distinguished from one another by their fondness for different kinds of wetness; so that some grow in marshes, others in lakes, others in rivers, others even in the sea . . . Some are water plants to the extent of being submerged, while some project a little from the water; of some again the roots and a small part of the stem are under the water, but the rest of the body is altogether above it.
>
> Theophrastus (370–*c*.285 BC)

FOR OVER two millennia, aquatic vegetation has fascinated philosophers and natural historians. Today we recognise the features common to aquatic plants across continents but we are also fascinated by their differences. Despite Australia's dryness as a continent, wetlands and aquatic vegetation occur in all regions. The range and extent of wetlands, permanent and temporary, saline and fresh, running and static, within all ecosystems gives plenty of scope for contrasts and comparisons of vegetation.

What makes a plant aquatic?

The ability to exist in an aqueous medium distinguishes aquatic plants from terrestrial: some plants are completely aquatic, others use both air and water, and some alternate between complete submersion and desiccation.

The central theme of this chapter is the survival of aquatic plants in space and time, in various aquatic habitats, of a range of wetland types, in different climatic regions of inland Australia. Water is examined as the distinctive medium of aquatic habitats. This approach contrasts with, and complements, previous reviews that have discussed wetland vegetation and plant communities from a more terrestrial viewpoint. Detailed reviews of vegetation types, community structure and species occurrence (with references prior to 1981) can be found in Beadle (1981), Briggs (1981) and Specht (1981). Reviews of Australian wetlands that contain useful comments and references on wetland vegetation include Jacobs (1983), Paijmans *et al.* (1985), McComb & Lake (1988), Finlayson & von Oertzen (1993) and Jacobs & Brock (1993).

The aqueous environment as a plant habitat

Wetlands are characterised by the aqueous rather than gaseous environment they provide for organisms. The properties of water largely define this environment. For aquatic life to persist, the water must allow for photosynthesis and other metabolic processes to occur.

Water has a greater density and viscosity than air and this allows many organisms to live in the aqueous medium without complex systems for support. The density of water is 775 times greater than air at standard temperature and pressure and the viscosity of water has about 100 times the frictional resistance to a moving particle as does air (Wetzel, 1975; Vogel, 1981). These properties help explain why many organisms live suspended in water but few live suspended in air. High density and viscosity also mean that moving water exerts huge forces on plants making rooted existence difficult in water that is flowing or subject to wave action. Unlike air, however, the aquatic medium at one place may be always flowing or always stationary, thereby allowing appropriately adapted plants to settle.

The carbon dioxide needed for photosynthesis and the oxygen needed for respiration are abundant in air but the capacity of water to dissolve these gases is much less. Aquatic vegetation must occupy habitats with sufficient dissolved gases for photosynthesis and respiration. Therefore many plants are excluded from anoxic waters.

Similarly, light, of the necessary wavelengths for photosynthesis, is more likely to be limiting for plants in water than in air because water absorbs and scatters light much more rapidly than air. The light climate of water can be further modified by dissolved and suspended sediments which again reduce light available for photosynthesis (see Kirk, 1983). Hence aquatic plants are absent from deep or turbid waters.

Relative to the air, water provides a tempered environment in which extreme fluctuations of water availability and temperature are ameliorated. Water is thermally a relatively stable environment for living things because of its heat-requiring and heat-retaining properties which are related to its high specific heat and the high latent heat of evaporation (Wetzel, 1975). Diurnal and seasonal temperature fluctuations occur more gradually and extremes are smaller in aquatic than terrestrial habitats. This means that plants living in habitats with alternating wet and dry phases must cope with the extremes of both aquatic and terrestrial life.

The ability of water to dissolve organic and inorganic substances influences aquatic plants. Dissolved oxygen is required for photosynthesis, carbon dioxide for respiration, and macro- and micro-nutrients for plant growth. Aquatic plants must live in an osmotically and chemically suitable environment; they are much more vulnerable to substances dissolved in water than their terrestrial counterparts are to substances diffusing in air. Increasing salinity and chemical pollution of water bodies are of major concern in human-modified systems as many aquatic plants cannot tolerate or escape these changes.

Thus the properties of both air and water are important to aquatic vegetation. Plants may occupy the water, the sediments under it, the atmosphere above it, or indeed any combination of these media. The air–water interface and the water–sediment interface are important parts of this environment; processes such as gas exchange and pollination occur at the air–water interface and much nutrient uptake occurs at the often anaerobic sediment surface. The relative positions of these interfaces change as water depth changes and this can be a crucial determinant of the biota present.

Marked seasonal or erratic rainfall usually combined with high evaporation rates characterise most Australian wetlands. Thus widely fluctuating water levels are typical and many wetlands actually dry completely for varying periods. With these fluctuations, the relative positions of the air, water and sediment change, thereby giving quite different habitat characteristics for the flora at different times.

Microvegetation and macrovegetation

The properties of the water environment allow it to support a suspended as well as a rooted vegetation, comprising organisms from a wide range of size classes and phylogenetic groups. Aquatic vegetation includes photosynthesisers from those groups which have always been aquatic (various algal groups), as well as some vascular plants of terrestrial ancestry which have re-adapted to the aquatic environment.

Often the smaller invisible groups of organisms are ignored. Micro-algae may form plant communities at least as complex as terrestrial vascular plant communities. One litre of lake water may contain millions of organisms from a variety of phyla, a huge range of sizes, and a full complement of trophic levels. To ignore these would give a false impression of the complexity and processes within the wetland ecosystem. Within the water body, at the microscopic scale, the phyto-plankton are the autotrophs surrounded by a range of herbivores, carnivores and detritivores.

Algae (micro and macro), bryophytes, ferns and the flowering plants (angiosperms) are all important components of aquatic vegeta-tion (Fig. 15.1). Yet the microflora of many Australian wetlands are either poorly described or have been ignored. Consideration of the vegetation of wetlands therefore must encompass both diversity and dynamics: of organisms, of habitats and of wetland ecosystems. This is a daunting task for one chapter, but the fascinating complexity of plant form and association in a range of wetland ecosystems is easier to convey.

Plants and their habitats in Australian wetlands

Australian wetlands: types and wetting frequency

Rainfall regularity rather than amount is one of the most important factors differentiating types of wetlands in Australia. Wetlands have seasonal or aseasonal (erratic) filling and drying patterns that reflect the rainfall patterns of the geographical region in which they occur (see Beadle, 1981). Both the tropical north, with its monsoonal sum-mer rainfall, and the temperate southern areas of Australia, with dominant winter rainfall, have seasonal climates. Even though seasonal rainfall is reliable, variability in the timing and length of the wet phase may be important to some species (Beadle, 1981; Taylor & Tulloch, 1985). In contrast to these seasonal climates, the arid central areas of Australia with low, erratic rainfall and areas with higher rainfall with neither a regular winter nor summer dominance of rainfall (e.g. Northern Tablelands of New South Wales) have aseasonal rainfall patterns with high variability within and between years.

Predictability or unpredictability of rainfall may be of crucial importance for the plant communities in wetlands. Seasonal environ-

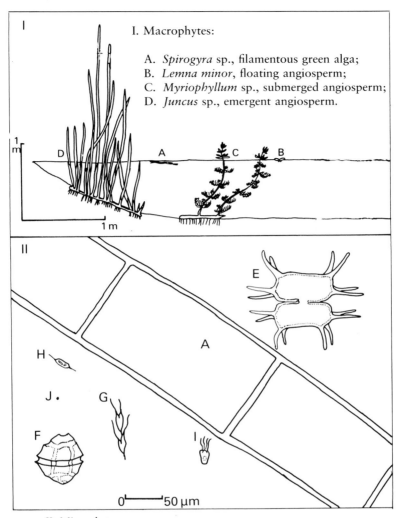

I. Macrophytes:

A. *Spirogyra* sp., filamentous green alga;
B. *Lemna minor*, floating angiosperm;
C. *Myriophyllum* sp., submerged angiosperm;
D. *Juncus* sp., emergent angiosperm.

II. Microphytes:

A. *Spirogyra* sp., cells from single filaments of green alga (see Macrophyte A);
E. *Xanthidium antilopaeum*, planktonic desmid;
F. *Peridinium gutwinskii*, planktonic dinoflagellate;
G. *Dinobryon bavaricum*, planktonic golden-brown alga;
H. *Centritractus belanophorus* planktonic yellow-green alga;
I. *Pyramimonas* sp., planktonic unicellular green alga;
J. Bacterium.

Fig. 15.1. Relative sizes and phylogenetic groups of some major components of the aquatic flora. (Adapted from Sainty & Jacobs, 1981; Ling & Tyler, 1986; Moss, 1988.)

Table 15.1. *Australian inland wetland types: characteristics and wetting frequencies (adapted from Paijmans et al., 1985)*

Characteristics

Lakes: open water generally >1 m; little or no persistent emergent vegetation. Wetting frequency: P, S, I or E.

Swamps: water, where present, <1 m; generally vegetated, persistent emergent vegetation. P, S, I, or E.

Land subject to inundation: water not present long enough for wetland vegetation to develop. S or I.

River and creek channels: water, when present, has a directional flow; development of vegetation depends on the intensity of flow and frequency and duration of wetting. P, S, I or E.

Wetting frequency

Permanent and near permanent (P): annual inflow exceeds minimum annual loss in 90 per cent of years.

Seasonal (S): alternately wet and dry every year according to season.

Intermittent (I): alternately wet and dry but less frequently and regularly than seasonal wetlands.

Episodic (E): annual inflow is less than minimum annual loss in 90 per cent of years; dry most of the time with rare and very irregular wet phases.

---

ments are predictable and may select for groups of plants and animals which require seasonal influences for some stage of their life cycle, whereas aseasonal environments are unpredictable in rainfall timing and amount and will select for a biota which can cope with erratically timed events.

Several wetland types occur within each Australian climatic region (Paijmans *et al.*, 1985). Four inland (non-coastal) wetland types are distinguished (lakes, swamps, land subject to inundation, and river and creek channels) and wetting frequency (permanent, seasonal, intermittent and episodic) can be defined within each (Table 15.1).

Plant habitats

The common environmental factor which creates habitats for wetland plants is the presence of water above the soil surface, at least temporarily (Paijmans *et al.*, 1985). Aquatic habitats cannot be sharply distinguished from terrestrial because of the marked seasonal or aseasonal fluctuations in the water table (Sculthorpe, 1967). This is particularly true of the many Australian wetlands which are subject

Table 15.2. *Plant habitats within a typical Australian wetland; the types of plant growth-form each habitat supports and the wetland types in which each occurs (L – lake, S – swamp, R – rivers and creeks, In – land subject to inundation)*

Habitat	Growth form	Wetland type
Edge	terrestrial plants, emergent aquatics	L,S,R,In
Aquatic–terrestrial ecotone	terrestrial plants, emergent and some submerged aquatics	L,S,R
Shallow water to 0.5 m	emergent, submerged and floating aquatics, suspended and attached microalgae	L,S,R
Deeper open water	suspended microalgae, submerged rooted aquatic plants and attached microalgae	L,R
Bottom sediments	rooted submerged aquatic plants, attached algae, fungi and bacteria	L,S,R
Air–water interface	floating aquatics, floating microflora of algae, pollen, seeds and spores	L,S,R,In
Sediment–water interface	attached algae, fungi and bacteria, seeds	L,S,R,In

to wide seasonal or erratic climatic fluctuations: some habitats with standing water dry seasonally, others do so only in dry years, just as some floodplains flood seasonally and others only in wet years.

Typically, a lake or lagoon (with little directional water flow) will have distinct habitats supporting suites of plant species with particular growth forms. Although these habitats form a continuum, seven are distinguished; the edge or littoral zone, the aquatic–terrestrial ecotone, shallow water, deeper open water, the bottom sediments, the air–water interface and the water–sediment interface (Table 15.2). In many systems microhabitats can be further defined within each habitat. Each habitat will have a characteristic flora made up from species of submerged, emergent or floating species of angiosperms, bryophytes, ferns, charophytes and macro- and micro-algae. Each habitat may not occur in all wetland types (Table 15.2).

Growth form

Growth form may indicate what parts of the aquatic system plants occupy. Floating (air–water interface), submerged and emergent (with photosynthetic parts above the water surface) are three often-used categories (Sculthorpe, 1967) which can be further divided to describe the use of habitats in Australian waters (Sainty & Jacobs, 1981). Figure 15.2 combines and extends these classifications to describe the range of macroflora and microflora found in Australian wetlands.

Aquatic plants of each growth form (Fig. 15.2) are found in some habitats and not others and are more common in some wetland types than others (Tables 15.1 & 15.2). Many habitats are reduced in running waters as strong unidirectional water flow prevents plant establishment. In many swamps, the extent of the open water and the air–water interface habitats are reduced because of the density of rooted emergent vegetation. Thus habitat availability is a dynamic product of interactions between the physico-chemical properties of the wetland and its vegetation.

Submerged and suspended growth forms occupy habitats which hold water long enough for their development and where the water is not too erosive for rooted plants and planktonic communities to develop. These habitats are least likely in areas of land subject to inundation and in flowing waters.

The microflora occurs either free and suspended (phytoplankton) or attached (periphyton) within the waterbody. Microflora will not develop in the edge or aquatic–terrestrial ecotone where drying is regular. The algal components of the flora are less tolerant of drying than those of terrestrial origin that have readapted to the aquatic environment.

Emergent aquatic plants are rooted in the bottom sediments and project through the water and the air–water interface to the aerial environment. Many Australian swamps, lakes and rivers have such emergent plants in their shallow water, aquatic–terrestrial ecotone and edge zones; shallow waters and wetland edges tend to be dominated by stands of one or two species, often from the genera *Typha*, *Phragmites*, *Baumea* or *Eleocharis*. Water level fluctuations in some aquatic–terrestrial ecotone habitats often maintain a species-rich suite of low growing herbs. Emergent trees and shrubs are more likely to occur in floodplains and land subject to intermittent or episodic, rather than more regular (or persistent), inundation.

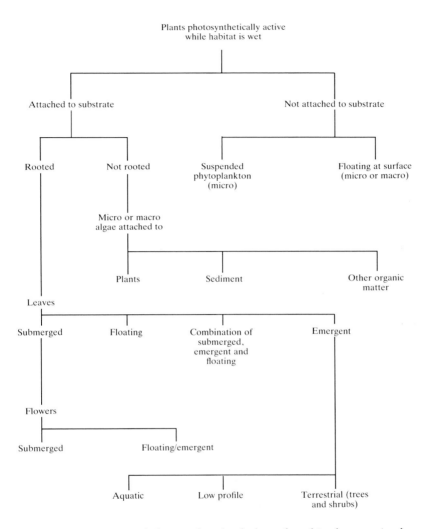

Fig. 15.2. Growth forms of wetland plants found in the aquatic phase of habitats in Australian wetlands.

Plants and plant groupings

Plants (both macro and micro) found in Australian wetlands come from a diverse range of plant groups and life-forms (Fig. 15.2). They range from algal unicells (a few microns) to the smallest angiosperm, the floating *Lemna* (a few millimetres), to large emergent river red gums *Eucalyptus camaldulensis* (several metres in trunk diameter). It is common to find all three in different habitats in the same floodplain (see Table 15.2): the unicellular algae suspended in the phytoplankton in the open water, the *Lemna* floating with other plant species at the air–water interface, and the river redgum, a permanent resident of the occasionally flooded aquatic–terrestrial ecotone. Many habitats will have species from a range of evolutionary groups.

Cyanobacteria or blue-greens

The kingdom Monera (prokaryotes) is represented by the Cyanobacteria, or blue-greens, and by other bacterial groups. Some are photosynthetic and others are detritivores which are important in nutrient cycling in wetlands. At least twelve common genera of blue-greens occur in Australian wetlands. Some are nitrogen-fixing and are important in soil and sediment processes (see Parbery, 1970). Other blue-greens form symbiotic associations; for example, the association of the blue-greens *Anabaena azollae* with the floating fern *Azolla filiculoides* where the fern provides a habitat for the blue-green which fixes nitrogen that is available to the fern. Unfortunately, some blue-greens have made a less welcome name for themselves as they can become dominant and form blooms in eutrophic wetlands. Some of these, particularly species of *Microcystis* and *Anabaena*, produce toxins which can poison other organisms (both aquatic and terrestrial).

Fungi

The kingdom Fungi, although not photosynthetic, is an important group of decomposers in wetlands. It is represented in Australian wetlands mainly by the aquatic group Hyphomycetes (Cowling & Waid, 1963; Thomas, Chilvers & Norris, 1989). These are often detected only as spores floating at the air–water interface or on decaying plants. The hyphomycetes and other aquatic fungi vary from unicellular to a complex mass of filaments generally saprophytic on submerged surfaces. They are mostly aerobic and occur in lakes and swamps as well as in running waters; polluted freshwaters are favoured habitats whereas they are not recorded from saline lakes. Their abundance and species-richness is probably higher than recorded as befits their major role as decomposers of leaf litter, plankton and other organisms. Mycorrhizal associations with wetland plants are poorly known and may well deserve further investigation particularly in temporary wetlands without anaerobic sediments (P. McGee, pers. comm.).

Algae

Algae of Australian inland wetlands come from various phyla of protists and plants. The microalgae are dominated by a diverse range of species including desmids (Chlorophyta), dinoflagellates (Dinophyta), diatoms (Bacillariophyta), euglenoids (Euglenophyta) and golden algae (Chrysophyta). A wide variety of species occurs, many of them cosmopolitan, at least at the generic level. However, many are locally or regionally specific or specific to particular types of water body (John, 1983; Ling & Tyler, 1986).

Microalgae occur in community groupings in several habitats and wetland types. Phytoplankton communities (suspended microalgae and cyanobacteria) are typical of open water habitats, and are more diverse and common in still than running waters. Different suites of algae, often dominated by diatoms, occupy the bottom sediments of lakes, swamps or streams, whilst other diatom-dominated communities of epiphytes use submerged surfaces such as macrophyte leaves as substrates suspended at an appropriate depth for photosynthesis.

Phytoplankton communities characteristically fluctuate in numbers, species composition and photosynthetic activity within a wetland. Diurnal, seasonal and year-to-year variations in productivity and species composition occur in most waterbodies. Variation in light, nutrients, turbidity, temperature and the associated zooplankton grazers in the community are some of the causes of change (Walker & Hillman, 1981; Ganf, 1982; Ganf & Oliver, 1982; Merrick & Ganf, 1988).

The macroscopic algae of inland waters are dominated by charophytes and filamentous green algae. These occur mostly in open water, often in association with a suite of aquatic flowering plants. The charophytes occur widely, submerged in permanent and temporary waters, both fresh and saline (Brock, 1986; Casanova & Brock, 1990; Brock & Casanova, 1991a). This situation contrasts with northern hemisphere reports of charophytes as species mainly from deep, permanent, clear waters. Charophytes are restricted to the habitats with free water as they cannot survive desiccation; they maintain themselves in temporary wetlands by resistant oospores in the seed bank.

The almost complete absence in inland waters of representatives of the marine red and brown macroalgae is noteworthy.

Aquatic plants of terrestrial origin

The 'terrestrial' groups of plants are relatively well represented in inland waters compared with marine waters. There are Australian aquatic mosses, liverworts, ferns and angiosperms but no true aquatic gymnosperms. These groups evolved in terrestrial environments and their aquatic members have adaptations for life in water that are modifications of previous terrestrial structures (see Sculthorpe, 1967).

Bryophytes, ferns and flowering plants occur in plant communities in permanent and temporary Australian inland lakes, swamps and rivers as well as on land subject to inundation.

The mosses are represented by some that grow in the deep clear waters of crater lakes around Mount Gambier in South Australia as well as by the better known *Sphagnum* species of alpine and some coastal wet heaths of Tasmania and the temperate mainland.

Liverworts of the genera *Riccia* and *Ricciocarpus* occur in mixed-species communities in open water, at the air—water interface, and stranded in the aquatic—terrestrial ecotone of freshwater lagoons. In the same systems ferns also occupy a variety of aquatic habitats. The floating fern *Azolla filiculoides* occupies the air—water interface and can also survive stranded in the aquatic—terrestrial ecotone, whereas *Isoetes muelleri* (one of the fern allies) occurs submerged in shallow or deeper water of both permanent and temporary systems. Whilst one species of nardoo, *Marsilea*, may be rooted in the sediment with floating leaves at the surface, others may grow completely submerged or as stranded plants.

The flowering plants are well represented in Australian wetland types and habitats by a wide variety of species from a diverse set of flowering plant families. Many aquatic plant genera are cosmopolitan; others, particularly many of the semi-aquatic herbs, derive from the tropics, Asia or Malaysia. Many of the woody groups (trees, shrubs and heaths) are endemic or confined to the southern hemisphere. Tropical derivation of plant groups is more obvious in northern Australia in contrast to endemism in the south (Sculthorpe, 1967; Briggs, 1981; Taylor & Dunlop, 1985; Dunlop & Webb, 1991). Only one genus, *Maundia* (Juncaginaceae), is endemic to Australia, and *Liparophyllum* (Menyanthaceae) is confined to Australia and New Zealand (Beadle, 1981). Many cosmopolitan genera have endemic or Australasian species, e.g. *Ruppia tuberosa* is endemic and *R. megacarpa* and *R. polycarpa* are Australasian.

Particular families dominate Australian wetland communities. Chenopodiaceae, Cyperaceae, Epacridaceae, Juncaceae, Myrtaceae, Poaceae, Proteaceae and Restionaceae dominate edge habitats, whereas a group of true aquatics from cosmopolitan families dominate open waters (Aston, 1973; Sainty & Jacobs, 1981, 1988). Some genera are confined to the tropics, e.g. *Aldrovandra*, *Caldesia* and *Nelumbo*, whilst other tropical genera extend down the east coast to temperate latitudes, e.g. *Aponogeton* and *Nymphaea*. Other genera are confined to the temperate south, e.g. *Damasonium* and *Villarsia*, and some to the southeast, e.g. *Alisma*. Some genera are widespread across the continent with different species in each region, e.g. *Myriophyllum, Ottelia, Potamogeton* and *Utricularia*, whilst a few species occur throughout the continent, e.g. *Lemna minor*. Isolation may at least partially account for the relative paucity of aquatic vascular plants in Tasmania and the southwest of Western Australia (Beadle, 1981).

The greatest floristic diversity is found in seasonal and intermittent temporary wetlands which otherwise do not experience environmental extremes (Paijmans *et al.*, 1985). The vegetation tends to become poorer in species and to exhibit more special adaptations as environmental factors in the habitat shift towards the extremes; for example, the group of plants occupying highly saline lakes consists of a very few angiosperms and algae with specialisations to cope with wide fluctuations of salinity and drying (Brock, 1983, 1986).

Permanent wetlands, particularly deeper habitats, often have plant communities dominated by only one or a few species which may reflect the small suite of species of angiosperms which have adapted to a fully aquatic life. This contrasts with the greater richness of species found in shallow and wet–dry ecotone habitats which are generally more extensive in swamps than rivers or lakes.

Aquatic and terrestrial wetland vegetation

Community structure

The structure of an aquatic plant community is determined by plant growth forms, water regime and hence the habitats available in the wetland. Climate, sediment type and topography interact with aquatic parameters to determine what plant types and hence what vegetation communities occur. Zonation of growth form is often described in relation to water depth. Australian systems are similar in aquatic plant community structure to those elsewhere in the world. However, fluctuations in water regime on various time scales probably plays a more significant role in Australian wetland systems.

'Terrestrial' wetland vegetation types with a forest, woodland, scrub, heath, shrub or grassland structure often occur with an understorey of aquatic sedges and herbs and microscopic suspended or attached aquatic organisms. In some wetlands, terrestrial vegetation formations exist without a well-developed aquatic community, just as some aquatic formations do not include terrestrial components. As each terrestrial vegetation formation has been considered separately in other sections of this book, only the general occurrence of these types in wetlands is considered here (Table 15.3). Details of the floristics of the vascular plants of Australian wetland communities are well described by Beadle (1981) and Briggs (1981).

These 'terrestrial' vegetation types dominate the edge and the aquatic–terrestrial ecotone habitats where water depth and its fluctuation influence community composition. Small changes in water level often make the difference between drought, waterlogging or flooding. The vegetation of these habitats may be subject to inundation episodically, intermittently, seasonally or semi-permanently.

Aquatic communities of macrophytes and microphytes dominate the habitats within the waterbody itself. Emergent plants dominate shallow water, submerged and rooted floating-leaved plants deeper water up to several metres (depending on light penetration), and floating plants the air–water interface. The rooted emergent and submerged and floating-leaved plants also use the bottom sediments and the sediment–water interface as part of their habitat. The microphytic aquatic communities can either be suspended (planktonic) or attached. The larger aquatic plants (hydrophytes) are common in permanent water and temporary water which persists long enough for aquatic communities to develop. Communities of hydrophytes are most common in lakes and swamps and generally less well developed in rivers (because of the unidirectional flow) and land subject to inundation (because duration of inundation often is insufficient for communities to develop and persist until reproduction).

Spatial and temporal changes in aquatic vegetation

Changes in aquatic vegetation can occur over a range of spatial and temporal scales. In temporary wetlands, inundated seasonally, intermittently or erratically, all habitats (Table 15.2) are potentially available for aquatic plants with all types of growth form (Fig. 15.2). However, habitats will vary in time and space depending on wetting frequency, extent, timing and duration; hence all habitats may not be available at once and one particular space may change its habitat type over time as a wetland dries or refills. These habitat changes, caused by water depth fluctuations, may select for differences in species composition or for morphological plasticity within a species. Examples of this selection process are widely reported in Australian and other aquatic plants where heterophylly (growing leaves whose

Table 15.3. *Occurrence of vegetation types and growth form in various wetland habitats, wetland types and water regimes in Australia (vegetation types after Briggs, 1981)*

Wetland vegetation type	Habitats	Plant form	Wetland type	Water regime
Swamp Forest	E,ATE	T	L,S,R,I	S,I,E
Swamp Woodlands	E,ATE	T	L,S,I	S,I,E
Swamp Scrub and heath	E,ATE	T	S,I	S,I,E
Swamp Shrubland	E,ATE	T	L,S,R,I	S,I,E
Sedgelands	E,ATE,SW	E	L,S,R,I	P,S
Swamp Grasslands	E,ATE	T,E,Fl	L,S,R,I	P,S,I,E
Swamp herbland of aquatic macrophytes	E,ATE,SW DW,B,AWI, SWI	E,S,F P,A	L,S,mainly R – sometimes I – rare	P,S,I,E
Aquatic microphytes suspended	SW,DW	P,A	L,S	P,S,I,E
attached	ATE,B, SWI,AWI	A,F	L,S	P,S,I,E

Habitats		Plant form		Wetland		Water regime	
E:	edge	T:	terrestrial	L:	lakes	P:	permanent
ATE:	aquatic– terrestrial ecotone	S:	submerged	R:	rivers	S:	seasonal
		F:	floating	S:	swamps	I:	intermittent
		E:	emergent	I:	land	E:	episodic
SW:	shallow water	P:	suspended (planktonic)		subject to inundation		
DW:	deep water	A:	attached				
B:	bottom sediments						
AWI:	air–water interface						
SWI:	sediment–water interface						

shapes differ below and above water) allows a plant to live in spatially different parts of the aquatic system (Sculthorpe, 1967; Aston, 1973; Sainty & Jacobs, 1981).

Some aquatic plants exist in the same space but cope with habitat change over time by morphological plasticity and reproductive flexibility. *Myriophyllum variifolium*, an aquatic plant common in permanent and temporary lagoons in the northern New South Wales, is a good example. It grows upright with underwater and above-water leaves in deep water, prostrate with entire leaves when stranded on mud and can change form within hours or days of water level change. It reproduces prolifically, both sexually and asexually, can exist for long periods as dormant seeds in dry sediments, and can germinate and establish under a range of conditions, thereby maintaining itself under the full range of water regime fluctuation (Brock, 1991; Brock & Casanova 1991*b*). Other species in these temporary freshwater lagoons and in saline lakes show similar flexibility of life-cycle characteristics (Brock 1986; Casanova & Brock, 1990; Brock & Casanova, 1991*a*).

Unpredictable environmental fluctuations help maintain a variety of semi-permanent, seasonal intermittent, and episodic temporary wetlands on the Northern Tablelands of New South Wales. Diverse aquatic communities are maintained in the various aquatic habitats when the lagoons are wet, yet when dry the lake beds may be colonised by a variety of native and introduced terrestrial species leaving the aquatics to persist as seed in the dry sediments. In Llangothlin Lagoon at 1500 m altitude and 31° S over 45 species of angiosperms, ferns, bryophytes and charophytes were recorded in the aquatic–terrestrial ecotone and shallow water habitats in January 1992 after a wet summer, yet four months previously at the end of a dry autumn and winter the aquatic–terrestrial ecotone was bare, cracked and dry sediment with only a few vegetative shoots of *Myriophyllum variifolium* struggling to survive. Fast recolonisation from germinating seeds occurred after rewetting. This type of water level fluctuation is typical and plants may experience flooding, waterlogging or drying at irregular intervals (from months to years to decades) and for irregular durations (from days to weeks, months or years). Plants survive as seeds or resistant vegetative propagules which form a seed bank in the sediments. These germinate and grow when the habitats rewet.

Factors other than water level also fluctuate on different spatial and temporal scales and wetland vegetation responds accordingly. The changes in distribution and abundance of co-existing aquatic species in one lagoon were attributed to both temporal and spatial variation in wind, wave, chemical and water level fluctuations which caused different reproductive and growth responses of each species (Yen & Myerscough, 1989 *a,b*). On the floodplains of the River Murray horizontal (wave action and currents) rather than vertical gradients (depth and change in water level) were important in defining the riparian vegetation (Roberts & Ludwig, 1991).

Even within the more predictable seasonally inundated floodplains of the monsoonal north of Australia, mosaics of aquatic vegetation do not have constant patterns in space and time. Monospecific populations of grasses and sedges, such as the annual, *Oryza meridionalis*, the tufted perennial *Fimbristylis tristachya*, and rhizomatous and tuber forming species of *Eleocharis*, change in distribution from wet season to wet season (Dunlop & Webb, 1991). Annual species regenerating from seed are particularly susceptible to variation between years. The lotus lily, *Nelumbo nucifera*, from permanent swamps, appears to have a cycle of establishment to moribundity to dormancy over several seasons. These cycles may be due, in part, to grazing by feral buffalo (Dunlop & Webb, 1991). Feral buffalo are also responsible for changes in the development of characteristic tropical floating mats in permanent swamp and billabongs. Floating mats first develop as rafts of floating aquatics (*Pistia stratiotes* and *Azolla pinnata*) which are immobilised by rooted vines (*Ludwigia adscendens* and *Ipomoea aquatica*) and then colonised by vines, shrubs and trees as detritus accumulates. These fragile mats may be periodically set adrift by floodwaters and are susceptible to damage by water buffalo (Hill & Webb, 1982; Dunlop & Webb, 1991).

Environmental fluctuation is a normal part of wetland function. Often these fluctuations are treated as disturbance (Sousa, 1984) and the response of the vegetation as one of regeneration and recolonisation. Many of the species in these wetlands are those with life-cycle patterns that maximise their fitness in changing environments. It also has been suggested that species richness increases under moderate levels of disturbance (Grime, 1979). The species richness in the temporary wetlands of southern Australia, described above and documented by McIntyre, Ladiges and Adams (1988), supports this contention of Grime.

Natural fluctuations of water regime have been modified by human activities in many Australian wetlands. These modifications have altered, or in some cases destroyed, the wetland vegetation. Drainage, dam construction and raising or lowering of water levels all serve to truncate the natural fluctuations. Wide acceptance of the value of wetlands and their vegetation has only occurred in the 1980s; recognition of the function of fluctuations of water regime in maintaining these systems and their vegetation has been even slower.

Special Australian wetland types and their vegetation

Some wetland types could be considered unique to, or characteristic of, parts of inland Australia. The vegetation of some ephemeral (temporary) wetlands, salt lakes and the River Murray are chosen for specific consideration.

Intermittent floodplain lakes

Most intermittent lakes are in the arid and semi-arid areas of Australia and many are parts of the floodplains of the large internal drainage basins. Most have lunettes (small dunes) on the eastern side which help create the lake basin (Paijmans *et al.*, 1985). Water regime and its long term and shorter term variations are critical in defining plant habitats in time and space. This has been demonstrated for the Macquarie Marshes (Knights, 1980) and Paroo (M. Maher, pers. comm.) systems.

These lakes commonly develop an aquatic flora when full; submerged aquatics commonly include *Myriophyllum verrucosum*, *Najas tenuifolia*, and species of *Nymphoides* and *Vallisneria* and charophytes. As the water level falls, other aquatics, such as species of *Marsilea* (nardoo), *Eleocharis*, *Cyperus* and *Damasonium*, grow and often complete their life cycle on the drying mud. The aquatic–terrestrial ecotone can contain a variety of emergent herbs, sedges and colonising species and may also contain shrubs or low trees such as lignum *Muehlenbeckia cunninghamii* and trees such as the river redgum *Eucalyptus camaldulensis* or the river oak *Casuarina cunninghamiana*. Many of these species survive by leaving long-lived resistant seeds in the sediments, but some, such as lignum, also sprout from old rootstocks which can survive prolonged inundation as well as several years without flooding (Beadle, 1981; Paijmans *et al.*, 1985; Sainty & Jacobs, 1990).

Salt lakes

The plant communities of saline lakes are largely determined by salinity and water regime and the level (amplitude) and timing (periodicity) of fluctuations of these factors. Australia has a wide range of saline wetlands which range from brackish to hypersaline (up to six times sea water) and from permanent to temporary (seasonal, intermittent or episodic). Even within a saline lake both salinity and water regime vary with season and from season to season. Salt-lake plant communities have characteristic floristics and structure over wide ranges of salinity, water regime and geographic region. A relatively small group of species, tolerant of these environmental extremes and fluctuations, dominate these communities.

The aquatic angiosperms that thrive in salt are relatively limited. The marine angiosperms or seagrasses include about 11 closely related genera, none of which occur in inland waters (Beadle, 1981). The genus *Ruppia* (Potamogetonaceae), which sometimes occurs in estuaries with these seagrasses, dominates Australian inland wetlands which range from fresh to hypersaline and permanent to very temporary. The three species of this genus in Australia occur in communities with species of *Lepilaena* (Zannichelliaceae) and with a single member of the charophytes, *Lamprothamnium papulosum*. Although the species composition of these shallow water aquatic communities is limited, their productivity can be high and the communities dense. Species tend to be limited by salinity and/or drying (Fig. 15.3), with freshwater species dropping out in the brackish end of the salinity range. The limit for growth of a submerged angiosperm community is a salt concentration three times that of sea water. The salt marsh communities which occupy the aquatic–terrestrial ecotone habitats of many of these saline wetlands are considered elsewhere (Adam, this volume).

The saline lake microflora is also limited by salinity and drying, in diversity but not necessarily in numbers. The most tolerant algal species is the unicellular green alga *Dunaliella salina*, which gives many salt lakes a red/pink appearance because it accumulates carotene pigments in high salinities. This species tolerates salinities up to five times sea water; the only organisms which tolerate salinities higher than this are a few species of halobacteria (Borowitzka, 1981).

The saline lake plants cope with fluctuations of salinity and drying by a variety of physiological, morphological and life-cycle adaptations. For example, species of *Ruppia* accumulate proline as an inert osmotic substance to prevent salt entering the cytoplasm (Brock, 1981); *Ruppia* varies its morphology in different environments and produces prolific numbers of both seeds and resistant asexual propagules which remain alive on the dry salt crust until the wetland refills (Brock, 1983). *Dunaliella salina* has a similar range of adaptations for salinity and drying. It produces glycerol as an osmotic protector

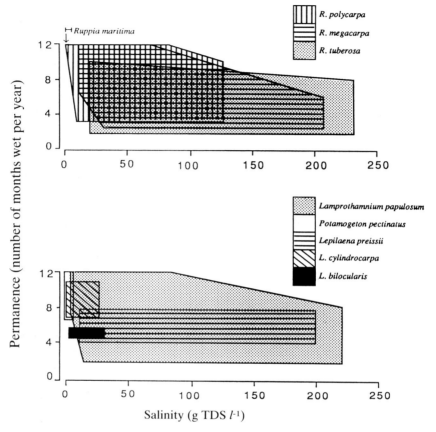

Fig. 15.3. Ranges of salinity and habitat permanence for macrophyte species in salt lakes in Australia: (*a*) *Ruppia* spp.; (*b*) other species. TDS 1^{-1} = total dissolved solutes per litre. (After Brock, 1986.)

and carotene probably serves as a protection against high light intensity. *Dunaliella salina* varies its morphology in different salinities and forms resistant spores which tolerate drying (Borowitzska, 1981). The feasibility of commercial harvesting of *Dunaliella* from salt lakes for extraction of carotenes (for food colouring) and possibly glycerol has been the subject of recent research (McComb & Lake, 1990).

The River Murray wetlands

The Murray–Darling catchment drains 1 000 000 square kilometres or one-seventh of Australia. The Murray is a large floodplain river 2530 km in length and its major tributary the Darling is 2740 km in length (Pressey, 1990). The need for management of river resources has led to recent efforts to coordinate our understanding of how this river system functions (see Mackay & Eastburn, 1990).

Floodplain rivers comprise three major wetland types: the mainstream, the intermittently wet floodplain, and more permanent floodplain waters such as billabongs (Hillman, 1986). The River Murray has an extensive floodplain of 9000 sq km which contains many billabongs (Pressey, 1986). This river has a history of human modification of the natural flow regime since the 1820s when clearing of vast areas of the catchment began. Since then, water diversion for urban uses and irrigation and the construction of reservoirs, locks and weirs have altered the natural flow regimes (Close, 1990). Changes to the frequency, timing and magnitude of flooding have altered many of the habitats in the river and billabongs, and on the floodplain (Walker, 1985). However, this river system contains characteristic floodplain and billabong communities which exemplify the interrelated functioning of different wetland types within the same catchment.

Billabongs

Billabongs are cut-offs or backwaters of the river system which are generally non-flowing and without directional flow except in times of flood. Their over-riding characteristic is their physico-chemical and hence biological variability over both space and time (Hillman, 1986). They are flooded periodically and then gradually dry, sometimes to dehydration. The biota must cope with this highly variable and often unpredictable environment.

The billabongs of the Murray form a heterogeneous mosaic of habitats and biological communities on the floodplain (Boon *et al.*, 1990). They are nutrient-rich environments that support a high primary productivity of submerged, emergent and floating hydrophytes as well as of planktonic and attached microflora which in turn support a complex community of heterotrophs (invertebrate and vertebrate) and detritivores. The full range of habitats (Table 15.2) are occupied by this diverse billabong flora. The species composition can vary in space and time both within and between billabongs. The factors determining such changes, although not completely understood, include fluctuations in water regime, sediment and water chemistry, turbidity and disturbance such as grazing (Boon *et al.*, 1990; Boon & Sorrell, 1991).

The aquatic vegetation of the billabongs is much more diverse than that of the adjacent river, and those plants that do occur in the river may be dependent on the billabong populations for re-establishment from seed or vegetative propagules. Whereas the river influences the billabongs in physical ways (hydrology and morphology), the billabongs influence the river in biological ways: as a major source of organic matter and of plants and animals.

Floodplain vegetation and river redgum forests

If a billabong does not receive sufficient scouring from river flow during floods, it may gradually move towards complete siltation (Hillman, 1986) and have characteristics similar to the intermittent floodplain. The floodplain connects the river channel and billabong systems. The River Murray floodplain is dominated by forests of river redgum (*Eucalyptus camaldulensis*) on the river banks and more frequently flooded areas (up to 25 km wide) and by black box (*Eucalyptus largiflorens*) in the outer floodplain and the mid-section of the river flowing through the semi-arid zone. Although the trees appear uniform the understorey varies both across the floodplain and along the river, depending largely on frequency and duration of flooding (Bren, 1990; Sainty & Jacobs, 1990; Smith & Smith, 1990).

The river redgum forests depend on the river water for their survival. In many floodplain areas, such as the Barmah Forest, with an annual average rainfall of 400 mm, the tree density is similar to areas with 1200 mm annual average rainfall; the surface flooding and/or aquifer recharge (which the trees can access) supplied by the river provides for the additional water needs (Bren & Gibbs, 1986).

The floodplain vegetation underneath the redgum forests is largely herbaceous with grasses (Poaceae), sedges (Cyperaceae) and daisies (Asteraceae) well represented. Some species are common along the length of the river floodplain (e.g. Warrengo summer grass *Paspalidium jubiflorum*, the bluebell *Wahlenbergia fluminalis*, and common sneezeweed *Centipeda cunninghamii*) but most are more restricted by the changes in climate, geomorphology and soils along the river. Those species typical of more flooded sites, such as Moira grass *Pseudoraphis spinescens*, and common spike-rush *Eleocharis acuta*, are more common in the upper riverine plain where there are extensive flood-prone areas than in the lower Murray mallee zone in South Australia (Smith & Smith, 1990).

Changes in both understorey and tree cover occur across the floodplain; minor differences in soil and topography affect the frequency and duration of flooding. With distance from the river the forest height and density decrease to form a woodland and the understorey floristics change. In parts of the Barmah Forest floodplain flooding is too frequent for trees to grow and a Moira grass plain has developed. Where flooding is even more prolonged, semi-permanent wetlands of emergent aquatics *Phragmites australis* and *Juncus ingens* grow to 2–3 m.

The regulation of river flows has changed the distribution of some species. The dampening of fluctuations in water level has favoured some species and disadvantaged others. *Phragmites australis* beds fringe the lower Murray where previously they were absent and *Pseudoraphis spinescens* is no longer recorded from the lower Murray regions where it once was common (Smith & Smith, 1990).

Management of wetland vegetation

Management problems

Alteration of water regime, river regulation, clearing of catchments, grazing, increasing salinity, eutrophication and the spread of introduced species are just some of the human-induced factors causing changes to wetland vegetation. Often we have not predicted or assessed the consequences of these changes before irreversible changes to the wetland ecosystems have been made. Hence they have become present and future management problems.

Increasing salinity

Increasing salinity of freshwater wetlands (of soils, groundwater and surface water) is a major management concern. Human activities such as land clearance and irrigation or weir construction have resulted in rising water tables, thereby bringing saline groundwater either to, or closer to, the surface. The resultant increase in wetland and soil salinity has killed or changed natural vegetation which in most cases has not been replaced by salt-tolerant species. The most striking examples are the once fresh lake systems of the wheatbelt in the southwest of Western Australia (Froend *et al.*, 1987; Zedler, Paling & McComb, 1990) and the salinised lands in the irrigated areas of the Murray–Darling basin (Macumber, 1990).

Eutrophication

Eutrophication is the process of nutrient enrichment of water bodies; it may occur naturally or as the result of human activities. Natural eutrophication is a slow, long-term process of accumulation of nutrients from the effects of climate on weathering and leaching of catchment soils. Nutrient-rich waters encourage plant growth and typically have dense macrophyte and/or phytoplankton growth.

Human-induced eutrophication is, in contrast, very rapid and has rapid biological consequences. Land clearance, addition of fertilisers to the catchment and raw or treated sewage are well-known sources of nutrients (nitrogen and phosphorus). Nutrient enrichment leads to increased plant growth (phytoplankton, filamentous algae and aquatic plants) and often reduction in oxygen availability which may cause mass death of aquatic organisms (McComb & Lake, 1990; Sullivan, 1990). Such blooms are often dominated by one or a few successful species and a decrease in plant diversity is often a consequence of eutrophication.

Blooms of cyanobacteria have been associated with such nutrient enrichment and some of these organisms produce toxins which have been implicated in mass death of stock and aquatic organisms. The major blooms in the Murray–Darling system in the summer of 1991–2 are related to increasing nutrients from clearance and fertilisers in the catchments combined with conditions of low river flow. The problems will not go away; we must recognise their magnitude, understand their causes and manage our waters for present and future uses.

Nuisance aquatics

Many introduced plant species are now naturalised in a great variety of our wetlands. Some have been intentional introductions, others accidental, but all have altered the floristics and structure of the natural vegetation and the characteristics of the wetlands. Aquatic weeds can be both an agricultural and a conservation problem. Some of our native species, such as the native grass *Diplachne fusca* in rice-fields, have become agricultural problems in artificial wetlands (McIntyre, Mitchell & Ladiges, 1989).

Those nuisance aquatics which have the ability to invade natural wetlands are perhaps of greatest concern: *Eichhornia crassipes* (water hyacinth), *Salvinia molesta* (floating fern), *Alternauthera philoxeroides* (alligator weed), *Mimosa pigra* and *Ludwigia peruviana* are of greatest concern into the 1990s (Jacobs & Sainty, 1987; Jacobs & Brock, 1993). Biological control programmes have targeted many of these with partial success. The weevil *Cyrtobagous singularis* has been dramatically successful in controlling *Salvinia molesta* in most areas but not in the tropical north. Other management options are discussed by Sainty & Jacobs (1981).

Many of these aggressive weeds outcompete the native species for space and light and thus alter the wetland environment. Many aquatic weed species are not killed or controlled merely by removing the growing parts because they have great potential to regenerate from the smallest fragment or from millions of long lived seeds in the sediment. Water hyacinth seeds have survived for at least 12 years on the dry Gingham watercourse and still retain their viability (J. Duggin, pers. comm.); *Ludwigia peruviana* adds hundreds of thousands of seeds per square metre to the seed bank each year in the Botany wetlands in Sydney (S. Jacobs, pers. comm.).

Perhaps the greatest potential threats of aquatic weeds to our native wetland vegetation lie in those species that we do not yet recognise as problems; illegal immigrants from the aquarium industry are one source.

Management for restoration and rehabilitation

The management of wetland vegetation is essential if we wish to maintain or rehabilitate our wetlands. We are only just beginning to realise the problems of predicting what sorts of wetland vegetation will regenerate when altered ecosystems are restored. Unfortunately it must be accepted that we cannot re-create the past vegetation by simply removing previous alterations. Changing water regimes back to those of 50 or 200 years ago will not do this. We must manage the present systems for our future needs and accept that introduced species, altered water regimes, eutrophication, salinity, algal blooms and multiple uses are all components and states we need to include.

Restoration of eutrophic wetlands by reducing phosphorus input from the catchment or planting fringing vegetation to trap nutrients which might enter the wetland has been suggested. Similarly, there is considerable interest in the use of artificial wetlands with aquatic plants to remove nutrients from wastewater. These ideas give some indication of the concern and interest in restoring wetland vegetation to help control eutrophication.

As our understanding of the natural resilience of wetland vegetation increases we will be in a better position to predict the consequences of various management and manipulation options. A better understanding of the biology of individual wetland species under different water regimes will further our basis for predictive management.

Management, conservation and appreciation of wetlands and their vegetation

The days of 'if it's wet, drain it' are almost gone. Our laws and politics now at least pay lip-service to the value of wetlands. There will always be threats from development and cries of 'don't touch it' from preservationists. But we must be realistic and accept that humans have to take the responsibility for managing all types of wetlands and their vegetation for a variety of purposes: some wetlands need to be preserved, others conserved, and most managed for multiple uses including conservation and human use combined.

Wetland vegetation, like other vegetation types, must be managed for human needs, including conservation, if we wish to see it survive in self-sustaining ecosystems into the next century. This means we need to understand better the functioning of wetland ecosystems and the role of their vegetation. The interactions and links between the catchment and all parts of the wetland and the interrelationships between wetlands are all important. Once we recognise and understand the causes and consequences of some of our wetland problems, we can hope to predict the consequences of future management decisions and thus manage wetlands and their vegetation for multiple future uses.

Acknowledgements

Information provided by Drs John Duggin (University of New England) on *Eichhornia* seedbanks, Surrey Jacobs (National Herbarium of NSW) on *Ludwigia* seedbanks, Peter McGee (University of Sydney) on mycorrhizal associations and Michael Maher (Western Lands Commission of NSW) on Paroo wetlands is acknowledged with thanks. Dr Peter Jarman is gratefully acknowledged for his comments on the manuscript and for drawing Figure 15.1. Rosemary Torbay and Barbara Blenman are thanked for producing this manuscript with the facilities of the Botany Department, University of New England.

References

Aston, H.I. (1973). *Aquatic Plants of Australia*. Melbourne University Press.
Beadle, N.C.W. (1981). *The Vegetation of Australia*. Cambridge University Press.
Boon, P., Frankenberg, J., Hillman, T., Oliver, R. & Shiel, R. (1990). Billabongs. In *The Murray*, eds N. Mackay, & D. Eastburn, pp. 180–200. Canberra: Murray–Darling Basin Commission.

Boon, P.I. & Sorrell, B.K. (1991). Biogeochemistry of billabong sediments. 1. The effect of macrophytes. *Freshwater Biology*, 26, 209–26.

Borowitzka, L.J. (1981). The microflora: adaptations to life in extremely saline lakes. *Hydrobiologia*, 81, 33–46.

Bren, L. (1990). Red gum forests. In *The Murray*, eds N. Mackay & D. Eastburn, pp. 231–45. Canberra: Murray–Darling Basin Commission.

Bren, L.J. & Gibbs, N.L. (1986). Relationships between flood frequency, vegetation and topography in a river red gum forest. *Australian Forest Research*, 16, 357–70.

Briggs, S.V. (1981). Freshwater wetlands. In *Australian Vegetation*, ed. R.H. Groves, pp. 335–60. Cambridge University Press.

Brock, M.A. (1981). Accumulation of proline in a submerged aquatic halophyte, *Ruppia* L. *Oecologia*, 51, 217–19.

Brock, M.A. (1983). Reproductive allocation in annual and perennial species of the submerged aquatic halophyte *Ruppia*. *Journal of Ecology*, 71, 811–18.

Brock, M.A. (1986). Adaptation to fluctuations rather than to extremes of environmental parameters. In *Limnology in Australia*, eds P. De Deckker & W.D. Williams, pp. 131–40. Dordrecht: W. Junk.

Brock, M.A. (1991). Mechanisms for maintaining persistent populations of *Myriophyllum variifolium* J. Hooker in a fluctuating shallow Australian lake. *Aquatic Botany*, 39, 211–19.

Brock, M.A. & Casanova, M.T. (1991*a*). Plant survival in temporary waters: a comparison of charophytes and angiosperms. *Verhandlungen Internationale Vereinigung für theoretishe und Angewandte Limnologie*, 24, 2668–72.

Brock, M.A. & Casanova, M.T. (1991*b*). Vegetative variation of *Myriophyllum variifolium* in permanent and temporary wetlands. *Australian Journal of Botany*, 39, 487–96.

Casanova, M.T. & Brock, M.A. (1990). Charophyte germination and establishment from the seed bank of an Australian temporary lake. *Aquatic Botany*, 36, 247–54.

Close, A. (1990). The impact of man on the natural flow regime. In *The Murray*, eds N. Mackay, & D. Eastburn, pp. 61–77. Canberra: Murray–Darling Basin Commission.

Cowling, S.W. & Waid, J.S. (1963). Aquatic Hyphomycetes in Australia. *Australian Journal of Science*, 26, 122–3.

Dunlop, C.R. & Webb, L.J. (1991). Flora and vegetation. In *Monsoonal Australia: landscape ecology and man in the northern lowlands*, eds C.D. Haynes, M.G. Ridpath & M.A. Williams, pp. 41–60. Rotterdam: A.A. Balkema.

Finlayson, C.M. & von Oertzen, I. (1993). Wetlands of Australia; Northern (tropical) Australia. In *Wetlands of the World*, ed. D. Whigham, pp. 195–243. Amsterdam: Elsevier.

Froend, R.H., Heddle, E.M., Bell, D.T. & McComb, A.J. (1987). Effects of salinity and waterlogging on the vegetation of Lake Toolibin, Western Australia. *Australian Journal of Ecology*, 12, 281–98.

Ganf, G.G. (1982). Influence of added nutrient on the seasonal variation of algal growth potential of Mount Bold Reservoir, South Australia. *Australian Journal of Marine and Freshwater Research*, 33, 475–90.

Ganf, G.G. & Oliver, R.L. (1982). Vertical separation of light and available nutrients as a factor causing replacement of green algae by blue-green algae in the plankton of a stratified lake. *Journal of Ecology*, 70, 829–44.

Grime, J.P. (1979). *Plant Strategies and Vegetation Processes*. Chichester: Wiley.

Hill, R. & Webb, G. (1982). Floating grass mats of the Northern Territory wetlands – an endangered habitat? *Wetlands*, 2, 45–50.

Hillman, T.J. (1986). Billabongs. In *Limnology in Australia*, eds P. De Deckker & W.D. Williams, pp. 457–70. Dordrecht: W. Junk.

Jacobs, S.W.L. (1983). Vegetation. In *Wetlands in New South Wales*, ed. C. Haigh, pp. 14–19. Sydney: New South Wales National Parks and Wildlife Service.

Jacobs, S.W.L. & Brock M.A. (1993). Wetlands of Australia: Southern (temperate) Australia. In *Wetlands of the World I: Inventory, Ecology and Management*. Handbook of Vegetation Science 15/2, eds D.F. Whigham, D. Dykyjová & S. Hejný, pp. 244–304. Dordrecht: Kluwer.

Jacobs, S.W.L. & Sainty, G.R. (1987). Water weeds into the third century. In *Proceedings of the 8th Australian Weeds Conference, Sydney*, eds D. Lemerle & A.R. Leys, pp. 148–51.

John, J. (1983). The diatom flora of the Swan River Estuary, Western Australia. *Bibliotheca Phycologia Bnd*, 64. Vaduz: J. Cramer.

Kirk, J.T.O. (1983). *Light and Photosynthesis in Aquatic Ecosystems*. Cambridge University Press.

Knights, P. (1980). *Macquarie Marshes: Wetlands and Management Requirements*. Sydney: Total Environment Centre.

Ling, H.U. & Tyler, P.A. (1986). A limnological survey of the Alligator Rivers Region. 11. Freshwater algae exclusive of diatoms. Supervising Scientist for the Alligator Rivers Region, Research Report No. 3. Canberra: Australian Government Publishing Service.

McComb, A.J. & Lake, P.S. (eds) (1988). *The Conservation of Australian Wetlands*. Chipping Norton: World Wildlife Fund.

McComb, A.J. & Lake, P.S. (1990). *Australian Wetlands*. Sydney: Angus & Robertson.

McIntyre, S., Ladiges, P.Y. & Adams, G. (1988). Plant species-richness and invasion by exotics in relation to disturbance of wetland communities on the Riverine Plain, NSW. *Australian Journal of Ecology*, 13, 361–73.

McIntyre, S., Mitchell, D.S. & Ladiges, P.Y. (1989). Germination and seedling emergence in *Diplachne fusca*: a semi-aquatic weed of rice fields. *Journal of Applied Ecology*, 26, 551–62.

Mackay, N. & Eastburn, D. (eds) (1990). *The Murray*. Canberra: Murray–Darling Basin Commission.

Macumber, P. (1990). The salinity problem. In *The Murray*, eds N. Mackay & D. Eastburn, pp. 111–27. Canberra: Murray–Darling Basin Commission.

Merrick, C.J. & Ganf, G.G. (1988). Effects of zooplankton grazing on phytoplankton communities in Mt Bold Reservoir, South Australia, using exclosures. *Australian Journal of Marine and Freshwater Research*, **39**, 503–24.

Moss, B. (1988). *Ecology of Freshwaters: Man and Medium*. Oxford: Blackwell Scientific Publications.

Paijmans, K., Galloway, R.W., Faith, D.P., Fleming, P.M., Haantjens, H.A., Heyligers, P.C., Kalma, J.D. & Loffler, E. (1985). *Aspects of Australian Wetlands*. CSIRO Division of Water and Land Resources Technical Paper No. 44.

Parbery, I.H. (1970). A survey of the algal flora of some semiarid and arid soils of N.S.W. M Litt thesis, University of New England.

Pressey, R.L. (1986). Wetlands of the River Murray below Lake Hume. River Murray Commission Report 86/1.

Pressey, R.L. (1990). Wetlands. In *The Murray*, eds N. Mackay & D. Eastburn, pp. 167–83. Canberra: Murray–Darling Basin Commission.

Roberts, J. & Ludwig, J.A. (1991). Riparian vegetation along current-exposure gradients in floodplain wetlands of the River Murray, Australia. *Journal of Ecology*, **79**, 117–27.

Sainty, G.R. & Jacobs, S.W.L. (1981). *Waterplants of New South Wales*. Sydney: New South Wales Water Resources Commission.

Sainty, G.R. & Jacobs, S.W.L. (1988). *Waterplants in Australia*. Sydney: Sainty & Associates.

Sainty, G.R. & Jacobs, S.W.L. (1990). Waterplants. In *The Murray*, eds N. Mackay & D. Eastburn, pp. 265–75. Canberra: Murray–Darling Basin Commission.

Sculthorpe, C.D. (1967). *The Biology of Aquatic Vascular Plants*. Königstein: Koeltz Scientific Books.

Smith, P. & Smith, J. (1990). Floodplain vegetation. In *The Murray*, eds N. Mackay & D. Eastburn, pp. 215–31. Canberra: Murray–Darling Basin Commission.

Sousa, W.P. (1984). The role of disturbance in natural communities. *Annual Review of Ecology and Systematics*, **15**, 353–91.

Specht, R.L. (1981). Major vegetation formations in Australia. In *Ecological Biogeography of Australia*, ed. A. Keast, pp. 163–297. The Hague: W. Junk.

Sullivan, C. (1990). Phytoplankton. In *The Murray*, eds N. Mackay & D. Eastburn, pp. 251–65. Canberra: Murray–Darling Basin Commission.

Taylor, J.A. & Dunlop C.R. (1985). Plant communities of the wet–dry tropics of Australia: the Alligator Rivers region, Northern Territory. *Proceedings of the Ecological Society of Australia*, **13**, 83–129.

Taylor, J.A. & Tulloch, D. (1985). Rainfall in the wet–dry tropics: extreme events at Darwin and similarities between years during the period 1870–1983 inclusive. *Australian Journal of Ecology*, **10**, 281–95.

Theophrastus (370–c. 285 BC). *Enquiry into Plants*. Translation by Sir A. Hort, London, (1916) as quoted by Sculthorpe C.D. (1967).

Thomas, K., Chilvers, G.A. & Norris, R.H. (1989). Seasonal occurrence of conidia of aquatic Hyphomycetes (Fungi) in Lees Creek, Australian Capital Territory. *Australian Journal of Marine and Freshwater Research*, **40**, 11–23.

Vogel, S. (1981). *Life in Moving Fluids: The Physical Biology of Flow.* Boston: Willard Grant Press.

Walker, K.F. (1985). A review of the ecological effects of river regulation in Australia. *Hydrobiologia*, **125**, 111–29.

Walker, K.F. & Hillman, T.J. (1981). Phosphorus and nitrogen loads in waters associated with the River Murray near Albury–Wodonga and their effects on phytoplankton populations. *Australian Journal of Marine and Freshwater Research*, **33**, 223–43.

Wetzel, R.G. (1975). *Limnology.* Philadelphia: W.B. Saunders Co.

Yen, S. & Myerscough, P.J. (1989a). Co-existence of three species of amphibious plants in relation to spatial and temporal variation: field evidence. *Australian Journal of Ecology*, **14**, 291–303.

Yen, S. & Myerscough, P.J. (1989b). Co-existence of three species of amphibious plants in relation to spatial and temporal variation: investigation of plant responses. *Australian Journal of Ecology*, **14**, 305–18.

Zedler, J.P., Paling, E. & McComb, A. (1990). Differential responses to salinity help explain the replacement of native *Juncus kraussii* by *Typha orientalis* in Western Australian salt marshes. *Australian Journal of Ecology*, **15**, 57–72.

16

Alpine and subalpine vegetation

R.J. WILLIAMS & A.B. COSTIN

THE DISTRIBUTION of high mountain vegetation in Australia is related primarily to summer temperatures, as in other alpine, arctic and subantarctic regions of the world (Daubenmire, 1954; Tranquillini, 1979). Tree growth is limited to areas where the mean temperature of the warmest month is 10 °C or greater; in areas with summer temperatures below this level trees do not survive. Such is the case for the alpine, treeless areas both in Tasmania and mainland Australia (Costin, 1967). The high mountain country comprises this treeless alpine zone and also the sparsely timbered subalpine zone which abuts the alpine region at a slightly lower altitude. Within this latter altitudinal belt, the vegetation consists of both woodland and treeless formations. The term 'alpine' may be used in both general and specific senses. Löve (1970) considers the subalpine tract to be a subset of the alpine tract, but that the belt within which treeless and wooded communities occur should be termed 'subalpine'. This chapter describes the vegetation in both these regions of Australia. Factors other than low summer temperature also contribute to the characteristic identities of these two regions, for example, frequent frosts, solifluction, low winter temperatures, persistent snow cover (of at least one month, but never lying permanently in Australia), and restricted biological production (Costin, 1954, 1973; Williams, 1987a, 1990b).

The alpine and subalpine areas (Fig. 16.1) are situated above an altitude of about 1370 to 1525 m on the mainland and above about 915 m in Tasmania. The morphology of treelines in Australia is complex, with few treelines well defined, especially in Victoria and Tasmania (Kirkpatrick, 1982; McDougall, 1982). The apparently low altitude of treeline in Tasmania is related to its cool oceanic summer climate (Costin, 1972; Kirkpatrick, 1982). On both the mainland and Tasmania, valley-bottom, or inverted treelines, may occur below the upper treeline. These are a consequence of unfavourable site factors, such as cold air drainage (Moore & Williams, 1976; Williams, 1987b; Paton, 1988), wet soils (Ashton & Hargreaves, 1983), exposure to strong winds (Ashton & Williams, 1989) or periodic fires.

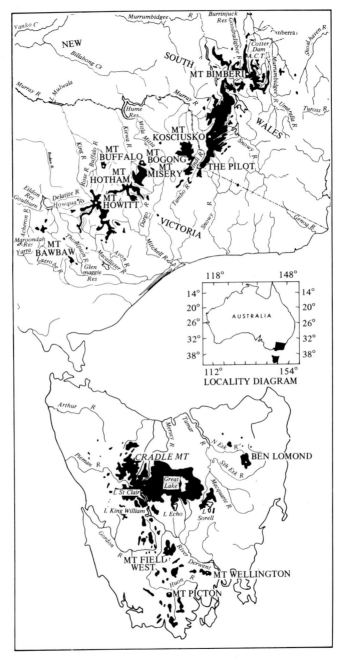

Fig. 16.1. The distribution of high mountain areas in southeastern Australia. (From Costin, 1981.)

The areas comprise about 6480 km² of Tasmania (about 10 per cent of that island state) and about 5180 km² of mainland Australia. The latter area occurs only in New South Wales, Victoria and the Australian Capital Territory and comprises about 0.07 per cent of the total mainland area. Together, they comprise approximately 0.15 per cent of the Australian land surface. Despite this apparently small overall area, the alpine and subalpine regions are of great significance to the use of land in Australia in terms of water yield, recreation and nature conservation (Barlow, 1986; Good, 1989).

In this chapter we wish to describe the alpine environment, and the composition, structure and dynamics of the major vegetation formations in these regions of Australia, and comment on the effects of humans on these different vegetation types.

Alpine and subalpine environments

The two most extensive areas of alpine and subalpine vegetation are the Central Plateau of Tasmania and the Snowy Mountains of New South Wales (including Mt Kosciusko, 2228 m in elevation, the highest peak). The high mountain areas of Victoria, of which the Bogong High Plains is the largest, have a less continuous distribution.

The alpine region of southeastern Australia is part of an uplifted plateau, the bedrocks of which include Devonian granites and gneiss, Carboniferous and Ordovician sediments, and Tertiary basalts. The Tasmanian mountains are predominantly Jurassic dolerites and Precambrian quartzites. The plateau is currently in various degrees of dissection, and often defined by relatively abrupt scarps. The palaeoplain is thought to be of Mesozoic age (Ollier, 1986). Periods of uplift which have given rise to the highlands of southeastern Australia may have commenced as early as 90 million years ago with three or more major episodes of uplift, through the Tertiary (Wellman, 1979; Jones & Veevers, 1982). The idea of a major 'Kosciusko Orogeny' at the end of the Tertiary is not supported by modern evidence (Ollier, 1986; Ollier & Wyborn, 1989). The morpho-tectonic history of Tasmania is probably similar to that of mainland southeastern Australia. Thus, physiographically, the mountains of southeastern Australia are considerably older than those of New Guinea or New Zealand. For these latter areas, the major orogenic events occurred during late Tertiary/ early Pleistocene times (Ollier, 1986).

In the Pleistocene, the Australian highlands were subjected to glacial and/or periglacial conditions, the effects of which contributed to the uniqueness of the high mountain environments. Glacial activity was widespread in Tasmania (Davies, 1965; Galloway, 1989) and led to the formation of the glacial lakes, cirques and moraines so apparent in the Tasmanian mountain landscape. On the mainland, activity was more localised, and restricted to the Kosciusko area (Peterson, 1971; Galloway, 1989) where there appears to have been only a single Quaternary glaciation (Ollier, 1986). Associated periglacial activity was widespread over both areas, and in Victoria, and produced extensive slope deposits. The limited evidence available (Costin, 1971, 1972) suggests that, on the mainland, the last period of periglacial and glacial conditions extended from about 32 000 to 15 000 BP and locally, perhaps until about 9000 BP. As in other areas of the world, the glacial maximum occurred at *c.* 18–20 000 BP (Galloway, 1989). A less severe cold interval was experienced about 3000 to 1500 years ago. The sequence of late Quaternary climates in Tasmania appears to have been broadly similar (Colhoun, 1978), although the Tasmanian highlands experienced two (or perhaps three) glaciations (Ollier, 1986).

The present climate is characterised by relatively low temperatures and relatively heavy precipitation. The latter occur because of the situation of the plateaux in the belt of westerly winds and the associated low pressure/cold front influences arising in the subantarctic region. The level of precipitation varies from about 800 mm at some of the relatively dry subalpine sites on the northeastern Snowy Mountains to more than 2500 mm on the Kosciusko Main Range and Mt Bogong. Most precipitation falls as winter snow, which normally persists from June until October, but may persist on sheltered southeastern slopes until January. In summer, thunderstorms are frequent. Temperatures are mild to low, with mean annual temperatures of the order of 5–8 °C (Costin, 1954, 1967). Frosts are frequent (100–200 per annum) and may occur at any time of the year. The undulating character of much of the high country is conducive to the development of cold air pondages and severe frosts (< −20 °C) may occur within such hollows (Leigh *et al.*, 1987; Williams, 1987*b*). However, subsoil freezing is rare (Costin & Wimbush, 1973).

Soils have been affected considerably by both past and present climatic conditions. There is a relatively wide range of soil types within the Australian Alps, many of which may be deep by world standards (Costin, 1989). Costin (1986) lists nine Great Soil Groups which occur in the alpine and subalpine tracts. These include lithosols on well-drained areas of glacial and peri-glacial weathering and erosion, peats in locally wet situations and alpine humus soils on better drained glacial and peri-glacial deposits. The latter soils are the most common in both alpine and subalpine tracts. Most soils are nutrient poor, high in organic matter (10–100 per cent), acidic (pH 3–4.5), very friable and have a high water-holding capacity (Costin *et al.*, 1959; Costin, 1962*b*; Costin, Wimbush & Cromer, 1964; Rowe, 1972). Similar soils occur in some Tasmanian alpine areas, such as Ben Lomond, but the alpine humus soils are absent from the mountains in the west, where very poor, shallow lithosols and acid peats predominate (Kirkpatrick, 1983).

Fire in alpine and subalpine environments was relatively rare in pre-European times with fire-free intervals being of the order of decades to centuries, depending on site. Two- to five-fold increases in the fire frequency occurred after the arrival of Europeans in the alps. Occasional catastrophic fires occur within the alps, such as occurred in 1900, 1926 and 1939. In the latter, some 3 million hectares of forest country between Melbourne and Canberra were burned (Banks, 1989). Carr & Turner (1959*a*) estimate that approximately three-quarters of the snowgum woodlands on the Bogong High Plains in Victoria were razed by this fire, although the extent of burning in the treeless vegetation is not known precisely. In Tasmania, 16 per cent of the treeless vegetation was burned between 1960 and 1980 (Kirkpatrick & Dickinson, 1984).

Vegetation

The diversity of present and past climates and microclimates and the range of soil types existing in the alpine and subalpine environments correlates well with the diverse pattern of vegetation types. The vegetation of the high country consists of a number of structural formations, primarily forest and woodland, heathland, grassland, herbfield and bog. At a continental scale, the mainland and Tasmanian alpine floras form a complex continuum, with the flora of the mainland southeastern mountains and that of southwestern Tasmanian mountains as the two extremes. Most of the floristic variation in the Australian alpine flora occurs within Tasmania (Kirkpatrick, 1982, 1989). At a landscape scale, the distribution of the various formations depends upon interactions between altitude, exposure to wind and frost, depth and persistence of snow, and type, depth and moisture content of the soil. The dynamics of these communities are varied and depend on controlling climatic and biotic factors as well as the life-history characteristics of the dominant plants (Williams, 1990*a*).

Plant growth is limited by low temperatures, low nutrient soils and variation in climatic conditions within seasons and from season to season. Above-ground productivity varies from <200 g/m^2/y to >800 g/m/y (Costin, 1967; Gibson, 1990; Williams, 1990b). Carr & Turner (1959a) listed conditions considered to be favourable for plant growth such as snow cover by early winter, thaw in mid-spring which is accompanied by rains, few frosts during both spring and summer, and rain every two to three weeks during the summer period. Variation in any of these features may lead to profound effects on the relative growth of species, as Wimbush & Costin (1979a,b,c) highlighted for a number of plant communities in the Kosciusko region.

Adaptations of plants

Alpine and arctic vegetation across the world shows convergent patterns of evolution in response to low temperatures (Costin, 1967; Billings & Mooney, 1968; Bliss, 1971). Plants of the tundra, both woody and herbaceous, are typically dwarf; taller woody species are restricted to the most favourable locations. Herbaceous species usually display rosette/tussock habits. This life-form effectively traps heat, enabling leaf temperatures to be >5 °C higher than ambient air (Körner & Cochrane, 1983). Many tundra plants have substantial carbohydrate reserves in underground organs (Bliss, 1971; Billings, 1974). This allows rapid mobilisation of resources at the onset of the growing season. In addition, many woody species have high concentrations of lipids in their shoots (Bliss, 1962; Hadley & Bliss, 1964). Numerous Australian species, e.g. *Phebalium* spp., form their floral buds during the autumn, and flower as soon as the snow melts. An extreme example is *Caltha introloba*, which flowers under the snow (Wardlaw, Moncur & Totterdell, 1989).

Drought in alpine systems may be locally severe, especially during winter when soil temperatures are at or below freezing (Billings & Mooney, 1968; Tranquillini, 1979). Studies of the water relations of snow gum (*Eucalyptus pauciflora*) indicate that severe water deficits are uncommon, with mid-winter shoot water potentials rarely falling below -1.4 MPa, which is considerably higher than the estimated lethal limit of -5.5 MPa (Cochrane & Slatyer, 1988). Where treeless vegetation is well covered by a protective blanket of snow throughout the winter, soils rarely freeze (Costin & Wimbush, 1973; Williams, 1987a).

Major vegetation formations

Extensive surveys of alpine and subalpine vegetation have been undertaken in New South Wales by Costin (1954, 1957), Costin *et al.* (1959), and McVean (1969); in Victoria by Carr & Turner (1959*a*), Costin (1962*a*), McDougall (1982), Walsh, Barley & Gullan (1984), Williams & Ashton (1987*a*); and in Tasmania by Jackson (1973), Kirkpatrick (1982, 1983, 1986, 1989), and Kirkpatrick, Minchin & Davies (1985). The following account is synthesised from these sources. The composition and distribution of the major alpine and subalpine formations are summarised in Table 16.1.

The subalpine woodlands occur between about 915 and 1200 m altitude in Tasmania and between about 1400 and 1900 m on the mainland; in Tasmania they are dominated by *Eucalyptus coccifera* and *E. gunnii* and on the mainland *E. pauciflora* ssp. *niphophila* (Table 16.1; Fig. 16.2). In Tasmania, as well as these *Eucalyptus*-dominated subalpine woodlands, distinctive rainforest thickets of deciduous beech (*Nothofagus gunnii*) and stands of *Athrotaxis cupressoides* and *A. selaginoides* occur (see Chapter 5). Whilst there is a good correlation world-wide between treeline and the 10° isotherm of the warmest month, establishment and survival of *E. pauciflora* seedlings may also be affected by competition from surrounding vegetation. In a long-term study, Ferrar, Cochrane & Slatyer (1988) found that survival rates of germinants at and above the treeline in the Kosciusko region are typically <1 per cent. In contrast, older seedlings experimentally transplanted up to 200 m above current treeline altitudes have survived to reproductive maturity, indicating that beyond a critical seedling size, snowgums can maintain positive carbon balance (by virtue of their greater metabolic reserves) at altitudes well above the current treeline.

The understorey of snowgum woodland may be grassy or heathy. At elevations of 1200–1500 m, the understorey is usually a shrub stratum, 3–5 m tall. At higher elevations, the proportion of shrub and grass in the understorey varies according to exposure, soil depth and rockiness. Shrubs 1–3 m tall predominate in more sheltered locations, especially on the steeper, rockier slopes. Predominant species include *Prostanthera cuneata*, *Bossiaea foliosa*, *Phebalium squamulosum*, *P. phylicifolium*, *P. ovatifolium* and *Orites lancifolia*. Grassy understoreys of tussock *Poa* species (e.g. *P. hiemata*) are common on more gentle slopes, especially where soils are deeper and less rocky. Shrubby understoreys are, in general, more common in Tasmania than on

Table 16.1. Major plant communities of Australian high mountain areas

Vegetation type	Main dominants	Distribution
Subalpine woodland	Eucalyptus pauciflora E. stellulata E. coccifera E. gunnii Athrotaxis spp. Nothofagus gunnii	Widespread subalpine community on mainland. Relatively deep alpine humus soils Widespread subalpine community in Tasmania
Heath Closed	Podocarpus lawrencei Oxylobium ellipticum Prostanthera cuneata Orites lancifolia Phebalium spp. Acacia alpina Tasmannia xerophila Bossiaea foliosa Kunzea muelleri Baeckea gunniana Callistemon sieberi	Widespread in rocky situations, especially in Tasmania. Lithosols, alpine humus soils. 25% of Bogong High Plains
Open	Hovea purpurea var. montana Leucopogon ericoides Grevillea australis Asterolasia trymalioides Pherosphaera hookeriana	Widespread on deeper alpine humus soils; slopes and flats. 30% of Bogong High Plains

Vegetation type	Characteristic species	Description
Tussock grassland/ Tall alpine herbfield	*Poa hiemata, P. costiniana* *Danthonia nudiflora* *Empodisma minus* *Themeda australis* (locally) *Celmisia* spp. *Helipterum albicans* ssp. *alpinum* *Chionochloa frigida* (locally) *Craspedia* spp. *Euphrasia* spp.	Widespread along valleys and in basins of cold air drainage, especially in subalpine areas. Alpine humus soils. The main alpine communities above tree line on mainland; more restricted in Tasmania. Mostly on freely drained, relatively deep alpine humus soils. 25% of Bogong High Plains
Fen	*Carex gaudichaudiana* *Danthonia nudiflora* *Festuca muelleri* *Eleocharis acuta* *Poa* spp. *Pratia surrepens* *Scirpus montivagus*	Widespread locally in wet, acid, almost level alpine and subalpine situations influenced by mineral soil. Acid fen peats, acid marsh soils. 1% of Bogong High Plains
Wet heath	*Epacris* spp. *Kunzea muelleri* *Baeckea gunniana*	Locally common in damp situations marginal to bog
Bog	*Carex gaudichaudiana* *Sphagnum cristatum* *Epacris* spp. *Callistemon sieberi* *Richea continentis* *Restio australis* *Carpha* spp. *Astelia* spp. *Empodisma minus*	Widespread locally in wet, acid valley situations and around hillside springs, both in alpine and subalpine areas; relatively little influence of mineral soil. Bog peats. 10% of Bogong High Plains

Table 16.1. (*Cont.*)

Vegetation type	Main dominants	Distribution
Bolster	*Abrotanella fosteroides* *Donatia novaezelandiae* *Phyllachne colensoi* *Pterygopappus lawrencii* *Astelia* spp. *Oreobolus* spp.	Wet alpine situations in Tasmania, little snow cover
Short alpine herbfield	*Plantago* spp. *Neopaxia australasica* *Caltha introloba* *Brachsycome stolonifera* *Ranunculus niphophilus* *Luzula* spp. *Carex hebes* *Viola betonicifolia* *Danthonia nudiflora* *Poa fawcettiae*	Local occurrence beneath alpine snow patches with persistent (>8 months) snow cover, especially on mainland. Snow patch meadow soils, acid fen peats. 1% Bogong High Plains
Fjaeldmark	*Coprosma* spp. *Colobanthus* spp. *Epacris petrophila* *Epacris microphylla* *Chionohebe densifolia* *Ewartia nubigena* *Drapetes tasmanicus* *Helipterum albicans* ssp. *alpinum*	Local alpine occurrences above persistent snow patches. Local very wind-exposed alpine sites

After Costin, 1954, 1981; Costin *et al.*, 1979; McDougall, 1982; Kirkpatrick, 1982, 1989; Walsh *et al.*, 1984.

Fig. 16.2. Woodland of *Eucalyptus pauciflora* (snow gum) in Kosciusko National Park, New South Wales. (Photo: C.J. Totterdell.)

the mainland. However, leguminous shrubs are virtually absent from Tasmania, whilst being common on the mainland.

Heath communities range in height from 0.5 to 2 m and in shrub cover from 20 per cent (open-heath) to more than 70 per cent (closed-heath). They may occur on well-drained alpine humus soils, waterlogged soils, lithosols, or on severely eroded alpine humus soils.

Closed-heaths occur on relatively sheltered, rocky sites. On the mainland they are dominated by *Prostanthera cuneata*, *Phebalium squamulosum*, *P. ovatifolium*, *Bossiaea foliosa* or *Orites lancifolia*, the cover of which is 70–100 per cent. Subdominants include *Oxylobium alpestre*, *Hovea longifolia*, *Grevillea australis*, *Helichrysum hookeri* and *Pimelea alpina*. The undergrowth consists of sparse grass, *Poa hiemata* or *P. hothamensis* and herbs such as *Craspedia*, *Celmisia*, and *Brachyscome* spp. On slopes of 10–20° shrubs grow downslope *en echelon* owing to the flattening effects of winter snow. Such shrubs generally layer from stems 20–30 cm behind tips where litter and soil are in close contact with stems (Williams & Ashton, 1987*a*). A distinctive closed-heath community is that dominated by mountain plum pine (*Podocarpus lawrencei*). This heath is characteristic of boulder fields and rock rivers, and occurs in most alpine and subalpine regions (Costin, 1954; McDougall, 1982; Walsh *et al.*, 1984). Floristic diversity is relatively low, but the community is an extremely important habitat for the rare mountain pygmy possum (*Burramys parvus*), for example in the Mount Hotham area (Mansergh & Scotts, 1986). *Podocarpus* is very long lived, and the fire frequency within the community is low – of the order of once a century or less (Costin *et al.*, 1979). *Podocarpus* occurs in Tasmania, but does not achieve the same degree of local dominance as on the mainland. There is a greater diversity of closed-heaths in Tasmania, including some which are restricted to Tasmania, e.g. coniferous heaths dominated by *Microcachrys tetragona* and *Diselma archeri*, and several heaths of diverse dominance restricted to the quartzitic mountains of western Tasmania (Kirkpatrick, 1989).

Open-heaths, with a shrub cover of 20–50 per cent, are ecotonal between closed-heath and more open forb/grass communities. The common shrubs within such heaths are *Grevillea australis*, *Hovea longifolia*, *Phebalium squamulosum*, *P. ovatifolium* and *Leucopogon montanus*. On the Bogong High Plains, *Grevillea* is the predominant shrub on alpine humus soils derived from metamorphic substrates; *Hovea* is more common on the basaltic soils (McDougall, 1982). Intershrub areas are occupied by tussock grasses (*Poa* spp.) and a wide range of major and minor forbs. The formation is relatively restricted in Tasmania, and best represented by a *Poa–Leucopogon* community (Kirkpatrick, 1989).

Grasslands and tall alpine herbfields are distinctive communities on alpine humus soils within the upper subalpine and alpine tracts of the mainland, but are less extensive in Tasmania. The dominant genera are *Poa* and *Celmisia*, and the associated herbaceous flora is rich, especially in composites such as *Gnaphalium*, *Craspedia*, *Helichrysum*, *Podolepis* and *Helipterum*. Dwarf shrubs such as *Asterolasia trymalioides* and *Grevillea australis* may also be found within the grasslands/herbfields (Williams & Ashton, 1987*a*). Past and present grazing by domestic livestock has undoubtedly altered the composition of these herbaceous communities (Costin *et al.*, 1959; Williams & Ashton, 1987*a*; Wahren, Papst & Williams, 1991). In Tasmania, the most extensive areas of highland grassy vegetation are on the basalt plains of the northwest and the Central Plateau; isolated patches also occur on carbonate rocks (Kirkpatrick & Duncan, 1987).

There are no extensive areas in Tasmania where the herbaceous composites attain the same degree of dominance as in the Kosciusko and
Bogong areas (Kirkpatrick, 1989).

Numerous studies of vegetation in the heath-grass complexes on
well-drained alpine humus soils have stressed the importance of complex topographic/exposure gradients as determinants of the shrub–
grass balance. Closed-heath dominates on steeper slopes where soils
are shallow with a higher fraction of surface and subsoil rocks. On
such sites, exposure to wind and extreme frosts is least. In contrast,
grassland occupies the more level slopes, where soils are deeper and
less rocky, especially in hollows, where exposure to both wind and
frost is high. Open grassy-heath occupies sites of intermediate topography and exposure (Costin, 1954; Costin *et al.*, 1959; Williams,
1987*b*; Williams & Ashton, 1987*a*). Thus, on the mainland, there
are definite limits to the distribution of heath, especially the taller,
closed-heaths, which are set primarily by exposure as affected by
topography and altitude. In Tasmania, the relative dominance of
heath over grassland/herbfield may be attributable to a more maritime
climate, and less persistent snow, both of which are likely to reduce
the competitive stature of herbs compared with woody plants (Kirkpatrick, 1989).

Wetlands occur on highly organic soils wherever the water table is
close to the surface, such as in valley bottoms and along stream
courses or on seepage areas on valley sides. They are floristically and
structurally very complex, and vary from wet heaths and grassland
fens dominated by *Carex*, to complex peat bogs rich in *Sphagnum*
moss and hydrophytic shrubs and herbs. They are extremely important hydrologically, and are highly susceptible to disturbance, especially the effects of trampling by domestic livestock (Costin, 1962*a,b*;
Farrell & Ashton, 1973; McDougall, 1982; Ashton & Hargreaves,
1983). The fens are dominated by *Carex gaudichaudiana*, in association with *Stackhousia pulvinaris*, *Pratia surrepens*, *Scirpus montiv-
agus* and *Myriophyllum pedunculatum*. They occur on level to gently
sloping situations, on peaty soils which contain some mineral matter.
Dwarf heath of species of *Epacris*, *Empodisma* and *Poa* occur on
shallow sheets of peat on broad, flat valley bottoms or saddles. The
bogs occur on the deeper peats, devoid of mineral matter. The bogs
are a complex mosaic of hummocks, hollows and pools rich in *Sphagnum*, *Carex*, *Myriophyllum*, and hydrophytic shrubs such as *Richea*,
Epacris and *Baeckea*, *Restio* and herbaceous species such as *Astelia
alpina*, *Drosera* and *Carpha*. In areas marginal to bogs, wet grasslands
or heaths may occur. The grasslands are dominated by *Poa costiniana*

and the rope-rush, *Empodisma minus*. The wet heaths consist of *Epacris* spp., *Baeckea gunniana* and *Kunzea muelleri*.

Both *Sphagnum* and peat have a high water-holding capacity, and the bog vegetation has a high surface roughness (Costin *et al.*, 1959; Wimbush & Costin, 1983). Consequently, water movement within the system is retarded and although channelled flow occurs, it is localised. Water tends to move on a wide front through the bog rather than over its surface. Thus at times of peak run-off following heavy rain or during the spring thaw the bogs can absorb large quantities of water, with a consequent delay in the loss of water from the catchment.

Bolster heaths are distinctively Tasmanian and consist of hard, cushion-like plants. They are widespread and occur largely on poorly drained ground where there is little winter snow. A dozen or so species – both dicots and monocots – may form cushions of sufficient hardness to be considered bolsters (Kirkpatrick *et al.*, 1985). Of these, *Abrotanella fosteroides*, *Donatia novae-zelandia*, *Dracophyllum minimum* and *Phyllachne colensoi* are the most common dominants. Dominance of any one species, however, is likely to be short-lived. The complex mosaic patterns of cushions is likely to be maintained by non-equilibrium processes (Gibson, 1990). Bolster heaths commonly occur in highly exposed situations, and the compact, cushion form minimises transpiration losses and damage to leaves/shoots arising from strong, often ice-laden winds (Turner & Hill, 1981; Kirkpatrick, 1983). The internal moistness may also minimise fire damage.

On leeward, steep sites where snow cover is prolonged (December – January) the vegetation is short sedge–herbfield. Such 'snow patch' vegetation may occur in both the subalpine and alpine tracts, with the largest snow patches occurring at the highest altitudes. The dominant genera are short ($<$ 10 cm) rosette forbs or tussock graminoids, and include genera such as *Plantago*, *Colobanthus*, *Coprosma*, *Celmisia*, *Neopaxia*, *Caltha*, *Carex*, *Luzula*, *Poa*, and *Danthonia*. Such herbfields typically occur at the base of snow patches, where snow melts earliest, and where irrigation (from snow melt upslope) is more or less constant. The deep and late-lying snow exerts a considerable shear force on the soil surface (0.2–37.9 Mpa), which is ample to account for the patterns of soil erosion which may be observed on such sites (Costin *et al.*, 1973). Consequently, these communities are particularly susceptible to trans-humant disturbances.

Fjaeldmark (= feldmark) is a low-growing community variously dominated by dwarf shrubs, low herbs, mosses, with substantial areas of bare ground between individual plants. A fjaeldmark of *Coprosma* and *Colobanthus* occurs on the drier, upper slopes of snowpatches, where snow persists the longest (Costin, 1954). Fjaeldmark dominated by *Epacris* spp. and *Chionohebe densifolia* occurs on highly exposed high-altitude, relatively snow-free alpine sites. On the mainland, these communities occur only at Kosciusko. Analogues of both the leeward and high-altitude fjaeldmarks in Tasmania have been described by Kirkpatrick (1980), Kirkpatrick & Harwood (1980) and Gibson & Kirkpatrick (1985). The formation does not occur in Victoria.

Vegetation dynamics

There have been numerous long-term studies of Australian alpine/ subalpine vegetation (Costin, 1954; Carr & Turner, 1959*b*; Carr, 1977; Wimbush & Costin, 1979*a,b,c*; van Rees *et al.*, 1985; Leigh *et al.*, 1987; Williams & Ashton, 1987*b*; Wimbush & Forrester, 1988; Wahren *et al.*, 1991), and the dynamics of these vegetation types are relatively well understood. These studies have emphasised interactions between site factors, unpredictable climatic fluctuations, biotic and abiotic disturbances, and species life-history characteristics as determinants of both the rate and direction of vegetation change.

Disturbance is integral to all plant communities (Pickett & White, 1985), and arctic and alpine vegetation in general may be subject to recurrent disturbance, because of the severe nature of the climate (Churchill & Hanson, 1958). Fire, wind, frost, extreme climatic events, and insect attacks are some of the natural disturbances within Australian alpine ecosystems (Williams, 1990*a*). In contrast to ecosystems at lower altitudes the alpine biota and soils have evolved largely in the absence of both frequent fire and substantial grazing by vertebrates (Costin, 1983; Banks, 1982, 1989). In response to differing disturbance regimes, both endogenous and exogenous, patterns of dominance in all formations may alternate between species, which may or may not be of the same life form (Ashton & Williams, 1989).

Interactions between woody and herbaceous species within the heath–grass complex have received considerable attention (Carr & Turner, 1959*b*; Wimbush & Costin, 1979*a,b,c*; Williams & Ashton, 1987*b*, 1988; Williams, 1987*a*, 1990*a,b*; Wahren *et al.*, 1991). Most shrubs are facultative seed regenerators, capable of resprouting following canopy damage or senescence. Within closed-heath and open-heath, seedlings of shrubs will only establish on bare ground, which may result from damage to the local ground cover, either woody or herbaceous. Shrub seedlings will not establish on fully vegetated micro-sites, or on thick, fixed grass litter. Endogenous disturbances which may lead to the formation of such patches of bare ground include wind and frost, fire, senescence of plant material, attack of *Poa* by hepialids, and infestation of shrubs by merimnetrids. Exogenous disturbances which cause bare ground include grazing and trampling by livestock, and an increased fire frequency. Subsequent to establishment, shrub growth depends upon species and site-exposure, with shrubs from more sheltered, closed-heath having greater growth rates than those from open-heath. As the canopy senesces, closed-heath species such as *Prostanthera* and *Orites* tend to resprout, and individual plants may survive for more than a century. Thus the understorey of such heaths remains typically sparse. Open-heath species such as *Grevillea* and *Asterolasia*, in contrast, are obligate seeders and tend to be replaced fully by a closed sward of *Poa* and other herbs during and after shrub senescence, which usually occurs after 20–40 years. Further disturbance leading to the formation of bare ground is necessary to initiate the establishment of a subsequent generation of seedlings of these shrubs.

The regeneration cycles of the bog communities have also been studied in some detail and involve formation and degeneration of pools, hollows and hummocks. *Sphagnum* (principally *S. cristatum*) is involved in the formation of hummocks and the pattern of hollows and hummocks is of the order of 1–2 m². The hollows are often water filled, and are dominated by *Carex gaudichaudiana*. *Sphagnum* may invade the hollows, forming hummocks because of the lateral and vertical expansion of the *Sphagnum*. The hummocks in turn may be colonised by the hydrophytic shrubs and herbs. Continued vertical expansion of the vegetation leads to the development of substantial layers of peat which may be metres thick and several thousand years old (Ashton & Hargreaves, 1983; Wimbush & Costin, 1983).

Fire can have substantial impacts on Australian alpine vegetation, given that its natural occurrence is relatively rare. Initial impacts include reduction in shrub cover within heathland and creation of intertussock patches within grassy vegetation (Leigh *et al.*, 1987). Frequent prescribed burning, as practised in the past by graziers, is likely to have lead to substantial long-term increases in the cover of shrubs, especially fast-growing, non-palatable species such as occur within closed-heath. Numerous fire-sensitive shrubs occur in Tasmanian alpine environments, and such species may take many decades to recover from fire (Kirkpatrick & Dickinson, 1984).

The potential for vegetation change in response to climatic changes associated with the greenhouse effect is open to speculation, as is the extent of any global warming associated with elevated levels of atmospheric CO_2. A number of scenarios have been described for upland southeastern Australia, including a rise in mean annual temperature of $1-2$ °C, reduced winter snow, increased summer rain and increased wind (Galloway, 1989). Whilst the likelihood of this scenario is difficult to determine, as the resolution of the climatic models is not fine enough for most biological predictions, should it occur, there are a number of implications for both the vegetation and land use.

First, there is the potential for expansion of the woody vegetation. The elevation of treeline may rise by *c.* 100 m per 1° rise in mean annual temperature (Galloway, 1989). Elevated summer temperatures may also be expected to increase the growth rates of extant shrubs (Williams, 1990*b*). Any expansion of woody vegetation is dependent, however, on the availability of patches of bare ground for establishment of juveniles (Williams & Ashton, 1987*b*). Such patches are not necessarily provided by climate change alone, and, given the high inertia of herbaceous alpine vegetation (Slatyer, 1989), then the potential for shrub expansion into grasslands and herbfields will be much greater if grazing by stock continues, given the undoubted increase in the cover of bare ground that such grazing causes. Vegetation 'responses' are therefore likely to lag behind climatic 'stimuli', as Graetz, Walker and Walker (1988) have argued for potential responses to climatic change in arid vegetation.

Secondly, if summer rainfall increases, then wetland vegetation will assume an even greater role in regulating catchment discharge. To achieve this, however, the wetlands must be in optimal condition, with a minimum of entrenched drainage channels. Given that most wetlands are currently degraded, or in the process of recovering from grazing-induced degradation, then the removal of grazing from the whole of the high country must occur in order to achieve optimal condition of the wetlands.

Origins of the flora

With increasing elevation, the proportions of the different phytogeographic elements change (Table 16.2). For example, the so-called 'southern' or Antarctic element in the flora increases proportionally from tableland to alpine communities and the proportions of both the autochthonous Australian and the 'tropical' elements decrease.

Table 16.2. *Geographical elements in floras of Monaro Region of New South Wales, according to elevation*

	Geographical element (%)			
	Southern	Cosmopolitan	Australian	Tropical
Tableland (610–915 m)	5	36	50	9
Montane (915–1525 m)	6	33	53	7
Subalpine (1525–1830 m)	14	48	36	2
Alpine (>1830 m)	22	46	31	1

From Costin, 1981.

There is a high proportion of the southern element, showing similarities with elements of the New Zealand, South American, Antarctic and South African floras.

The flora is also characterised by a high degree of endemism at the species level (Costin, 1981), especially in the genera *Abrotanella*, *Athrotaxis*, *Chionochloa*, *Colobanthus*, *Craspedia*, *Diselma*, *Donatia*, *Microcachrys*, *Microstrobos*, *Plantago* and *Ranunculus*. There is also a considerable degree of regional endemism, as around Mt Kosciusko, where at least 10 per cent of the alpine plants are endemic to that area (Costin *et al.*, 1979). No genera are endemic to mainland alpine areas; however, some of the characteristic insects and crustaceans are also endemic and show southern affinities (Costin, 1981), possibly indicating a co-evolution of some sections of the fauna and flora. Endemism within the Tasmanian alpine flora is more pronounced than within the mainland alpine flora, and increases strongly from east to west (Kirkpatrick, 1982; Kirkpatrick & Brown, 1984).

The alpine flora of Australia exists on upland 'islands' in the 'sea' of the old continental landscape. Unlike tropical rainforest, where current patchiness is related to the operation of relatively old relictual processes since the beginning of the Tertiary, the alpine flora is relatively young, with many species derived from sources at lower elevations (Barlow, 1989). Smith (1986) concluded that the Gondwanan floristic elements were not of major significance to the origin of the Australian alpine flora. The relatively high content (*c.* 40 per cent) of bi-hemispheric genera indicated an important role for immigrants, via distance dispersal from cool, extra-Australian sources. Barlow (1989 and Chapter 1), however, argued for a greater contribution from autochthones in the evolution of the flora. Evidence from Tasmania in particular (e.g. Hill & Gibson, 1986; Hill, 1987) indicates the existence of macrofossils of microphyllous (and therefore cold-adapted) species, including *Nothofagus*, from Oligocene/Miocene deposits. McPhail (1986), in analysing pollen assemblages from upland Tasmania, described palynofloras of the last glacial which were similar to those of the modern alpine flora. Two elements of this flora were evident – an older, autochthonous one, and a younger immigrant one.

One factor in the evolution of a pre-Pleistocene cold-adapted flora may have been the long history of response of the lowland flora to periods of aridity, as physiological adaptation to drought may also confer a degree of frost tolerance. For example, lowland/semi-arid species of *Scleranthus* possess morphological features which are more primitive than those of the alpine species, thereby indicating that the alpine species are derived from lowland elements; they have radiated into highland areas only comparatively recently (West & Garnock-Jones, 1986). Areas available for the establishment of the evolving alpine flora are likely to have been provided by large-scale disturbances associated with glaciation (Smith, 1986). The questions of the origins of the flora remain open, with the possibility that evolution from lowland Australian sources, long-distance dispersal and Gondwanan relics have all contributed to the existing alpine flora.

Human impacts on alpine/subalpine vegetation

Use by the Aborigines

Human association with the alpine regions began with the Aborigines, although it is probable that the cold, inhospitable glacial and peri-glacial conditions prevented them from using the high country at least until the last few thousand years. They seem to have migrated annually to the mountains from lower areas, for instance, to feast on Bogong moths, *Agrotis infusa* (Flood, 1980). On the mainland, these moths swarm in late spring in large numbers to the mountains where they rest during the summer among rocks and in crevices. The Aborigines collected and cooked them each summer, before both the moths and their Aboriginal hunters migrated to lower altitudes with the onset of autumn. This appears to have been the only use of the land prior to the time of European settlement. The low intensity of use by

Aborigines and the virtual absence of large indigenous mammals meant that the pastoral practices of Europeans had the potential to modify greatly the alpine/subalpine vegetation, ill adapted as it was to utilisation by domestic herbivores.

Effects of European settlement

Grazing

With the introduction of domestic livestock in the middle of the nineteenth century came substantial changes to the soils and vegetation of the high country. By the late nineteenth century, the practice of summer grazing of sheep, cattle and occasionally horses for a period of more than 50 years, combined with the associated practice of 'burning off' the high mountain pastures to improve their palatability, had already changed the condition of much of the vegetation. Plant cover was reduced, soil erosion promoted, and some plant species, such as *Aciphylla glacialis*, *Chionochloa frigida* and *Ranunculus anemoneus*, became rare and disappeared locally under this combined regime of grazing, trampling and burning.

Grazing by stock has affected the composition of the native alpine vegetation because domestic animals are free ranging, and are highly selective in both the communities in which they graze, and the species within the communities upon which they feed. Cattle prefer to graze within the open communities, such as grassland, open-heaths and snow patches, as the availability of the more palatable species is greater in these communities than in the closed-heaths (McDougall, 1982; van Rees & Hutson, 1983). Palatable species include grasses, e.g. the snow grasses (*Poa* spp.), major herbs, e.g. silver snow daisy (*Celmisia*) and billy buttons (*Craspedia*), and some small- to medium-sized shrubs, e.g. *Asterolasia* and *Grevillea*. In contrast, most shrubs such as the prostrate *Hovea longifolia* and the taller *Prostanthera cuneata* are non-palatable. Monitoring of reference plots both at Kosciusko and on the Bogong High Plains has shown clearly that, in the long-term absence of grazing, the composition of all communities alters dramatically, with a greater abundance of palatable major herbs (Carr & Turner, 1959*b*; Wimbush & Costin, 1979*a,b,c*; Wahren *et al.*, 1991).

The structure and composition of some areas of open-heath and grassland on the Bogong High Plains has changed dramatically following four decades of exclusion from cattle. At the Pretty Valley experimental exclusion plots to the west of Cope Hut, the cover of palatable herbs such as *Craspedia* and *Celmisia* increased substantially between 1946 and 1989. The cover of the palatable shrub *Aster-*

olasia trymalioides also increased over this period (Fig. 16.3). However, this species, and other palatable shrubs such as *Grevillea australis* senesce, are likely to be replaced by herbaceous and grassy species as described above. It is clear that open-heath is a very dynamic formation and capable of increasing shrubbiness or grassiness depending on the age of the stand, the life history characteristics of the dominant species and variations in climate and cattle activity.

Disturbance by cattle is an important aspect of the present composition and structure of grasslands and grassy patches within heathland (Williams & Ashton, 1987a,b). Tussocks are scalded by urine and may be damaged by trampling and grazing or obliterated by droppings. Such areas may be converted to gaps and the soil bared by frost or wind. Gaps may remain largely bare or may be recolonised by graminoids or seedling shrubs. If cattle are removed from the Bogong High Plains it is highly probable that the palatable, obligate seed-regenerating shrubs (such as *Asterolasia* and *Grevillea*) which have established on the bare sites within the grassland will increase in both cover and density, at least for the next 20 to 30 years, before senescing. On the other hand, the cover of shrubs such as *Hovea longifolia*, *Kunzea muelleri*, *Phebalium squamulosum* and *Prostanthera cuneata* is unlikely to be inhibited by cattle activity, as these shrubs are non-palatable and more tolerant of trampling than either *Grevillea* or *Asterolasia*. Indeed, the continued increase in the cover of taller, denser, non-palatable shrubs on all experimental plots in grassland and heathland subject to grazing between 1945 and 1989, compared with downward trends in shrub cover on ungrazed plots (Wahren *et al.*, 1991), shows clearly that cattle have done little or nothing to reduce the standing crop of the taller shrubs. Given that these species are the most flammable (Good, 1982), then there is little evidence to support the contention that grazing in alpine areas reduces the risk of fire.

It has long been recognised that all wetland communities are particularly vulnerable to physical damage, especially that due to trampling by domestic livestock (Costin, 1954; Anon., 1957; Carr & Turner, 1959a; Costin *et al.*, 1959; Wimbush & Costin, 1979a,b,c, 1983). Although bogs are not preferred grazing communities by cattle (van Rees & Hutson, 1983), they enter them to drink and graze palatable species such as *Carex gaudichaudiana*, *Empodisma minus*, *Poa costiniana* and *Astelia alpina* (van Rees, 1984). Cattle trample bogs and create lines of enhanced drainage, for example by joining up naturally discreet hollows into a continuous drainage line. Grazing and trampling also decrease the surface roughness of the vegetation,

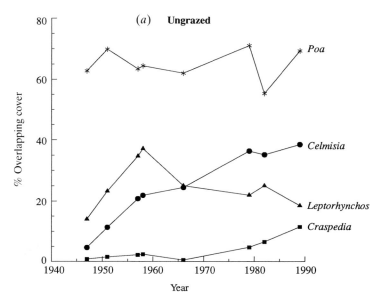

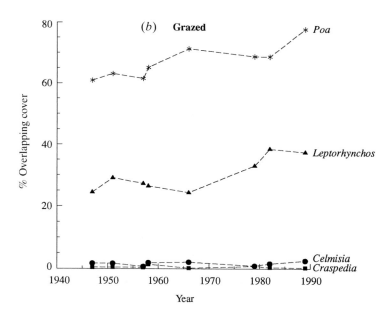

Fig. 16.3. Long-term vegetation changes, Pretty Valley experimental grassland plots, Bogong High Plains, Alpine National Park, Victoria. The plots were established by S.G.M. Carr and J.S. Turner in 1946. Changes in percentage overlapping cover of: (a,b) the dominant herbs

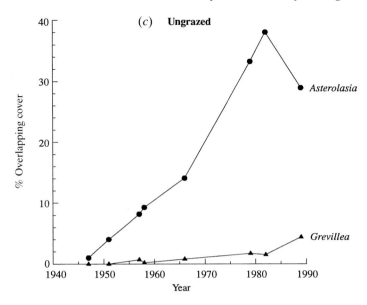

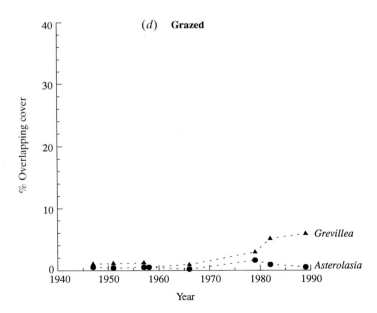

Fig. 16.3. (*continued*) *Poa hiemata*, *Celmisia* spp., *Leptorhynchos squamatus* and *Craspedia* spp.; (c,d) the shrubs *Asterolasia trymalioides* and *Grevillea australis*, on plots grazed or ungrazed by domestic cattle (from Wahren *et al.*, 1991).

which leads to accelerated overland flow (Wimbush & Costin, 1983). The combined effects of these processes is to dry out the *Sphagnum* and peat, and increase incision and erosion in the peat by accelerated erosion, eventually leading to entrenched drainage lines. These may continue to erode through the peat until large portions of the rock pavement at the base of the bog are exposed. Erosion of both peat and stream banks may continue for decades after grazing ceases (Wimbush & Costin, 1979*b*, 1983).

Over a period of the last 20 or more years, since sheep and cattle grazing and associated burning off ceased over the whole of the Kosciusko National Park, there have been significant improvements to the condition of the landscape in the alpine and subalpine regions (Wimbush & Costin, 1979*a*,*b*,*c*, 1983). Established snowgrass tussocks (of *Poa* spp.) and/or shrubs have spread onto adjacent bare ground, and formerly bare ground and intertussock spaces have also been colonised by various minor and major herbs. There has been an increase in the number and abundance of species of palatable major forbs, as co- and subdominants in the widespread snowgrass communities. Snowgum has regenerated around existing trees, but not on deforested areas. Partial regeneration of *Sphagnum* bogs has occurred, except where the bogs are incised by gullies and drainage lines.

Water

On mainland Australia the high mountain catchments contribute about 25 per cent of the average annual yield of the largest Australian water catchment, the River Murray system. As early as the 1920s, concern for the state of the vegetative cover and soils of these alpine and subalpine regions was expressed. Water from the Western slopes of Kosciusko was first collected by the Hume Dam, situated on the Murray River and built at this time, whence it is reticulated for subsequent use in the irrigation areas in New South Wales, Victoria and South Australia. This purpose was also part of the concept of the Snowy Mountains Scheme, begun in 1948 and completed in 1972, as well as the purpose of generating peak-load hydro-electric power. In Victoria and Tasmania the catchments are used mainly for hydro-electric power generation, for the development of which there is still further scope.

Results of studies of catchment hydrology in the Australian alps have shown that maintenance of optimum catchment condition depends upon maximising vegetative cover (Costin *et al.*, 1959, 1960, 1964; Costin, 1966), and that the potential for soil erosion increases substantially with increasing slope (van Rees & Boston, 1986). No single type of plant cover (i.e. herbs, shrubs or trees) was found to be ideal for all catchment requirements. The most effective control of surface run-off and soil loss was provided by a virtually continuous herbaceous cover. But this cover was found to be a relatively in-efficient collector of wind-blown snow, rain, fog and cloud, and had limited value in delaying snowmelt. These aspects of water collection were best provided by trees, preferably as fairly open stands, or denser groups of trees with small clearings or strips between them. The extra water collected by the trees was not offset by greater evapotranspir-ation which was found to be much the same in all non-ground-water vegetation. Quality and continuity of yield were further improved by bogs and similar ground-water communities.

In summary, an open-forest or woodland, underlain by dense her-baceous cover, with ground-water communities along seepages and drainage lines gives water of optimum quality and quantity. Long-term protection of the vegetation from both livestock and grazing and fires helps achieve this condition.

Recreation and tourism

The alpine and subalpine areas are all close to the main centres of population in southeastern Australia, and the use of these regions for tourism has increased exponentially in the last 40 years. The building of the roads and hotels for tourist purposes commenced in the early 1900s and continues to the present day. In places, such as the ski resort complexes at Thredbo and Perisher (New South Wales), Falls Creek and Mount Hotham (Victoria), tourism has had and will con-tinue to have a major effect on land use, in the area itself and in adjoining areas, especially as it conflicts with nature conservation. Research into revegetation of disturbed sites, such as ski-slopes, requires particular support. The potential of native herbs and shrubs for slope stabilisation is high (Good, 1980) and numerous broad-acre trials are being conducted in the Victorian ski resorts by the Victorian Department of Conservation & Environment.

The network of roads initially built in conjunction with the con-struction of hydro-electric power schemes, has opened up large areas of alpine and subalpine country for scenic driving, camping sites, tourist villages and ski resorts. The multifarious uses usually involve some destruction and disturbance of the vegetation, both directly and from the accelerated run-off of rain water and melted snow, as does the use of four-wheel drive and over-snow vehicles. Even foot traffic along walking tracks is causing local damage to some of the most restricted plant communities (Wimbush & Costin, 1973). Management plans for the Alps (e.g. Anon., 1989) must continue to address these issues so as to provide maximum protection to the native biota.

National parks and nature conservation

To a large extent, preservation of the scientific values of the high mountain areas is compatible with their use for such activities as wilderness-based recreation and for water harvesting. These uses are incompatible, however, with grazing and sophisticated, highly intensive tourism, such as ski resorts.

Moves to preserve adequate areas of the Alps within National Parks date from the 1920s and 1930s. In 1944, a small part of the mainland alpine area, that around Mt Kosciusko itself, was withdrawn from grazing. Subsequently, all areas in Kosciusko National Park above about 1370 m were similarly protected, and in 1967 the Kosciusko State Park (later National Park) was declared. In Victoria, the Alpine National Park, contiguous with the Kosciusko National Park, was declared in 1989. However, in Victoria, non-compatible land uses such as cattle grazing will continue over the majority of the alpine and subalpine area (Land Conservation Council, 1983) despite clear evidence that grazing and nature conservation are incompatible land uses.

Considering Australia as a whole, the high mountain ecosystems appear to be reserved more adequately than any other (Newsome, 1973). In Tasmania and New South Wales most of the high mountain country is reserved in national parks. In Victoria, the 'multi-use' strategy is most unsatisfactory. Whilst preservation of a large and representative sample of alpine and subalpine vegetation in Australia seems assured, there is no room for complacency when it comes to managing these communities for fauna and flora conservation (Wimbush & Costin, 1973; Good, 1989). Herein lies the challenge for nature conservation and its practitioners in the immediate future.

Other uses

Commercial forestry has no significance in the alpine and subalpine regions of Australia as yet. Intermittent mining, such as for gold at Kiandra, New South Wales, predominantly in the 1860s, has had only local effects on the vegetation and is of minor significance overall at present.

Conclusions

This chapter has emphasised the uniqueness of the alpine and sub-alpine environments and their vegetation in relation to those of some other, more widespread vegetation types. Their occurrence is restricted, and the patchy distribution of the major vegetation types reflects the influence of complex climatic and geomorphic processes, both present and past. The dynamics of the vegetation depends on the occurrence of both average and extreme conditions, and in this context, rare combinations of conditions and other non-equilibrium processes may have substantial and long-lasting effects on the vegetation. The vegetation is fragile, and its condition has undoubtedly been affected adversely by disturbances associated with European land uses such as livestock grazing. Maintenance of optimal vegetation condition, and the minimisation of any potential deleterious effects associated with global warming require that grazing be discontinued.

Australia's alpine systems have been much studied and a seemingly adequate sample has been preserved for the future. Conflicts in the use of this land have been and are acute, largely because its uses for water harvesting, recreation and for nature conservation are mostly incompatible with grazing, highly intensive tourism and some hydro-electric developments. At the national level, these conflicts have sometimes been resolved in State parliaments. The recently declared Memorandum of Understanding signed by the governments of New South Wales, Victoria and the Commonwealth, addresses the issue of future conservation of the Australian Alps. Conflicts in land use still arise, and plans of management (National Parks and Wildlife Service of New South Wales, 1974; Anon, 1989) attempt to minimise such conflicts and maximise the benefits to the different categories of users through zoning the land, with appropriate management for each zone.

Acknowledgements

We thank David Ashton, Jenny Read and Robert Flower for critical comments on earlier drafts of this chapter.

References

Anon. (1957). *A Report on the Condition of the High Mountain Catchments of New South Wales and Victoria.* Canberra: Australian Academy of Science.

Anon. (1989). *Proposed Management Plans, Alpine National Park.* Melbourne: Department of Conservation, Forests and Lands.

Ashton, D.H. & Hargreaves, G.R. (1983). Dynamics of subalpine vegetation at Echo Flat, Lake Mountain, Victoria. *Proceedings of the Ecological Society of Australia,* **12,** 35–60.

Ashton, D.H. & Williams, R.J. (1989). Dynamics of the subalpine vegetation in the Victorian Region. In *The Scientific Significance of the Australian Alps,* ed. R. Good, pp. 143–65. Canberra: Australian Alps National Parks Liaison Committee/Australian Academy of Science.

Banks, J. (1982). The use of dendrochronology in the interpretation of the dynamics of the snow gum forest. PhD thesis, Australian National University.

Banks, J.C. (1989). A history of forest fire in the Australian Alps. In *The Scientific Significance of the Australian Alps,* ed. R. Good, pp. 265–80. Canberra: Australian Alps National Parks Liaison Committee/Australian Academy of Science.

Barlow, B.A. (ed.) (1986). *Flora and Fauna of Alpine Australasia.* Melbourne: CSIRO/Brill.

Barlow, B.A. (1989). The alpine flora: autochthones and peregrines. In *The Scientific Significance of the Australian Alps,* ed. R. Good, pp. 69–78. Canberra: Australian Alps National Parks Liaison Committee/Australian Academy of Science.

Billings, W.D. (1974). Adaptations and origins of alpine plants. *Arctic & Alpine Research,* **6,** 129–42.

Billings, W.D. & Mooney, H.A. (1968). The ecology of arctic and alpine plants. *Biological Review,* **43,** 481–529.

Bliss, L.C. (1962). Caloric and lipid content in alpine tundra plants. *Ecology,* **43,** 753–75.

Bliss, L.C. (1971). Arctic and alpine life cycles. *Annual Review of Ecology & Systematics,* **2,** 405–438.

Carr, S.G.M. (1977). *Report on inspection of the Bogong High Plains.* Melbourne: Land Conservation Council of Victoria.

Carr, S.G.M. & Turner, J.S. (1959a). The ecology of the Bogong High Plains. I. The environmental factors and the grassland communities. *Australian Journal of Botany,* 7, 12–33.

Carr, S.G.M. & Turner, J.S. (1959b). The ecology of the Bogong High Plains. II. Fencing experiments in grassland C. *Australian Journal of Botany,* 7, 34–63.

Churchill, E.D. & Hanson, H.C. (1958). The concept of climax in arctic and alpine vegetation. *Botanical Review,* **24,** 127–91.

Cochrane, P.M. & Slatyer, R.O. (1988). Water relations of *Eucalyptus pauciflora* near the alpine treeline in winter. *Tree Physiology*, **4**, 45–52.

Colhoun, E.A. (1978). The Late Quaternary environment in Tasmania as backdrop to man's occupance. *Records of the Queen Victoria Museum, Launceston*, No. 61.

Costin, A.B. (1954). *A Study of the Ecosystems of the Monaro Region of New South Wales*. Sydney: Government Printer.

Costin, A.B. (1957). The high mountain vegetation of Australia. *Australian Journal of Botany*, **5**, 173–89.

Costin, A.B. (1962*a*). Ecology of the High Plains. I. *Proceedings of the Royal Society of Victoria*, **75**, 327–37.

Costin, A.B. (1962*b*). Soils of the High Plains. *Proceedings of the Royal Society of Victoria*, **75**, 291–9.

Costin, A.B. (1966). Management opportunities in Australian high mountain catchments. In *Forest Hydrology*, eds W.E. Sopper & H.W. Lull, pp. 565–77. Oxford & New York: Pergamon Press.

Costin, A.B. (1967). Alpine ecosystems of the Australian region. In *Arctic and Alpine Environments*, eds H.E. Wright & W.H. Osburn, pp. 57–87. Bloomington: Indiana University Press.

Costin, A.B. (1971). Vegetation, soils, and climate in Late Quaternary south-eastern Australia. In *Aboriginal Man and Environment in Australia*, eds D.J. Mulvaney & J. Golson, pp. 26–37. Canberra: Australian National University Press.

Costin, A.B. (1972). Carbon-14 dates from the Snowy Mountains area, south-eastern Australia, and their interpretations. *Quaternary Research*, **2**, 579–90.

Costin, A.B. (1973). Characteristics and use of Australian high country. In *The Lake Country of Tasmania*, ed. M.R. Banks, pp. 1–23. Hobart: Royal Society of Tasmania.

Costin, A.B. (1981). Vegetation of the high mountains of Australia. In *Ecological Biogeography in Australia*, ed. A. Keast, pp. 717–32. The Hague: W. Junk.

Costin, A.B. (1983). Mountain lands in the Australasian region. *Proceedings of the Ecological Society of Australia*, **12**, 1–13.

Costin, A.B. (1986). Genesis of Australian alpine soils. In *Flora and Fauna of Alpine Australasia*, ed. B.A. Barlow, pp. 37–44. Melbourne: CSIRO/Brill.

Costin, A.B. (1989). The alps in a global perspective. In *The Scientific Significance of the Australian Alps*, ed. R. Good, pp. 7–19. Canberra: Australian Alps National Parks Liaison Committee/Australian Academy of Science.

Costin, A.B., Gray, M., Totterdell, C.J. & Wimbush, D.J. (1979). *Kosciusko Alpine Flora*. Melbourne: CSIRO & Collins.

Costin, A.B., Jennings, J.N., Bautovich, B.C. & Wimbush, D.J. (1973). Forces developed by snow action, Mt Twynham, Snowy Mountains, Australia. *Arctic & Alpine Research*, **5**, 121–6.

Costin, A.B. & Wimbush, D.J. (1973). Frost cracks and earth hummocks at Kosciusko, Snowy Mountains, Australia. *Arctic & Alpine Research*, **5**, 111–20.

Costin, A.B., Wimbush, D.J. & Cromer, R.N. (1964). Studies in catchment hydrology in the Australian Alps. V. Soil moisture characteristics and evapotranspiration. *CSIRO Division of Plant Industry, Technical Paper* No. 20.

Costin, A.B., Wimbush, D.J. & Kerr, D. (1960). Studies in catchment hydrology in the Australian Alps. II. Surface run-off and soil loss. *CSIRO Division of Plant Industry, Technical Paper* No. 14.

Costin, A.B., Wimbush, D.J., Kerr, D. & Gay, L.W. (1959). Studies in catchment hydrology in the Australian Alps. I. Trends in soils and vegetation. *CSIRO Division of Plant Industry, Technical Paper* No. 13.

Daubenmire, R. (1954). Alpine timberlines in the Americas and their interpretation. *Butler University (Indianopolis, USA) Botanical Studies,* **11,** 119–36.

Davies, J.L. (1965). Landforms. In *Atlas of Tasmania,* ed. J.L. Davies, pp. 19–22. Hobart: Department of Lands & Survey.

Farrell, T.P. & Ashton, D.H. (1973). Ecological studies on the Bennison High Plains. *Victorian Naturalist,* **90,** 286–98.

Ferrar, P.J., Cochrane, P.M. & Slatyer, R.O. (1988). Factors influencing germination and establishment of *Eucalyptus pauciflora* near the alpine treeline. *Tree Physiology,* **4,** 27–43.

Flood, J. (1980). *The Moth Hunters. Aboriginal Prehistory of the Australian Alps.* Canberra: Australian Institute of Aboriginal Studies.

Galloway, R.W. (1989). Glacial and periglacial features of the Australian Alps. In *The Scientific Significance of the Australian Alps,* ed. R. Good, pp. 55–67. Canberra: Australian Alps National Parks Liaison Committee/Australian Academy of Science.

Gibson, N. (1990). The environments and primary production of cushion species at Mt Field and Mt Wellington, Tasmania. *Australian Journal of Botany,* **38,** 229–43.

Gibson, N. & Kirkpatrick, J.B. (1985). Vegetation and flora associated with localized snow accumulation at Mt Field West, Tasmania. *Australian Journal of Ecology,* **10,** 91–9.

Good, R.B. (1980). Native species in revegetation of alpine herbfields. Technical Note. Sydney: New South Wales National Parks & Wildlife Service.

Good, R.B. (1982). Effects of prescribed burning in the subalpine area of Kosciusko National Park. MSc thesis, University of New South Wales.

Good, R.B. (ed) (1989). *The Scientific Significance of the Australian Alps.* Canberra: Australian Alps National Parks Liaison Committee/Australian Academy of Science.

Graetz, R.D., Walker, B.H. & Walker, P.A. (1988). The consequences of climatic change for seventy percent of Australia. In: *Greenhouse: Planning for Climatic Change,* ed G.I. Pearman, pp 399–420. Melbourne: CSIRO.

Hadley, E.B. & Bliss, L.C. (1964). Energy relationships of alpine plants on Mt Washington, New Hampshire. *Ecological Monographs,* **34,** 331–57.

Hill, R.S. (1987). Tertiary *Isoetes* from Tasmania. *Alcheringa,* **2,** 157–62.

Hill, R.S. & Gibson, N. (1986). Macrofossil evidence for the evolution of the alpine and subalpine vegetation of Tasmania. In *Flora and Fauna of Alpine Australasia,* ed. B. Barlow, pp. 205–17. Melbourne: CSIRO/Brill.

Jackson, W.D. (1973). Vegetation of the Central Plateau. In *The Lake Country of Tasmania*, ed. M.R. Banks, pp. 61–86. Hobart: Royal Society of Tasmania.

Jones, J.G. & Veevers, J.J. (1982). A Cainozoic history of Australia's southeast highlands. *Journal of the Geological Society of Australia*, **29**, 1–12.

Kirkpatrick, J.B. (1980). Tasmanian high mountain vegetation. I. A reconnaissance survey of the Eastern Arthur Range and Mt Picton. *Papers & Proceedings of the Royal Society of Tasmania*, **114**, 1–20.

Kirkpatrick, J.B. (1982). Phytogeographical analysis of Tasmanian alpine floras. *Journal of Biogeography*, **9**, 255–71.

Kirkpatrick, J.B. (1983). Treeless plant communities of the Tasmanian high country. *Proceedings of the Ecological Society of Australia*, **12**, 61–77.

Kirkpatrick, J.B. (1986). Tasmanian alpine biogeography and ecology and interpretation of the past. In *Flora and Fauna of Alpine Australasia*, ed. B. Barlow, pp. 229–42. Melbourne: CSIRO/Brill.

Kirkpatrick, J.B. (1989). The comparative ecology of mainland Australia and Tasmanian alpine vegetation. In *The Scientific Significance of Australian Alps*, ed. R. Good, pp. 127–42. Canberra: Australian Alps National Parks Liaison Committee/ Australian Academy of Science.

Kirkpatrick, J.B. & Brown, M.J. (1984). A numerical analysis of Tasmanian higher plant endemism. *Botanical Journal of the Linnaean Society*, **88**, 165–83.

Kirkpatrick, J.B. & Dickinson, K.J.M. (1984). The impact of fire on Tasmanian alpine vegetation and soil. *Australian Journal of Botany*, **32**, 613–29.

Kirkpatrick, J.B. & Duncan, F. (1987). Tasmanian high altitude grassy vegetation: its distribution, community composition and conservation status. *Australian Journal of Ecology*, **12**, 73–86.

Kirkpatrick, J.B. & Harwood, C.E. (1980). The vegetation of an infrequently burned Tasmanian mountain region. *Proceedings of the Royal Society of Victoria*, **91**, 67–86.

Kirkpatrick, J.B., Minchin, P.R. & Davies, J.B. (1985). Floristic composition and macroenvironmental relationships of Tasmanian vegetation containing bolster plants. *Vegetatio*, **63**, 89–96.

Körner, Ch. & Cochrane, P. (1983). Influence of plant physiognomy on leaf temperature on clear midsummer days in the Snowy Mountains, south-eastern Australia. *Acta Oecologia, Oecologia Plantarum*, **4**, 117–24.

Land Conservation Council. (1983). *Final Report. Alpine Study Area*. Melbourne: Land Conservation Council.

Leigh, J.H., Wimbush, D.J., Wood, D.H., Holgate, M.D., Slee, A.V., Stanger, A.G. & Forrester, R.I. (1987). Effects of rabbit grazing and fire on a subalpine environment. I. Herbaceous and shrubby vegetation. *Australian Journal of Botany*, **35**, 433–64.

Löve, D. (1970). Subarctic and subalpine: where and what? *Arctic & Alpine Research*, **2**, 63–73.

McDougall, K.I. (1982). *The Alpine Vegetation of the Bogong High Plains*. Melbourne: Victorian Ministry of Conservation, Environmental Studies Publication No. 357.

McPhail, M.K. (1986). Over the top: Pollen based reconstructions of past alpine floras and vegetation in Tasmania. In *Flora and Fauna of Alpine Australasia*, ed. B. Barlow, pp. 173–204. Melbourne: CSIRO/Brill.

McVean, D.N. (1969). Alpine vegetation of the Central Snowy Mountains of New South Wales. *Journal of Ecology* 57, 67–87.

Mansergh, I.M. & Scotts, D.J. (1986). Winter occurrence of the Mountain Pygmy-possum, *Burramys parvus* (Broom) (Marsupialia: Burramidae) on Mt Higginbotham, Victoria. *Australian Mammalogy* 9, 35–42.

Moore, R.M. & Williams, J.D. (1976). A study of a subalpine woodland-grassland boundary. *Australian Journal of Ecology*, 1, 145–53.

National Parks & Wildlife Service of New South Wales (1974). *Kosciusko National Park Plan of Management*. Sydney: National Parks & Wildlife Service.

Newsome, A.E. (1973). The adequacy and limitations of flora conservation for fauna conservation in Australia and New Zealand. In *Nature Conservation in the Pacific*, eds A.B. Costin & R.H. Groves, pp. 93–110. Canberra: Australian National University Press.

Ollier, C.D. (1986). The origin of alpine land forms in Australasia. In *Flora and Fauna of Alpine Australasia*, ed. B. Barlow, pp. 2–36. Melbourne: CSIRO/Brill.

Ollier, C.D. & Wyborn, D. (1989). Geology of alpine Australia. In *The Scientific Significance of the Australian Alps*, ed. R. Good, pp. 35–53. Canberra: Australian Alps National Parks Liaison Committee/Australian Academy of Science.

Paton, D.M. (1988). Genesis of an inverted treeline associated with a frost hollow in south-eastern Australia. *Australian Journal of Botany*, 36, 655–63.

Peterson, J.A. (1971). The equivocal extent of glaciation in the south-eastern uplands of Australia. *Proceedings of the Royal Society of Victoria*, 84, 207–11.

Pickett, S.T.A. & White, P.S. (eds) (1985). *The Ecology of Natural Disturbance and Patch Dynamics*. New York: Academic Press.

Rees, H. van (1984). *Behaviour and Diet of Free-ranging Cattle on the Bogong High Plains, Victoria*. Melbourne: Victorian Department of Conservation, Forests & Lands, Environmental Studies Publication No. 409.

Rees, H. van & Boston, R.C. (1986). Evaluation of factors affecting surface runoff on alpine rangeland in Victoria. *Australian Rangelands Journal*, 8, 97–102.

Rees, H. van & Hutson, G. (1983). The behaviour of free-ranging cattle on an alpine range in Australia. *Journal of Range Management*, 36, 740–3.

Rees, H. van, Papst, W.A., McDougall, K. & Boston, R.C. (1985). Trends in vegetation cover in the grassland community on the Bogong High Plains, Victoria. *Australian Rangelands Journal*, 7, 93–8.

Rowe, R.K. (1972). *A Study of the Land in the Catchment of the Kiewa River*. Melbourne: Soil Conservation Authority.

Slatyer, R.O. (1989). Alpine and valley bottom treelines. In *The Scientific Significance of the Australian Alps*, ed. R. Good, pp. 169–84. Canberra: Australian Alps National Parks Liaison Committee/Australian Academy of Science.

Smith, J.M.B. (1986). Origins of Australasian tropic alpine and alpine floras. In *Flora and Fauna of Alpine Australasia*, ed. B. Barlow, pp. 109–28. Melbourne: CSIRO/Brill.

Tranquillini, W. (1979). *Physiological Ecology of the Alpine Timberline.* Berlin: Springer-Verlag.

Turner, A.L. & Hill, R.S. (1981). Cushion plants. In *Vegetation of Tasmania*, ed. W.D. Jackson, pp. 169–74. Hobart: University of Tasmania, Department of Botany.

Wahren, C.H., Papst, W. & Williams, R.J. (1991). *The Ecology of Grasslands and Heathlands on the Bogong High Plains – An Analysis of 45 years of Vegetation Records.* Melbourne: Department of Conservation & Environment.

Walsh, N.G., Barley, R.H. & Gullan, P.K. (1984). *The Alpine Vegetation of Victoria (Excluding the Bogong High Plains).* Melbourne: Victorian Department of Conservation, Forests & Lands, Environmental Studies Publication No. 376.

Wardlaw, I.F., Moncur, M.W. & Totterdell, C.J. (1989). The growth and development of *Caltha introloba* F. Muell. I. The pattern and control of flowering. *Australian Journal of Botany*, **37**, 275–89.

Wellman, P. (1979). On the Cainozoic uplift of the south-eastern Australian highlands. *Journal of the Geological Society of Australia*, **26**, 1–9.

West, J.G. & Garnock-Jones, P.J. (1986). Evolution and biogeography of *Scleranthus* (Caryophyllaceae) in Australasia. In *Flora and Fauna of Alpine Australasia*, ed. B.A. Barlow, pp. 435–50. Melbourne: CSIRO/Brill.

Williams, R.J. (1987a). *The Dynamics of Heathland and Grassland Communities in the Subalpine Tract of the Bogong High Plains, Victoria.* Melbourne: Victorian Department of Conservation, Forests & Lands, Environmental Studies Publication No. 413.

Williams, R.J. (1987b). Patterns of air temperature and accumulation of snow in subalpine heathland and grassland communities on the Bogong High Plains, Victoria. *Australian Journal of Ecology*, **12**, 153–63.

Williams, R.J. (1990a). Cattle grazing within subalpine heathland and grassland communities on the Bogong High Plains: Disturbance, regeneration, and the shrub-grass balance. *Proceedings of the Ecological Society of Australia*, **16**, 255–65.

Williams, R.J. (1990b). Growth of subalpine shrubs and snowgrass following a rare occurrence of frost and drought in south-eastern Australia. *Arctic and Alpine Research*, **22**, 412–22.

Williams, R.J. & Ashton, D.H. (1987a). The composition, structure and distribution of heathland and grassland communities in the subalpine tract of the Bogong High Plains, Victoria. *Australian Journal of Ecology*, **12**, 57–71.

Williams, R.J. & Ashton, D.H. (1987b). The effects of disturbance and grazing by cattle on the dynamics of heathland and grassland communities on the Bogong High Plains, Victoria. *Australian Journal of Botany*, **35**, 413–31.

Williams, R.J. & Ashton, D.H. (1988). Cyclical patterns of regeneration in heathland communities on the Bogong High Plains, Victoria. *Australian Journal of Botany*, **36**, 605–19.

Wimbush, D.J. & Costin, A.B. (1973). *Vegetation mapping in relation to ecological interpretation and management in the Kosciusko alpine area.* CSIRO, Australia, Division of Plant Industry, Technical Paper No. 32.

Wimbush, D.J. & Costin, A.B. (1979*a*). Trends in vegetation at Kosciusko. I. Subalpine grazing trials 1957–1971. *Australian Journal of Botany*, **27**, 741–87.

Wimbush, D.J. & Costin, A.B. (1979*b*). Trends in vegetation at Kosciusko. II. Subalpine range transects 1959–1978. *Australian Journal of Botany*, **27**, 789–831.

Wimbush, D.J. & Costin, A.B. (1979*c*). Trends in vegetation at Kosciusko. III. Alpine range transects 1959–1978. *Australian Journal of Botany*, **27**, 833–71.

Wimbush, D.J. & Costin, A.B. (1983). Trends in drainage characteristics in the subalpine zone at Kosciusko. *Proceedings of the Ecological Society of Australia*, **12**, 143–54.

Wimbush, D.J. & Forrester, R.I. (1988). Effects of rabbit grazing and fire on a subalpine environment. II. Tree vegetation. *Australian Journal of Botany*, **36**, 433–64.

17

Coastal dune vegetation

P.J. CLARKE

COASTAL dune vegetation covers about half the length of Australia's shoreline and about one-quarter of the area of coastal land systems (Galloway *et al.*, 1980). Coastal dune vegetation varies both in floristic composition and structure around the coast, e.g. in Queensland and in New South Wales; heaths, open-forest, swamp forests, and rainforest all occur on coastal dune land systems. Hence many of the vegetation types found on coastal dunes are covered elsewhere in this book.

Broadly speaking, dune vegetation can be separated into three major zones that grade into each other; the strandline zone, the foredune zone, and the hind-dune zone (Fig. 17.1). The latter may extend many kilometres inland and often consists of a mosaic of vegetation types on different dune land systems. This chapter is mainly concerned with the vegetation that is found on the strandline and foredune. It draws together information on the floristic composition, vegetation patterns, vegetation dynamics and the changes wrought by people on coastal dune vegetation.

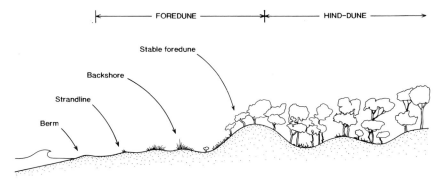

Fig. 17.1. Coastal dune nomenclature and zonation sequence.

501

Foredune floristics

The total number of native species that grow on the strandline and foredune is conservatively estimated to be about 250 species, most of which grow on the foredune. Species richness is high along the southeastern coast and apparently decreases towards the tropics (Fig. 17.2). The flora is diverse in its composition, with more than 70 families being represented. Only the Amaranthaceae, Asteraceae, Chenopodiaceae, Cyperaceae, Epacridaceae, Fabaceae, Goodeniaceae, Mimosaceae, Myrtaceae, and Poaceae have more than a few genera represented, of which only *Scaevola* (Goodeniaceae), *Spinifex* and *Sporobolus* (Poaceae) have a continent-wide distribution on foredunes (Fig. 17.3).

A distinctive feature of the coastal dune flora is the difference between the southern temperate and northern tropical regions. On the temperate and subtropical coasts a large proportion of the dune flora is endemic to Australia, whereas on the tropical coasts an increasing proportion of species has pantropical or Indo-Pacific

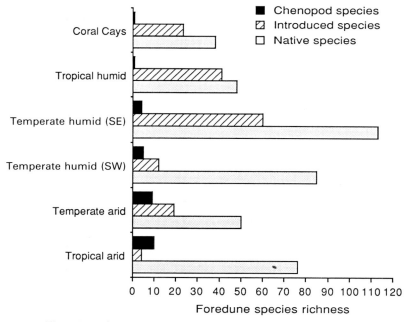

Fig. 17.2. Plant species richness on foredunes from the major coastal climatic regions of Australia.

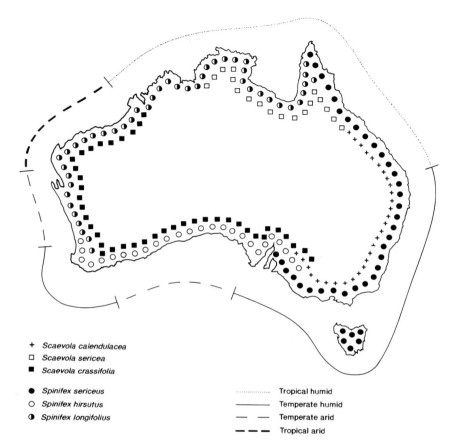

+   *Scaevola calendulacea*
□   *Scaevola sericea*
■   *Scaevola crassifolia*

●   *Spinifex sericeus*   ················   Tropical humid
○   *Spinifex hirsutus*   ——————   Temperate humid
◑   *Spinifex longifolius*   — — —   Temperate arid
                                                   ▬ ▬ ▬   Tropical arid

Fig. 17.3. Major coastal climatic regions of Australia and the coastal distribution of the genera *Spinifex* (Poaceae) and *Scaevola* (Goodeniaceae).

ranges (Fig. 17.4). Another broad difference is that between arid and humid coasts. On the dry and seasonally dry dunes of Western Australia and South Australia the endemic species have primarily inland distributions, which is reflected by the number of inland Chenopodiaceae found there (Fig. 17.2). Conversely, the endemic foredune flora of the humid southeastern coasts has primarily a coastal fidelity (Fig. 17.5).

Differences in the floristic composition of the foredune appear to correlate with four simplistic coastal climatic regions; temperate arid, tropical arid, tropical humid and temperate humid (Davies, 1977) (Fig. 17.3). It is not clear, however, how closely these correspond

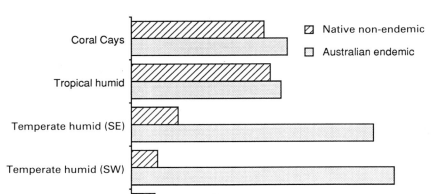

Fig. 17.4. The proportion of native foredune plant species that are endemic to the major coastal climatic regions of Australia.

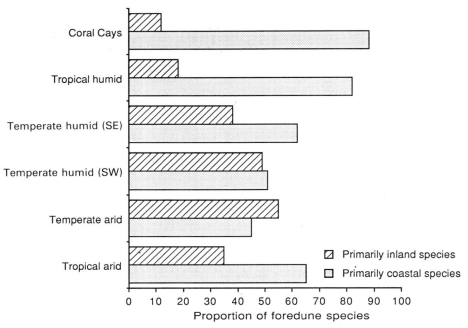

Fig. 17.5. The proportion of native foredune species that have predominantly inland or coastal distributions among the major coastal climatic regions of Australia.

with floristic regions because our knowledge of the dune flora is incomplete, particularly in northern Australia.

Gradual changes in floristic composition between major climatic regions occur latitudinally along the west and east coasts. In Western Australia less than 10 per cent of the 100 foredune species are shared between the arid and the humid regions but there is a gradual change between these extremes (Sauer, 1965). Similarly, there is considerable change in composition between southern Queensland (Batianoff & McDonald, 1980) and Victoria (Turner *et al.*, 1962). The gradational nature of this change can be seen in New South Wales where elements of both the 'endemic' temperate and 'cosmopolitan' tropical floras overlap (Clarke, 1989). Shifts in the zonal location of species are also found on the east coast, e.g. the scrambler *Hibbertia scandens* is often found on the seaward face of the foredune in southern Queensland, but on the central and southern coast of New South Wales it is confined to the back of the foredune and in Victoria it is found only in the hind-dune forests.

Zonation

Floristic variation occurs universally in a zonal pattern across the major environmental gradient from the strand landwards (Doing, 1985). Common species found across the foredune from the major climatic regions are tabulated in Table 17.1 after Beadle (1981). One of the difficulties in describing the niche of dune species is the scattered distribution of species and varied form of the foredune (see Hesp (1988) for the plethora of terms used to describe the morphology of the foredune). Comparison of the relative position of species across the foredune can be made by calculation of an index of the relative position of species. Such an 'outpost index' was used by Sauer (1965) in a pioneering study of the Western Australian foredunes; no such indices have been published for other sections of the coast.

Species richness increases from the strand through the foredune and usually into the hind-dune, though few studies have quantitatively examined the entire gradient. In New South Wales, Clarke (1989) measured species richness in 400 m² quadrats on 224 foredunes; the most seaward beach 'mound' zone had an average of seven species (range 2–20), whilst the windward face of the foredune had 14 (range 3–30) and the sheltered face had 20 (range 3–41) species.

Structural differences in zonation are also present in different regions. On temperate humid coasts grassland through to low

Table 17.1. *Common species found on the foredune in the major climatic sections of the coast*

	Form	Tropical arid	Temperate arid	Temperate humid SW	Temperate humid SE	Tropical humid	Coral cays
Strandline							
Salsola kali	A	*	*				
Cakile maritima[a]	A	*	*	*	*		
Cakile edentula[a]	A	*	*	*	*		*
Backshore							
Ipomoea pes-caprae	H	*				*	*
Sesuvium portulacastrum	H	*				*	*
Euphorbia atoto[a]	H	*				*	*
Spinifex longifolius	G	*		*		*	
Atriplex isatidea	S	*	*	*			
Sporobolus virginicus	G	*	*	*	*	*	*
Carpobrotus spp.	H	*	*	*	*	*	*
Spinifex hirsutus	G		*	*	*		
Euphorbia paralias[a]	H		*	*	*		
Isolepis nodosa	G		*	*	*		
Arctotheca populifolia[a]	G		*	*	*		
Atriplex cinerea	S		*	*	*		
Ammophila arenaria[a]	G			*	*		
Festuca littoralis	G			*	*		
Hydrocotyle bonariensis[a]	H			*	*		

Species	Form						
Spinifex sericeus	G	*	*	*	*		
Stackhousia spathulata	H			*			
Gladiolus gueinzii[a]	H			*			
Lepturus repens	G	*	*				
Ischaemum triticeum	G	*	*				
Thuarea involuta	G	*	*				
Stable foredune							
Acacia binervosa	S						*
Acacia translucens	S						*
Aerva javanica[a]	S						*
Triodia pungens	G						*
Whiteochloa airiodes	G						*
Crotalaria cunninghamii	S						*
Threlkeldia diffusa	H					*	*
Acacia rostellifera	S					*	*
Nitraria billardieri	S					*	*
Scaevola crassifolia	S				*	*	*
Olearia axillaris	S			*	*	*	*
Einadia baccata	S			*		*	*
Agonis flexuosum	T				*		
Lepidosperma gladiatum	G			*	*	*	
Myoporum insularae	S			*	*		
Alyxia buxifolia	S			*	*		
Calocephalus brownii	S			*	*	*	
Leucopogon parviflorus	S			*	*	*	

Table 17.1. (Cont.)

	Form	Tropical arid	Temperate arid	Temperate humid SW	Temperate humid SE	Tropical humid	Coral cays
Acacia sophorae	S		*		*		
Banksia integrifolia	S				*		
Leptospermum laevigatum	S				*		
Chrysanthemoides monilifera[a]	S				*		
Lomandra longifolia	G				*		
Wedelia biflora	S				*	*	*
Casuarina equisetifolia	T					*	*
Pandanus spp.	T					*	*
Hibiscus tiliaceus	T					*	*
Thespesia populneoides	T					*	*
Argusia argentia	S					*	*

A = annual herb, H = perennial herb, G = graminoid, S = shrub, T = tree. [a] = recent introduction
Based mainly on Beadle (1981).

shrubland, open heath and closed scrub, together have an aerodynamic structure. Along the Queensland coast this sequence is truncated with woodland abutting the strandline herbfield; this is because of the sheltering effect of the Great Barrier Reef. Similarly, on the sheltered shores of northern Australia the backshore rapidly grades into woodland and monsoon vine thicket. Not all zonations have increasing structural complexity from sea to land, for on the arid coasts around the Pilbara the hindunes are covered by an open hummock grassland (Craig, 1983).

The zonal patterns of species have often been interpreted as a seral sequence, with one zone replacing another through time, but as outlined in the section on dune building and dynamics, such models should be interpreted cautiously. Differences in species composition and form across the foredune correlate with the rapid environmental changes that occur across the foredune. Major environmental variables include wind shear, sand blasting, salt spray, shifting substrates, water stress and nutrition. Not surprisingly, given the large number of physical, chemical and biological variables, studies of the cause of zonation in Australia have been limited to correlative accounts of environmental gradients and species turnover (e.g. Pidgeon, 1940). There have been some studies of the physical and chemical factors that influence the distribution, abundance and performance of species on dunes (e.g. Parsons & Gill, 1968; Maze & Whalley, 1990). However, biological factors such as dispersal ability (Heyligers, 1985), mycorrhizae (Logan, Clarke & Allaway, 1989), and competition (Clarke, 1989) may also have important roles in influencing vegetation patterns.

Vegetation descriptions

Tropical arid coasts

In northern Australia, dune vegetation is found on the open coast and on sheltered storm beaches and shelly ridges (cheniers) behind mangrove and saltmarsh vegetation. In many places, the vegetation types intermingle and the distinction between mangrove, saltmarsh and dune vegetation becomes blurred (Craig, 1983). The seaward edge of the foredune is dominated by *Spinifex longifolius* together with tropical vines which extend down to the strandline. The foredunes are covered with low open shrubland of *Acacia translucens* with *Whiteochloa airoides* as a ground layer (Table 17.1). In Sauer's (1965) survey of the coast around Port Headland, only about seven species are reported as being common, although the profile diagrams of Craig (1983) indicate the presence of additional chenopod species, especially in more protected environments. Beyond the foredune the hind-dune is covered by *Triodia* hummocks and *Acacia* shrubs. These form an open shrubland, which according to Davies (1977), is 'astonishingly green and well vegetated' when viewed from the air.

Temperate arid coasts

Western Australia

The coastal dune communities of the central west and southern coasts of Western Australia are influenced by relative aridity and high levels of offshore wave activity. The dune systems in this region often extend inland and consist of carbonate sediments that have been lithified in the older Pleistocene systems (Davies, 1977). The strandline vegetation throughout the region is mainly composed of *Cakile maritima*, *Salsola kali* and *Arctotheca populifolia*. Higher on the beach, *Spinifex longifolius* is common on the central west coast, whereas *S. hirsutus* is found around the Great Australian Bight (Fig. 17.3). The foredune proper is covered with an open shrubland and grades into the sand plain heath and shrublands (Sauer, 1965) (Table 17.1). On the driest coasts of the central west, *Acacia rostellifera* thickets are common in hind-dune situations (Beadle, 1981). The sand plain areas support extensive and floristically rich heaths and shrublands called 'Kwongan'; these are described in Pate & Beard (1984) and more recently in Beard (1990).

South Australia

The strand vegetation of semi-arid South Australia is similar to that of Western Australia. The backshore is covered with tussock grasslands of *Spinifex hirsutus* and *Isolepis nodosus* whilst the chenopod shrub *Atriplex cinerea* can also commonly occur near the strandline. The calcareous foredunes are covered with low shrubland dominated by *Calocephalus brownii* and *Olearia axillaris*, whereas a woodland of *Melaleuca lanceolata* and *Allocasuarina verticilliata* is often found on the hind-dune in the driest areas (Specht, 1972; Buckley & Fotheringham, 1987).

Tropical humid coasts

Capricorn Coasts

Extensive areas of dune vegetation occur on the siliceous parallel, parabolic and transgressive dunes on the Tropic of Capricorn (Batian-off & McDonald, 1980). Along this section of the coast, and further north, the foredune zonation is telescoped because of the low wave energy (Sauer, 1985). The strand is typically dominated by *Spinifex sericeus* and *Ipomoea pes-caprae* which grade into a low woodland of *Casuarina equisetifolia* on the foredune (Table 17.1). Behind this zone, a narrow band of vine thicket is often found. These develop into vine forest (closed-forest) in sheltered locations free from fire. On hind-dune ridges sclerophyllous woodlands and forests cover extensive areas. *Melaleuca* spp. and *Livistona decipiens* usually domi-nate the swales, whilst *Eucalyptus tessellaris*, *E. intermedia* and *Allo-casuarina littoralis* are common on dunes. Parabolic dunes are similarly covered in woodland and forests, but myrtaceous and epacri-daceous heaths also occur.

Cape York

Both the east and west sides of Cape York have extensive areas of dune vegetation much of which remains to be described in detail. The strand typically consists of tropical grasses and vines, though the endemic *Spinifex sericeus* and *S. longifolius* are common, the former being restricted to the east coast and the latter to the west coast. The foredune is not very wide and the strand is often overhung by a woodland of *Casuarina equisetifolia*. Semi-deciduous vine forest is distributed patchily on the hind-dune as are open forest and heath (Pedley & Isbell 1971; Specht, Salt & Reynolds; 1977; Pyke & Jackes, 1981; Tracey, 1982; Lavarack & Goodwin, 1987).

Coral cays

Coral cays along the Great Barrier Reef support a range of tropical vegetation that grows on calcareous sand and rubble. Cribb & Cribb (1985) describe three generalised zones that often form concentric rings around cays. The most seaward consists of tropical strand herbs and grasses, followed by a wind-pruned shrub zone, and finally a closed forest often dominated by *Pisonia grandis*. Considerable vari-ation in this pattern exists (e.g. Elsol, 1985) but as yet no overall vegetation analysis has been attempted. Heatwole (1991) has, how-ever, analysed the floristics of these islands, whilst Buckley (1988) provides an account of temporal changes in strand vegetation.

Northern Territory

The northern coasts have relatively small dune systems and frequently they form chenier ridges behind a fringe of mangroves. On open coasts the tropical beach creepers cover the strand whilst tropical shrubs and trees such as *Casuarina equisetifolia* and *Hibiscus tiliaceus* form foredune woodlands. Eucalypt woodland is found on adjacent dunes interspersed with monsoon vine thickets which are described in Russell-Smith & Dunlop (1987) and McKenzie & Kenneally (1983) for the Kimberley in Western Australia.

Temperate humid coast

Southwest

The coastal dune vegetation of the southwest corner of Western Australia has many similarities with the more arid coastal vegetation; however the foredunes are more continuously covered with a wider range of species. *Spinifex hirsutus* is ubiquitous on the backshore as are a number of introduced species (*Ammophila arenaria, Euphorbia paralias* and *Trachyandra divaricata*). The foredune is often covered with open heath, whilst the hind-dune is covered with distinctive *Agonis* thickets (closed scrub) that develop into woodland on well-drained dunes. On low relief dunes *Acacia decipiens* forms a shrub-land, whilst Tuart (*E. gomphocephala*) forest occurs on old dunes in a narrow band north from Busselton (Beadle, 1981). Cresswell & Bridgewater (1985) describe the dune vegetation of the Swan Coastal Plain; several foredune community-types occur within the *Olearia axillaris – Rhagodia baccata* and *Tetragonia implexicoma* complexes. Smith (1985) provides a general account of the coastal vegetation of the southwest and remarks that the coastal dune scrub has a paucity of Fabaceae, Proteaceae and Myrtaceae, which are notably abundant in the sandplain flora away from the coast.

Southeast

A wide range of coastal dune vegetation occurs along the southeastern coast of Australia and there is little floristic similarity between vegetation found in Tasmania and Victoria, and that of the sand islands of southern Queensland. Nevertheless, a common feature of this region is the development of dune sclerophyllous forests and woodlands on the predominantly siliceous sands. The nature of this vegetation in relation to the Quaternary land systems is elaborated in the section on Quaternary deposition and vegetation patterns.

Tasmanian and Victorian dune vegetation

The temperate strandline shores are currently colonised by the introduced sea rocket (*Cakile maritima*), though in sheltered locations the native shrub *Atriplex cinerea* is occasionally found on the strand. The backshore was mainly covered by *Festuca littoralis* and *Spinifex sericeus*, but these have often been displaced by the introduced *Ammophila arenaria* or *Thinopyrum junceum*. In Tasmania, many swales behind the foredune are below the watertable and saline herbfields of *Schoenus nitens* and *Samolus repens* develop (Harris, 1991). Stable foredunes are mostly covered by a mosaic of graminoids, herbs and low shrubs to form an open heath. Two major foredune complexes can be distinguished; those on calcareous deposits of the seasonally dry coast and those found on the more humid siliceous deposits east of Wilson's Promontory. Adjacent to the foredune, thickets of *Melaleuca*, *Acacia* and occasionally *Leucopogon* are found on calcareous dunes, whilst *Leptospermum laevigatum* and *Banksia integrifolia* form dense thickets in eastern Victoria (Barson & Calder, 1981). Eucalypt woodlands and forests are commonly found on parallel dune ridges that can extend several kilometres inland. Bowden & Kirkpatrick (1973) describe an open forest of *Eucalyptus globulus* grading into *E. viminalis* on the east coast of Tasmania, whereas on the west coast *E. nitida* woodland is found on older parabolic dunes. In eastern Victoria *E. viminalis* and *Banksia integrifolia* form dune woodlands. Other eucalypts commonly found on dunes in eastern Gippsland include *E. botryoides*, *E. baxteri*, *E. radiata*, *E. willisii*, *E. bridgesiana* and *E. obliqua* (Barson & Calder, 1981).

New South Wales and southern Queensland

The dune vegetation along the warm temperate high energy coastline of New South Wales and southern Queensland is found on a range of dune landsystems described by Thom, Polach & Bowman (1978). Numerous localised studies have recorded the vegetation in detail, e.g. Blake (1968), Austin & Sheaf (1976), Durrington (1977), Clifford & Specht (1979) and Heyligers *et al.* (1981), whilst summary accounts are given in Clarke (1989) and Myerscough & Carolin (1986). Three latitudinal zones grading into each other have been distinguished in the backshore and stable foredune complexes in New South Wales (Clarke, 1989). The ephemeral strandline is covered with *Cakile edentula* or *C. maritima* with a zone of *Spinifex sericeus* immediately behind. Sedges, mainly *Carex pumila* and *Isolepis nodosus*, dominate soaks and the occasional moist dune swale. Stable foredunes are usually covered in open heath, but frequent fires can result in foredune herbfields and grasslands. *Acacia sophorae* and *Banksia integrifolia* are common shrub dominants with additional species co-occuring depending on the latitude (Table 17.1). On many foredunes the introduced *Chrysanthemoides monilifera* forms closed shrublands that are replacing the native vegetation (Clarke, 1989). In southern New South Wales the hind-dune area rapidly grades into eucalypt woodland and open forest of *E. botryoides*, *E. pilularis* and *Banksia serrata*. On the central and north coast of New South Wales, however, sheltered hind-dunes often support a thicket where mesic species such as *Cupaniopsis anacardioides* and *Acmena smithii* are common. In sheltered locations free from fires, low closed forests (dune littoral rainforests) are found in isolated pockets north from Jervis Bay. Beadle (1981) describes the various eucalypt forest alliances and suballiances found on Holocene and Pleistocene dunes. In general, *E. pilularis*, *E. gummifera* and *Angophora costata* are found on the central coast of New South Wales, whilst *E. pilularis*, *E. intermedia* and *E. signata* are found on the dune systems of southern Queensland. Extensive heaths are also found on dune land systems throughout the east coast; those that occur on Pleistocene dunes are briefly described in the following section.

Quaternary deposition and vegetation patterns

The Quaternary dune systems between central New South Wales and southern Queensland support extensive vegetation complexes that can be related to the age and mode of deposition of siliceous sands through the Quaternary. These dunes form a chronosequence showing increasing weathering, soil formation and reduction in nutrient content through to the oldest Pleistocene dunes (Thompson, 1981). The oldest well-drained dunes carry heath dominated by *Banksia aemula* whilst on poorly drained dunes wet heaths dominated by *Banksia oblongifolia* and *B. robur* occur (Beadle, 1981). Forests and woodlands dominated by *Eucalyptus gummifera* and *Angophora costata* (in the south) and *E. intermedia* and *E. signata* (in the north) with a heathy understorey are found on sequentially younger well-drained Pleistocene dunes (Myerscough & Carolin, 1986; Heyligers *et al.*, 1981). The youngest transgressive dunes of Holocene age with incipient podsol formation support tall sclerophyllous open forests. These forests are typically dominated by *Eucalyptus pilularis* with *Banksia serrata* replacing *B. aemula* in the understorey (Myerscough & Carolin, 1986). Nearer to the foredune, forest stature decreases to a thicket, but there is an increasing number of species with meso-morphic leaves present.

Myerscough & Carolin (1986) have drawn together four strands of evidence that 'make a reasonable argument for linking depletion of nutrient stocks to change of vegetation type with increasing age of the sand surface in wind-blown dunes'. The type and structure of the vegetation, as outlined above, is one. The other factors are: regen-eration attributes of species on differing systems (Myerscough & Carolin, 1986), nutrient stocks in the system (Westman & Rogers, 1977) and the relative podzolisation of the sands (Thompson, 1981).

Dune building and vegetation dynamics

The interdependence of vegetation and the development of dunes has long been of interest in regions of dune instability. The important processes of dune building and vegetation development have received little attention in Australia, however. Several recent studies describe dune initiation and stabilisation in relation to the strandline-backshore zone (Hesp, 1984; Heyligers, 1985; Bird & Jones, 1988). There is scant information, however, on the vegetation dynamics of the more stable foredune vegetation or the formations that occur behind.

The causal factors leading to the initiation of dunes in eastern Australia have aroused considerable debate (Hesp, 1984); neverthe-less, most authors agree that dune building is an interrelated process of aeolian sand deposition with plant establishment and growth. The major mode of contemporary dune initiation involves the germination of propagules in the strand litter and the ensuing plants form the nuclei for wind-blown sand deposition. The shape of the dune formed appears to depend on the form, growth and spacing of strand species present; thus *Ammophila* with its tussock form and rapid vertical growth forms steep hummocks, whereas *Spinifex* with its sprawling horizontal growth builds lower rounded dunes (Heyligers, 1985). Hesp (1984) documents the building of 'incipient' foredunes and defines several modes of formation depending on the type and density of strand species. Most 'incipient' foredunes are rarely stable long enough for compositional change in the vegetation as there are few beaches where progradation is found.

It is tempting to interpret the zonal sequence of increasing species richness and increasing structural complexity across the foredune and into the hind-dune as a seral sequence, where one zone precedes another through time. Along much of the coast, however, there is a disjunction (often exceeding several thousand years), in the age of the established foredune and the recently stabilised backshore (Thom *et al.*, 1978). Thus, Clarke (1989) suggests caution in the seral interpretations of zones because there are few long-term observations of vegetation change on prograding dunes.

Compositional changes across the foredune are viewed by Clarke (1989) as a response to a number of environmental gradients with limited directional succession. Cyclic changes in population numbers and composition were related to irregular disturbance events such as storms, slumping and fires. In southeastern Australia a high fire frequency of low intensity results in foredunes covered in graminoids and bracken; conversely, dunes burnt infrequently by intense fires are often dominated by *Leptospermum* thickets (Harris, 1991). Along much of the eastern coast the foredune is scarped from storms in the 1970s. The scarp face has been colonised from the established foredune *via* slippage and clonal growth, rather than by the strandline and backshore stabilisers. Our understanding of these processes is limited, however, and more quantitative and experimental evidence is needed to test predictive models about dune vegetation dynamics.

Introduced flora

A wide array of introduced species has become naturalised on the strand and foredunes, especially on temperate urbanised coasts where more than 100 species have been recorded (Clarke, 1989). The lowest numbers of introductions are found in the northwestern arid coast (Craig, 1983) and the humid tropical shores (Fig. 17.2). Families commonly represented include the Asteraceae, Aizoaceae, Brassicaceae, Euphorbiaceae, Fabaceae, Liliaceae and Poaceae. Three modes of introduction have occurred; accidental introduction, ornamental escapes, and deliberate introduction.

Ballast from early shipping is thought to have provided the main source of accidentally introduced strand plants. The northern hemisphere *Cakile maritima*, *C. edentula* and *Euphorbia paralias* are found on temperate strandlines, whereas higher on the beach *Arctotheca populifolia*, *Gladiolus gueinzii*, *Hydrocotyle bonariensis*, and *Tetragonia decumbens* are introductions from the southern hemisphere (Heyligers, 1985).

Commonly encountered ornamental plants that have escaped on to the foredune in southeastern Australia include *Polygala myrtifolia*,

Senecio elegans, *Coprosma repens*, *Protasparagus densifolius* and *Chrysanthemoides monilifera*. The latter was first introduced as a garden plant in the 1800s. Between 1940 and 1970 it was deliberately used for dune stabilisation until the pernicious nature of its spread was realised. Weiss & Noble (1984) have shown unequivocally that *C. monilifera* displaces the endemic shrub *Acacia sophorae*, whilst recent surveys of the New South Wales coast show that it has become dominant along much of the coast.

Marram grass (*Ammophila arenaria*) has been the main deliberate introduction for dune stabilisation, although the wheat grasses *Thinopyrum junceum* and *T. distichum* have also been used. The first two species appear to affect the formation of foredunes in southeastern Australia as they build larger hummocky dunes than the native grasses (Heyligers, 1985). *Ammophila arenaria* may also have displaced the native tussock grass *Festuca littoralis* in parts of Tasmania, Victoria and New South Wales.

Human impacts

The changes wrought by human impact on coastal dune vegetation have been widespread and profound. As previously described, the naturalised flora is large and for the most part still spreading. This prompted Heyligers (1985) to say that the chance of finding a foredune with exclusively native plants was remote in southeastern Australia. Fortunately, many of the widespread introduced strand plants are relatively benign, but some introduced foredune plants have displaced extensive areas of native vegetation.

Sand mining for heavy minerals has occured on barrier dunes, high transgressive dunes and frontal dunes at more than 100 locations between Broken Bay in New South Wales and Gladstone in Queensland (Chapman, 1989). Mined areas have been revegetated by conventional soil conservation methods, but the re-establishment of original vegetation has been more problematic. Buckney & Morrison (1992) have followed changes in the floristic composition of mined and unmined dunes at Myall Lakes from 1982 to 1990 and found differences in species composition and a decreasing similarity between mined and unmined sites with age since mining. They conclude that the goal of complete reconstruction of the dune community as it was before mining is probably unrealistic.

Human disturbance has caused the loss of foredune vegetation and the destabilisation of dunes over extensive areas. In New South Wales about 10.5 per cent of the dune lands are unvegetated drifts and 4.7 per cent are in a semi-stable state (Chapman, 1989). The semi-stable dunes are a direct result of 4-wheel drive vehicles, pedestrians, and heavy camping use. The drift sands themselves are, however, mostly natural or in some cases may have resulted from grazing or aboriginal burning. There has been a reduction in the larger areas of drift through revegetation by soil conservation authorities throughout Australia and better management of foredunes on urbanised coasts. A recent report of the House of Representative Standing Committee on Environment, Recreation and the Arts about the state of the Australian coastline (1991) highlights the fragmented approach to management and the lack of knowledge about coastal processes.

Acknowledgements

The assistance of P. Adam, R.C. Carolin, P.C. Heyligers, J.B. Kirkpatrick, V.S. Logan and P.J. Myerscough in the preparation of this manuscript is gratefully acknowledged.

References

Austin, M.P. & Sheafe, J. (1976). Vegetation survey of the South Coast study area. *CSIRO, Division of Land Use Research, Technical Memorandum*, 76/15.

Barson, M.M. & Calder, D.M. (1981). The vegetation of the Victorian coast. *Proceedings of the Royal Society of Victoria*, 92, 55–65.

Batianoff, G.N. & McDonald, T.J. (1980). Capricorn Coast sand dune and headland vegetation. *Queensland Department of Primary Industries, Botany Branch, Technical Bulletin*, 6, 1–71.

Beadle, N.C.W. (1981). *The Vegetation of Australia*. Cambridge University Press.

Beard, J.S. (1990). *Plant Life of Western Australia*. Kenthurst, N.S.W.: Kangaroo Press.

Bird, E.C. & Jones, D.J.B. (1988). The origins of foredunes on the coast of Victoria, Australia. *Journal of Coastal Research*, 4, 181–92.

Bowden, A.R. & Kirkpatrick, J.B. (1974). The vegetation of Rheban Spit, Tasmania. *Proceedings of the Royal Society of Tasmania*, 108, 199–210.

Blake, S.T. (1968). The plant communities of Fraser, Moreton and Stradbroke Islands. *Queensland Naturalist*, 19, 106–13.

Buckley, R. (1988). Plant succession under repeated disturbance: establishment and growth of strandline forbs and creepers on unstable beaches in the wet tropics. *Proceedings of the Ecological Society of Australia*, 15, 307–11.

Buckley, R.C. & Fotheringham, D.G. (1987). *Terrestrial Vegetation of the Eyre*

Coast. Adelaide: Department of Environment and Planning, South Australia.

Buckney R.T. & Morrison, D.A. (1992). Temporal trends in plant species composition on mined sand dunes in Myall Lakes National Park. *Australian Journal of Ecology*, 17, 241–54.

Chapman, D.M. (1989). *Coastal Dunes of New South Wales, Status and Management.* University of Sydney, Coastal Studies Unit, Technical Report 89/3.

Clarke, P.J. (1989). *Coastal Dune Vegetation of New South Wales.* Coastal Studies Unit, University of Sydney, Technical Report 89/4.

Clifford, H.T. & Specht, R.L. (1979). *The Vegetation of North Stradbroke Island.* St Lucia: University of Queensland Press.

Craig, G.F. (1983). *Pilbara Coastal Flora.* Perth: Western Australian Department of Agriculture.

Cresswell, I.D. & Bridgewater, P.B. (1985). Dune vegetation of the Swan Coastal Plain, Western Australia. *Journal of the Royal Society of Western Australia*, 67, 137–48.

Cribb, A.B. & Cribb, J.W. (1985). *Plant Life of the Great Barrier Reef and Adjacent Shores.* St Lucia: University of Queensland Press.

Davies, J.L. (1977). The coast. In *Australia, A Geography*, ed. D.M. Jeans, pp. 134–51. Sydney: University of Sydney Press.

Doing, H. (1985). Coastal fore-dune zonation and succession in various parts of the world. *Vegetatio*, 61, 65–75.

Durrington, L.R. (1977). Vegetation of Moreton Island. *Queensland Department of Primary Industries, Botany Branch, Technical Bulletin*, 1, 1–44.

Elsol, J.A. (1985). Vegetation of an eastern Australian coral cay – Lady Musgrave Island, Great Barrier Reef. *Proceedings of the Royal Society of Queensland*, 96, 33–48.

Galloway, R.W., Story, R., Cooper, R. & Yapp, G.A. (1980). *Coastal Lands of Australia.* CSIRO, Division of Land Use Research, Technical Memorandum 80/24.

Harris, S. (1991). Coastal vegetation. In *Tasmanian Native Bush, A Management Handbook*, ed. J.B. Kirkpatrick, pp. 130–47. Hobart: Tasmanian Environment Centre.

Heatwole, H. (1991). Factors affecting the number of species of plants on islands of the Great Barrier Reef, Australia. *Journal of Biogeogaphy*, 18, 213–21.

Hesp, P. (1988). Morphology, dynamics and internal stratification of some established foredunes in southeast Australia. *Sedimentary Geology*, 55, 17–41.

Hesp, P.A. (1984). Foredune formation in southeast Australia. In *Coastal Geomorphology in Australia*, ed. B.G. Thom, pp. 69–97. Sydney: Academic Press.

Heyligers, P.C. (1985). The impact of introduced plants on foredune formation in south-eastern Australia. *Proceedings of the Ecological Society of Australia*, 14, 23–41.

Heyligers, P.C., Meyers, K., Scott, R.M. & Walker, J. (1981). An ecological reconnaissance of the Evans Head training area, Budjalung National Park, New South Wales. *CSIRO, Division of Land Use Research, Technical Memorandum*, 81/35.

House of Representatives Standing Committee on Environment, Recreation and the Arts (1991). *The Injured Coastline, Protection of the Coastal Environment*. Canberra: Australian Government Publishing Service.

Lavarack, P.S. & Goodwin, M. (1987). Rainforests of northern Cape York Peninsula. In *The Rainforest Legacy*, Volume 1, pp. 201–22. Canberra: Australian Government Publishing Service.

Logan, V.S., Clarke, P.J. & Allaway, W.G. (1989). Mycorrhizas and root attributes of plants of coastal sand-dunes of New South Wales. *Australian Journal of Plant Physiology*, 16, 141–6.

Maze, K.M. & Whalley, R.D.B. (1990). Resource allocation patterns in *Spinifex sericeus* R.Br.: a dioecious perennial grass of coastal sand dunes. *Australian Journal of Ecology*, 15, 145–53.

McKenzie, N.L. & Kenneally, K.F. (1983). Wildlife of the Dampier Peninsula, Part 1, Background and Environment. *Wildlife Research Bulletin of Western Australia*, 11, 1–15.

Myerscough, P.J. & Carolin, R.C. (1986). The vegetation of the Eurunderee sand mass, headlands and previous islands in the Myall Lakes area, New South Wales. *Cunninghamia*, 1, 399–466.

Parsons, R.F. & Gill, A.M. (1968). The effects of salt spray on coastal vegetation at Wilsons Promontory, Victoria, Australia. *Proceedings of the Royal Society of Victoria*, 81, 1–10.

Pate, J.S. & Beard, J.S. (1984). *Kwongan: Plant Life of the Sandplain*. Nedlands: University of Western Australia Press.

Pedley, L. & Isbell, R.F. (1971). Plant communities of Cape York Peninsula. *Proceedings of the Royal Society of Queensland*, 82, 51–74.

Pidgeon, I.M. (1940). The ecology of the central coastal area of New South Wales. III. Types of primary succession. *Proceedings of the Linnean Society of New South Wales*, 65, 221–49.

Pyke, K. & Jackes, B. (1981). Vegetation of the coastal dunes at Cape Bedford and Cape Flattery, North Queensland. *Proceedings of the Royal Society of Queensland*, 92, 37–42.

Russell-Smith, J. & Dunlop, C. (1987). The status of monsoon vine forests in the Northern Territory: a perspective. In *Australian Rainforest Legacy*, Volume 1, pp. 227–88. Canberra: Australian Government Publishing Service.

Sauer, J.D. (1965). Geographic reconnaissance of Western Australia seashore vegetation. *Australian Journal of Botany*, 13, 36–9.

Sauer, J.D. (1985). How and why is Australian seashore vegetation different? *Proceedings of the Ecological Society of Australia*, 14, 17–22.

Smith, G.G. (1985). *A Guide to the Coastal Flora of South Western Australia*, 2nd edition. Perth: Western Australia Naturalists Club.

Specht, R.L. (1972). *The Vegetation of South Australia*. (2nd edn). Adelaide: Government Printer.

Specht, R.L., Salt, R.B. & Reynolds, S.T. (1977). Vegetation in the vicinity of Weipa, North Queensland. *Proceedings of the Royal Society of Queensland,* **88**, 17–38.

Thom, B.G., Polach, H.A. & Bowman, G.M. (1978). Holocene age structure of coastal sand barriers in New South Wales, Australia. Geography Department, Faculty of Military Studies, University of New South Wales.

Thompson, C.H. (1981). Podzol chronosequences on coastal dunes of eastern Australia. *Nature,* **291**, 59–61.

Tracey, J.G. (1982). *The Vegetation of the Humid Tropical Region of North Queensland.* Melbourne: CSIRO.

Turner, J.S., Carr, S.G.M. & Bird, E.C.F. (1962). The dune succession at Corner Inlet, Victoria. *Proceedings of the Royal Society of Victoria,* **75**, 17–33.

Weiss, P.W & Noble, I.R. (1984). Interaction between seedlings of *Chrysanthemoides monilifera* and *Acacia longifolia. Australian Journal of Ecology,* **9**, 107–15.

Westman, W.E. & Rogers, R.W. (1977). Nutrient stocks in a subtropical eucalypt forest, North Stradbroke Island. *Australian Journal of Ecology,* **2**, 447–60.

PART 4

CONSERVATION OF VEGETATION

18

Biodiversity and conservation

R.L. SPECHT

The Australian flora

The plant species collected in Australia by the early explorers created great interest when examined by European botanists of the eighteenth and nineteenth centuries. Much attention was focussed on the floras of heathlands and dry sclerophyll forests of the Hawkesbury Sandstone around Sydney, New South Wales, and on the flora of the sandplain and lateritic soils centred on Perth and Albany, Western Australia. The great diversity of colourful and unusual flowering species in these floras made them particularly exciting. Although many regarded the flora as typically Australian, it was realised that it was closely related to the heathland (fynbos) flora of southwestern Cape Province on the other side of the Indian Ocean.

Similarly, the *Nothofagus* flora of southeastern Australia was seen to be closely related to the cool temperate, closed-forest floras of New Zealand and of southern Chile, and later this genus was found in New Caledonia and New Guinea (Specht, 1981*a*).

The tall open-forest vegetation, which grew in the Otway Basin of southeastern Australia during the Late Cretaceous, contained austral conifers (*Podocarpus, Dacrydium, Dacrycarpus*) and proteaceous taxa (*Knightia, Gevuina, Macadamia*). Today, these taxa form a minor component of the upland closed-forests of tropical northeastern Australia and New Caledonia (Specht, Dettmann & Jarzen, 1992).

The relict pockets of tropical/subtropical closed-forest around the coast of northern and eastern Australia were clearly related to the Gondwanan rainforest floras of Africa and South America, at the family and generic levels. Even the remnants of the rainforest flora within the eucalypt forests on the sandstone outcrops of northern Australia are similar, at the species level, to the flora of the sandstones of the western peninsula of India (Specht, 1988*b*). When the Australian Continental Plate collided with the Sundaland Plate during the Miocene, many rainforest taxa migrated northwards across this Plate, up into the Malay Peninsula (Specht, 1981*b*).

The grass *Themeda triandra*, widespread across Australia, is essentially the same as the taxon in Africa. Tropical and subtropical Australian grass genera appear to be Gondwanan in origin, extending from northern Australia to Africa, and sometimes into South America (Clifford & Simon, 1981).

The evidence strongly points to the conclusion that the Australian flora is but part of the angiosperm flora which developed and became widespread across the super-continent of Gondwanaland before it broke up into the present-day continental plates. New Guinea, New Caledonia and New Zealand were part of the Australasian tectonic plate and show close floristic relationships. The flora of the Australasian plate became isolated from the other Gondwanan continents early in the Tertiary (Specht, 1981*b*; Specht, Dettmann & Jarzen, 1992). It was not until the mid-Miocene that the New Guinea section of the Australasian tectonic plate came in contact with the island archipelago (the Sundaland Plate) extending from southeast Asia. Migration then occurred northward from the Australasian plate at least up into the Malay Peninsula, but apparently little further into Asia. At the same time, a number of Sundaland plants invaded New Guinea and northern Australia. Contact between the two land masses was relatively brief, however, and occurred at a time when the climate of the Australian continent was becoming more arid; grasses, chenopods and acacias began to appear in the fossil records of central and eastern Australia, and there was a virtual explosion in speciation of the eucalypts (Specht, 1993*b*).

The continent of Australia can thus be considered as a 'floating museum' of Gondwanan flora, containing many genera and species which retain extremely primitive structures in both floral and leaf morphology (Specht, Roe & Boughton, 1974; Melville, 1975). Examples of these primitive plants are scattered throughout the whole continent; plants showing primitive floral characteristics are particularly prominent in the closed-forests and related vegetation of northeastern Australia (Fig. 18.1*a*); plants with primitive leaf morphology are characteristic of the sclerophyll (heath) flora, with greatest concentrations in southwest Western Australia and near Sydney (Fig. 18.1*b*).

Until the mid-Miocene, the climate of Australia was 5 to 10 °C warmer and more humid than at present (Specht, Dettmann & Jarzen, 1992; Specht, 1981*b*). Seasonal fluctuations in rainfall led to the

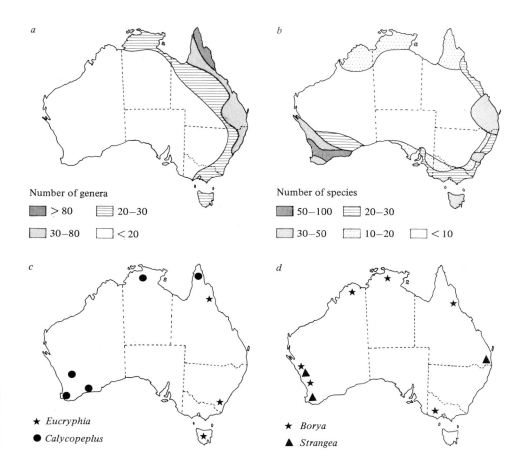

Fig. 18.1. Distribution of: (*a*) 126 genera (592 species) of angiosperms retaining primitive floral characteristics (after Specht *et al.*, 1974); (*b*) 381 species of angiosperms retaining primitive morphological characters (after Specht *et al.*, 1974); (*c*) *Eucryphia* (Eucryphiaceae) and *Calycopeplus* (Euphorbiaceae) (after Specht, 1981*a*); and (*d*) *Borya* (Xanthorrhoeaceae) and *Strangea* (Proteaceae) (after Specht, 1981*a*) in Australia.

development of extensive areas of lateritic soils across the continent. At least in southern Australia where fossil evidence is available, the floral elements were widespread across the continent. It was only in the mid-Miocene that some evidence of aridity appeared in the fossil records and this dryness has continued until the present day with periods of extreme intensity. Aridity induced widespread destruction;

the formerly continuous flora survived in disjunct areas often separated by thousands of kilometres (Fig. 18.1*c*, *d*).

Today, remnants of the primitive Gondwanan flora, after considerable speciation in certain areas, survive in relict pockets of diverse vegetation across the continent; restricted and rare species abound.

Of the 20 000 native plant species which are estimated to occur in Australia, 3329 species (17 per cent) are considered to be *rare* or *threatened* (Briggs & Leigh, 1988; updated from Leigh, Briggs & Hartley, 1981 and Specht, Roe & Boughton, 1974; also Pryor, 1981). Approximately 43 per cent (1442) of the Australian species are listed from Western Australia, the majority being found in the Mediterranean-type climate of southwest Western Australia – the centre of angiosperms with primitive morphological characters (Fig. 18.1*b*). The next most endangered flora is in the tropical, northeastern part of the continent – the centre of angiosperms with primitive floral characters (Fig. 18.1*a*), where more than half (633 species) of the endangered flora of Queensland (1059 species, or 32 per cent of the

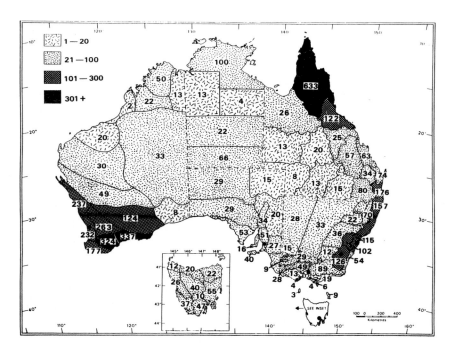

Fig. 18.2. Total number of rare or threatened plant species for each geographical region of Australia (from Briggs & Leigh, 1988).

Australian total) are located. The numbers of rare or threatened plant species are much less in the other Australian States and Territories, with 16 per cent in New South Wales and 4–6 per cent each in the Northern Territory, South Australia, Tasmania and Victoria. The number of rare or threatened plant species is listed for each geographical region of Australia in Figure 18.2 (Briggs & Leigh, 1988).

Only one-third of Australia lies in the humid and subhumid climatic zones (the rest lying in the semi-arid and arid zones), and this part of the continent must support the ever-increasing pressures of urbanisation, tourism, recreation, farming, forestry, water conservation, etc. The greatest number of rare or threatened plants are located in these areas. Before long, even the most common plants may be included in the list of 'plants at risk'.

Plant formations and alliances

There are many logistic problems in acquiring and maintaining satisfactory conservation areas to ensure the survival of the 3329 plant species listed as rare or threatened (Briggs & Leigh, 1988), plus the countless numbers of plants which may become endangered over the next century.

Australian ecologists, working co-operatively during the International Biological Programme (Peterken, 1967), recommended the establishment of a *National System of Ecological Reserves* (Specht, Roe & Boughton, 1974; Fenner, 1975) which would conserve representative examples of all Australian ecosystems. A network of reserves, based on all distinctive ecosystems, would conserve a large proportion of the plant and animal species recorded in Australia. The network would focus long-term conservation objectives on those areas which would provide the greatest diversity of ecosystems, a wide range of plant and animal species, as well as a wide variety of heterogeneous habitats in which evolution could proceed far into the future. In effect, this approach would channel limited manpower and resources into those conservation areas likely to achieve the greatest long-term benefit, rather than dissipate these energies on the emotional issues of conserving a single 'rare and endangered' species.

The definition of major plant communities proved to be an insoluble problem in the IBP Conservation Survey (Specht, Roe & Boughton, 1974). Plant ecological surveys in Australia have been made haphazardly across the continent, often with markedly different degrees of resolution depending on the nature of the survey. On the information available, a committee of ecologists and naturalists in each state/territory produced a list of major and minor plant communities recorded in the region. There was little attempt to co-ordinate the ecological studies on a national basis, even to rationalise the plant communities defined on either side of state/territory borders. Some of the units recognised were very broad in concept, others were subdivisions of larger units. The broadly defined plant communities were often only recognised by the dominant tree or shrub species of the overstorey, with little consideration of the contribution made by the many associated understorey species. To overcome some of this difficulty, the same alliance was often recorded in a number of structural formations; e.g. *Eucalyptus marginata* (the jarrah of Western Australia) may be found as a dominant in an open-forest, a low open-forest, a woodland, or even in a shrubby heathland, with a gradient of changing understorey species. Detailed studies would sub-divide these broad alliances into several subdivisions.

In order to achieve a National System of Ecological Reserves, the major and minor vegetation formations and alliances were defined, using all available ecological surveys, for each State and Territory of Australia (Specht, 1970; Specht, Roe & Boughton, 1974; Specht, 1981*d*) and a map of the natural vegetation of Australia was prepared (Carnahan, 1976, updated 1990).

As the Fosberg (1967) classification of plant formations of the world (adopted for the International Biological Programme) proved unsatisfactory for Australian vegetation, a structural classification was devised, based on the photosynthetic capacity of the plant community (the foliage canopy of each stratum) and the ultimate growth potential of the community (expressed as life form and height of each stratum). The major Australian plant formations were defined on a two-way table (Specht, 1970) relating Foliage Projective Cover and height/life form of the tallest stratum in the plant community (Table 18.1). The area of Australia covered by each formation is shown as a percentage on this Table (Specht, Roe & Boughton, 1974).

After the somewhat subjective approach in devising the IBP list of major and minor plant communities, it was essential to place the list of major plant communities on an objective basis. The lists of plant species recorded in almost 4000 plant communities in over 1500 ecological surveys throughout Australia have been collated into several large databanks. These databanks have been analysed by the polythetic-divisive classificatory program TWINSPAN (Hill, 1973). Over 400 floristic groups have been defined by this technique for the whole continent of Australia (Specht, in prep.). Probably another 40

Table 18.1. *Structural formations in Australia and their percentage area (from Specht et al., 1974)*

Life form and height of tallest stratum	Foliage projective cover of tallest stratum			
	100–70%	70–30%	30–10%	<10%
Trees[a] >30 m	Tall closed-forest	Tall open-forest 1.64%	(Tall woodland)	(Tall open-woodland)
Trees 10–30 m	Closed-forest	Open-forest	Woodland 17.84%	Open-woodland
Trees <10 m	Low closed-forest 0.92%	Low open-forest 3.61%	Low woodland 4.81%	Low open-woodland 5.80%
Shrubs[a] >2 m	Closed-scrub	Open-scrub 7.69%	Tall shrubland 9.89%	Tall open-shrubland 0.22%
Shrubs 25 cm–2 m[b]	Closed-heath	Open-heath 1.47%	Low shrubland 4.00%	Low open-shrubland 14.39%
Shrubs <25 cm[b]	—	—	Fellfield	Open-fellfield
Hummock grasses <2 m	—	—	Hummock grassland 3.95%	Open-hummock grassland 14.71%
Graminoids, herbs, ferns, etc.	Closed-grassland	Grassland	Open-grassland 6.69%	Ephemeral herbland

1.13%

Littoral complex 0.37%; Alpine complex 0.37%; Salt lakes 0.57%.
[a]A *tree* is defined as a woody plant usually with a single stem; a *shrub* is a woody plant with many stems arising at or near the base.
[b]These categories may be subdivided into (1) sclerophyllous and (2) semi-succulent/succulent.

Floristic Groups will be defined for the rainforest remnants of the Northern Territory (Russell-Smith, 1991), southeast Queensland, (Forster *et al.*, 1991) and northern and central Queensland. Sixteen TWINSPAN floristic groups have been defined in the alpine vegetation of south-eastern Australia and Tasmania (Kirkpatrick, 1989).

Closed-forests (rainforests and monsoon forests)	59
Semi-deciduous closed-forests (NT and southeast Qld only)	32
Open-forests and woodlands	190
Heathy shrublands	14
Heathlands and shrubby heathlands	22
Grasslands	11
Mallee vegetation	12
Desert *Acacia* vegetation	19
Hummock grass understorey	20
Chenopod (± low trees) vegetation	10
Forested wetland vegetation (including brigalow and bull oak)	25
Freshwater swamp vegetation	15
Coastal dune vegetation	11
Coastal wetland (mangrove and saltmarsh) vegetation	12
Alpine vegetation	16

Most Australian ecologists have adopted, sometimes with modification, the structural classification of vegetation developed for the IBP Conservation Survey, shown in Table 18.1. To describe each stratum within a plant community, it is possible to combine the two structural parameters – life form and Foliage Projective Cover – with a code for the major plant genus found in that stratum. A triplet notation (of three code characters) will then provide a coded picture of each stratum within the plant community (Beard & Webb, 1974; Carnahan, 1976 revised 1990; Specht, 1993*a*). For example, a picture of the savanna open-forest in Brisbane Forest Park, southeastern Queensland, is readily given by the triplet formula, eM3. wL1. yG3. In this plant community *Eucalyptus* (e) trees (medium height, M) dominate the upper stratum, with a Foliage Projective Cover (3) of 60%; a few scattered (1) *Acacia* (wattle, w) trees (low, L) are found in the mid-stratum: the ground stratum is composed of a dense (3) stand of the kangaroo grass (G), *Themeda triandra* (coded y).

Over the last 20 years, an intensive vegetation mapping program has been undertaken in all States and Territories:

New South Wales: Hayden (1971), Fox (1991), Pickard (in prep.).
Northern Territory: Wilson *et al.* (1990).
Queensland: Boyland (1984), Elsol *et al.* (1976–79), Elsol (1991), Neldner (1984, 1991), Young & McDonald (1989), the tropical rainforest and the Cape York survey (in progress).
South Australia: Specht (1972*b*), Laut *et al.* (1977).
Tasmania: Kirkpatrick & Dickinson (1984).
Victoria: Land Conservation Council of Victoria (regional reports), Frood & Calder (1987).
Western Australia: Beard (1974–80).

Each of the mapping units defined during these local vegetation surveys has been reclassified into the major Floristic Groups analysed by TWINSPAN (Specht, in prep.); all minor plant communities with very restricted distributions have been listed.

At the same time, considerable effort has been made to define the limits of rare or threatened plant species throughout Australia (Briggs & Leigh, 1988).

Biogeographical regions

The recent vegetation maps listed above, together with older land systems surveys conducted by CSIRO in northern Australia, have enabled all the major plant communities (TWINSPAN Floristic Groups) to be tabulated for every 1° latitude × 1° longitude grid-square throughout the continent of Australia.

This large database, the first comprehensive study in Australia (compare Udvardy, 1975; Beard, 1981; Barlow, 1984, 1985), has enabled biogeogeographical regions to be defined objectively, based on the total flora of the continent (Specht, Specht & Allen, 1993). The resultant botanical regions of Australia are shown in Figure 18.6.

Any TWINSPAN Floristic Group which spreads across two or more biogeographical regions should be conserved, with associated Floristic Groups, in several conservation reserves along the environmental gradient.

(*a*)

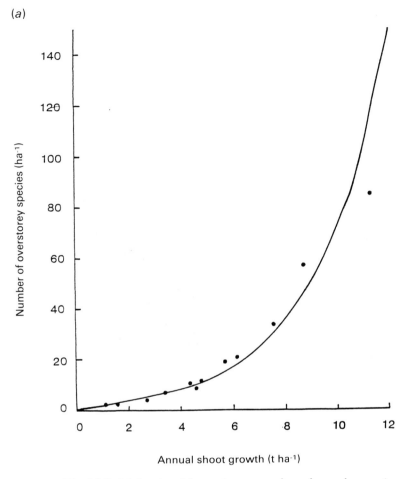

Fig. 18.3. (*a*) Species richness (mean number of vascular species per hectare) of the overstorey of Australian plant communities in relation to annual productivity (biomass in tonnes dry weight per hectare per year) of the foliage canopy (A. Specht & Specht, 1993); (*b*) Species richness (mean number of vascular species per hectare) in the overstorey and the understorey of perhumid plant communities from tropical to cool temperate Australia, plotted against the mean solar radiation (MJ/m²/day) falling on a horizontal surface (at ground level) in each climatic region (A. Specht & Specht, 1993); (*c*) influence of foliage projective cover of the overstorey on the species richness (number of vascular species per 10 m²) of the understorey along the soil-nutrient gradient from heath to heathy open-forest on North Stradbroke Island, Queensland (Specht & Morgan, 1981).

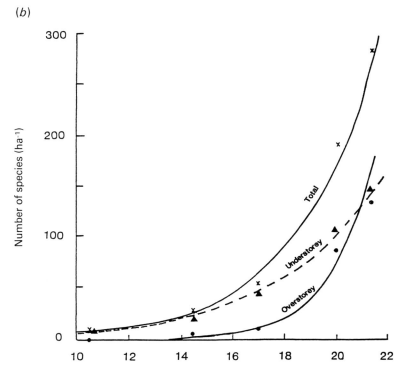

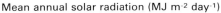

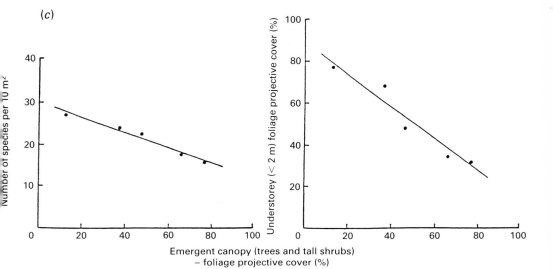

Biodiversity of the Australian vegetation

Alpha diversity (species richness)

The number of species of vascular plants per hectare (alpha diversity, or species richness) of Australian plant communities is the product of the balance between the dominance of the overstorey and the response of the understorey to the shading of the overstorey.

The annual productivity of leaves and stems in the canopy of the plant community is shown to be limited by the requirements of photosynthesis, in particular the availability of light, water and mineral nutrients, and the length of the growing season. For all climatic regions and zones, the species richness of the overstorey of the plant community is shown to be exponentially related to the annual shoot growth (and the Leaf Area Index) of the overstorey canopy, until the latitudinal or altitudinal tree line is reached (Fig. 18.3a, after A. Specht and Specht, 1993). In perennial vegetation, it is probable that this exponential correlation has developed over many centuries; the equilibrium values of species richness develop more rapidly in annual vegetation (Specht, Grundy & Specht, 1990).

With latitudinal increase outside the tropics, overstorey canopies of forest communities absorb increasingly more of the incident solar radiation, thereby markedly reducing the growth rates and the species richness of the understorey strata (Fig. 18.3b). If, in any area, the overstorey cover is reduced, say by water-stress induced by extreme drought or waterlogging, less of the solar radiation is absorbed by the widely spaced trees or tall shrubs, thus enabling the species richness of the understorey to be increased (Fig. 18.3c; Specht & Morgan, 1981).

The species richness of small mammals appears to parallel the increase in the species richness of vascular plants along the climatic gradient from the semi-arid to the humid zone in both the Mediterranean-type climate of southern Australia and the tropical climate of northern Australia (Fig. 18.4, after Specht, 1988 a). The information collated on the species richness of other vertebrate groups in southern Australia shows the same relationship for amphibia, resident birds and snakes as that demonstrated for small mammals and vascular plants. Species richness of lizards decreases as the overstorey of the plant community becomes denser along the climatic gradient from the semi-arid to the humid zone (Specht, 1993c).

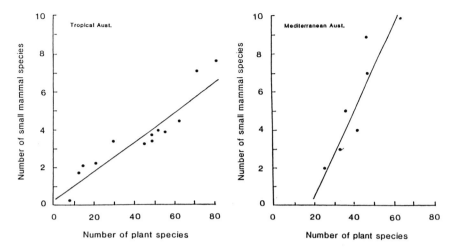

Fig. 18.4. Number of small mammals recorded in plant communities in tropical Australia and in the Mediterranean-type climate of southern Australia, plotted against the number of vascular plant species recorded in the same ecosystems (Specht, 1988*a*; 1993*c*).

Community diversity

The number of plant communities (as defined by the polythetic-divisive classificatory program TWINSPAN), which are found in any part of the continent, seems to be associated with the same environmental factors (solar energy, water and nutrient stress, and length of the growing season) which control the species richness of each plant community in the location. The TWINSPAN communities recorded in each 1° latitude × 1° longitude grid-square have been collated throughout the continent by Specht & Specht (1993).

In all climatic regions, maximum community diversity is found in the humid zone, and diversity is minimal in the arid zone (Fig. 18.5). Community diversity is highest in the tropical region where solar radiation is greatest, lowest in southern Australia where solar radiation is least (Fig. 18.5). Canopy shoot growth is markedly reduced in the plant communities growing on the nutrient-poor, lateritic peneplains of northern Australia and southwest Western Australia; both species richness per hectare and community diversity per latitude–longitude grid-square are reduced.

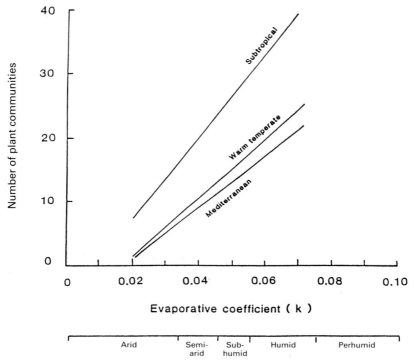

Fig. 18.5. Relationship between the number of major TWINSPAN plant communities recorded in grid-squares (of 1° latitude × 1° longitude) along the climatic gradient (evaporative coefficient) from the arid to the perhumid zones (Specht & Specht, 1993).

Beta diversity

Within any one of the TWINSPAN communities, many species are widespread throughout its distribution (e.g. 82 species of the serpentinite vegetation are common from tropical to subtropical Queensland, Batianoff & Specht, 1992); it is these species which distinguish one TWINSPAN floristic group from another. The other species have a more restricted distribution within the TWINSPAN community, may be unique to the plant community, or occur only in isolated pockets. Although the species richness (number of vascular plant species per hectare) may be essentially the same throughout the TWINSPAN floristic group, the species-composition of the plant community will vary from site to site. Certain areas, such as southwest Western Australia and northeastern Queensland, are centres of speciation where the floristic variation (beta diversity) is very high within the same TWINSPAN community; these centres of speciation also show the highest numbers of rare or threatened plants (Fig. 18.2).

Conservation reserves

On March 1st 1872, the Congress of the United States of America formally proclaimed the marvels of the Yellowstone area as a National Park. The area was set aside as 'pleasuring-ground for the benefit and enjoyment of the people' and the act provided for 'the preservation, from injury or spoliation, of all timber, mineral deposits, natural curiosities or wonders . . . and their retention in their natural condition'.

The conservation movement, begun in the United States, soon swept the world. Australia did not lag behind and, in fact, established the world's second national park, Royal National Park, south of Sydney, in 1879, only seven years after Yellowstone National Park was proclaimed (Goldstein, 1979). National Park, Belair, near Adelaide, followed in 1891, and Tower Hill National Park in Victoria in 1892. Western Australia had a Parks and Reserves Act in 1895.

In Australia, responsibility for conservation policies is shared between State and Federal authorities. In general, regional policies are the responsibility of the States; national and international policies are developed by the Federal Authorities in consultation with State Authorities. Considerable variation existed between State Acts dealing with conservation, although some integration of policies has been achieved by regular meetings of the former Council of Nature Conservation Ministers (CONCOM).

In general, four major conservation categories have been adopted by the States (see J.G. Mosley, in Specht, Roe & Boughton, 1974; Mobbs, 1989; AUSLIG, 1989).

1. *National Parks* protect and preserve indigenous plants and animals, and features of scenic, scientific or historical interest; maintain the existing environment of the area; and provide for the education and enjoyment of visitors to the area and encourage and control such visitors. National Parks can only be revoked or altered by an Act of Parliament.
2. *Wildlife Reserves, Fauna Sanctuaries, Nature Reserves, Conservation Parks,* etc. are for the protection and care of wildlife, the propagation of wildlife and the promotion and study of wildlife.

Compared with National Parks, Wildlife Reserves may be used as scientific reference areas, for the development of habitat management techniques, maintenance of habitat for rare species and for formal education. In most states, these reserves are not as secure as National Parks, being subject to change by Governor-in-Council or by gazettal.

3. *Flora Reserves, Forest Reserves* have been established under the Forestry Act in New South Wales, Queensland and Victoria where sample areas representative of the main forest communities have been dedicated.

4. *Marine Parks, Aquatic Parks, Fisheries Habitat Reserves* are being established by governmental marine authorities to conserve samples of marine ecosystems. Some reserves still preserve the right of the public to fish, gather worms, oysters and other molluscs.

The UNESCO World Heritage Convention came into force in 1975. The Convention aims to ensure international co-operation to protect the world's irreplaceable cultural and natural heritage. Under the Convention, a *World Heritage List* has been established, on which seven Australian properties have been inscribed (Mobbs, 1989).

Conservation legislation and the declaration of reserves in Australia, although initiated a century ago, have had a checkered history. Progress in the declaration of conservation reserves has depended largely on the enthusiasm of public conservation organisations and the pressure campaigns which these organisations initiated.

In most cases the actions were not seen in a state-wide or national perspective. The action of the Australian Academy of Science in initiating, in 1959, a survey of national parks and reserves throughout Australia enabled planners to see the State and national conservation problems in perspective (Specht & Cleland, 1961, 1963; Day, 1968; Frankenberg, 1971; Specht, Roe & Boughton, 1974; Fenner, 1975; Specht, 1975*a*, 1978; Davies, 1982; Frood & Calder, 1987; Forestry Commission of NSW, 1981, 1989). As a nation, we accepted our *'evolutionary responsibility'* (Frankel, 1970, 1974) and are now in a position to plan for a *National System of Ecological Reserves* (Specht, Roe & Boughton, 1974; Fenner, 1975).

Table 18.2 shows the proportion of each Australian State or Territory proclaimed as Conservation Reserves in 1971, 1978 and 1988. The conservation of the major and minor plant communities, which have been defined in Australia (Specht, in prep.), has improved greatly over the last 20 years.

Table 18.2. *Area of National Parks (NP) and other Conservation Reserves (CR) proclaimed by the States and Territories of Australia by 1971 (Specht, Roe & Boughton, 1974), by 1978 (Specht, 1981, first edition of this chapter), and by 1988 (Mobbs, 1989)*

State/territory:– area (km^2)	Conservation category	Conservation reserves area (km^2)			% of state
		1971	1978	1988	1988
Australian Capital Territory (2400)	NP	–	–	94	3.9
	CR	47	98	18	0.8
New South Wales (801 600)	NP	9427	17 346	31 038	3.9
	CR	1920	3449	7084	0.9
Northern Territory (1 346 200)	NP	2258	12 003	1412	0.1
	CR	45 131	45 516	38 824	2.9
Queensland (1 727 200)	NP	9767	21 927	35 221	2.0
	CR	–	639	1416	0.1
South Australia (984 000)	NP	1655	4650	26 484	2.7
	CR	33 341	36 949	84 687	8.6
Tasmania (67 800)	NP	3775	6253	8511	12.6
	CR	5222	725	1158	1.7
Victoria (227 600)	NP	2051	4281	12 021	5.3
	CR	861	4283	6278	2.7
Western Australia (2 525 500)	NP	13 957	44 631	47 572	1.9
	CR	91 436	75 208	104 949	4.1
Total (7 682 300)	NP	42 890	111 091	162 353	2.1
	CR	177 958	166 867	244 414	3.2

A national system of ecological reserves

Following the conservation survey conducted by Australian ecologists as part of the International Biological Programme, the Australian Academy of Science sponsored a symposium entitled 'A National System of Ecological Reserves in Australia' (Fenner, 1975).

The basic premises proposed during the symposium are listed below.

1. Samples of all major ecosystems in Australia should be conserved in reserves which could be termed 'ecological reserves'.
2. As most animals depend on particular plant communities for food, shelter, and nest-sites and since plant communities can be studied relatively easily and should be conserved for their own sake, one can readily substitute major plant community for major ecosystem in the first premise.
3. Replicate samples are needed to cover the variation observed throughout the geographical range of major ecosystems, especially if the ecosystem extends across two or more biogeographical regions (Fig. 18.6).
4. The replicate samples of major ecosystems should provide a satisfactory spatial system of habitats for migratory animals.
5. The 'ecological reserve' should be of sufficient size (Slatyer, 1975; Kitchener *et al.*, 1980) so that:
   (1) there is sufficient space for the perpetuation of larger animal and plant species;
   (2) man's unplanned environmental impact on the ecosystems within the 'ecological reserve' is minimal;
   (3) diverse environments (necessary for mobile animals and for the continued conservation and evolution of both plants and animals in response to environmental stress) are included in the reserve.

The various conservation reserves already proclaimed in the States and Territories of Australia provided the framework for this proposed network of 'ecological reserves'. It was necessary, therefore, to determine what ecosystems were included in the existing reserves, how well they were conserved, and how well the current network of reserves covered all the major and minor ecosystems in Australia. Such a conservation survey would indicate the *gaps* in the conservation network, and encourage positive efforts to overcome these deficiencies.

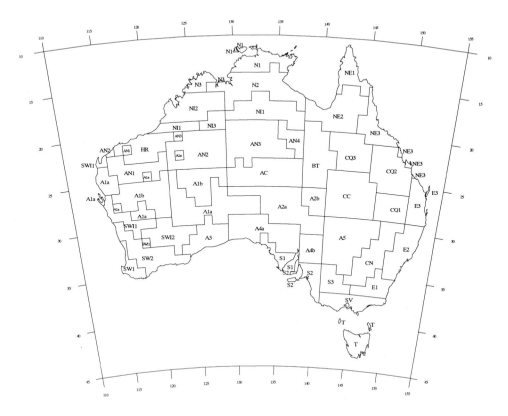

Figure 18.6. Biogeographic regions in Australia derived from a Kulczynski Symmetric pattern analysis of the TWINSPAN Floristic Groups recorded in 1° latitude × 1° longitude grid-squares throughout Australia.

A = Arid Southern (A1a, A1b, A2a, A2b, A3, A4a, A4b, 5); AC = Arid Central; AN = Arid Northern (AN1, AN2, AN3, AN4); BT = Barkly Tableland; CC = Channel Country; CN = Central New South Wales; CQ = Central Queensland (CQ1, CQ2, CQ3); E = Eastern (E1, E2, E3); HR = Hamersley Range; N = Northern (N1, N2, N3); NE = North Eastern (NE1, NE2,. NE3); NI = Northern Inland (NI1, NI2, NI3); S = Southern (S1, S2, S3); SV = Southern Victoria; SW = South Western (SW1, SW2); SWI = South Western Inland (SWI1, SWI2); T = Tasmania.

Almost 1200 major and minor alliances were recognised throughout Australia during the IBP Conservation Survey of the continent (Specht, Roe & Boughton, 1974). In 1971, the conservation status of about half of the alliances could be considered to be reasonable, though not necessarily ideal. Small States such as Tasmania had a good conservation record (based both on area and on the diversity of ecosystems conserved). The large State of Queensland, which had conserved almost a quarter of the total area (42 890 km²) of National Parks declared for the whole of Australia, lagged behind other States in the diversity of ecosystems conserved.

Many Australian plant communities (formations and alliances) were (in 1971) reasonably well conserved in the network of conservation reserves proclaimed throughout Australia. However, the following major Australian plant formations were virtually absent from or poorly conserved in the network. The second Conservation Survey of Australia (Specht, in prep.) shows that this position has been little improved over the last 20 years, though considerable effort is now (1990–92) being made in Queensland to conserve examples of these ecosystems.

1. Tropical and subtropical tussock grasslands (*Astrebla, Dichanthium*) in the coastal and semi-arid zones of northern Australia, and in southeastern Queensland (Everist & Webb, 1975).
2. *Acacia aneura* (mulga) and related *Acacia* tall shrubland communities of the semi-arid zone of Queensland, New South Wales, Northern Territory, South Australia and Western Australia (Sattler, 1986*b*).
3. Low shrubland (shrub steppe) communities dominated by *Atriplex* spp. and *Maireana* spp. in semi-arid southern Australia.
4. Temperate tussock grasslands (*Themeda, Danthonia, Lomandra*) of western Victoria and South Australia (Groves, 1979).
5. *Acacia harpophylla* (brigalow) open-forests in central and south eastern Queensland (Bailey, 1984).
6. Savanna woodland communities (dominated by many *Eucalyptus* spp.) in the wheat belt of south eastern Queensland, New South Wales, Victoria, South Australia and Western Australia.
7. The mallee open-scrub, *Eucalyptus socialis* alliance, in the wheat belt of northwestern Victoria and the adjacent Murray Lands of South Australia.

The first three formations listed above form a vital part of the pastoral industry in the semi-arid zone of Australia; large, but degraded, conservation reserves could still be acquired. The last four formations lie within the wheat belt/improved pasture zone of Australia. Only small 'islands', plus roadside corridors, still remain. Research programs aimed at assessing the conservation potential of the remnants of the original ecosystems and developing a conservation plan are essential (see Kitchener *et al.*, 1980, for a discussion of the wheat belt of southwest Western Australia).

Selection of nature conservation reserves

A biologically based method for the selection of nature conservation reserves in the pastoral districts of arid and semi-arid Australia has been developed, based on the criteria of *diversity, representativeness, naturalness* and *effectiveness* (including *size*) as a conservation unit (Bolton & Specht, 1983; Purdie, 1987).

Existing vegetation or land system maps are the major source of data. These maps are digitised for use as a flexible computerised data base; each cell of a regular grid laid over the map surface is the basic unit ('record') of the data base. Various indices are calculated for these squares; contour plots of each index are made over the whole map surface. The map squares are also classified into groups of squares with similar contents. Adjacent map squares with high priorities for conservation are treated as potential conservation reserves. These potential reserves are further investigated and ranked for suitability for conservation.

Figure 18.7 shows potential conservation reserves in part of the mulga study-area in southwestern Queensland. Category 'A' areas are the first choice for representing each selection group. These areas comprise only 6 per cent of the map area, yet they sample 88 per cent of the ecosystem variability and all of the major ecosystems in the mapping area. Any remaining ecosystems can be readily pinpointed, and cases for smaller reserves developed.

It must be stressed that all the potential conservation reserves shown on Figure 18.7 contain a *mosaic of ecosystems*, each with a distinctive complement of plant and animal species which cover the biodiversity of the region (Fig. 18.3, 18.4 and 18.5). In such diverse conservation reserves, evolution has some chance of proceeding within the multitude of microhabitats of the reserve, if and when environmental change occurs. As well, a series of ecosystems is available to cope with the different feeding and nesting habits of some of the resident fauna.

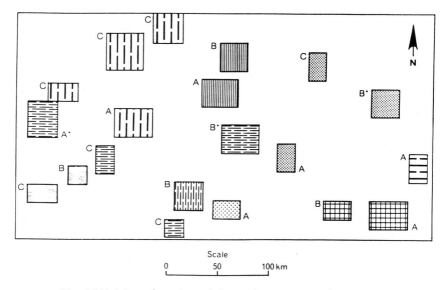

Fig. 18.7. Map of portion of the study area in southwestern Queensland showing potential conservation reserves (Bolton & Specht, 1983). Selected areas representing each typical landscape group are marked 'A'. Alternative areas, which could be selected to represent each landscape group if the category 'A' areas are not available, are marked 'B' and 'C'.

Conservation management

Once a satisfactory system of 'ecological reserves' has been established, the greatest difficulties faced by conservation authorities is the maintenance of the conserved ecosystems for posterity. In the creation of conservation reserves, we have accepted '*evolutionary responsibility*' as indicated by Frankel (1970, 1974). Ecosystems are dynamic systems, which change continually with time. In order to maintain one ecosystem in a 'conservation island' surrounded by a miscellany of alternative land-use systems needs far-sighted, long-range planning. For example, the mound building mallee-fowl (*Leipoa ocellata*) needs a good supply of *Acacia* seed as food for much of the year (Frith, 1962); as seed supply is poor in young stands of mallee regenerating after fire, and also in senescing stands, careful management is necessary to promote a series of age-classes of mallee vegetation. This mosaic of age-classes will ensure a reasonable supply of *Acacia* seed for the mallee-fowl in the small 'island' reserves which will eventually be all that will remain of the formerly widespread mallee communities.

Some of the management problems faced by 'island' reserves may be classed as external, such as hydrology, atmospheric and water pollution, soil erosion, wind damage, weeds, vermin, pathogens, fire, etc. On a continent as dry as Australia, the hydrological relations of the landscape determine the nature of the various ecosystems found in the region. Water movement from adjacent areas will influence water tables which affect the development of closed-forests, coastal dune vegetation, mangrove vegetation, forested wetlands, wet-heathlands, etc. (Specht, 1972a; Specht, Salt & Reynolds, 1977). Changes in adjacent land-use, outside the control of the Conservation Authority, may affect the water supply to the conservation area. Unless controlled by an overall-planning authority, little co-operation between adjacent land owners can be expected in dealing with the wide range of external environmental impacts on the conservation area.

The global increases in carbon dioxide and methane in the atmosphere, the changes in the 'ozone hole', are now increasing exponentially at such a rate that meteorologists are predicting an increase in global temperatures of $2-4\,°C$ within the next century. Such increases in global temperature have taken thousands, if not millions, of years in the geological past; now they are expected in less than a hundred years! There is no possibility that ecosystems can move across the landscape of the earth in such a short period of time – even if the impeding areas of alternative land-use did not exist. The ecosystems within conservation 'islands' will have to adjust rapidly, and many ecophysiological problems, will be associated with such rapid changes in global warming (Specht, 1988b). Research to solve some of these problems on the conservation of biodiversity should be an urgent priority.

In the case of internal impacts, the conservation authorities have a little more control. Many of the serious environmental impacts appear to be solved by controlling humans and their domestic animals. Problems of compaction, erosion, litter, weeds, pathogens, vermin, fire, etc. are in this category. However, there is always someone who flaunts the controls. It is thus necessary to initiate research programmes which will solve the problems created by the minority. For example, small amounts of litter strewn along pathways through sclerophyll (heath) vegetation usually lead to an increase in nutrient level along the paths and the invasion of weeds and herbs which out-compete the native flora (Specht, 1963; Heddle & Specht, 1975; Specht, 1990). In order to maintain the conservation objectives of these reserves, it is necessary to reduce the nutrient level, thus changing the balance back in favour of the sclerophyllous plants. It is believed that this can be achieved by the application of kaolinitic clays along the tracks (Specht, 1975*b*, 1981*c*).

Considerable ecological research is necessary to provide a background for wise long-range management programs. Part of this background can be developed by intensive studies of a single ecosystem over a long period of time (see Specht, Rayson & Jackman, 1958 and Specht, Clifford & Rogers, 1984, for heathland vegetation; Hall, Specht & Eardley, 1964, for arid zone vegetation; Ashton, 1976, for tall open-forest; Groves, 1979, for various grassland communities; Specht & Morgan, 1981, for examples from the humid to the arid zone; Clifford & Specht, 1986 for tropical plant communities).

Many of the changes which may occur within a conservation reserve due to human impact are likely to be solved, however, by the combined knowledge of the long-term dynamics of the ecosystem and of the community-physiological principles which maintain the equilibrium interface between an ecosystem and its neighbours (Specht, 1981*e*). Until recently, there were many areas of natural or semi-natural vegetation where ecological studies could be undertaken. Much of this landscape has now disappeared, or has been incorporated into National Parks where research, in most States, is not permitted. It is thus essential to create a series of research reserves (such as the *Biosphere Reserves* of UNESCO, 1974) where an adequate research background, on management problems, can be developed.

Twelve Biosphere Reserves have been declared in Australia (Mobbs, 1989). The primary objectives of Biosphere Reserves are:

1. To conserve for present and future use the diversity and integrity of biotic communities of plants and animals within natural ecosystems, and to safeguard the genetic diversity of species on which their continuing evolution depends.
2. To provide areas for ecological and environmental research including, particularly, baseline studies, both within and adjacent to such reserves, such research to be consistent with the first objective.
3. To provide facilities for education and training.

Research in Biosphere Reserves will provide the answers to the pressing problems on *Biodiversity and Conservation* being tackled internationally by the Scientific Committee on the Problems of the Environment (SCOPE). During the 1990s, two central questions are being addressed: –

1. Is biodiversity important in the maintenance of system processes, and if so, how will it be affected by anticipated global changes such as those of climate, land use and plant-animal invasions?
2. Is the stability and resilience of systems affected by biodiversity, and will global changes alter these relationships?

Now that humans have acquired *evolutionary responsibility*, answers to these questions are vital to the survival of the biodiversity of the planet.

References

Ashton, D.H. (1976). The development of even-aged stands of *Eucalyptus regnans* F. Muell. in central Victoria. *Australian Journal of Botany*, **24**, 397–414.

AUSLIG, Australia (1989). *Nature Conservation Reserves*. Fourth Edition. (1:5 million scale). Canberra: Australian Surveying and Land Information Group.

Bailey, A. (ed.) (1984). *The Brigalow Belt of Australia*. Brisbane: The Royal Society of Queensland.

Barlow, B.A. (1984). Proposal for delineation of botanical regions in Australia. *Brunonia*, 7, 195–201.

Barlow, B.A. (1985). A revised natural regions map for Australia. *Brunonia*, 8, 387–92.

Batianoff, G.N. & Specht, R.L. (1992). Queensland (Australia) serpentine vegetation. In *The Vegetation of Ultramafic (Serpentine) Soils*, eds A. J. M. Baker, J. Proctor & R.D. Reeves, pp. 109–28. Andover, Great Britain: Intercept Ltd.

Batianoff, G.N., Specht, R.L. & Reeves, R.D. (1991). The serpentinite flora of the humid subtropics of eastern Queensland. *Proceedings of the Royal Society of Queensland*, 101, 137–58.

Beard, J.S. (1974–80). *Vegetation Survey of Western Australia 1:1 000 000 series*. Nedlands: University of Western Australia Press.

Beard, J.S. (1981). The history of the phytogeographical region concept in Australia. In *Ecological Biogeography of Australia*, ed. A. Keast. pp. 335–53. The Hague: W. Junk.

Beard, J.S. & Webb, M.J. (1974). The vegetation survey of Western Australia: its aims, objects and methods. In *Vegetation Survey of Western Australia. Part 1 of Explanatory Notes to Sheet 2. Great Sandy Desert*. Nedlands: University of Western Australia Press.

Bolton, M.P. & Specht, R.L. (1983). *A Method for Selecting Nature*

Conservation Reserves. Canberra: Australian National Parks and Wildlife Service Occasional Paper No. 8.

Boyland, D.E. (1984). *Vegetation Survey of Queensland. South Western Queensland.* Brisbane: Queensland Department of Primary Industries, Queensland Botany Bulletin No. 4.

Briggs, J.D. & Leigh, J.H. (1988). *Rare or Threatened Australian Plants.* Revised Edition. Canberra: Australian National Parks and Wildlife Service Special Publication No. 14.

Carnahan, J.A. (1990). *Atlas of Australian Resources. Third Series.* Vol. 6. *Vegetation.* Canberra: Australian Surveying and Land Information Group.

Clifford, H.T. & Simon, B.K. (1981). The biogeography of Australian grasses. In *Ecological Biogeography of Australia*, ed. A. Keast, pp. 537–54. The Hague: W. Junk.

Clifford, H. T. & Specht, R. L. (eds) (1986). *Tropical Plant Communities. Their Resilience, Functioning and Management in Northern Australia.* St Lucia (Brisbane): Department of Botany, University of Queensland & Utah Foundation.

Davies, R. J-P. (1982). *The Conservation of Major Plant Associations in South Australia.* Adelaide: Conservation Council of South Australia.

Day, M.F. (1968). *National Parks and Reserves in Australia.* Canberra: Australian Academy of Science Report No. 9.

Elsol, J.A. (1991). *Vegetation Description and Map: Ipswich, South-Eastern Queensland.* Brisbane: Queensland Department of Primary Industries, Queensland Botany Bulletin No. 10.

Elsol, J.A., Dowling, R.M., McDonald, W.J.F., Sattler, P.S. & Whiteman, W.G. (1976–79). *Moreton Region Vegetation Map Series.* Brisbane: Queensland Department of Primary Industries, Botany Branch.

Everist, S.L. & Webb, L.J. (1975). Two communities of urgent concern in Queensland: Mitchell grass and tropical closed-forests. In *A National System of Ecological Reserves in Australia*, ed. F. Fenner, pp. 39–52. Australian Academy of Science Report No. 19.

Fenner, F. (ed.) (1975). *A National System of Ecological Reserves in Australia.* Canberra: Australian Academy of Science Report No. 19.

Forestry Commission of New South Wales (1981 & 1989). *Forest Preservation in State Forests of New South Wales.* First & Second Editions. Sydney: Forestry Commission of New South Wales, Research Note No. 47.

Forster, P.I., Bostock, P.D., Bird, L.H. & Bean, A.R. (1991). *Vineforest Plant Atlas for South-East Queensland.* Brisbane: Queensland Government, Queensland Herbarium.

Fosberg, F.R. (1967). A classification of vegetation for general purposes. In *Guide to the Check Sheet for IBP Areas*, ed. G.F. Peterken, pp. 73–120. International Biological Programme (IBP) Handbook No. 4. Oxford: Blackwell Scientific Publications.

Fox, M.D. (1991). The natural vegetation of the Ana Branch – Mildura 1:250 000 map sheet (New South Wales). *Cunninghamia*, 2, 443–93.

Frankel, O.H. (1970). Variation – the essence of life. *Proceedings of the Linnean Society of New South Wales*, 95, 158–69.

Frankel, O.H. (1974). Genetic conservation: our evolutionary responsibility. *Genetics*, 78, 53–65.

Frankenberg, J. (1971). *Nature Conservation in Victoria*, ed. J.S. Turner. Melbourne: Victorian National Parks Association.

Frith, H.J. (1962). *The Mallee Fowl: The Bird That Builds an Incubator*. Sydney: Angus & Robertson.

Frood, D. & Calder, M. (1987). *Nature Conservation in Victoria. Study Report.* 2 vols. Melbourne: Victorian National Parks Association Report.

Goldstein, W. (ed.) (1979). Australia's 100 years of National Parks. *Parks and Wildlife*, vol 2 (3–4), pp. 1–160. Sydney: National Parks and Wildlife Service.

Groves, R.H. (1979). The status and future of Australian grasslands. *New Zealand Journal of Ecology*, 2, 76–81.

Hall, E.A.A., Specht, R.L. & Eardley, C.M. (1964). Regeneration of the vegetation on Koonamore Vegetation Reserve, 1926–1962. *Australian Journal of Botany*, 12, 205–64.

Hayden, E. (1971). *Natural Vegetation of New South Wales*. Bathurst, NSW: Department of Decentralisation and Development of New South Wales.

Heddle, E.M. & Specht, R.L. (1975). Dark Island heath (Ninety-Mile Plain, South Australia). VIII. The effect of fertilizers on composition and growth, 1950–1972. *Australian Journal of Botany*, 23, 151–64.

Hill, M.O. (1973). Reciprocal averaging: an eigen vector method of ordination. *Journal of Ecology*, 61, 237–49.

Kirkpatrick, J.B. (1989). The comparative ecology of mainland Australian and Tasmanian alpine vegetation. In *The Scientific Significance of the Australian Alps*, ed. R. Good, pp. 127–42. Canberra: Australian Alps National Parks Liaison Committee/Australian Academy of Science.

Kirkpatrick, J.B. (ed.) (1991). *Tasmanian Native Bush: A Management Handbook*. Hobart: Tasmanian Environment Centre.

Kirkpatrick, J.B. & Dickinson, K.J.M. (1984). *Vegetation of Tasmania*. (Scale 1:500 000). Hobart: Forestry Commission of Tasmania.

Kitchener, D.J., Chapman, A., Dell, J., Muir, B.G. & Palmer, M. (1980). Lizard assemblage and reserve size and structure in the Western Australian wheat belt – some implications for conservation. *Biological Conservation*, 17, 25–62.

Laut, P., Heyligers, P.C., Keig, G., Löffler, E., Margules, C., Scott, R.M. & Sullivan, M.E. (1977). *Environments of South Australia, Provinces 1–8 & Handbook*. Canberra: CSIRO Australia, Division of Land Use Research.

Leigh, J., Briggs, J. & Hartley, W. (1981). *Rare or Threatened Australian Plants*. Canberra: Australian National Parks and Wildlife Service Special Publication No. 7.

Melville, R. (1975). The distribution of Australian relict plants and its bearing on angiosperm evolution. *Botanical Journal of the Linnaean Society*, 71, 67–88.

Mobbs, C.J. (1989). *Nature Conservation Reserves in Australia (1988)*. Canberra: Australian National Parks and Wildlife Service Occasional Paper No. 19.

Neldner, V.J. (1984). *Vegetation Survey of Queensland. South Central*

Queensland. Brisbane: Queensland Department of Primary Industries, Queensland Botany Bulletin No. 3.

Neldner, V.J. (1991). *Vegetation Survey of Queensland. Central Western Queensland*. Brisbane: Queensland Department of Primary Industries, Queensland Botany Bulletin No. 9.

Peterken, G.F. (ed.) (1967). *Guide to the Check Sheet for IBP Areas*. International Biological Programme (IBP) Handbook No. 4. Oxford: Blackwell Scientific Publications.

Pryor, L.D. (1981). *Australian Endangered Species: Eucalypts*. Canberra: Australian National Parks and Wildlife Service Special Publication No. 5.

Purdie, R.W. (1987). Selection of key area networks for regional nature conservation – The revised Bolton and Specht Method. *Proceedings of the Royal Society of Queensland*, 98, 59–71.

Russell-Smith, J. (1991). Classification, species richness, and environmental relations of monsoon rainforest in northern Australia. *Journal of Vegetation Science*, 2, 259–78.

Sattler, P.S. (1986a). Nature conservation in Queensland: planning the matrix. *Proceedings of the Royal Society of Queensland*, 97, 1–21.

Sattler, P.S. (ed.) (1986b). *The Mulga Lands*. Brisbane: The Royal Society of Queensland.

Slatyer, R.O. (1975). Ecological reserves: size, structure and management. In *A National System of Ecological Reserves in Australia*, ed. F. Fenner, pp. 22–38. Canberra: Australian Academy of Science Report No. 19.

Specht, Alison & Specht, R.L. (1993). Species richness and canopy productivity of Australian plant communities. *Biodiversity and Conservation*, 2, 152–67.

Specht, R.L. (1963). Dark Island heath (Ninety-Mile Plain, South Australia). VII. The effect of fertilizers on composition and growth, 1950–1960. *Australian Journal of Botany*, 11, 67–94.

Specht, R.L. (1970). Vegetation. In *The Australian Environment*, ed. G.W. Leeper, 4th edn (rev.), pp. 44–67. Melbourne: CSIRO & Melbourne University Press.

Specht, R.L. (1972a). Water use by perennial, evergreen plant communities in Australia and Papua New Guinea. *Australian Journal of Botany*, 20, 273–99.

Specht, R.L. (1972b). *The Vegetation of South Australia*. Second Edition. Adelaide: Government Printer.

Specht, R.L. (1975a). The report and its recommendations. In *A National System of Ecological Reserves in Australia*, ed. F. Fenner, pp. 11–21. Canberra: Australian Academy of Science Report No. 19.

Specht, R.L. (1975b). A heritage inverted: our flora endangered. *Search*, 6, 472–7.

Specht, R.L. (1978). *In Wildness is the Preservation of the World*. Sixth Romeo Watkins Lahey Memorial Lecture. Brisbane: National Parks Association of Queensland.

Specht, R.L. (1981a). Major vegetation formations in Australia. In *Ecological Biogeography of Australia*, ed. A. Keast, pp. 163–297. The Hague: W. Junk.

Specht, R.L. (1981*b*). Evolution of the Australian flora: some generalisations. In *Ecological Biogeography of Australia*, ed. A. Keast, pp. 783–806. The Hague: W. Junk.

Specht, R.L. (1981*c*). Conservation: Australian heathlands. In *Ecosystems of the World*. Vol. 9B. *Heathlands and Related Shrublands. Analytical Studies*, ed. R.L. Specht, pp. 235–40. Amsterdam: Elsevier Scientific Publishing Company.

Specht, R.L. (1981*d*). Structural attributes – foliage projective cover and standing biomass. In *Vegetation Classification in the Australian Region*, eds A.N. Gillison & D.J. Anderson, pp. 10–21. Canberra: CSIRO & Australian National University Press.

Specht, R.L. (1981*e*). Ecophysiological principles determining the biogeography of major vegetation formations in Australia. In *Ecological Biogeography in Australia*, ed. A. Keast, pp. 299–332. The Hague: W. Junk.

Specht, R.L. (1981*f*). Developments in terrestrial ecology in Australia. In *Handbook of Contemporary Developments in World Ecology*, eds E.J. Kormondy & J.F. McCormick, pp. 387–415. Westport, Connecticut: Greenwood Press.

Specht, R.L. (1988*a*). Climatic control of ecomorphological characters and species richness in mediterranean ecosystems in Australia. In *Mediterranean-type Ecosystems. A Data Source Book*, ed. R.L. Specht. pp. 149–55. Dordrecht: Kluwer.

Specht, R.L. (1988*b*). Origin and evolution of terrestrial plant communities in the wet-dry tropics of Australia. *Proceedings of the Ecological Society of Australia*, **15**, 19–30.

Specht, R.L. (1988*c*). Geosphere-biosphere interaction in terrestrial ecosystems. In *Global Change*, ed. K.D. Cole. pp. 169–76. Canberra: Australian Academy of Science.

Specht, R.L. (1990). Changes in the eucalypt forests of Australia as a result of human disturbance. In *The Earth in Transition: Patterns and Processes of Biotic Impoverishment*, ed. G.M. Woodwell, pp. 177–97. Cambridge University Press.

Specht, R.L. (1993*a*). Vegetation. In *Land Degradation Processes in Australia*, eds G. McTainsh & W. Boughton, in press. Sydney: Longman Cheshire Pty Ltd.

Specht, R.L. (1993*b*). The influence of soils on the evolution of the eucalypts. In *Nutrition of The Eucalypts*, eds P.M. Attiwill & M.A. Adams, in press. Melbourne: CSIRO Australia Publications.

Specht, R.L. (1993*c*). Species richness of vascular plants and vertebrates in relation to canopy productivity. In *Plant–Animal Interactions in Mediterranean-type Ecosytems*, eds M. Ariannoutsou & R. H. Groves pp. 15–23. Dordrecht: Kluwer.

Specht, R.L. & Cleland, J.B. (1961). Flora conservation in South Australia. 1. The preservation of plant formations and associations recorded in South Australia. *Transactions of the Royal Society of South Australia*, **85**, 177–96.

Specht, R.L. & Cleland, J.B. (1963). Flora conservation in South Australia. 2. The preservation of species recorded in South Australia. *Transactions of the Royal Society of South Australia*, **87**, 63–92.

Specht, R.L. & Morgan, D.G. (1981). The balance between the foliage projective covers of overstorey and understorey strata in Australian vegetation. *Australian Journal of Ecology*, 6, 193–202.

Specht, R.L. & Specht, A. (1989a). Canopy structure in *Eucalyptus* dominated communities in Australia along climatic gradients. *Acta Oecologia, Oecologia Plantarum*, 10, 191–213.

Specht, R.L. & Specht, A. (1989b). Species richness of overstorey strata in Australian plant communities – the influence of overstorey growth rates. *Australian Journal of Botany*, 37, 321–36.

Specht, R.L. & Specht, A. (1989c). Species richness of sclerophyll (heathy) plant communities in Australia – the influence of overstorey cover. *Australian Journal of Botany*, 37, 337–50.

Specht, R.L. & Specht, A. (1993). Diversity of plant communities in Australia (in a 1° latitude × 1° longitude grid). *Australian Journal of Botany*, 41, in press.

Specht, R.L., Rayson, P. & Jackman, M. E. (1958). Dark Island heath (Ninety-Mile Plain, South Australia). VI. Pyric succession: changes in composition, coverage, dry weight, and mineral nutrient status. *Australian Journal of Botany*, 6, 59–88.

Specht, R.L., Roe, E.M. & Boughton, V.H. (eds) (1974). *Conservation of Major Plant Communities in Australia and Papua New Guinea*. Australian Journal of Botany Supplement No. 7.

Specht, R.L., Salt, R.B. & Reynolds, S. (1977). Vegetation in the vicinity of Weipa, north Queensland. *Proceedings of the Royal Society of Queensland*, 88, 17–38.

Specht, R.L., Clifford, H.T. & Rogers, R.W. (1984). Species richness in a eucalypt open-woodland on North Stradbroke Island, Queensland. The effect of overstorey and fertilizer, 1965–1984. In *Focus on Stradbroke*, eds R.J. Coleman, J. Covacevitch & P. Davie, pp. 267–77. Brisbane: Boolarong Press.

Specht, R.L., Grundy, R.I. & Specht, A. (1990). Species richness of plant communities: Relationships with community growth and structure. *Israel Journal of Botany*, 39, 465–80.

Specht, R.L. Clifford, H.T., Arianoutsou, M., Bird, L.H., Bolton, M.P., Forster, P.I., Grundy, R.I., Hegarty, E.E. & Specht, A. (1991). Structure, floristics and species richness of plant communities in southeast Queensland. *Proceedings of the Royal Society of Queensland*, 101, 27–78.

Specht, R.L., Dettmann, M.E. & Jarzen, D.M. (1992). Community associations and structure in the Late Cretaceous vegetation of southeast Australasia and Antarctica. *Palaeogeography, Palaeoclimatology, Palaeoecology*, 94, 283–309.

Specht, A., Specht, R.L. & Allen, C. (1993). Biogeographical regions of Australia – an objective analysis. *Australian Journal of Botany*, (in press).

Udvardy, M.D.F. (1975). *A Classification of the Biogeographical Provinces of the World*. Morges, Switzerland: IUCN Occasional Papers No. 18.

UNESCO (1974). *Criteria and Guidelines for the Choice and Establishment of Biosphere Reserves. Final Report*. Paris: UNESCO Programme on Man and the Biosphere (MAB) No. 22.

Wilson, B.A., Brocklehurst, P.S., Clark, M.J. & Dickinson, K.J.M. (1990). *Vegetation Survey of the Northern Territory, Australia.* Darwin: Conservation Commission of the Northern Territory, Land Conservation Unit Technical Report No. 49.

Young, P.A.R. & McDonald, T.J. (1989). *Vegetation Map and Description: Warwick, South-Eastern Queensland.* Brisbane: Queensland Department of Primary Industries, Queensland Botany Bulletin No. 8.

INDEX

Aborigines *see* humans, aboriginal
Acacia open-forests, woodlands and
    shrublands: of central and
    western Australia, 274–87
  with hummock grasses, 283–6; of
    A. ancistrocarpa, 284; of kanji
    (*A. pyrifolia*), 283–4; of pindan
    wattles (*Acacia* spp.), 284
  with tussock grasses, 274–83; of
    bastard mulga (*A. stowardii*),
    280; of mulga (*A. aneura*),
    274–80; of sandhill mulga
    (*A. ramulosa*), 281–2; of
    snakewood (*A. xiphophylla*),
    282; of turpentine mulga
    (*A. cibaria*), 281; of witchetty
    bush (*A. kempeana*), 281
  their utilisation, 286–7
  of northeastern Australia, 259–71:
    on deep fine-textured alkaline
    soils, 263–70; of blackwood
    (*A. argyrodendron*), 269; of
    boree (*A.tephrina*), 269; of
    brigalow (*A. harpophylla*),
    264–7; of georgina gidgee
    (*A. georginae*), 269–70; of gidgee
    (*A. cambagei*), 267–9; of prickly
    acacia (*A. nilotica*), 270
  on shallow coarse-textured acid
    soils, 261–3; of bendee
    (*A. catenulata*), 262–3; of
    lancewood (*A. shirleyi*), 262
  their utilisation, 270–1
  of southern Australia, 271–3
  of myall (*A. pendula*), 272–3; of

western myall (*A. papyrocarpa*),
    272
  their utilisation, 273
algae, aquatic, 447, 455–6
alien flora, 60–5
alien plants
  algae, 60
  definitions, 57–8
  distribution in Australia, 72–4
  ferns and fern allies, 61–2
  in coastal dune vegetation, 516–17
  in pollen record, 49
  in saltmash and mangrove
    vegetation, 407–8
  in wetlands, 460–1
  invasions of natural ecosystems,
    66–7
  prickly acacia (*A. nilotica*), 270
  origins, history and spread, 67–72
  proportions in flora, 61, 63–4
  uncertainties in origin, 59–60
  variation in, 74–5
alpine and subalpine vegetation
  conservation, 492
  distribution, 467–9
  dynamics of, 481–3
  environments, 469–71
  grazing of, 486–90
  human impacts on: aboriginal,
    485–6; European, 486–92
  hydrology of, 490–1
  major formations: bogs, 475,
    479–80; bolster heaths, 476,
    480; closed heaths, 474, 477–8;
    fens, 475, 479; fjaeldmark, 476,

480–1; open heaths, 474, 478;
short alpine herbfields, 476, 480;
subalpine woodlands, 473, 477;
tall alpine herbfields, 475,
478–9; tussock grasslands, 475,
478–9; wet heaths, 475, 479–80
origins of, 23–9, 483–5
plant adaptations, 472
recreation and tourism, 491–2
relationships with New Zealand,
26–8
aquatic vegetation
acquatic plants, 437–40, 447–9
community structure, 449–50
dynamics of, 450–4
management, 459–62; for
conservation, 462; problems,
459–61; restoration and
rehabilitation, 461
micro- and macro- vegetation,
439–40, 441
spatial and temporal changes,
450–1
special types of: billabongs, 457–8;
intermittent floodplain lakes,
454; river Murray wetlands,
456–7; river redgum (*Eucalyptus
camaldulensis*) forests, 458–9;
salt lakes, 454–6
wetlands, 440–9; associations,
445–9, 451; growth forms of
plants, 444–5; habitats, 442–3;
types, in relation to wetting
frequency, 440–2
arid zone vegetation
historical changes in, 20–3, 48

biodiversity
alpha diversity, 536–7
beta diversity, 538–9
community diversity, 537–8
of heathlands, 324–5
species richness, 534–7
biogeographical regions, 533–5, 543
of northern rainforest, 111–14

charcoal in pollen deposits, 49, 187
Chenopod shrublands

community relationships, 352–4
'dieback' in, 362
distribution, 345–6
economically important species:
black bluebush (*Maireana
pyramidata*), 351; bladder
saltbush (*Atriplex vesicaria*), 349;
cotton bush (*M. aphylla*), 350;
oldman saltbush (*A.
nummularia*), 348–9; pearl
bluebush (*M. sedifolia*), 351–2;
southern bluebush
(*M. astrotricha*), 351
floristics, 345–6
in relation to: grazing: effects of,
356; mineral content and forage
value, 357–61; palatability,
354–5; fire, 361
occurrence and present condition,
347–8
closed-forests *see* northern *and/or*
southern rainforests
coastal dune vegetation:
alien plants in, 515–17
dynamics of, 515–16
floristics, 502–5, 509–14; of
foredunes, 502–5; coral cays,
502, 504, 506–8, 511; temperate
arid, 502, 504, 506–8, 510;
temperate humid, 502, 504,
506–8, 512–14; tropical arid,
502, 504, 506–8, 509–10;
tropical humid, 502, 504, 506–8,
511–12
human impacts on, 517–18
nomenclature, 501
Quaternary deposition and
patterning, 514–15
zonation, 501, 505–9
cool temperate rainforests *see*
southern rainforests
conservation
in relation to management, 546–9
of alpine and subalpine vegetation,
492
of aquatic vegetation, 462
of Australian flora, 525–9
of ecological reserves, 542–5

conservation – cont'd.
  of *Eucalyptus* scrubs and
    shrublands, 312
  of formations and alliances,
    529–33
  of rare and threatened species,
    528–9
  reserves, 539–41; their selection,
    545–6
continental displacement, 4–6, 132
Council of Nature Conservation
  Ministers (CONCOM), 539
Cretaceous period, 4–5, 116
Cyanobacteria, 446

derived grasslands, 270–1
  see also natural and derived
    grasslands
dry sclerophyll forests *see* open-forests

environmental weeds, 66
Eucalyptus scrubs and shrublands
  conservation of, 312
  definition, 291
  evolutionary history, 294–5
  floristics, 297–9
  general ecology: of 'core' areas in
    eastern Australia, 295–301; in
    Western Australia, 301–3
  mallee growth habit, 293–4
  production ecology, nutrient cycling
    and hydrology, 303–6, 307
  regeneration: community studies,
    311–12; germination, 308;
    lignotuber regrowth, 310–11;
    mortality and age structure, 309;
    seed availability, 306–8; seed
    predators and soil storage, 308;
    seedling establishment, 308–9
eutrophication, of wetlands, 459–60
exotics *see* alien plants

fire
  in alpine and subalpine vegetation,
    471
  in Chenopod shrublands, 361
  in heathlands, 338–9
  in southern rainforests, 146–7

in tall open-forests, 169–70,
    186–8
  in woodlands, 238–9
  use by Aborigines, 48–9
fungi, aquatic, 446

geographic information system, for
    woodlands, 229
global change, likely effects of,
    151–2, 483, 547
Gondwanan flora, 4–5, 7, 14–17,
    25–6, 116, 121, 132, 526
grazing
  of *Acacia* open-forests, woodlands
    and shrublands, 270–1, 273,
    286–7
  of alpine and subalpine vegetation,
    486–90
  of Chenopod shrublands, 354–61

halophytes *see* saltmarsh and
    mangrove
heathlands
  attributes: bradysporous fruits,
    328; leaf characteristics, 327; life
    forms, 326–7; root systems,
    327–8; seeds, 328–9
  biodiversity, 324–5
  bolster, alpine, 476, 480
  closed, alpine, 474, 478
  community structure, 322–4
  distribution, 329–35
  ecological relationships: animals,
    339–40; climate, 335–6; fire,
    338–9; soil nutrients, 336–8
  floristic classification, 330–4
  open, alpine, 474, 478
  wet, alpine, 475, 379–80
humans
  aboriginal, 121; use of rainforest
    plants, 121–2
  effects on herbivores, 38; on
    landscape, 38, 48–9; on alpine
    and subalpine vegetation,
    485–92
  movement of alien plants, 70
  use of fire, 48–9
hummock grasslands

in *Acacia* vegetation, 283–6; *see also* natural and derived grasslands
hydrology: of alpine and subalpine vegetation, 490–1
of tall open-forests, 178

International Biological Programme (IBP), 529, 530, 532, 542

lignotubers, in *Eucalyptus*, 293, 310–11
litter
in mangroves, 422
in open-forests, 212
in tall open-forests, 171–2

mallee eucalypts, mixed with *Acacia*, 284–6
see also scrubs and shrublands
mangrove *see* saltmarsh and mangrove
marlocks, 294, 310
Miocene epoch, 5, 11–12, 21–2
monsoon vegetation, 18–20

National System of Ecological Reserves, 529, 530, 540, 542–5
natural and derived grasslands
classification: arid hummock, 369–70; arid tussock, 369; coastal, 369; derived, 370–1; subhumid, 369–70
climatic variability, 378–80
conservation biology, 373–4
distribution, in northern Australia, 375
floristics, in northern Australia, 377
monitoring, 384–6
plant production, 374–8: GRASP model for, 375–7, 379
spatial heterogeneity, 380–3: within land systems, 382; within land units, 383
naturalised plants: definitions, 57–8
northern rainforests
as bird habitat, 119–20
biogeographic significance, 110–15

boundaries and ecotones, 117–18
community: rank, 95–7; types, 114–15
conservation status, 121–4
definition, 88–91
discontinuity, 91
distribution and area, 89, 92–4
dynamics, 115–17
ecological characteristics, 88–92
environmental relationships, 99–110; of core areas, 100–3
floristics: classification, 95–9; elements, 97–9; patterns, 111–14
interspersions, 117
relationships with New Guinea, 15–18
role of refugia, 115–17
segregation, from sclerophyll vegetation, 92
social values, 121–4
use for forestry, 122
structure: classification, 94–5; patterns, 110–11
vagility, 118–19
nutrient cycling
in *Eucalyptus* scrubs and shrublands, 306, 307
in intertidal wetlands, 422–3
open-forests, 210–13
in tall open-forests, 171–4

open-forests
concept, 197–9
conservation, 218
distribution: in southwest Western Australia, 203–4; in Sydney Basin, 199–201; in Victorian Eastern Highlands, 201–3; patterns, 199–204
fire in, 213–15
floristics, 204–10, 304–5; of understorey, 208–10
growth and seed dispersal, 215–17
nomenclature, 198
nutrient cycling, 210–13
role of soil phosphate, 200–1, 203
utilisation, 217–19

palynology, 22, 40, 50–1
pests and diseases
  in southern rainforests, 149–50
phytogeography
  alpine region, 23–9
  arid zone, 20–3
  floristic elements and subelements:
    Antarctic, 7; Autochthonous
    (Australian), 6, 7, 8, 11–13, 26;
    Cosmopolitan, 7, 8, 20;
    Gondwanan, 7, 8, 9–13, 25–6;
    Indomalayan, 6–7, 116;
    Intrusive, 7, 14–17; Neoaustral,
    8; Relict, 7, 8, 9–11; Tropical, 8,
    14–17
  invasion theory, 4, 6–7
  landform history, 3–6
  monsoon forests, 18–20
plate tectonics *see* continental
    displacement
pliocene epoch, 5–6, 20
pollen analysis, 50–1

Quaternary period
  data base, 40–2
  dune systems, 514–15
  Early, 42–5
  fauna, 42
  in relation to: fire, 48–50; human
    impact, 48–50
  Late, 45–8
  major events, 37–9
  pollen analysis, 50–1
  vegetation history, 37–51

rainforests, 9–11
  see also northern *and/or* southern
    rainforests

salinisation, 459
saltmarsh and mangrove
  as bird habitat, 423
  composition and biogeography:
    introduced species, 407–8;
    mangroves, 400–3; non-vascular
    flora, 405–7; saltmarsh, 403–5
  distribution and abundance, 395–6
  ecosystem aspects, 421–4

  in relation to: coastal physiography,
    420–1; environment, 408–16;
    tidal regimes, 408–11
  litter, 422
  management issues, 424–7
  related vegetation types, 396–9
  zonation, 416–20
savanna *see* woodlands
scleromorphy, 13–14
Scientific Committee on Problems of
    the Environment (SCOPE), 549
southern rainforests
  botanical characteristics, 136–8
  conservation status, 150–1
  distribution, 133, 135–6
  dynamics, 146–7; of populations,
    146
  habitat, 142–4
  in relation to: environment, 137,
    142–4; fire, 146–7, 148–9;
    global change, 151–2; human
    impacts, 148
  life cycle attributes, 144–6
  pests and diseases, 149–50
  present status, 148–50
  species composition, 138–42
  vegetation: history, 46–7, 132,
    134; variation in, 138–42
structuregrams, for woodlands, 230,
    240

tall open-forests
  climate, 167–9
  environment, 167–75
  evolution, 186–8
  fire in, 169–70, 186–8
  floristics, of understorey, 159–61,
    167
  forest form, 176–8
  growth and development, 176–86
  insects, 175
  nutrient cycling, 171–4
  regeneration, 178–81
  succession, 162–3, 181–6
tertiary period, 9, 11–12, 13, 37,
    134
TWINSPAN floristic groupings, 530,
    532–3, 537–9

tussock grasslands
  alpine, 475, 478–9; mixed with
    Acacia, 274–83; *see also* natural
    and derived grasslands

UNESCO Biosphere Reserves, 548

weeds *see* alien plants
wetlands *see* aquatic vegetation
wet sclerophyll forests *see* tall
  open-forests
woodlands
  adaptive traits, 236–9
  bioclimatic provinces, 232–4

definition, 228
descriptive framework, 229
distribution, 232–4
floristics, 234–5
subalpine, 473–7
types, 230–2: low (*Melaleuca
  viridiflora, M. nervosa*), 245–6;
  low (deciduous) (*Lysiphyllum,
  Terminalia*), 249–50; medium
  (*Eucalyptus populnea*), 243–4;
  tall (*E. moluccana*), 241–3; very
  low (*E. pruinosa, E. brevifolia*),
  246–9; very tall (*E. miniata,
  E. tetrodonta*), 239–40